Quality Management for Organizations Using Lean Six Sigma Techniques

Quality Management for Organizations Using Lean Six Sigma Techniques

Erick C. Jones

CRC Press
Taylor & Francis Group
Boca Raton London New York

CRC Press is an imprint of the
Taylor & Francis Group, an **informa** business

CRC Press
Taylor & Francis Group
6000 Broken Sound Parkway NW, Suite 300
Boca Raton, FL 33487-2742

First issued in paperback 2017

© 2014 by Taylor & Francis Group, LLC
CRC Press is an imprint of Taylor & Francis Group, an Informa business

No claim to original U.S. Government works

Version Date: 20131025

ISBN 13: 978-1-4398-9782-9 (hbk)
ISBN 13: 978-1-138-07512-2 (pbk)

Library of Congress Cataloging-in-Publication Data

Jones, Erick C.
    Quality management for organizations using lean Six Sigma techniques / Erick C. Jones.
      pages cm
    Includes bibliographical references and index.
    ISBN 978-1-4398-9782-9 (hardback)
    1. Six sigma (Quality control standard) 2. Quality control. I. Title.

TS156.17.S59J66 2014
658.5'62--dc23   2013040935

Visit the Taylor & Francis Web site at
http://www.taylorandfrancis.com

and the CRC Press Web site at
http://www.crcpress.com

Contents

Preface...xxvii
Acknowledgments..xxix
Introduction xxxi
Author..xxxv

Section I: Modern quality and Lean Six Sigma overview

Chapter 1   Overview of modern quality..3
1.1   Introduction...3
1.2   What is quality? 3
       1.2.1   How is quality defined?...3
                1.2.1.1   Global approach 3
                1.2.1.2   Product-based approach 3
                1.2.1.3   User-based approach..4
                1.2.1.4   Manufacturing-based approach 4
                1.2.1.5   Value-based approach 4
1.3   What do organizational departments or functions think about quality?.................6
       1.3.1   Functional perspectives on quality..6
                1.3.1.1   Engineering perspective on quality....................................6
                1.3.1.2   Operations perspective on quality 6
                1.3.1.3   Systems perspective on quality 6
                1.3.1.4   Executive management perspective on quality..................7
                1.3.1.5   Marketing perspective on quality 8
                1.3.1.6   Financial perspective on quality 8
                1.3.1.7   Human resources perspective on quality 10
1.4   Total quality management and quality overview 10
1.5   Who is taught and trained to attain quality by an organization?...........11
       1.5.1   The implementation of quality...11
                1.5.1.1   The integrated approach..11
                1.5.1.2   Three spheres of quality 11
1.6   What if quality is too expensive to justify?...12
       1.6.1   "As-needed" implementation of quality 12
1.7   Limitation of quality measures 12
1.8   How can the COQ be justified? 13
       1.8.1   Cost of quality..13
       1.8.2   Traditional COQ 13

1.9   Who started quality and what ideas did they contribute? 14
         1.9.1   Quality philosophers, gurus, and contributors 14
                    1.9.1.1   William E. Deming 14
                    1.9.1.2   Joseph M. Juran 14
                    1.9.1.3   Philip B. Crosby 15
                    1.9.1.4   Armand V. Feigenbaum 15
                    1.9.1.5   Garvin 15
                    1.9.1.6   Ishikawa 15
                    1.9.1.7   Taguchi 15
                    1.9.1.8   Japanese contributions to quality 15
1.10   How is quality rewarded and enforced globally? 16
         1.10.1   Quality awards 16
                    1.10.1.1   The Deming Prize 16
                    1.10.1.2   Malcolm Baldrige National Quality Award 16
                    1.10.1.3   European Quality Award 17
1.11   Quality standards 18
         1.11.1   ISO 9000 18
         1.11.2   ISO 14000 and 14001 18
1.12   Integrating QM with LSS techniques 19
1.13   Will LSS really work? 19
1.14   How can I integrate LSS into the operations throughout my supply chain? 20
References 20
Further Reading 21

Chapter 2   Quality gurus and philosophies 23
2.1   Introduction 23
2.2   William E. Deming 23
2.3   Joseph M. Juran 24
2.4   Philip B. Crosby 24
2.5   Armand V. Feigenbaum 24
2.6   Garvin 25
2.7   Ishikawa 25
2.8   Taguchi 25
2.9   Japanese contributions to quality 25
References 26
Further Readings 26

Chapter 3   Quality awards and standards 27
3.1   The Deming Prize 27
         3.1.1   Categories 28
         3.1.2   Deming Prize and TQM 28
         3.1.3   Deming Prize in detail 29
3.2   The Malcolm Baldrige Award 30
         3.2.1   Baldrige model for quality 30
         3.2.2   Six Sigma method 31
         3.2.3   Baldrige model and Six Sigma 31
                    3.2.3.1   Implementing Six Sigma within the Baldrige model 32
                    3.2.3.2   Full integration 33

      3.2.4   A quality alliance 34
3.3   ISO 9000 34
      3.3.1   Purpose 35
      3.3.2   ISO 9000—Performance...36
3.4   Other quality awards and standards.....................................36
      3.4.1   Strategic quality planning......................................36
      3.4.2   European Quality Award...37
      3.4.3   Quality Systems..38
References 38
Further Reading 39
Appendix 3A: A relationship between Six Sigma and Malcolm Baldrige
          Quality Award 40
3A.1   Introduction...40
3A.2   The Baldrige model for quality..40
3A.3   Six Sigma methodology 41
3A.4   The Baldrige model and Six Sigma 42
      3A.4.1   Implementing Six Sigma within the Baldrige model.....43
3A.5   Integrating Six Sigma with the Baldrige model.......................44
      3A.5.1   The case for Motorola 45
      3A.5.2   Leadership and Six Sigma.......................................46
      3A.5.3   Human resource management and Six Sigma.............46
      3A.5.4   Strategic planning and Six Sigma 46
3A.6   Conclusion 47
References 47
Further Readings...48
Appendix 3B: Comparing quality management practices between the United States
          and Mexico..48
3B.1   Introduction...48
3B.2   Literature review 49
3B.3   Theoretical framework: Comparative management....................50
3B.4   Hypotheses 51
3B.5   Research methodology 51
      3B.5.1   Survey instrument...51
      3B.5.2   Data collection...53
References 54
Further Readings...56

Chapter 4   Six Sigma without Lean..**57**
4.1   Introduction...57
      4.1.1   The Six Sigma thinking in this text 57
4.2   Background 58
      4.2.1   Which Six Sigma points are retained from TQM? 59
      4.2.2   Which Six Sigma point changed from TQM? 59
      4.2.3   Six Sigma roots...59
            4.2.3.1   From product quality program to firm-wide
                    decision-making strategy 59
4.3   Corporates support Six Sigma popularity 60
      4.3.1   Industries other than manufacturing......................60
4.4   Challenges of Six Sigma 61

4.5   Six Sigma and engineers...62
        4.5.1   Six Sigma and engineering education...62
        4.5.2   Controversy between engineers education and green/black belts.........62
4.6   The statistics of Six Sigma 63
4.7   The 3.4 defects per million or long-term shift confusion.............................64
4.8   Methodologies..65
        4.8.1   The Juran Six Sigma approach..65
        4.8.2   Different methodologies...65
        4.8.3   Six Sigma statistical tools 66
4.9   Conclusion 66
References 66
Further Readings..67

Chapter 5   Lean 71
5.1   Introduction..71
5.2   Lean enterprise 71
5.3   Ohno seven sources of waste 72
5.4   Lean elements of manufacturing 72
5.5   Lean tools..73
5.6   Value stream mapping 73
5.7   Lean success 74
5.8   Conclusion 75
References 75
Further Readings..76

Chapter 6   Lean Six Sigma...79
6.1   Introduction..79
        6.1.1   History of Lean Six Sigma..79
        6.1.2   What is the real difference?...79
References 81
Further Readings..82

Section II: Deciding on firm qualities and Lean Six Sigma initiatives

Chapter 7   The cost approaches and engineering economics of Six Sigma projects......87
7.1   Introduction..87
7.2   Cost approach...88
7.3   Value approach...88
7.4   Application approach...89
7.5   LSS/Quality decision matrix..91
7.6   Determining economic justification of engineering projects.....................91
7.7   Previous economic studies...92
        7.7.1   Impact of Six Sigma on performance of a company.....................92
7.8   Critical factors in Six Sigma implementation 93
7.9   Methodology 94
        7.9.1   Baseline 94
        7.9.2   Identifying engineering economics metric....................................94

7.9.3   Comparison 94
7.9.4   Evaluation 96
7.9.5   Sensitivity 96
7.10   Case study 96
7.10.1   Sample calculation 96
7.11   Results and discussions 98
7.11.1   Decision comparison based on EBIT growth rate and IRR 98
7.12   Conclusion 98
References 98

Chapter 8   Lean Six Sigma and firm performance: A managerial perspective 101
8.1   Introduction 101
8.2   What is special about Six Sigma? 102
8.3   Process management 103
8.3.1   Understanding organizational processes 104
8.3.2   Process management and firm performance 105
8.3.3   Process management and innovation 106
8.4   Final notes 108
References 108
Further Readings 110

Chapter 9   Benchmarking high-performing Six Sigma companies 113
9.1   Introduction 113
9.1.1   Lean Six Sigma criteria testing model 113
9.1.1.1   Identification of Six Sigma companies 114
9.2   Lean Six Sigma priority evaluation models 117
9.2.1   Estimation of MARR 118
9.2.2   Discount for inflation 118
9.2.3   Calculate performance of each company 118
9.3   Non-LSS projects in the LSS organization—How do we manage non-LSS
         projects? Evaluation and comparison of practices between high- and
         low-performing companies 119
9.3.1   Study factors critical for Six Sigma implementation 119
9.3.2   Examine and compare failure factors for companies under our study 122
9.3.3   Difference between traditional and high performance organization ... 125
9.4   Lean Six Sigma/no Lean Six Sigma management integration strategy 128
References 128
Appendix 9A: Developing an instrument for measuring Six Sigma implementation 129
9A.1   Introduction 129
9A.2   Background and literature review 130
9A.2.1   Six Sigma methodology 130
9A.2.2   Literature review 130
9A.3   Construct development 131
9A.4   Methodology 132
9A.5   Analysis 132
9A.5.1   Black belt roles 132
9A.5.2   DMAIC versus DMADV 133
9A.5.3   Plan 134

9A.5.4	Do	134
9A.5.5	Check	135
9A.5.6	Act	136
9A.5.7	Financial responsibility	136
9A.5.8	Executive support	137
9A.5.9	Final reduced questionnaire	137

9A.6   Summary and conclusions 137
      9A.6.1   General findings 139
      9A.6.2   Limitations 140
      9A.6.3   Future research 140
      9A.6.4   Implications for managers 140
References 140
Further Readings 141
Appendix 9B: Six Sigma deployment success according to shareholder value 141
9B.1   Introduction 141
      9B.1.1   *Good to Great* 142
      9B.1.2   Six Sigma 142
      9B.1.3   Lean Six Sigma 143
      9B.1.4   Use of Lean Six Sigma for company growth and compliance 143
      9B.1.5   Previous study 144
9B.2   Study methodology 144
      9B.2.1   Population sample (representative companies) 144
      9B.2.2   Measurement criteria 145
      9B.2.3   Criteria for success 147
      9B.2.4   Identify characteristics 147
9B.3   Conclusion 147
References 148

Chapter 10   Six Sigma and HR: Getting the right talent and industrial engineers in key Six Sigma roles **149**
10.1   Introduction 149
10.2   LSS in education (graduate and undergraduate) 150
      10.2.1   Six Sigma and engineers 150
         10.2.1.1   Six Sigma and engineering education 150
         10.2.1.2   Controversy between engineers education and green/black belts 150
      10.2.2   Aspects of Lean Six Sigma 150
      10.2.3   Roles of Lean Six Sigma 151
         10.2.3.1   Industrial roles 151
         10.2.3.2   Six Sigma–IE interface framework 151
10.3   LSS integration into the Malcolm Baldrige Award 153
      10.3.1   Baldrige model for quality and Six Sigma method 153
      10.3.2   Six Sigma within Baldrige model 154
      10.3.3   Full integration 155
      10.3.4   Leadership 155
      10.3.5   Human resource/work force management 156
      10.3.6   Strategic planning 156
10.4   LSS integration into ISO 9000 156

10.5   LSS integration into logistics...156
References...157
Appendix 10A: Quality control measurement of an engineering productivity index 157
10A.1   Introduction...157
10A.2   Background 159
10A.3   Application of quality control principles 160
10A.4   Use of control charts for cognitive turnover....................160
         10A.4.1   Chart run rules 161
10A.5   Chart development...161
10A.6   Results 162
10A.7   Conclusions 165
References 165
Further Readings...166
Appendix 10B: The value of industrial engineers in Lean Six Sigma organizations........166
10B.1   Introduction..166
10B.2   Background 166
10B.3   Aspects of Lean Six Sigma 167
10B.4   Roles in Lean Six Sigma..167
10B.5   Academic discipline 168
         10B.5.1   Aspects of the Lean Six Sigma..........................168
         10B.5.2   Roles in Lean Six Sigma......................................168
10B.6   Industrial roles...168
10B.7   Six Sigma–IE interface framework 168
10B.8   Implementation of the framework 170
10B.9   Proposed study 171
10B.10   Conclusion...171
References 172

Section III: The science and engineering of Lean Six Sigma

Chapter 11 LSS process, steps, tools, and statistics **175**
11.1   Introduction...176
11.2   Hypothetical statements 178
11.3   Define 179
        11.3.1   Chartering a Six Sigma project............................179
        11.3.2   The assignments of the team 181
        11.3.3   Methods to collate information 182
11.4   Measure..182
        11.4.1   Data collection and accuracy 183
        11.4.2   Process description...183
        11.4.3   Focus/prioritization tools 184
        11.4.4   Quantifying and describing variation 184
        11.4.5   Generating and organizing ideas.........................185
11.5   Analyze 185
11.6   Improve 186
11.7   Control...188

11.8   DFSS-R introduction 190
         11.8.1   Define 190
         11.8.2   Measure...190
         11.8.3   Analyze 191
         11.8.4   Design 191
         11.8.5   Identify..191
         11.8.6   Optimize..191
         11.8.7   Verify 192
References 192

Chapter 12 Fundamental statistics and basic quality tools...........................193
12.1   Introduction...193
12.2   Basic definitions 193
12.3   Module 1—Empirical methods...194
         12.3.1   Frequency distribution 194
         12.3.2   Histogram..194
         12.3.3   Other graphical methods 198
                   12.3.3.1   Scatter diagram 198
                   12.3.3.2   Stem and leaf diagram 199
                   12.3.3.3   Box and whisker plot...201
12.4   Module 2—Statistical models for describing populations 202
         12.4.1   Binomial distribution...202
         12.4.2   Poisson distribution 203
         12.4.3   Normal distribution..204
         12.4.4   Application of central limit theorem 205
                   12.4.4.1   Central limit theorem..205
         12.4.5   Hypothesis testing 205
                   12.4.5.1   US legal system hypothesis tests................................206
                   12.4.5.2   Why and when are hypothesis tests used? 206
                   12.4.5.3   Null hypothesis 206
                   12.4.5.4   Alternate hypothesis 206
                   12.4.5.5   Translation 207
         12.4.6   Regression..208
                   12.4.6.1   Simple linear regression model 209
                   12.4.6.2   The method of least squares.......................................209
                   12.4.6.3   Residuals 210
                   12.4.6.4   Multiple regression...211
         12.4.7   Analysis of variance...212
                   12.4.7.1   *P*-value 212
                   12.4.7.2   Tukey's method 212
                   12.4.7.3   Fisher's least significant difference 212
12.5   Module 3—Quality tools..212
         12.5.1   Flowcharting 212
         12.5.2   Cause-and-effect diagram...213
         12.5.3   Check sheets..213
         12.5.4   Pareto diagrams..214
         12.5.5   Control charts 215
                   12.5.5.1   Variable and attribute process control charts 215

		12.5.5.2	Understanding the process of implementing the chart	215
		12.5.5.3	$\bar{x}$ and R charts	216
		12.5.5.4	Median charts	216
		12.5.5.5	$\bar{x}$ and s charts	217
		12.5.5.6	Attribute process control charts	218

References 220
Further Readings..220

Chapter 13 Define: Identifying and organizing Six Sigma projects for success 221

13.1	Introduction	222
13.2	Project definition worksheet	224
13.3	Voice of the customer	228
	13.3.1 Customer identification	228
	13.3.2 Internal customers	229
	13.3.3 External customers	232
	13.3.4 Customer service	232
	13.3.5 Customer retention and loyalty	232
	13.3.6 Customer metrics selection	233
	13.3.7 Customer data collection	234
	13.3.8 Customer surveys	234
	13.3.9 Customer data analysis	235
	13.3.9.1 Line graph	235
	13.3.9.2 Control charts	235
	13.3.9.3 Matrix diagrams	236
	13.3.10 Voice of customers	236
	13.3.10.1 Methods of hearing the VOC	236
	13.3.10.2 Process to understand VOC	236
	13.3.10.3 Critical to quality	237
13.4	Project charter and teams	237
	13.4.1 Setting project boundaries	238
	13.4.2 Project charter	239
	13.4.2.1 Content of the charter	239
	13.4.2.2 Financial impact	240
	13.4.3 Goal statement	240
	13.4.3.1 Step 1: Identify	240
	13.4.3.2 Step 2: Exploit	241
	13.4.3.3 Step 3: Subordinate	241
	13.4.3.4 Step 4: Elevate	241
	13.4.3.5 Step 5: Go back to Step 1	241
	13.4.4 Milestones or deliverables	241
	13.4.5 Charter negotiation	242
	13.4.6 Project management and its benefits	242
	13.4.7 Project measures	243
	13.4.8 Teams	244
	13.4.8.1 Team leader roles and responsibilities	244
	13.4.8.2 Team formation and evolution	245
	13.4.8.3 Types of teams	246
	13.4.8.4 Implementing teams	247

13.5   Project tracking 250
          13.5.1   Work breakdown structure (WBS)..251
                       13.5.1.1   Advantages of manual methods....................................252
                       13.5.1.2   Advantages of computer software 252
          13.5.2   Milestones reporting...252
          13.5.3   Project report...253
                       13.5.3.1   Document archiving..253
13.6   Yellow belt...253
          13.6.1   SMART objective 253
          13.6.2   SIPOC 254
          13.6.3   Voice of the customer in yellow belt...254
                       13.6.3.1   Critical to quality 256
13.7   Green belt...257
          13.7.1   Pareto diagrams..257
          13.7.2   Benchmarking..258
          13.7.3   Thought process map..259
                       13.7.3.1   Value stream mapping 259
                       13.7.3.2   Process mapping 261
          13.7.4   Brainstorming..262
13.8   Black Belt..262
          13.8.1   Kano model 262
                       13.8.1.1   Dissatisfiers (basic requirements)............................. 262
                       13.8.1.2   Beijing Olympics 262
                       13.8.1.3   Delighters.. 262
          13.8.2   Quality function deployment...263
          13.8.3   Affinity diagram..266
          13.8.4   Interrelationship diagram 269
References 270

Chapter 14 Measure: Identifying obtainable data for realistic use 279
14.1   Introduction...280
14.2   Critical characteristics..281
          14.2.1   Critical to satisfaction 282
          14.2.2   Critical to quality...282
          14.2.3   Critical to process 282
14.3   Flow down process..283
          14.3.1   Construction of a flow down process...283
          14.3.2   Characteristics of good "Little y"...284
14.4   Two dimensions 285
14.5   Determination of targets and specifications...285
          14.5.1   The need for targets and specifications...286
14.6   Sponsorship level..287
          14.6.1   Balanced scorecard 287
14.7   Yellow belt...287
          14.7.1   Process mapping..288
          14.7.2   Data collection...290
                       14.7.2.1   Control charts 291
                       14.7.2.2   Check sheets 291

14.7.3 Waste and variation analysis 292
    14.7.3.1 First S—Sort (organization) 292
    14.7.3.2 Second S—Stabilize (orderliness) 293
    14.7.3.3 Third S—Shine (cleanliness) 293
    14.7.3.4 Fourth S—Standardize (promote adherence) 293
    14.7.3.5 Fifth S—Sustain (self-discipline) 293
    14.7.3.6 Value analysis 293
14.8 Green belt 294
  14.8.1 Pareto analysis 294
  14.8.2 Data collection 295
    14.8.2.1 Measurement scales 295
    14.8.2.2 Types of sampling 296
    14.8.2.3 Ensuring data accuracy 296
  14.8.3 Data collection methods 296
    14.8.3.1 Check sheets 297
    14.8.3.2 Checklists 298
  14.8.4 Types of data 298
  14.8.5 Measurement methods 299
    14.8.5.1 Measurement system analysis 299
    14.8.5.2 Range method 300
    14.8.5.3 Analysis of variance 301
    14.8.5.4 Measurement correlation 303
    14.8.5.5 Enterprise measurement systems 306
    14.8.5.6 Calibration 307
  14.8.6 Graphical methods 308
    14.8.6.1 Boxplots 308
    14.8.6.2 Scatter diagrams 308
    14.8.6.3 Stem and leaf plots 310
    14.8.6.4 Run (TREND) charts 310
    14.8.6.5 Histograms 310
14.9 Black belt 311
  14.9.1 Control charts 311
    14.9.1.1 Understanding the process for implementing the chart 312
    14.9.1.2 $\bar{x}$ and R charts 313
    14.9.1.3 Median charts 315
    14.9.1.4 $\bar{x}$ and s charts 315
  14.9.2 Attribute process control charts 315
    14.9.2.1 P-charts for proportion defective 316
    14.9.2.2 np charts 316
    14.9.2.3 C and u charts 317
References 317
Further Reading 318

Chapter 15 Analyze: Evaluating data to determine root causes 321
15.1 Introduction 322
15.2 Establishing baseline 322
  15.2.1 Baseline performance metrics 324
    15.2.1.1 Counting "defects" or "defectives" 324
    15.2.1.2 Measuring characteristic or variable 324

            15.2.1.3   Baseline process metrics 324

   15.2.2   Discrete outputs...325

            15.2.2.1   Process capability for binomial process outputs...................331

            15.2.2.2   Process capability for Poisson process output.......................331

   15.2.3   Continuous outputs..332

            15.2.3.1   Process capability analysis for normal process outputs 333

   15.2.4   Potential process capability..334

            15.2.4.1   Short-term variation 335

15.3   Yellow belt..337

   15.3.1   Cause-and-effects diagram (fishbone diagram)....................................337

   15.3.2   Fault tree analysis...338

            15.3.2.1   Fault tree symbols..339

   15.3.3   Root cause analysis 340

            15.3.3.1   Subjective tools...341

            15.3.3.2   Analytical tools 341

            15.3.3.3   5 Whys 341

            15.3.3.4   5 Ws and H..342

15.4   Green belt..342

   15.4.1   Entitlement 342

            15.4.1.1   Types of errors..342

   15.4.2   Hypothesis testing 345

            15.4.2.1   US legal system hypothesis tests 345

            15.4.2.2   Why and when are hypothesis tests used? 345

            15.4.2.3   Null hypothesis...346

            15.4.2.4   Alternate hypothesis...346

            15.4.2.5   Translation 346

   15.4.3   Chi-squared test..349

   15.4.4   Analysis of variance 349

            15.4.4.1   *P*-value..349

            15.4.4.2   Tukey's method 350

            15.4.4.3   Fisher's least significant difference 350

   15.4.5   *X–Y* process map..350

   15.4.6   Failure mode and effect analysis..351

            15.4.6.1   Failure mode and effect analysis process steps......................353

            15.4.6.2   Risk assessment and risk priority number 354

            15.4.6.3   Types of FMEA...354

15.5   Black belt..356

   15.5.1   Multi-vari studies 356

            15.5.1.1   Advantages of multi-vari chart...357

            15.5.1.2   Characteristics of multi-vari chart 357

   15.5.2   Regression..357

            15.5.2.1   Fitted line plot 360

            15.5.2.2   Residuals 362

            15.5.2.3   Multiple regression..362

            15.5.2.4   Simple linear regression model 364

            15.5.2.5   The method of least squares..365

            15.5.2.6   Logistics regression 366

   15.5.3   Process capability analysis..368

            15.5.3.1   Process capability for normal distribution..............................369

      15.5.3.2 Process capability indices 370

      15.5.3.3 Process performance 370

      15.5.3.4 Process performance metrics 371

      15.5.3.5 Rolled throughput yield 372

      15.5.3.6 Chi-squared analysis of contingency tables 372

References 373

Chapter 16 Improve: Utilizing data to predict implementation success **375**

16.1   Introduction 376

16.2   New process architecture 376

16.3   Green belt 379

     16.3.1   Benchmarking 379

        16.3.1.1 Purpose of benchmarking 380

        16.3.1.2 Types of benchmarking 380

        16.3.1.3 Benefits of benchmarking 381

        16.3.1.4 Different standards for comparison 381

        16.3.1.5 How is benchmarking done? 381

     16.3.2   SWOT analysis 383

        16.3.2.1 Strengths and weaknesses 383

        16.3.2.2 Opportunities and threats 384

     16.3.3   Theory of constraints 385

        16.3.3.1 Methods for implementing theory of constraints 386

        16.3.3.2 Drum buffer rope 386

16.4   Black belts 387

     16.4.1   DOE terminology 387

     16.4.2   Design selection guidelines 390

     16.4.3   Various examples for performing DOE 391

        16.4.3.1 One factor, multiple levels 391

        16.4.3.2 Planning experiments 399

References 401

Further Readings 402

Chapter 17 Control: Using data to maintain success **403**

17.1   Introduction 404

17.2   Validating the plan 405

     17.2.1   Purpose 405

     17.2.2   Objective 405

     17.2.3   Deliverable 405

17.3   Control inputs and monitor outputs 406

     17.3.1   Purpose 406

     17.3.2   Objective 406

     17.3.3   Deliverable 406

17.4   Controlling x's 407

17.5   Green Belt 408

     17.5.1   5S (workplace organization) 408

        17.5.1.1 Details of the 5S program 409

17.6   Kaizen 411

     17.6.1   Kaizen blitz 411

17.7 Total Productive Maintenance 412
    17.7.1 Designing for maintainability and availability 413
    17.7.2 TPM metrics 414
    17.7.3 Benefits of TPM 414
    17.7.4 Steps to implement TPM 415
    17.7.5 Autonomous TPM small group activities 415
17.8 Black belts 415
    17.8.1 Statistical process control 415
        17.8.1.1 Selection of variables 416
        17.8.1.2 Rational subgrouping 416
17.9 Control chart selection 417
    17.9.1 Control chart analysis 417
    17.9.2 Trends 418
    17.9.3 Jump-in-process level 418
    17.9.4 Recurring cycles 419
    17.9.5 Points near or outside limits 419
    17.9.6 Lack of variability 419
17.10 Poka-yoke (mistake proofing) 420
17.11 Kanban pull 422
References 424

Chapter 18 Lean integration into DMAIC **425**
18.1 Introduction 425
18.2 The marriage of Lean and Six Sigma 426
18.3 Enterprise-wide deployment 426
    18.3.1 Challenges with deployment 426
18.4 Systems and processes 427
References 428

Chapter 19 Nontraditional tools for LSS: Work measurement and time studies **429**
19.1 Introduction 429
19.2 Motion and time study in a Lean framework 429
19.3 History of time and motion study 430
19.4 Continuous improvement 431
    19.4.1 Problem definition 432
    19.4.2 Analysis of the problem 432
    19.4.3 Search for possible solutions 432
    19.4.4 Evaluation of alternative 432
    19.4.5 Recommendation for action 432
    19.4.6 Continuous monitoring of the action 433
19.5 Time study 433
    19.5.1 Time study equipment 433
    19.5.2 Important definitions 434
    19.5.3 Making the time study 435
        19.5.3.1 Recording information about the operation and the operator 435
        19.5.3.2 Dividing the operation into its elements 435

          19.5.3.3 Recording the time 435
          19.5.3.4 Number of cycles to be timed 437
          19.5.3.5 Rating an operator 437
          19.5.3.6 Determining the allowances and standard time 437
      19.5.4   Types of time study..438
19.6   Methods-time measurement..438
19.7   Time studies and human factors 439
19.8   Challenges with time study 439
19.9   Economic tools 439
19.10   Evaluating quality costs..441
      19.10.1 Cost of good quality..441
      19.10.2 Cost of poor quality 441
      19.10.3 Total quality costs...442
19.11   Conclusion 442
References 442
Further Readings...443

Chapter 20 DFSS methods and uses for product- and process-oriented research
           and development..445
20.1   Introduction...445
20.2   Background 445
20.3   Operational Six Sigma and design for Six Sigma 446
      20.3.1   DMAIC: OSS...447
      20.3.2   Design for Six Sigma..447
      20.3.3   Design for Six Sigma research 449
References..451
Further Readings...451
Appendix 20A: Using Design for Six Sigma—To Develop Real-World Testing
           Environments for RFID Systems..452
20A.1   Introduction...452
      20A.1.1 RFID system details 452
      20A.1.2 Six Sigma methodology 452
20A.2   3P's theoretical model 453
20A.3   Plan 453
      20A.3.1 Define 453
      20A.3.2 Measure..454
20A.4   Predict 454
      20A.4.1 Analyze 454
      20A.4.2 Design 455
20A.5   Perform..455
      20A.5.1 Optimize 456
          20A.5.1.1 Factors and levels 456
      20A.5.2 Verify 457
20A.6   Conclusion 458
Further Readings...460
Appendix 20B: A framework for effective Six Sigma implementation 460
20B.1   Introduction...460
20B.2   Six Sigma methodology 461

20B.3   Literature review 462
20B.4   Development of constructs and their relationships 463
20B.5   The conceptual model for Six Sigma implementation 465
20B.6   Summary..467
References..467

Section IV: Applying and using Lean Six Sigma

Chapter 21 Using Six Sigma to evaluate radio frequency technologies at NASA.......471
21.1   Introduction...471
21.2   Methodology: Integrated DFSS for research 471
         21.2.1   Plan..471
                   21.2.1.1   Define..471
                   21.2.1.2   Measure 472
         21.2.2   Predict 472
                   21.2.2.1   Analyze 472
                   21.2.2.2   Design...472
                   21.2.2.3   Identify 472
         21.2.3   Perform...472
                   21.2.3.1   Optimize 472
                   21.2.3.2   Verify 472
21.3   Define 473
         21.3.1   Project charter...473
                   21.3.1.1   Problem statement 473
                   21.3.1.2   Business case 473
                   21.3.1.3   Scope...473
                   21.3.1.4   Deliverable 473
                   21.3.1.5   Test requirements...474
                   21.3.1.6   Obstacles 474
                   21.3.1.7   Define acceptable hardware vendors 475
                   21.3.1.8   Application strategy 475
21.4   Measure...475
         21.4.1   Read range testing...475
                   21.4.1.1   Read range experiment procedure 475
         21.4.2   Orientation sensitivity testing.................................475
                   21.4.2.1   Orientation sensitivity experiment procedure 475
         21.4.3   Read accuracy testing 476
                   21.4.3.1   Read accuracy experiment procedure 476
21.5   Analyze 476
         21.5.1   Read range analysis...476
         21.5.2   Orientation sensitivity analysis...............................477
         21.5.3   Read accuracy analysis...477
21.6   Develop 478
         21.6.1   Smart shelf...478
         21.6.2   Door tracking...478
         21.6.3   Waste tracking 478
         21.6.4   Sensor active tag 479

21.7 Optimize 480
    21.7.1 Implementation..480
    21.7.2 Cost analysis...480
        21.7.2.1 Fixed cost..480
21.8 Conclusion 481
References 481

Chapter 22 Using Six Sigma to evaluate automatic identification technologies to optimize broken case warehousing operations...483
22.1 Introduction..483
    22.1.1 RAID laboratory 483
    22.1.2 The National Science Foundation–International Research Experiences for Students 483
22.2 Background 484
    22.2.1 Company background 484
    22.2.2 Problem statement 484
22.3 Research objective 484
    22.3.1 Research question..484
    22.3.2 Relationship to business problem 484
    22.3.3 Research hypothesis..484
        22.3.3.1 Test metrics..485
        22.3.3.2 Null hypothesis...485
        22.3.3.3 The rejection criteria......................................485
22.4 Research methodology 485
    22.4.1 Research methodology 485
    22.4.2 Research approach 486
    22.4.3 DFSSR—Steps and description..486
        22.4.3.1 Scenario 1 489
        22.4.3.2 Scenario 2 490
        22.4.3.3 Scenario 3 490
    22.4.4 Summary 490
    22.4.5 Identify 491
22.5 Results 491
22.6 Discussion...491
    22.6.1 Rejection of hypothesis..491
22.7 Limitation 492
22.8 Conclusion 492
Further Readings..492

Chapter 23 Using Six Sigma to implement RFID automation at an automotive plant...493
23.1 Research problem and literature search...493
23.2 Research overview and objectives 494
23.3 Research method and approach 495
    23.3.1 Measure...497
    23.3.2 Design 499
    23.3.3 Identify...501

23.3.4 Results 501
      23.3.4.1 Benefit assumptions of using RFID technologies 502
      23.3.4.2 Final recommendation 502
23.4 Conclusion 502
23.5 Gantt chart and project timeline 502
Acknowledgment 503
References 503

Chapter 24 Using Six Sigma to evaluate using automated inventory tracking to reduce in processing food product shortages at an international food processing plant 505
24.1 Introduction 505
    24.1.1 RAID laboratories 505
    24.1.2 The NSF and the IRES 505
24.2 Background 506
    24.2.1 Business problem statement 506
24.3 Research objective 506
    24.3.1 Research question 506
    24.3.2 Relation to business problem 506
    24.3.3 Research hypothesis 507
24.4 Research methodology 507
    24.4.1 Research laboratory methodology with steps 507
    24.4.2 Data collection techniques 508
        24.4.2.1 Tools of DFSS-R 508
    24.4.3 Research approach 508
        24.4.3.1 DFSS-R steps and description 508
24.5 Results 511
    24.5.1 DFSS-R approach 511
        24.5.1.1 Define 511
        24.5.1.2 Measure 512
        24.5.1.3 Analyze 512
        24.5.1.4 Design 514
        24.5.1.5 Identify 514
    24.5.2 Return-on-investment findings 514
24.6 Discussion 515
    24.6.1 Rejection of hypothesis 515
    24.6.2 Selection of options 515
24.7 Limitations 515
24.8 Conclusion 515
Acknowledgments 516
Further Readings 516

Chapter 25 Using Six Sigma to improve trucking 517
25.1 Introduction 517
    25.1.1 Research problem and literature review 517
25.2 Research overview and objectives 518
25.3 Research method and approach 518
    25.3.1 Task 1—Flowchart 518
    25.3.2 Task 2—Web-based software solution 519

25.3.3 Task 3—Investigate RFID technologies for automating data capture .. 519
25.3.4 Define 519
25.3.4.1 Define—Results 519
25.3.5 Measure 519
25.3.6 Analyze 519
25.3.7 Design 520
25.3.7.1 Solution 1 520
25.3.7.2 Solution 2 520
25.3.7.3 Results 521
25.3.8 Benefits of RFID 521
25.4 Web order entry 521
25.5 Tracking and tracing 521
25.6 Carrier selection 521
25.7 Information exchange 521
25.8 Mapping 522
25.9 Automated paperless logs 522
25.10 Gantt chart and project timeline 522
25.11 Conclusion 523
Acknowledgments 523
Reference 523
Further Readings 524

Chapter 26 Using Six Sigma logistics to optimize a city's supply chain
inventory supplies 525
26.1 Introduction 525
26.2 Background 526
26.3 Capacitated plant location model 527
26.4 Network modeling steps incorporated into a Six Sigma service project ... 527
26.4.1 Define 528
26.4.2 Measure 528
26.4.3 Analyze 528
26.4.4 Improve 528
26.4.5 Control 529
26.5 Case description 529
26.5.1 Organizational description 529
26.5.2 Project description 529
26.5.3 Lessons learned 531
26.6 Implications for the technical manager 532
26.7 Conclusions 532
References 533
Further Readings 533

Section V: Nontraditional Lean Six Sigma and modern quality trends

Chapter 27 Lean Six Sigma certification and belt levels 537
27.1 Introduction 537
27.2 Executive level 537

	27.2.1	Sponsorship information	537
	27.2.2	Belt information	538
		27.2.2.1 Master black belt	538
		27.2.2.2 Yellow belt	539
		27.2.2.3 Green belt	539
		27.2.2.4 Black belt	539

27.3   ASQ certified Six Sigma black belt 539
27.4   IIE BB certification 539
         27.4.1 Institute of industrial engineers 540
27.5   Summary 542
27.6   Lesson learned 542
27.7   ISCEA LSS pathways to excellence 542
         27.7.1 Executive LSS yellow belt—Invitation only 542
         27.7.2 Certified LSS yellow belt 543
         27.7.3 Certified LSS green belt certification program 543
         27.7.4 Certified LSS black belt certification program 543
27.8   Company certifications 543
References 544

Chapter 28 Lean Six Sigma practitioners, consultants, and vendors 545
28.1   Introduction 545
28.2   Champion 545
28.3   Black belt 545
28.4   Green belt 545
28.5   Current major vendors 546
28.6   Vendor selection model 546
28.7   Consultants and integrators 547
         28.7.1 Master black belt 547
References 547

Chapter 29 Six Sigma project management: What to do after the Six Sigma
**   sponsorship phase 549**
29.1   Introduction 550
29.2   Project definition worksheet 551
29.3   Voice of the customer 553
         29.3.1 Customer identification 553
         29.3.2 Internal customers 556
         29.3.3 External customers 557
29.4   Project charter and teams 559
         29.4.1 Setting project boundaries 560
         29.4.2 Project charter 561
                  29.4.2.1 Content of the charter 561
                  29.4.2.2 Negotiation of the charter 565
                  29.4.2.3 Project management and its benefits 565
                  29.4.2.4 Project measures 566
         29.4.3 Teams 567
                  29.4.3.1 Team leader roles and responsibilities 567
                  29.4.3.2 Team formation and evolution 568
                  29.4.3.3 Team rules 569

		29.4.3.4	Types of teams	569
		29.4.3.5	Implementing teams	570
	29.4.4	Meeting management	571
	29.4.5	Conflict resolution in teams	572
29.5	Project tracking	573
	29.5.1	Work breakdown structure	574
	29.5.2	Milestone reporting	575
	29.5.3	Project report	576
	29.5.4	Document archiving	576

References 577
Further Readings...577

Index 579

Preface

With the book *Quality Management for Organizations Using Lean Six Sigma Techniques*, we hope to establish the concepts and principles by which students, quality and Six Sigma practitioners, and corporate quality managers will learn about Lean Six Sigma (LSS) and its origins in quality, total quality management (TQM), and statistical process control and how it can be integrated into manufacturing, logistics, and health-care operations. This is a comparative evaluation of LSS against other quality initiatives such as ISO 9000, Malcolm Baldrige Awards, Deming Prizes, and other programs that industry professionals seek to implement.

The integration of LSS principles into other quality initiatives is described from best practices, failed integration, and successful academic and industrial case studies. Also, a description of the job market and expectations at different levels from current Six Sigma practitioners, program sponsors, and executives are provided. This guide will provide an unbiased description of the good, bad, and ugly in integrating LSS into modern operations as they seek value-based quality initiatives.

Moreover, the description of this modern LSS and other quality initiatives' histories, current use, and future application will serve to educate student, academia, and organizations that focus on quality. We also discuss the use of design for Six Sigma for start-up operations or companies as a means for product development. Lastly, we hope to provide self-help for organizations, contractors, individuals, and corporate managers who wish to improve and modernize quality initiatives with LSS techniques.

LSS is a fast-growing approach in the organizational quality field that has come of age. Other quality initiatives have been useful in improving quality over the years such as TQM, ISO 9000, and Malcolm Baldrige Awards, which have been adopted and have improved worldwide quality in the past. LSS is the modern method for quality to improve organizational performance. Although LSS is a popular trend, there are numerous challenges for using LSS in lieu of, in conjunction with, or integrated with other quality initiatives. Thus, there is a need for a new book in this area.

Many practitioners, contractors, and researchers do not have a good reference on the workings of modern LSS technique such as design of experiments (DOEs), multifactor analysis, earnings before interest and taxes (EBIT) calculations, Pareto analysis, benchmarking, and house-of-quality techniques. These are not clearly understood and are presented in their historical use context. In this book, we will present the backgrounds on quality and LSS techniques and tools, the previous history of LSS in manufacturing, current applications of LSS in operations such as logistics and health care, as a decision model for choosing whether to do LSS or other quality initiatives, which projects should be selected and prioritized, what to do with non-LSS projects, an integration model for integrating and developing integrated LSS and other quality initiatives, and common mathematical techniques that practitioners can use for performing LSS statistical calculations.

Further, this book describes the methods to attain the different Six Sigma certifications and general processes for industry and organizations such as the American Society for Quality (ASQ) and the Institute of Industrial Engineers (IIE). We close with the future directions of LSS and quality.

The purpose of this book includes the following:

- It allows students to learn more about LSS.
- It allows quality and LSS practitioners, consultants, and contractors to learn and use the most modern LSS techniques, and understand the body of knowledge of Six Sigma with respect to certification and testing.
- It allows practitioners to learn and integrate process improvement and quality initiatives in operations.

The features of this book are as follows:

- It presents a decision model for choosing LSS as an organizational initiative.
- It presents an economic comparison model for implementing LSS and quality initiatives.
- It presents a useful way to use LSS techniques for process improvement at work.
- It demonstrates how to integrate LSS with other quality initiatives in quality organizations.
- It has a history of modern popular quality initiatives such as ISO 9000.
- It demonstrates how to integrate LSS with future complex organizations such as health care.
- It discusses the case studies of LSS experienced by Dr. Erick C. Jones as University CSSBB program chair.
- It provides case studies from academic researchers.
- It presents LSS statistics and quality tools with classroom theory and problems.

MATLAB® is a registered trademark of The MathWorks, Inc. For product information, please contact:

The MathWorks, Inc.
3 Apple Hill Drive
Natick, MA 01760-2098 USA
Tel: 508 647 7000
Fax: 508-647-7001
E-mail: info@mathworks.com
Web: www.mathworks.com

Acknowledgments

This book is the product of many long hours, hard work, and persistence of my personal support team.

I thank God for giving me the strength and desire; my family—Ranita, Erick Jr., Chelsey, and Morgan—for giving me the motivation; my graduate students for providing me the inspiration through their many questions; and my undergraduate and graduate students for their enthusiasm toward this project and their comments and ideas on possible uses in the future.

I want to personally thank Harrison Davis Armstrong for his tireless work, without his Herculean efforts this book would not have been finished over the last year. His immense talent has allowed the updated concepts of Six Sigma to be articulated for his generation.

I especially thank my mother for her never-ending support and love and care. She inspires me in good and hard times to find the best in myself.

I thank all of the Six Sigma organizations that have supported these ideas including the American Society of Quality (ASQ), the Institute of Industrial Engineers (IIE), and the International Supply Chain Education Alliance (ISCEA). I also thank Goodyear Tire, Kelloggs, TRW, Novartis, Interdeli (Mexico), and Werner Trucking who have had their students participate in some of these Six Sigma programs over the past 10 years. I thank the University of Nebraska–Lincoln and the University of Texas at Arlington that allowed these programs to exist.

I express my special thanks to Harrison Armstrong, my student, who worked almost as hard as I did on this book. His persistence to make sure the chapters are perfect is commendable. I also thank Drs. Ida Lumintu, Vettrivel Gnaneswaran, and Maurice Cavitt and my students Chendur M. Anand, Karthika, Harini, and Boobathy who were part of my National Science Foundation Student Research in Queretero, Mexico, including Stan Ugogi, Juan Robles, Erick C. Jones Jr., Stan Chidebe Ugoji, Mohammad Siddiqui, Nithin Dainiel, Shin-Chiann Han, Walter Muflur, Cinthya Vinuez Garcia, and Mackenzie Dacres. Dr. Beatriz Murrieta-Cortes at Tecnológico de Monterrey, Campus Querétaro (ITESM QRO) supported us while in Mexico. Finally, I thank Drs. Mahour Parast, Vettrivel Gnaneswaran, John Priest, and Edmund Prater for their physical and conversational contributions to the book.

Erick C. Jones

Introduction

The radio frequency identification (RFID) and auto-identification (Auto-ID) or RAID laboratories were established in 2011 at the University of Texas at Arlington (UTA). The laboratories are located in two distinct facilities, 413 Woolf Hall (RFID laboratories) and 309 Engineering Labs Building (Auto-ID laboratories), and provide faculty and graduate and undergraduate students access to equipment for conducting research. Previous research performed by the principal investigator (PI) from NASA, the Department of Transportation, and the National Science Foundation (NSF) has provided additional industrial equipment that allows for testing of unique industrial environments. Software that the laboratories utilize include SAS, Minitab, MATLAB®, and Labview. Often to ensure that the equipment is in working condition, the laboratories at a minimum biannually provide a working tour to K–12 students and educators. The mission of the laboratories "to provide integrated solutions in logistics and other data driven environments through automatic data capture, real world prototypes, and analysis" is met through support for faculty and students to continue in this research area (www.uta.edu/rfid). The laboratories hosts RFID and Auto-ID equipment including handheld readers, antennas and antenna portals, and 3-D real-time location systems. For developing themes in health care and logistics, hospital beds and mannequins, working roller conveyors, and industrial grade warehouse portal stands are in the laboratories. The RAID laboratories of UTA have a vision: "Everything will be tracked wirelessly in 10 years."

Minitab Free-Trial Offer

A free 30-day trial of the Minitab Statistical Software is available at http://www.minitab.com/products/minitab/free-trial.aspx.

About the RAID laboratories

The managing quality text came out of Dr. Erick C. Jones' experience as a consultant, as a teacher and researcher in quality systems, and as a creator of the academic Six Sigma online program. As an additional motivator, the use of Design for Six Sigma has taken place while conducting his world-class research activities currently consisting of logistics, RF and Auto-ID, and automated inventory control theory. More details of Dr. Jones' background are provided in the following sections including his current research facilities in which Six Sigma is both utilized, applied, and researched.

RFID laboratories (413 Woolf Hall)

RFID equipment (general)

RFID 3-D real-time locating system (RTLS) equipment

RFID health-care team equipment

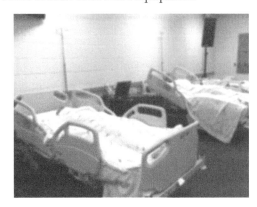
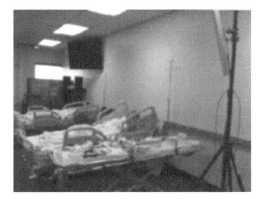

RFID logistics and warehouse equipment

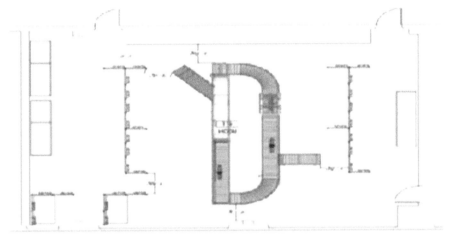

Auto-ID laboratories (309 Engineering Lab Building)

Author

Dr. Erick C. Jones, PE, American Society of Quality Certified Six Sigma Black Belt (ASQ CSSBB), is an associate professor in the Department of Industrial and Manufacturing Systems Engineering of the College of Engineering and Computer Science at the University of Texas, Arlington. In his current appointment, he is a prominent professor primarily in the field of Six Sigma, supply chain management, and radio frequency identification (RFID).

Six Sigma background

Dr. Jones brings industry experience in quality implementations as a former consultant. He is an American Society for Quality (ASQ) Certified Six Sigma Black Belt. He originated and directed a university-level Six Sigma black belt program for several years. The program expanded to a state-level program with an industry and university certification board. He is the chairman for the International Supply Chain Education Alliance (ISCEA)—Industry Technology Board, Indian Institute of Technology Bombay (IITB) that certifies the training programs, exams and review process for the ISCEA technology RFID supply chain management (RFIDSCM), and Six Sigma programs.

Academic background

Dr. Jones graduated from Texas A&M University with a degree in industrial engineering. He later earned a master's degree and a PhD in industrial engineering from the University of Houston while concurrently working in industry.

Industry background

Dr. Jones has held positions in industry that include engineering specialist, engineering director, engineering consultant, project manager, and executive manager for United Parcel Service (UPS), Tompkins Associates, Academy Sports and Outdoors, and Arthur Andersen.

He managed teams and operations as small as 3 people and as large as 500 people. He has managed projects implementing warehouse management systems (WMS) and enterprise resources planning (ERP) system, and designing and constructing new facilities and reengineering Fortune 1000 organizations. Operations managed include strategic systems deployment, teams of large-scale distribution operation, and human resources at an executive level. He is an expert in the field of supply chain optimization, distribution logistics, and inventory control. His unique background led him to originate one of the first and largest academic RFID laboratories in the country in 2003. He continues this research today, which includes a research project that led to RFID being shipped and utilized at NASA on the International Space Station (ISS) in 2013. Some astronauts have tweeted from space of the usefulness of the technology.

Current key responsibilities and scholarly accomplishment

Dr. Jones has several key responsibilities aside from his current academic appointment as an associate professor at UTA. He is currently the deputy director of the UTA Homeland Security focused Security Advances via Applied Nanotechnology (SAVANT) Center (www.uta.edu/rfid), director of the RAID laboratories (www.uta.edu/raid) at UTA, and campus site director of the Alfred P. Sloan Foundation Minority PhD Program.

Academically, Dr. Jones has received significant funding for academic research and projects from organizations such as the National Science Foundation, the Department of Transportation, various US State Departments, and NASA. He has published over 117 manuscripts including journal and magazine articles and book chapters. He has written two college textbooks on RFID and edited two industry texts on the subject. He has graduated over 26 master's and PhD students since 2003.

section one

Modern quality and Lean Six Sigma overview

chapter one

Overview of modern quality

> Let all things be done decently and in order.
>
> **—1 Corinthians 14:40**

1.1   Introduction

Most economists would agree that for a country such as the United States to remain a strong economic power, its manufacturing sectors must remain competitive. Often domestic economic conditions such as the national trade deficit are linked to a country's manufacturing competitiveness. Given that trade is driven by customers' willingness to pay for products and services, and many studies suggest that customers would pay more for a better quality of product, quality is an important factor for driving trade in the United States. In this chapter, we hope to provide an overview on the definition of quality. Section I of this book deals with the "Quality Trail" as given below. The diagram in Figure 1.1 shows that everything that has been established in quality management (QM) will lead to Lean Six Sigma (LSS).

1.2   What is quality?

1.2.1   How is quality defined?

Merriam-Webster defines quality in several ways, including the following: (1) peculiar and essential character, (2) degree of excellence, (3) superiority in kind, (4) distinguishing attribute, and (5) acquired skill. These definitions range from describing products to describing people. Our focus will be mainly on describing products and services. We will define quality by taking several approaches that are consistent with historical engineering themes. Though there may be other approaches that can be applied to defining quality, Garvin's (1984) approach is generally accepted in engineering. The five approaches to quality include global or transcendent, product-based, user-based, manufacturing-based, and value-based approaches.

1.2.1.1   Global approach

The global approach to quality is a very generic view. It is often difficult for companies to analyze and/or measure and be subjective. As described by customers, quality is generally described as "good enough" or it "gets the job done." Often this type of quality is further defined by companies as good enough not to get sued. In other words, if no one is defining this as bad quality, then it must be good quality.

1.2.1.2   Product-based approach

The product-based approach to quality relates to defining a product's attributes. The attributes or features that customers find more desirable would define the product as having higher quality. Often companies determine the higher quality attribute by defining

Figure 1.1 The quality trail.

what they are willing to pay more money for. For example, a car with leather seats and a high-end radio system would be considered higher quality than a car with cloth seats and no radio system.

1.2.1.3   User-based approach

The user-based approach to quality is related to users defining the features and attributes they consider important. One of the more important quality gurus, Joseph Juran (1998), defines quality as "fitness for use" as defined by the user. In other words, the users and their expectations of the product determine quality. This is further clarified that different users may use the product in different ways, and a high-quality product must possess multiple elements that "fit" different uses effectively.

1.2.1.4   Manufacturing-based approach

The manufacturing-based approach to quality is related to meeting engineering specifi-cations in order to meet quality standards. This is generally the most comfortable defi-nition for engineers in defining quality in that it is specific, measurable, and objective. Unfortunately, meeting engineering specifications does not always ensure higher sales from finicky customers. This approach is generally attributed to Crosby's (1979) "confor-mance to requirements" concept. The concept describes a set of requirements that must be met in a manner required by specifications or standards. Generally, if these specifications are not met, then the product is considered noncompliant and defective. Every distinct noncompliant attribute is considered a defect.

1.2.1.5   Value-based approach

The value-based approach to quality is related to the market-driving quality. It builds on the fact that consumers pay higher prices for better quality products. The general idea is that market value determines the quality of the product. There are attributes that are associated with the value-based approach. These attributes are generally associated with Garvin (1984) and include (1) performance, (2) features, (3) reliability, (4) conformance, (5) durability, (6) serviceability, (7) aesthetics, and (8) perceived quality.

1.2.1.5.1   Performance.   The performance attribute refers to the primary operating characteristics of the product. It describes the ability of a product to meet its intended pur-pose. For example, the ability for a car to operate on a freeway by accelerating into and out of traffic as opposed to the ability of the car's radio to play. The primary purpose of the car is its performance on the road.

1.2.1.5.2   Features.   The feature attribute describes the "nice to have" or the "bells and whistles" of a product. An example would be the extra food and drinks for first-class flight passengers, hands-free phone in sync inside a car, and automatic shut-off motion sensors on big screen televisions. Often these features mean higher quality to consumers and drive the sales of many products.

1.2.1.5.3   Reliability.   The reliability attribute describes the probability of the failure of a product within a specified period of time. Often this attribute is associated with a product's brand name. For example, American cars were perceived to have the propensity of not performing consistently for their designed useful life and were thus considered to be of lower quality than Japanese cars. Reliability in manufacturing literature is generally measured in the following terms: mean time to first failure (MTFF) and mean time between failures (MTBFs).

1.2.1.5.4   Conformance.   The conformance attribute is described as the degree to which a product's design and operating characteristics meet pre-established standards.
It involves the following:

1. Internal elements (inside the factory): Conformance is measured commonly by the number of defects (proportion of all units that fail to meet specifications and so require rework or repair).
2. External elements (outside the factory): Conformance is measured commonly by the number of repairs under warranty.

1.2.1.5.5   Durability.   The durability attribute is described as the measure of product life, defined in two dimensions:

1. Technically: Durability is the amount of use one gets from a product before it deteriorates physically and repair is not possible (i.e., after many hours of use, the filament of a light bulb burns up and the bulb should be replaced).
2. Economically: Durability is the amount of use one gets from a product before it breaks down and repair is required. The product's life is determined by repair costs, personal valuation of time and inconvenience, loss due to downtime, and other economic variables.

1.2.1.5.6   Serviceability.   The serviceability attribute is described as the speed, courtesy, and competence of the repair service. Objective and subjective views play a role in defining *serviceability.*

1.2.1.5.7   Aesthetics.   The aesthetic attribute is described as to how a product looks, feels, sounds, tastes, or smells (appearance and impression): *quality* is viewed as the combination of these attributes that best match consumer preferences.

1.2.1.5.8   Perceived quality.   The perceived quality attribute is described as how consumers feel when they use the product. The concepts of durability and reliability are related to how consumers feel about a product. It is implied that a product that fails frequently (low durability) is likely to be scrapped earlier than the one that is more reliable. Consumers, if financially capable of making a choice, would choose a product with higher durability as a means of choosing a product with higher quality.
These different attributes also further define the differences among the five traditional approaches to *quality,* as explained previously. Each of the approaches focuses implicitly on a different attribute and dimension of quality: For example, the product-based

approach encompasses the concepts of performance, features, and durability; the user-based approach describes aesthetics and perceived quality; and the most common concept of Six Sigma in the manufacturing-based approach includes and focuses on the ideas of conformance and reliability. The efforts toward improvement of these measures are normally viewed as translating directly into quality gains (objective measures of quality). They are less likely to reflect individual preferences.

1.3   What do organizational departments or functions think about quality?

1.3.1   Functional perspectives on quality

Organizational quality may mean different things to different organizational functions. Each function has marked the specific tasks for its function and then translates how quality impacts those tasks. Typical functions that are impacted by quality include engineering, operations, executive management, marketing, finance, and human resources. Each function has a different view of quality and its effect on its functional goals in the organization.

1.3.1.1   Engineering perspective on quality

A critical perspective, and one that our book focuses on, is generally the engineering perspective. Engineers generally seek to apply scientific and mathematical problem-solving skills and models to business and industrial problems. They generally seek solutions related to product and process design and/or redesign. Product design engineering involves activities associated with developing a product from concept to final design and implementation. This also includes product redesign that involves continuously improving current products. In product design and redesign, quality is an important component and engineers seek to design high quality into the products. Process design is generally inferred in product redesign and new product design. The idea of integrating quality into the process is generally there in most quality theories. This means that if you have processes that support high-quality components and subcomponents, then the products will have high quality (Figure 1.2).

1.3.1.2   Operations perspective on quality

The operations planning and management view of quality is rooted in the engineering approach. Like engineers, operations managers, planners, and schedulers are tasked with optimizing product and process design. In contrast to traditional engineering, operations researchers and some specialized industrial engineers focus on decision modeling rather than on the technical (traditional engineering) aspects of these activities; operations concentrates specifically on the best decisions and policies for the management to execute these activities. Operations management has developed into an integrative field, combining concepts from engineering, operations research, organizational theory, organizational behavior, and strategic management.

1.3.1.3   Systems perspective on quality

Operations management, in conjunction with industrial and systems engineers, utilizes a systems view to address quality problems that support modern QM thinking. The systems view involves the understanding that product quality is the result of the interaction of several variables such as machines, labor, procedures, planning, and management (Figure 1.3).

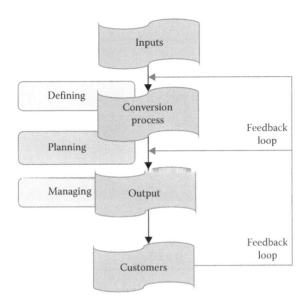

Figure 1.2 Engineering perspective of quality.

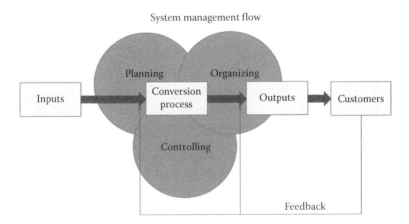

Figure 1.3 Systems perspective of quality.

1.3.1.4   Executive management perspective on quality

Executive management is tasked with managing the investments that support organizational strategic objectives. Strategy is defined as the planning processes used by an organization to achieve a set of long term goals. The strategic objectives must be understandable and executable in terms of goals, policies, and plans to achieve quality improvement. In some organizations, it is the tradition for quality-related strategic planning to be treated separately from firm-level strategic planning. It is more common for organizations to integrate strategic planning and include QM in the firm's business practices (Figure 1.4).

The ultimate goal of strategic quality planning is to aid an organization to achieve sustainable competitive advantage, market share, and returns for its investors. Research shows that quality is still one of the major concerns for CEOs of large organizations.

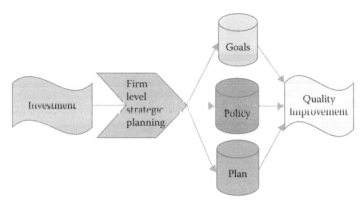

Figure 1.4 Executive management perspective of quality.

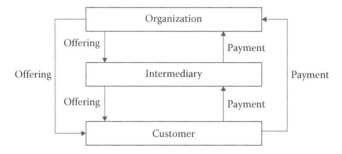

Figure 1.5 Marketing perspective of quality.

1.3.1.5 *Marketing perspective on quality*

Marketing efforts are often focused on managing the perception of quality by the public. This may include focusing on customer relationship management, including paying attention to delivering value to the customer. The tools for influencing customer perceptions of quality are primarily price and advertising. The marketing perspective primarily focuses on the end customer's perspective of the product. Customer service surveys are important mediums for assessing quality (Figure 1.5).

1.3.1.6 *Financial perspective on quality*

One of the most commonly asked questions about QM is "Will it pay us financial benefits?" The financial perspective relies more on quantified, measurable, and results-oriented thinking. W. Edwards Deming made the first theoretical attempt to link quality improvements to financial results through the "Deming Value Chain" (Foster 2010). Later, Juran (1998) mentioned that management will always ask the question "Will it make us money?" This has led to the propounding of the concept in quality described as the law of diminishing marginal returns (Figure 1.6).

According to this law, there is a point at which investments in quality improvement become uneconomical. According to the quadratic economic quality-level model, higher levels of quality will result in higher expenditures. This view is at odds with the ethic of continual improvement. Some suggest that this is the main distinction between traditional quality program concepts and LSS initiatives. The focus of deriving the firm return or the intrinsic value that will be received by performing quality as opposed to a firm seeking to do continual improvement on an ongoing basis has been a point of distinction (Figure 1.7).

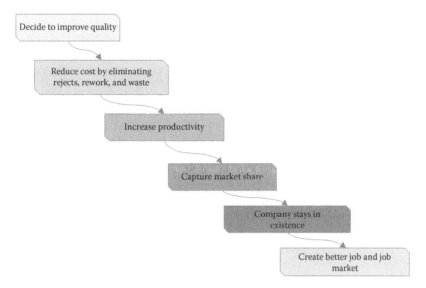

Figure 1.6 Deming value chain.

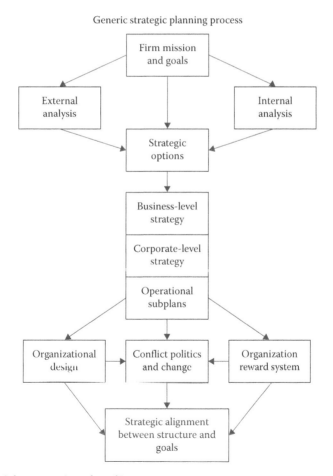

Figure 1.7 Financial perspective of quality.

1.3.1.7 Human resources perspective on quality

Understanding the human resources perspective on quality is difficult and necessary because implementation of quality without the commitment of employees brings disastrous results. This perspective of quality generally encompasses several concepts that include but are not limited to employee empowerment, organizational design, job analysis, and 360 degree evaluation:

1. *Employee empowerment:* Empowering employees involves moving decision making to the lowest level in the organization.
2. *Organizational design:* Human resource managers are involved in many aspects of organizational design, such as the design of rewards system, pay system, organizational structure, compensation, training mechanisms, and employee grievance arbitration.
3. *Job analysis:* This involves collecting detailed information about a particular job.
4. *360-degree evaluation:* This is a performance measurement system in which an employee's peers, supervisors, and subordinates are involved in evaluating the worker's performance.

1.4 Total quality management and quality overview

There are many different terms used to refer to an integrative philosophy of management that seeks to continuously improve the quality of the firm's products and internal processes which include total quality management (TQM), QM, and corporate quality, to name a few (Ahire 1997). The most prominent of these, the author suggests, is TQM.

TQM also promotes the idea that the organization and all of its employees support continuous improvement, and this integrative approach translates into high-quality products. The idea is prescribed that the quality of products and processes is the responsibility of everyone in the organization including the management, the workforce, and the suppliers. Cua et al. (2001) identified the following nine common TQM practices:

1. Cross-functional product design
2. Process management
3. Supplier QM
4. Customer involvement
5. Information and feedback
6. Committed leadership
7. Strategic planning
8. Cross-functional training
9. Employee involvement

TQM concepts and practices were developed by quality gurus in the United States. An interesting fact about these gurus or consultants was that they initially had few believers for these concepts in the United States. On the other hand, Japan was struggling with the perception of low-quality goods originating from that country as a whole at the time these gurus initiated the concepts. Only after the Japanese companies greatly succeeded in employing these principles did many of these gurus become well known. In fact, the highest Japanese honor for a company is named after one of these US gurus. The premier annual prize for manufacturing excellence in Japan is the Deming Award. The US gurus, who were recognized for their ideas in Japan, were mainly W. Edwards Deming, Joseph M. Juran, and Armand V. Feigenbaum. There are others, but they are the ones who made the main contributions that also extend into LSS concepts. Other contributors to LSS methodologies were Philip B. Crosby (1979) and Kaoru Ishikawa.

1.5   Who is taught and trained to attain quality by an organization?

1.5.1   The implementation of quality

Generally, there are two ways of implementing quality organizationally. One approach is to take the attitude that quality is integral to all functions, policies, and procedures of the firm. Another approach is to implement quality on an "as-needed" basis. The "as-needed" approach seeks to add quality as a means to improve profits through a highly specialized team. Generally, companies perform both approaches. They maintain acceptable quality throughout their operations so as to put acceptable products in the market. Later, they use quality teams, LSS initiatives, and Six Sigma teams to improve quality when it becomes necessary to make products more competitive in the marketplace.

1.5.1.1   The integrated approach

The integrated approach to quality is generally associated with a concept called the three spheres of quality. It is implied that the conceptual areas of quality are integrated throughout a firm's policies, procedures, and products.

1.5.1.2   Three spheres of quality

The three spheres of quality originated by Juran (1998) refer to quality control (QC), quality assurance (QA), and QM. Often activities in one sphere could also be completed in another sphere, and this is subject to the interpretation of how the activity supports the firms' needs. The tasks associated with these spheres are described in next page (Figure 1.8):

1.5.1.2.1   Quality control.   QC focuses on providing analysis and identifying relationships and causes of variation that create lower quality. QC generally refers to the following activities:

* Monitoring process capability and stability
* Measuring process performance
* Reducing process variability

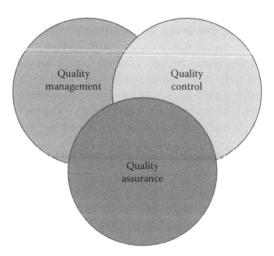

Figure 1.8 The three spheres of quality.

- Optimizing processes to nominal measures
- Performing acceptance sampling
- Developing and maintaining control charts

1.5.1.2.2 Quality assurance. QA focuses on testing and researching activities guarantees that products meet specifications and perform as advertised. QA activities are related to the following.

- Failure mode and effects analysis
- Concurrent engineering
- Experimental design
- Process improvement
- Design team formation and management
- Off-line experimentation
- Reliability/durability testing of product

1.5.1.2.3 Quality management. QM describes the management processes that overarch and tie together the completion, control, and transitioning of quality activities. QM activities include

- Planning for quality improvement
- Creating a quality organizational culture
- Providing leadership and support
- Providing training and retraining
- Designing an organizational system to reinforce quality ideals
- Providing employee recognition
- Facilitating organizational communication

1.6 What if quality is too expensive to justify?

1.6.1 "As-needed" implementation of quality

The "as-needed" implementation of quality throughout the firm includes the following concept: "What does the customer perceive as quality? And if the customer does not want to pay for it, then excessive quality is not necessary." This concept drives the thought that activities that do not support the customers' view of quality are nonvalue activity or wasted effort in the firm. Minimizing these activities or making processes "Lean" is supported by this implementation approach. Also, using specialized teams such as those of Six Sigma to identify a process that the customer and/or financial motivations have suggested as needing improvement yields a higher quality and thereby improves the firm's bottom-line.

Additionally, this implementation also is influenced by international markets and customers who may have differing preferences between cultures and countries. This drives how countries influence quality.

1.7 Limitation of quality measures

Quality is difficult to measure and often organizationally difficult to justify in traditional accounting terms such as return on investment (ROI). The timing between quality initiatives and cost saving creates difficulty in justifying quality investments.

Generally, hidden costs that usually are not accounted for are the main drivers of quality investment. Understanding the cost of quality (COQ) is one of the main reasons why modern LSS techniques have replaced other quality initiatives. Six Sigma focuses on the COQ.

1.8 How can the COQ be justified?

1.8.1 Cost of quality

After World War II, many quality departments emerged globally because infrastructure was destroyed in several countries due to the war. In order to make competitive products, these companies had to change their image as suppliers of poor quality products to even improve their export goods. Quality initiatives in various organizations across the world began when Deming started helping Japanese companies. Many companies discovered that quality was sometimes the root cause to their bad economic performance. For instance, even though IBM instilled a zero-defect policy, their economic performance did not improve. Juran (1999) addressed the concept of quality costs in the first edition of *Quality Control Handbook*. It describes that losses in defects were equal to the COQ control.

1.8.2 Traditional COQ

Conformance costs are those that are incurred to ensure manufactured products or delivered services conform to specifications. Conformance costs are made up of the following:

1. Prevention costs or activities relating to preventing defects
2. Appraisal costs or activities relating to measuring and evaluating the product or service
3. Nonconformance costs are the prices paid when the product or service does not conform to a customer's requirement. Nonconformance costs are made up of internal failure costs, failures incurred prior to shipment to customer, external failure costs, and costs discovered after shipment or service delivery.
   a. Internal failure costs can be remedied by using rework, identifying the amount of scrap produced, reinspection, and retesting, to name a few.
   b. External failure costs can be remedied by investigating customer complaints, inspecting warranties, and repairs.

Caveats are the direct tradeoffs between conformance and nonconformance expenditures that are economically difficult to measure. Some of the main issues with the hypothetical COQ concepts include the following:

- Cost-of-quality reports do not provide specific actions.
- Cost-of-quality calculations do not capture all of the costs, and multiple tasks may count the same savings many times leading to inaccuracies in corporate accounting.
- Accounting conventions such as capital spending and rules for defining period and product cost render COQ of little use for evaluating a quality program.

1.9 Who started quality and what ideas did they contribute?

1.9.1 Quality philosophers, gurus, and contributors

1.9.1.1 William E. Deming

Deming is often considered the father of modern quality. Deming, a US quality consultant who had limited success with American businesses, introduced quality principles to the more-accepting Japanese manufacturers on a large scale post-World War II. Many of Deming's principles and ideals are the foundation of quality initiatives today. He utilized some of the concepts identified with Shewart, such as control charts, into his philosophies in his research (Foster 2010). Other notable ideas include Keiretsu partnerships with suppliers who are precursors to vendor-managed inventory and partnerships today. He also introduced the concepts of common and special causes of manufacturing defects. He described the idea of seeking profound knowledge to make breakthrough improvements in quality and manufacturing efficiency. Moreover, he had organizational understanding and appreciated what we describe as "system thinking," theories on process variation, and organizational knowledge management and information retention. The idea of understanding the psychology of the employee was a major contribution. This concept is realized in his overarching philosophy described as the "Fourteen Points." Deming's Fourteen Points is based on three principles: constancy of purpose, continual improvement, and cooperation between factions. Deming believed in the following Fourteen Points of quality:

1. Create constancy of purpose
2. Adopt the new philosophy
3. Cease inspection, require evidence
4. Improve the quality of supplies
5. Continuously improve production
6. Train and educate all employees
7. Supervisors must help people
8. Drive out fear
9. Eliminate boundaries
10. Eliminate use of slogans
11. Eliminate numerical standards
12. Let people be proud of their work
13. Encourage self-improvement
14. Commit to ever-improving quality

Another contribution that directly relates to Six Sigma techniques was that Deming was the first guru to theoretically attempt to link quality improvements to financial results. It was described as the "Deming Value Chain."

1.9.1.2 Joseph M. Juran

Juran (1998) followed Deming in Japan and is mainly known for using the Quality Trilogy (planning, control, and improvement). He also originated the three spheres of quality: QC, QA, and QM. Juran (1998) gave the definition of quality as "fitness for use" as defined by the user. In other words, the users and the expectations that they have of the product determines quality. This is further clarified to mean that different users may use the product in different ways and a high-quality product must possess multiple elements that "fit" different uses effectively.

He also contributed to the LSS and Six Sigma ideas of financially quantifying quality improvements. To support this idea, he introduced ideas such as decreasing chronic waste, limiting rework, and minimizing the COQ. He is known for stating that management will always ask the following question: "Will it make us money?" This has led to the concept in quality described as the law of diminishing marginal returns that extended the idea of COQ.

1.9.1.3   Philip B. Crosby
Crosby (1979) focused on improving preventive approaches that would reduce quality appraisal tradeoff. He is also widely recognized for his creation of the management maturity grid. This approach is generally attributed to Crosby's "conformance to requirements" concept. The concept describes a set of requirements that must be met in a manner required by specifications or standards. Generally, if these specifications are not met, then the product is considered noncompliant and defective.

1.9.1.4   Armand V. Feigenbaum
Armand Feigenbaum joined General Electric in 1944. He defines total quality as an excellence-driven rather than a defect-driven concept. He also feels that the quality philosophy extends beyond the factory floor to include all of the functions in the organization. He believed that an organization should be excellence-driven rather than defect-driven as well as for adjusting the view of the organization to match that of the customer or excellence-driven, rather than defect-driven view defined by the customer (Rao 1996).

1.9.1.5   Garvin
Garvin's (1984) approach is generally accepted in the field of engineering. The five approaches to quality include global or transcendent, product-based, user-based, manufacturing-based, and value-based approaches.

1.9.1.6   Ishikawa
Kaoru Ishikawa graduated from the University of Tokyo in 1939 with a degree in applied chemistry. He believed that all divisions and all employees in the organization should be involved in studying and promoting quality control by learning seven statistical tools. He perfected the cause-and-effect diagram. He also believed the customer defines the quality that the organization believes in. He was also a pioneer in quality circles and teamwork to improve quality (Rao 1996).

1.9.1.7   Taguchi
Genichi Taguchi, formerly an employee of Nippon Telephone and Telegraph, has had significant influence on the quality movement in Japan. His prime focus was in making statistics practical. He viewed quality as an issue for the entire company and focused on the use of statistical methods to improve quality, particularly in the area of product design. He created loss function and focused on improving quality in design with reduction of noise in the form of system design, parameter design, and tolerance design (Rao 1996).

1.9.1.8   Japanese contributions to quality
Japan has many contributions to the art of QM. The most notable is LSS production. Two views emerge that pertain to LSS. The first view of LSS is a philosophical view of waste reduction. This view asserts that anything in the process that does not add value for the customer

should be eliminated. The second view of LSS is a systems view stating that just in time (JIT) is a group of techniques or systems focused on optimizing quality processes. Other contributions include visibility, in-process inspection, 5S methodology, and emphasizing teamwork and quality circles.

Japanese quality is the belief and method in quality that the Japanese used post-World War II to the present day. These are as follows:

Fitness to Standard: ingrained in management to have quality over profit
PDCA Style: plan–do–check–act (Shewart charts)
Fitness to Use: understand customer needs
Fitness to Cost: high quality, low cost
Fitness to Latent Requirement: discover the needs of the customers before they do
Kano Diagram: model of a set of ideas for planning a new product, service, or process
Hoshin Kanri: vertical deployment of TQM strategy

1.10   How is quality rewarded and enforced globally?

1.10.1   Quality awards

Quality awards and standards are driven by how the country improved through its business enterprises and how these businesses are perceived globally. The awards are usually named after a famous individual who contributed to the business success of companies in that country.

1.10.1.1   The Deming Prize

The Deming Prize for quality was established in 1951 by the Japanese Union of Scientists and Engineers (JUSE). There are three categories of awards: application prize for division, application prize for small business, and QC award for factory. Judging is based on several criteria: policy, organization, and operations, collecting and using information, analysis, planning for the future, education and training, QA, quality effects, standardization, and control. In contrast to the Malcolm Baldrige Award, there is no limit on the number of companies that can receive the Deming Award in a given year. The Deming Prize is more focused on processes than the Baldrige Award.

1.10.1.2   Malcolm Baldrige National Quality Award

The Malcolm Baldrige National Quality Award (MBNQA) was created by the federal government to award companies and organizations that held the highest standard for quality success. The award is open to small and large firms in the manufacturing and service sectors. It is, however, not open to public sector and nonprofit entities. Each year, two winners are selected for each category, limiting the number of winners to six per year. The award is based on seven criteria: senior executive leadership; information and analysis; strategic quality planning; human resource development and management; management process of control, quality, and operational results; customer focus; and satisfaction. The criteria focus on business results. Companies must show outstanding results in the listed areas to win. The means of obtaining the results do not follow any given format and are adaptive to each hopeful organization (Figure 1.9).

The Baldrige selection process is thorough and exact, so as to ensure that the most deserving and qualified firm is selected. The first step is eligibility determination. The completed application is sent to the National Institute of Standards and Technology (NIST).

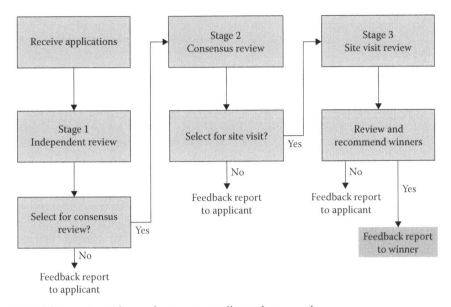

Figure 1.9 Baldrige criteria for performance excellence framework.

It is reviewed by examiners and then by judges, who determine whether it will be given a consensus score by the examiners. Firms that are granted Baldrige site visits sometimes refer to themselves as "Baldrige Qualified." The site visit consists of a team of four to six examiners visiting a company over a period not exceeding a week. One of the most important outcomes is examiner feedback. For more information on the Malcolm Baldrige Award, visit www.NIST.gov

1.10.1.3   European Quality Award

Much like the MBNQA, Europe issues the European Quality Award every year to the most accomplished applicant. There is also a second-level award given to all the organizations that meet the award criteria (Figure 1.10).

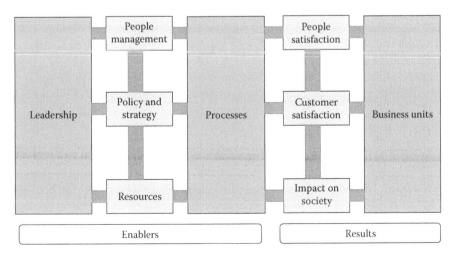

Figure 1.10 European Quality Award.

1.11   Quality standards

1.11.1   ISO 9000

International Organization for Standardization (ISO) 9000 is a quality initiative that is widely used across the world. There are 12 member countries that work together to eliminate separate standards. ISO 9000 is currently used in 91 countries, and this number is growing on a yearly basis. Additionally, there are three levels of audit robustness: ISO 9001, 9002, and 9003. ISO 9000 is the basic set of requirements for a TQM system (TQMS).

The current ISO standard is ISO 9000:2008, 2008 being the year of the latest update in the standardization. The focus of ISO 9000:2000 is for companies to document their quality systems in a series of manuals to facilitate trade through supplier conformance. There are two standards that follow ISO 9000, which are ISO 9001 and ISO 9004. ISO 9001 is used for internal implementation and contractual purposes. ISO 9004 is used for standardizing continuous improvement and enhancing overall performance. ISO is based on several underlying QM principles: customer focus, leadership, involvement of people, process approach, systems approach to management, continual improvement, factual approach to decision making, and mutually beneficial supplier relationship. Registering for ISO 9000:2008 is thorough and the steps are listed as follows:

1. Inquiry—Client contacts registrars to investigate the terms for registration and makes a selection.
2. Contract with registrar—Registration steps are determined and a price is negotiated. A preassessment or gap analysis may be done.
3. Phase 1 audit—Registrar performs an onsite audit of the documented quality system against the applicable standard.
4. Certification audit—Every element of the ISO 9000 standard is audited several times during the registration process. During each 3-year period, 100% of the organization is audited.
5. Process audits (optional)—The client may choose business processes for auditing to the applicable standard, allowing the client to learn and experience the registrar's auditing methods.
6. Final certification audit—Once the client's documented quality system has met the applicable standard, the registrar will conduct an audit to determine the system's effective implementation.
7. Rolling certification (surveillance) audits—The registrar returns either on a 6-month or on an annual cycle. Sometimes they are referred to as surveillance audits.

1.11.2   ISO 14000 and 14001

ISO 14000 is a series of standards that provide guidelines and compliance standards for environmental compliance. ISO 14001 is the compliance standard for ISO 14000 which focuses on environmental management systems. Both of the standards use the same approach as ISO 9000:2008 in several of the focus areas but direct the standardization to environmental compliance. To find out more about the ISO, visit www.iso.org

1.12   Integrating QM with LSS techniques

Many organizations emphasize quality as a means of staying competitive in the marketplace in the long run. They view that having a reputation of high quality is like representing future market share for new customers and maintaining the market share for existing customers over their lifetime. Further, improving quality can provide long-term financial savings, such as scrap and rework reduction. We associate these quality savings as long-term savings that are difficult to quantify. One method to quantify quality is the initiative known as Six Sigma. The label "Six Sigma" originates from statistical terminology, wherein sigma represents the standard deviation. The probability of falling within plus or minus Six Sigma on a normal curve is .9999966, which is more commonly represented as a defect rate of 3.4 parts per million (quoted in Narasimhan and Jones 2004; Zhang et al. 2012). This level of quality is seen as the goal in most Six Sigma initiatives.

LSS is a methodology that allows organizations to "maximize shareholder value by the fastest rate of improvement in customer satisfaction, cost, quality, process, speed, and invested capital." It is a combination of well-known waste elimination and process improvement techniques of LSS manufacturing and Six Sigma (Zhang et al. 2012). LSS is a well-structured theory-based methodology to improve performance, develop effective leadership, gather customer satisfaction, and gain bottom-line results (Jones et al. 2010a; Zhang et al. 2012). Together, LSS manufacturing and Six Sigma become more powerful and eliminate the constraints of each approach (Zhang et al. 2012). There are several approaches that can be utilized by LSS such as define, measure, analyze, improve, and control (DMAIC); define, measure, analyze, design, and verify (DMADV); or identify, design, optimize, and validate (IDOV). The DMAIC methodology utilized by Six Sigma is employed to reduce the resources wasted within existing processes. To design a new product to be of Six Sigma quality, the design for Six Sigma (DFSS) approach is needed; this process operates on either the DMADV or IDOV methodology (Jones et al. 2010b).

1.13   Will LSS really work?

As various QA methods are being developed and discarded, TQM is becoming popular through Six Sigma. The goal of TQM and Six Sigma is to identify the poor quality immediately during the production process, rather than spending time to inspect the finished product. The quality of the manufacturing process determines the quality of a finished product. In the supply chain, it is not always possible to control the manufacturing process for incoming materials, especially from outside suppliers. In this instance, quality can only be measured by the percentage of defective goods received from the suppliers and in order to manage the supply chain more effectively, companies must choose suppliers that produce quality materials without a substantial price tag.

Sigma stands for the Greek symbol σ that designates a standard deviation in statistics. Six refers to the number of standard deviations from the mean of the specifications. Begun at Motorola in 1982 as an effort to reduce costs and improve quality, it now involves planning, organization, training, human resources planning, and pay-for-knowledge. It requires both organizational and individual cooperation to achieve a goal.

It is a process that is so well understood and controlled that six standard deviations will fit between the average output and the specific limit. The main purposes of using the Six Sigma process are listed below:

- To deal with a world of declining product prices
- To compete successfully with the best companies in the world
- To establish standard language and approaches across functions and across businesses
- To develop the next generation of leaders

Honeywell invites key customers to participate in their black belt and other educational programs. GE provides on-site services to assist customers in solving their problems. GE customers feel the difference of implementing the Six Sigma. They have more than 3000 Six Sigma projects under way in the airline industry which will achieve $400 million in savings for the customers. This service to the customers will improve long-term relationships.

1.14 How can I integrate LSS into the operations throughout my supply chain?

Operations phase is expected to begin once LSS methodologies are implemented, that is, the change in processes, network, and outlook of the organization has been accepted. The following steps could be followed to integrate LSS into the operations (Devane 2004):

1. Executives must accept their roles and responsibilities and be able to set priorities.
2. Leaders must consider newly established expectations while evaluating performance. Those who ignore these new values must be discarded from the functional team.
3. Periodic review of efforts toward improvement and sustainability in the new framework with accurate data representing financial impact must be recorded.
4. Active management must support the project lead by the new perspective of LSS with proper training and feedback sessions.
5. Gain in cash flows need not be excessively emphasized as the net result of improvement. Encouragement through awards for quality and recognition is best for creating motivation in employees.
6. All goals must tally with the organization's overall mission and strategic plan for steering into the future successfully.
7. Cross-functional team exchanges are encouraged while keeping boundaries of each team intact. This is considered to be the job of the manager.
8. The organizational human resource structure must be flexible enough to incorporate such transformations.
9. The new methods must be publicized by the management who must be supportive of the change to encourage other employees to accept them.
10. Financial contribution must always be sought after.

References

Ahire, S. L. (1997). "Management science—Total quality management interfaces: An integrative framework." *Interfaces,* 27(6): 91–105.

Crosby, P. B. (1979). *Quality Is Free: The Art of Making Quality Certain.* New York: McGraw-Hill.

Cua, K. O., McKone, K. E., and Schroeder, R. G. (2001). "Relationships between implementation of TQM, JIT, and TPM and manufacturing performance." *Journal of Operations Management,* 19(6): 675–694.

Devane, T. (2004). *Integrating Lean Six Sigma and High Performance Organizations Leading the Charge toward Dramatic, Rapid and Sustainable Improvement*. San Francisco, CA: Pfeiffer, an imprint of Wiley.

Foster, S. T. (2010). *Managing Quality Integrating the Supply Chain*, 4th edn. Upper Saddle River, NJ: Pearson Prentice Hall.

Garvin, D. A. (1984). "What does product quality really mean?" *Sloan Management Review*, 26(1): 25–43.

Jones, E. C., Garza, A., and Exstrom, L. (2010a). "Six-Sigma deployment success according shareholder value." *Proceedings of the 2010 Institute of Industrial Engineers Annual Conference*, June 5–9, 2010, Cancún, Mexico.

Jones, E. C., Riley, M. W., and Battieste, T. (2010b). "The value of industrial engineers in Lean Six Sigma organizations." *Proceedings of the 2010 Industrial Engineering Research Conference*, June 5–9, 2010, Cancún, Mexico.

Juran, J. M. and Gryna, F. M. (1998). *Juran's Quality Control Handbook*. 5th edn. New York: McGraw-Hill.

Juran, J. M. and Godfrey, A. B. (1999). *Juran's Quality Handbook*. Vol. 2. New York: McGraw Hill.

Narasimhan, J. and Jones, E. C. (2004). "Reduction in the variation of welding process using the operational Six-Sigma methodology." *Proceedings of the 2004 Industrial Engineering Research Conference*, June 13–15, 2005, Portland, OR.

Rao, S. S. (1996). *Engineering Optimization: Theory and Practice*. 3rd edn. New York: Wiley.

Zhang, Q., Irfan, M., Khattak, M. A. O., Zhu, X., and Hassan, M. (2012). "Lean Six Sigma: A literature review." *Interdisciplinary Journal of Contemporary Research in Businesss*, 3(10): 599–605.

Further Reading

David, A. G. (1987). "Competing on the eight dimensions of quality." *Harvard Business Review*, November–December, pp. 101–109.

chapter two

Quality gurus and philosophies

You cannot inspect quality into the product; it is already there.

—W. Edwards Deming

2.1 Introduction

The modern era of development of quality started during the 1980s. Newer methodologies have been continuously added since then. It has encouraged growth in finding improved management practices. The foundation was provided by quality gurus of yesteryears such as W. Edwards Deming, Kaoru Ishikawa, Joseph M. Juran, Walter A. Shewart, and Genichi Taguchi. They have laid the foundations of Six Sigma principles.

2.2 William E. Deming

Deming is often considered the father of modern quality. Deming, a US quality consultant who had limited success with American businesses, introduced quality principles to the more-accepting Japanese manufacturers on a large scale post-World War II. Many of Deming's principles and ideals are the foundation of quality initiatives today. He utilized some of the concepts identified with Shewart such as control charts into his philosophies in his research. Other notable ideas include Keiretsu partnerships with suppliers who are the precursors of vendor-managed inventory and partnerships today. He also introduced concepts of common and special causes of manufacturing defects. He described the idea of seeking profound knowledge to make breakthrough improvements in quality and manufacturing efficiency. Moreover, he made organizations understand and appreciate what we describe as "system thinking," theories on process variation, and organizational knowledge management and information retention. The idea of understanding the psychology of the employee was a major contribution. This concept is realized in what is his overarching philosophy, described as the Fourteen Points. Deming's Fourteen Points is based on three principles, constancy of purpose, continual improvement, and cooperation between factions. Deming believed in Fourteen Points of quality:

1. Create constancy of purpose
2. Adopt the new philosophy
3. Cease inspection, require evidence
4. Improve the quality of supplies
5. Continuously improve production
6. Train and educate all employees
7. Supervisors must help people
8. Drive out fear
9. Eliminate boundaries
10. Eliminate use of slogans
11. Eliminate numerical standards

12. Let people be proud of their work
13. Encourage self-improvement
14. Commit to ever-improving quality

Another contribution that directly relates to Six Sigma techniques was that Deming was the first guru to theoretically attempt to link quality improvements to financial results. It was described as the "Deming Value Chain."

2.3   Joseph M. Juran

Juran followed Deming in Japan and is mainly known for using the quality trilogy (planning, control, and improvement). He also originated the three spheres of quality: quality control (QC), quality assurance (QA), and quality management (QM). Juran (1995) described quality's "fitness for use" as that which is defined by the user. In other words, the users and the expectations that they have of the product determines quality. This is further clarified to mean that different users may use the product in different ways, and a high-quality product must possess multiple elements that "fit" different uses effectively.

He also contributed to the Lean and Six Sigma ideas of financially quantifying quality improvements. To support this he introduced ideas such as decreasing chronic waste, limiting rework, and minimizing the cost of quality. He is known for mentioning that management will always ask the question "will it make us money." This has led to the concept described as the law of diminishing marginal returns that extended the idea of cost of quality.

2.4   Philip B. Crosby

Crosby focused on improving preventive approaches that would reduce quality appraisal tradeoff. He is also widely recognized for his creation of the management maturity grid. This approach is generally attributed to Crosby's (1979) "conformance to requirements" concept. The concept describes a set of requirements that must be met in a manner required by specifications or standards. Generally, if these specifications are not met then the product is considered noncompliant and defective.

2.5   Armand V. Feigenbaum

Feigenbaum believed that an organization should be excellence driven rather than defect driven as well as adjusting the view of the organization to match that of the customer. Excellence-driven, rather a defect-driven, view is defined by the customer. Feigenbaum had 19 steps to QM that the entire company should follow. The 19 steps are as follows:

1. Total quality control (TQC) is defined as a system for improvement.
2. Big Q quality (companywide commitment to TQC) is more important than little q quality (improvements on the production line).
3. Control is a management tool with four steps.
4. QC requires integration of uncoordinated activities.
5. Quality increases profits.
6. Quality is expected, not desired.
7. Humans affect quality.
8. TQC applies to all products and services.

9. Quality is a total life-cycle consideration.
10. Control the process.
11. Total quality system involves the entire companywide operating work structure.
12. There are many operating and financial benefits of quality.
13. The costs of quality are a means for measuring QC activities.
14. Organize for QC.
15. Managers are quality facilitators, not quality cops.
16. Strive for continuous commitment.
17. Use statistical tools.
18. Automation is not a panacea.
19. Control quality at the source.

2.6 Garvin

Garvin's (1984) approach is generally accepted in engineering. The five approaches to quality include global or transcendent, product-based, user-based, manufacturing-based, and value-based approaches.

2.7 Ishikawa

Ishikawa perfected the cause-and-effect diagram. He also believed that the customer defines the quality definition the organization believes in. He was also a pioneer in quality circles and teamwork to improve quality. Ishikawa spent his life perfecting quality in Japan. He believed a company should hold everyone accountable for the statistics and output of the company.

2.8 Taguchi

Taguchi created loss function and focused on improving quality in design with reduction of noise in the form of system design, parameter design, and tolerance design.

2.9 Japanese contributions to quality

Japan has made many contributions to the art of QM. Most notably, Lean production. Two views emerge that pertain to Lean. The first view of Lean is a philosophical view of waste reduction. This view asserts that anything in the process that does not add value for the customer should be eliminated. The second view of Lean is a systems view stating the JIT is a group of techniques or systems focused on optimizing quality processes. Other contributions include visibility, in-process inspection, 5S, and emphasizing teamwork and quality circles.

Japanese quality is the belief and methods of quality the Japanese used post-World War II to the present day. These are the following:

Fitness to Standard: ingrained in management to have quality over profit
PDCA Style: plan–do–check–act (Shewart charts)
Fitness to Use: understand customer needs
Fitness to Cost: high quality, low cost
Fitness to Latent Requirement: discover customer needs before they do
Kano Diagram: model of a set of ideas for planning a new product, service, or process
Hoshin Kanri: vertical deployment of TQM strategy

The history of QM, from mere "inspection" to total quality management (TQM), and its modern branded interpretations such as Lean Six Sigma (LSS), has led to the development of essential processes, ideas, theories, and tools that are central to organizational development, change management, and performance improvements that are generally desired from individuals, teams, and organizations (Department of Trade and Industry 2012).

During the early days of work and processes, inspection by another individual was the main method of QM. The inspecting individual would determine if the product was acceptable or needed to be rejected. As companies expanded, the role of the inspector became a full-time job and employees were hired to do only inspection of products. This method is highly flawed due to the amount of human error. Several issues surfaced with the designation of inspectors for defects: they were not sufficiently trained, there were learning curve effects, carelessness, and different standards adopted by them were a few of the many errors discovered.

This led to the creation of a chief inspector, for all inspectors to report to and standardize expectations during inspection. The introduction of chief inspectors greatly reduced the errors of random human inspectors. As chief inspectors began to improve in management, in-house initiatives were developed. As other companies viewed the decrease of defects and waste of the initiative-driven companies, the same initiatives spread.

The quality improvement revolution was slow to catch on in Western companies, as Japanese production systems advanced in improving quality techniques. With the growth of LSS in Japan, Western manufacturing companies began to identify the substantial benefits of managing quality. Many companies simply adapted Japanese ideals but some branched off and decided that creation of their own initiatives would fit the Western style of production better. This introduced initiatives such as Six Sigma.

Fast-forward to now and many initiatives are implemented in many companies. As stated previously, the biggest strategies will be discussed thoroughly in the forthcoming chapters.

References

Crosby, P. B. (1979). *Quality Is Free: The Art of Making Quality Certain*. New York: McGraw-Hill.

Department of Trade and Industry. (2013) "The History of Quality." Department of Trade and Industry, United Kingdom Government, September 28. Retrieved from http://www.dti.gov.ph/splash.php

Garvin, D. A. (1984). "What does product quality really mean?" *Sloan Management Review*, 26(1): 25–43.

Juran, J. M. (1995). "A history of managing for quality." *Quality Progress*, 28(8): 125–130.

Further Readings

Foster, S. T. (2007). *Managing Quality: Integrating the Supply Chain*. Upper Saddle River, NJ: Pearson Prentice Hall.

Grover, M. P. (2007). *Work Systems and the Methods, Measurement, and Management of Work*. Upper Saddle River, NJ: Pearson Prentice Hall.

chapter three

Quality awards and standards

> The right quality and uniformity are foundations of commerce, prosperity and peace.
>
> —Deming Medal inscription

3.1   The Deming Prize

Increased costs to producers, customers, and nations due to poor quality have fostered renewed appreciation of the quality assurance function. Japan initiated a quality revolution in the 1970s and has since received worldwide recognition for its achievements. United States joined the quality race in the mid-1980s and has also made rapid advances. More recently, Europeans have launched cooperative efforts to improve quality (Izadi et al. 1996).

Japan has successfully transformed from a country whose economy and industrial base was in shambles after World War II to become a true "economic superpower." Japanese goods, once equated with poor craftsmanship and quality, now inundate the US market and have driven many US firms out of business. A number of different theories—ranging from macro to micro to cultural—have been advanced to explain the success of Japanese companies in penetrating foreign markets (Dooley et al. 1990) Japanese corporations used "superior quality" to capture, hold, and build market share (Ghobadian and Woo 1996). A key theme that has emerged is the relentless pursuit of quality by most of the successful Japanese companies. The emphasis on quality permeates throughout Japanese organizations and is embodied in the widespread adoption of TQC principles and statistically based methods within firms and across industries. Positive reinforcement is provided in the form of a national award—the Deming Prize (Dooley et al. 1990).

In 1951, the Deming Prize was established in Japan in honor of Dr. W. Edwards Deming by the Union of Japanese Scientists and Engineers (JUSE). It is an avenue for disseminating knowledge of successful methods for improvement. Its purpose is to award companies that continually apply companywide quality control (CWQC), based on statistical quality control, and are likely to continue doing so (Izadi et al. 1996). The Deming Prize is an annual award presented to an organization that has implemented total quality management (TQM) suitable for its management philosophy, scope/type/scale of business, and management environment. Regardless of the type of business, any organization can apply for the Prize under certain conditions, be it public or private, large or small, domestic or overseas, or part of the entire organization. There is no limit to the number of potential recipients of the Prize each year. All organizations that score the passing points or higher on examination are awarded the Deming Prize (Dooley et al. 1990). The Deming Prize can be awarded to Japanese companies, overseas companies, small enterprises, divisions, and factories; it makes no distinction between private and public institutions and between manufacturing and service organizations (Dooley et al. 1990).

3.1.1   Categories

There are three award categories (Ghobadian and Woo 1996):

1. The Deming Prize for the individual—this is awarded to individuals who have contributed to the understanding and application of CWQC/TQC;
2. The Deming Application Prizes; and
3. The Quality Control Award for Factories.

The latter two prizes are awarded for the attainment of distinctive performance improvement through the application of CWQC/TQC and statistical methods. The Deming application prize is open to corporations or their subsidiaries. The Quality Control Award for Factories is only open to manufacturing sites.

3.1.2   Deming Prize and TQM

The Deming Prize Committee defines TQM as follows (SaferPak 2012):

> TQM is a set of systematic activities carried out by the entire organization to effectively and efficiently achieve company objectives so as to provide products and services with a level of quality that satisfies customers, at the appropriate time and price.

1. "Systematic activities" mean organized activities to achieve the company's mission (objectives) that are led by strong management leadership and guided by established clear mid- and long-term vision and strategies as well as appropriate quality strategies and policies.
2. The phrase "carried out by the entire organization to effectively and efficiently achieve" means to involve everyone at all levels and all parts of the company so as to achieve the business objectives speedily and efficiently with minimal management resources. This is accomplished through an appropriate management system that has a quality assurance system at its core, and it integrates other cross-functional management systems such as cost, delivery, environment, and safety. The respect for human values encourages the company to develop human resources which uphold its core technology, speed, and vitality. The company maintains and improves its processes and operations, and uses appropriate statistical techniques and other tools. Based on facts, the company manages its business by rotating the management cycle of plan–do–check–act (PDCA). The company also rebuilds its management system by utilizing appropriate scientific methods and information technology.
3. "Company objectives" refer to securing appropriate profit for the long term by satisfying customers consistently and continuously. Also, they encompass improving the benefit to all stakeholders including employees, society, suppliers, and stockholders.
4. "Provide" refers to activities from producing "products and services" to handing them off to customers, including surveys, research, planning, development, design, product preparation, purchasing, manufacturing, installation, inspection, order-taking, sales and marketing, maintenance, after-sales services, and after-usage disposal and recycling.

5. "Products and services" include manufactured products (finished products and parts and materials), systems, software, energy, information, and all other benefits that are provided to customers.
6. "Quality" refers to usefulness (both functional and psychological), reliability, and safety. Also, in defining quality, influence on the third parties, society, environment, and future generations must be considered.
7. "Customers" include buyers as also users, consumers, and beneficiaries.

- For any company, the shortest way to win the Deming Application Prize is to manage its business in the manner most appropriate to the company. It is undesirable to conduct unnecessary activities for its fundamental business just for the sake of the examination. Such activities will not help the company with its examination; rather they may negatively affect the examination.
- The emphasis of the examination is on whether the company has or not developed a unique brand of TQM suitable for its business and scale. It does not require all applicant companies to uniformly follow the same brand of TQM.
- If the company just copies the format of TQM from others or if it prepares rules and standards more than necessary under the name of TQM, such activities will not help it to win the Prize.
- Some people think that advanced statistical methods must be used to pass the examination. This is a misunderstanding.
- New activities suitable for the applicant company's business and scale are highly respected.
- Nonprofit organizations should read "companies" as "institutions" or "organizations."

3.1.3   Deming Prize in detail

The Deming Prize is given to organizations that have exerted an immeasurable influence directly or indirectly on the development of quality control/management in Japan (Dooley et al. 1990). Its primary purpose was to spread the quality gospel by recognizing performance improvements flowing from the successful implementation of company-wide or total quality control (CWQC or TQC) based on statistical quality control techniques (Ghobadian and Woo 1996). Historically, the Deming Prize, especially the Deming Application Prize that is given to companies, has exerted an immeasurable influence directly or indirectly on the development of quality control/management (SaferPak 2012). Hence, the Deming Application Prize is discussed in more detail in this book.

The steps involved in the application for the Deming Prize are (Dooley et al. 1990):

1. Application and initial screening,
2. A documented "Description of QC practices" document, and for those companies which pass this stage,
3. Site examinations.

The site examinations consist of three phases, in the following order (Dooley et al. 1990):

1. Schedule A, where the applicant presents the main points of their quality program and tour operations sites,
2. Schedule B, where the inspectors perform on-site investigations of locations of their choice, and interrogate people at those sites, and
3. An interview with the CEO.

The Prize guidelines do not specify scoring procedures; the judges' own expectations and experience weigh heavily in this regard. The final score for each inspected unit is computed and the company is awarded the Deming Prize if they receive (Ghobadian and Woo 1996):

1. A minimum of 70 points for the executive session and the head office examination
2. A minimum aggregate score of 70 for other facets of the examination
3. A minimum score of 50 for each unit examined

It would be wrong to assume that the Deming Prize is simply awarded for the successful application of the Deming theory (Dooley et al. 1990). Previous winners of the prize share that the emphasis is mainly on (1) commitment to continuous improvement, (2) collection of data, (3) clarification of responsibility for action, (4) feedback information from customers, and (5) using statistics as a practical tool (Dooley et al. 1990). Unlike other popular quality awards like the European Quality Award and the Baldrige Award, Deming Prize is not competitive. That is, any number of companies meeting the above criteria may be awarded the prize in any one year (Ghobadian and Woo 1996).

3.2   The Malcolm Baldrige Award

The Malcolm Baldrige National Quality Award has evolved from a way to recognize the best quality management practices to a comprehensive framework for world-class performance; its criteria have been widely accepted and used as a model for process improvement. Organizations strive to produce higher quality products and services and enhance profitability by implementing the Baldrige criteria. These criteria are well represented in the Six Sigma process improvement method (SSPIM). Whereas the Baldrige criteria and SSPIM have followed different development paths, they have similar roots. Congress established the Baldrige Award in 1987 to enhance US competitiveness by promoting quality and business excellence among US companies, and to publicize the successful performance of these companies. Motorola developed SSPIM in the late 1980s as an improvement initiative focused on quality. Both quality activities originated from management philosophies that attempted to enhance performance, increase profitability, and improve quality. The Baldrige framework can be used as a generic model for process improvement in an organization. We refer to SSPIM from a management point of view—as independent projects in a firm—that focus on improving operational and business performance. We believe it is possible to implement Six Sigma philosophies within the Baldrige framework and integrate them with the overall quality system of the organizations.

3.2.1   Baldrige model for quality

Described as a "badge of honor" and "the most important catalyst for transforming American business," the Baldrige Award is much more than a quality award for an organization. Juran argued that it is a helpful model for achieving world-class quality.

The Baldrige model consists of seven categories (Mellat-Parast et al. 2007a):

1. Leadership
2. Strategic planning
3. Customer/patient and market focus
4. Measurement, analysis, and knowledge management

5. Workforce focus
6. Process management
7. Results

The award is used not only as a model for quality management implementation, but also as a self-system for overall organizational performance management. It has also been used as a guideline for organizations to achieve higher levels of quality.

3.2.2 Six Sigma method

Six Sigma is defined by Feld as a data-driven philosophy used to influence management decisions and spark action across an organization (quoted in Jones et al. 2010). Six Sigma proponent Caulcutt says Six Sigma reduces waste, increases customer satisfaction, and improves processes with a considerable focus on financially measurable results (quoted in Jones et al. 2010). For the purpose of this book, Six Sigma is defined (definition by Rasis et al.) as the relentless pursuit of process variation reduction and breakthrough improvements that increase customer satisfaction and impact the bottom line (quoted in Jones et al. 2010). The standard framework for implementing a Six Sigma method is define, measure, analyze, improve, and control (DMAIC) process.

In this approach, key process input variables are narrowed to a vital few. Having control of a vital few allows for good control of the whole picture. DMAIC is widely used when a product or process already exists but performs inadequately. This management strategy seeks to make an organization more effective and efficient.

DMAIC focuses on eliminating unproductive steps, developing and applying new metrics, and using technology to drive improvement. Six Sigma has a strategic component aimed at not only developing employees' commitment to it, but also actively involving a higher level of management.

The strategic component requires management to identify key processes in its organization, measure process effectiveness and efficiency, and initiate improvements to the worst-performing processes.

While the Baldrige model tends to address system or enterprisewide improvement, Six Sigma attempts to produce microlevel or specific improvements (Jones et al. 2010). Firms can achieve higher levels of performance in the Baldrige criteria by implementing Six Sigma projects in the company that focus on the key Baldrige criteria.

3.2.3 Baldrige model and Six Sigma

The goal of SSPIM is to reduce variation in the product, service, or process in dimensions (characteristics) essential for customer satisfaction (Mellat-Parast et al. 2007b). To do so, firms need to analyze their product, process, or service, and determine the variables that affect customer satisfaction. Motorola's experience with Six Sigma helped the company receive the Baldrige Award in 1988. SSPIM laid the foundation for Motorola to be the first company to receive the Baldrige Award, which suggests a strong link between the two.

The advantage of the Baldrige model over other quality improvement systems (e.g., TQM or the PDCA cycle) revolves around the systems approach in the Baldrige model. The Baldrige model specifically addresses the areas companies need to concentrate on to improve their performance.

In addition, the Baldrige model is a self-assessment model. Companies can use the Baldrige model to assess their performance over time, identify strengths and weaknesses, and focus on areas that need improvement. This valuable self-assessment mechanism in the Baldrige model makes it attractive to companies as a tool to improve their performance over time and continue the never-ending journey for excellence in quality.

Companies can benefit from integrating Six Sigma and Baldrige. Others have explained how Six Sigma and Baldrige are related and how "Baldrige provides the framework, Six Sigma the methodology." Six Sigma proponents say that winning quality awards, improving quality, and increasing customer satisfaction is achieved through Six Sigma.

Whereas the Baldrige model is typically applied at the enterprise level, Six Sigma is deployed as a number of independent projects—throughout the enterprise. Accordingly, by implementing Six Sigma as independent projects within the Baldrige model, firms can achieve higher operational performance and efficiency through Six Sigma projects while operating under the guidelines and the framework of the Baldrige criteria.

In that regard the two approaches are complementary. The DMAIC cycle of the SSPIM can be employed as a practical tool for addressing the requirement in each category within the Baldrige framework.

3.2.3.1 Implementing Six Sigma within the Baldrige model

Our approach for implementing the SSPIM within the Baldrige model is consistent with other approaches taken to relate Six Sigma with ISO 9001:2000. According to one approach, the implementation of the Baldrige model would mean:

- A focus on performance excellence for the entire organization in an overall management framework.
- Identifying and tracking all important organizational results, customer satisfaction, product/service, financial objectives, human resources, and organizational effectiveness.

While the Baldrige model does not provide any specific method, tool, or technique that can be incorporated, it does offer general guidelines. As such, SSPIM is a practical, appropriate, and useful tool that can be used to transform the goals of the Baldrige model into reality.

Accordingly, we might apply Six Sigma to improve the performance of each of the seven Baldrige categories. SSPIM serves as a mechanism to determine how much progress we have made in each Baldrige category. In this respect, whereas companies are using Baldrige criteria as an overarching model for business excellence, improving enterprisewide quality level could be achieved through different Six Sigma projects. In other words, Six Sigma projects facilitate process improvement in a firm through focus on Baldrige criteria.

The major difference is that Six Sigma projects might not be independent; rather, they are all focused on achieving higher levels of quality set by Baldrige criteria. Figure 3.1 shows the integration of Six Sigma into Baldrige.

In this approach, SSPIM is used to improve each Baldrige category. Depending on the constraints or requirements of the organization, SSPIM can be implemented separately or simultaneously on each category.

The proposed idea of integrating SSPIM and Baldrige is based on the following principles:

- Six Sigma is linked with the Baldrige model. In fact, it becomes part of Baldrige—not separate from it—for achieving performance excellence. The top management sets such a performance requirement. After establishing goals—which need to be aligned with Baldrige requirements—SSPIM can be used to improve processes and meet quality objectives.

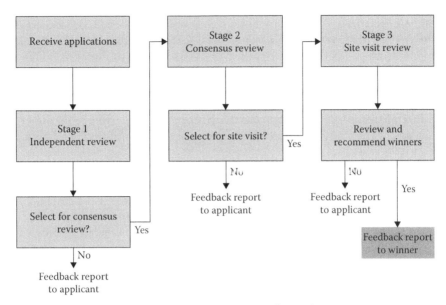

Figure 3.1 Malcolm Baldridge Award selection process and criteria.

- SSPIM can be applied to all types of projects, processes, and products. The top management directs the selection, administration, and control mechanisms.
- Baldrige self-assessment includes both the assessment of the seven Baldrige categories, as well as the efficiency and effectiveness of SSPIM. This approach toward quality ensures the company is benefiting from implementing SSPIM.
- The proposed model for the integrative Six Sigma–Baldrige framework is capable of addressing the core values of the Baldrige model. Through the focus of SSPIM on management by fact (data analysis and statistics) and goal setting (focus on results), the Baldrige core values—including visionary leadership, customer-driven excellence, organizational and personal learning, valuing employees and partners, agility, focus on the future, managing for innovation, social responsibility, and systems perspective—can be addressed.

3.2.3.2   Full integration

Currently, Six Sigma and Baldrige are linked, but not fully integrated. With reference to the Baldrige criteria, Six Sigma is directly related to customer focus and satisfaction, process management and information analysis, and knowledge management. The challenge is implementing Six Sigma in areas such as leadership, human resource/work force management, and strategic planning to achieve full integration. We offer the following recommendations for full integration of Six Sigma within the Baldrige model.

3.2.3.2.1   Human resource/workforce management.   Feedback from the Baldrige self-assessment can be used to identify potential Six Sigma projects focusing specifically on human resource development. In addition, SSPIM enhances the organizational knowledge base through training and systematic learning. But the impact of SSPIM on human resource development goes beyond that.

"There are hundreds of human resource issues that should be addressed when you begin Six Sigma," says David Silverstein, CEO and President of Breakthrough Management Group, a Six Sigma consulting firm based in Longmont, CO. This includes developing Six Sigma training delivery plan, preparing human resource teams for their role in Six Sigma, and creating and implementing a communication strategy to keep the organization informed about Six Sigma achievements.

"The human resource team controls the history and the culture of the company," Silverstein says. "(The team is) a vital part of the system that supports Six Sigma."

3.2.3.2.2 Strategic planning. Honeywell was another company that successfully implemented SSPIM in the entire organization. Honeywell developed a new generation of the Six Sigma approach called Six Sigma Plus. This initiative was primarily a quality strategy developed through the merger of Allied Signal and Honeywell in 1999.

With Six Sigma Plus, Honeywell focused on key strategic objectives providing more value to its customer by empowering the employees to improve its processes, products, and services, and capitalizing on e-business. In fact, Honeywell's Six Sigma Plus program focused on implementing high-impact projects consistent with its strategic planning. It employed rigorous project selection processes in line with the company's overall strategic plan.

3.2.4 A quality alliance

Organizations try to implement the Baldrige model as a means to achieving business excellence. The seven Baldrige categories are integrated and related with the purpose of addressing business challenges so that companies can be competitive in a dynamic business environment. However, the Baldrige model does not provide any tools or techniques that can be used by the firm for achieving the desired performance and quality objectives.

Six Sigma is a practical method for reducing cost, improving quality, and fostering continuous improvement in the product or process. It has been widely used by companies looking to achieve higher levels of customer satisfaction and profitability. Six Sigma's acceptance by many companies, its ability to produce bottom-line savings, and its foundations in quality make it a strong addition to the Baldrige model.

Integrating SSPIM into the Baldrige model not only helps firms achieve higher levels of performance and customer satisfaction in each Baldrige category, but also provides firms with a useful method for pursuing quality and a performance level set by the Baldrige model.

3.3 ISO 9000

In order to best fulfill customer needs, requirements, and expectations, effective organizations create and utilize quality systems. Effective quality management systems are dynamic and able to adapt and change to meet the needs, requirements, and expectations of their customers. Continued growth in international trade revealed the need for a set of quality standards to facilitate the relationship between suppliers and purchasers (Summers 2005). Standardization of business management practices through international management system standards has accelerated tremendously in the last decade. This boom was largely created by the publication of standards in different areas of a company's operations, including quality, environment, safety, information security, supply chain security, and social responsibility (Karapetrovic et al. 2010).

The ISO 9000 quality assurance standards, consisting of ISO 9000, 9001, 9002, 9003, and 9004, were first issued in 1987 and revised in 1994 by the International Organization for Standardization (ISO) (Tummala and Tang 1996). ISO certification stands for certain minimum quality standards that organizations should meet, and is said to assure a consistent quality of products, services, and processes (Singels et al. 2000). ISO 9000 is a series of international standards that sets out requirements and recommendations for design and assessment of management systems. ISO 9000 is grounded on the "conformance to specification" definition of quality (Buttle 1996). These standards are based on the concept that certain minimum characteristics of a quality management system could be usefully standardized, giving mutual benefit to suppliers and customers (Tummala and Tang 1996).

An important difference compared with earlier quality standards lies in the fact that ISO is focused on quality control systems in general, "from the process of product design to process design and from production process to service after-sale." Further, ISO is based on the notion that specific minimum characteristics of quality systems can be standardized, which can give mutual benefits for organizations and their suppliers because each of them knows that they both meet certain requirements concerning quality systems. It must be clear that ISO certification is not a standardized package that can be applied in the same way in every organization (Singels et al. 2000).

3.3.1   Purpose

ISO 9000 describes the guidelines for use of a particular standard whereas ISO 9004 describes the guidelines for establishing an internal quality management system within the broad and general context of TQM. The other three standards, ISO 9001, 9002, and 9003, are the generic standards containing minimum requirements for establishing and maintaining a documented quality system to instill confidence in customers that the intended products or services meet customer requirements (Tummala and Tang 1996).

There is always a misconception that ISO would mandate higher levels of product quality. ISO certification gives no guarantee that the quality of products or services of an organization is better than the quality of other organizations. Thus, ISO-certified organizations do not automatically have a good product quality. In fact it is possible that the products or services of a registered organization are not of such good quality, but of constant quality (Singels et al. 2000).

The major purpose of these standards is to provide an effective quality system reflecting a company's practices of producing goods and services that conform to specified requirements in order to enhance and facilitate trade (Tummala and Tang 1996). ISO standards facilitate the multinational exchange of products and services by providing a clear set of quality system requirements. The standards provide a baseline against which an organization's quality system can be judged. The foundation of this baseline is the achievement of customer satisfaction through multidisciplinary participation in quality-improvement efforts, documentation of systems and procedures, and other basic structural elements necessary to quality systems. Many companies use ISO 9000 as the foundation for their continuous improvement efforts (Summers 2005). The drivers behind the adoption of ISO come from the supply and demand side. The adoption of ISO reduces managerial inefficiency, rather than being a new production factor (Lafuente et al. 2009).

The ISO 9000 requirements describe what a company must accomplish in order to meet customer expectations. However, how these goals are accomplished is left to the particular company. In 2000, the ISO 9000 standards were revised significantly so that their structure more closely resembles the way organizations are managed. It tells us how quality levels can be obtained and is a process-oriented systems approach focusing on the customer for holistically improving relationships.

Eight key principles have been included in the ISO 9000.2000 standards (Summers 2005):

1. Customer-focused organization
2. Leadership
3. Involvement of people
4. Process approach
5. Systems approach to management
6. Continual improvement
7. Factual approach to decision making
8. Mutually beneficial supplier relationships

3.3.2   ISO 9000—Performance

As discussed earlier, firms adopt ISO 9000 certification as an effort of continuous process improvement. However, after implementing ISO 9000 standards the performance solely depends on the type of that firm. There are significant differences in the performance of manufacturing and service industry. The factors that affect the performance of a firm that has implemented ISO 9000 are (Lafuente et al. 2009):

- Implementation of just-in-time (JIT)
- Labor intensity
- Technology
- Ownership structure

In a nutshell, ISO 9000, altogether, increases labor productivity and revenue to total assets (ROA), thereby increasing the overall productivity of the firm.

3.4   Other quality awards and standards

3.4.1   Strategic quality planning

Strategic quality planning is an important process to be undertaken. It starts by designing the quality service with the help of creating the following (Foster 2010):

1. *The value chain*: The value chain consists of a network of *value chains* and value chain activities are the tasks that add value for the customers. Nonvalue chain activities typically have costs but create no effect on customers. The value chain can be considered as the linkage of *customers and suppliers.*
2. *Supplier partnering*: There is a trend toward developing closer relationships with fewer suppliers and this has been inspired by JIT supplier–partner relationships that have emerged. Options such as single or dual sourcing are now preferred. Suppliers are evaluated for the quality in their supply using sole-source filters such as:

   a. Baldrige
   b. ISO 9000:2000
   c. Vendor management inventory (VMI)
3. *Applying the contingency perspective to supplier partnering*: One variable which affects what customers want from their suppliers is the position in the supply chain. Suppliers are only marginally more interested in relationships than customers are.
4. *A supplier development program*: Using the following measures we plan to achieve supplier development:
   a. Identify critical products and services
   b. Identify critical suppliers
   c. Meet with supplier top-management
   d. Form cross-functional teams
   e. Identify key projects
   f. Define details of agreement
   g. Monitor status and modify strategies
   Let us look at some supplier development programs.
   a. Supplier relationship management system (SRMS) is one such supplier development program which is much like customer relationship management systems (CRMSs). It has the following components:
      i. Spend analytics
      ii. Sourcing execution
      iii. Procurement execution
      iv. Payment
      v. Supplier score-carding
      vi. Performance monitoring
   b. The ISO/TS 16949 standard applies only to automotive companies and it was developed by the International Automotive Task Force (IAFT). Members of this task force include Ford, GM, and Daimler Chrysler.
      ISO/TS 16949 Incorporates the following features a quality management system that comes with responsibility to manage resources for realization of product. It identifies the need for improvement through measurement and analysis.
5. *Acceptance sampling and statistical sampling techniques*: Acceptance sampling is done when dealing with unproven suppliers; during start-ups and when building new products; when products can be damaged during shipment, and when problems with a certain supplier have been noticed.

Operating characteristic (OC) curve is a sampling technique that is useful for assessment of the probabilities of acceptance given the existing quality of the shipment. It helps to identify the risk taken by the consumer.

3.4.2   *European Quality Award*

Much like the MBNQA, Europe issues the European Quality Award every year to the most accomplished applicant. There is also a second-level award given to all organizations that meet the award criteria.

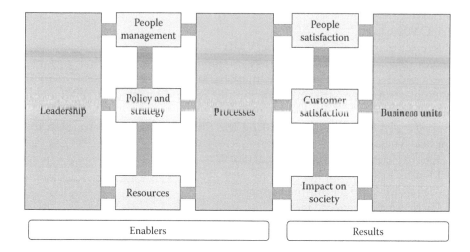

3.4.3   Quality Systems

Continuous learning under modern quality systems is required for development of sustainable systems (GMP Training Systems 2012). These systems have a necessity for identifying the weak points of a given process. Timely identification and having contingency measures can help avoid risks. One must allocate (qualified) resources for assessing the risk incurred, accuracy of reports, and the current quality in order to create carefully structured standardized processes and a quality control unit (QCU).

A good quality system adopted by an organization will help in meeting the goals of the organization. Using modern quality initiatives the risks have been allayed through robust design, automation of processes, optimal testing strategies, and other business initiatives to mitigate variations.

Let us look at some quality systems being used in major corporations:

- QS 9000: It encompasses two quality assessment standards. They are:
  - Customer-specific requirements which include FMEA, SPC, and others
  - ISO 9001 which is a requirement for suppliers for automotive components (Summers 2005).
- ISO 14000: It is an environmental management standard for preservation of value for environmental concerns. If an organization wishes to go "green" by taking steps to counter the negative effects of its activities on the environment and also eliminate waste to procure profits they must compete for this certification. It follows with a series of standards that may be process or product oriented and increases their responsibility toward global quality. Examples are ISO 14001, ISO 14014, ISO 14022, ISO 1402X, ISO 14050, etc.

There are many more standards adopted by different sectors of industries like TL 9000 for the telecom industry.

References

Buttle, F. (1996). "ISO 9000: Marketing motivations and benefits." *International Journal of Quality and Reliability Management*, 14(9): 936–947.

Dooley, K. J. et al. (1990). "The United States' Baldrige Award and Japan's Deming Prize: Two guidelines for total quality control." *Engineering Management Journal*, 2(3): 9–16.

Foster, S. T. (2010). *Managing Quality Integrating the Supply Chain*, 4th edn.

Ghobadian, A. and Woo, S. H. (1996). "Characteristics, benefits and shortcomings of four major quality awards." *International Journal of Quality and Reliability Management*, 13(2): 10–44.

GMP Training Systems. (2012). http://www.gmptrainingsystems.com/files/u1/pdf/Friedman.pdf.

Izadi, M., Kashef, A. E., and Stadt, R. (1996). "Quality in higher education: Lessons learned from the Baldrige Award, Deming Prize, and Iso 9000 registration." *Journal of Industrial Teacher Education* 33(2): 60–76.

Jones, E. C., Parast, M. M., and Adams, S. G. (2010). "A framework for effective Six-Sigma implementation." *Total Quality Management and Business Excellence*, 21(4): 415–424.

Karapetrovic, S., Fa, M. C., and Saizarbitoria, I. H. (2010). "What happened to the ISO 9000 lustre? An eight-year study." *Total Quality Management*, 21(3): 245–267.

Lafuente, E., Bayo-Moriones, A., and Garcia-Cestona, M. (2009). "ISO-9000 certification and ownership structure: Effects upon firm performance." *British Management Journal*, 21(3): 649–665.

Mellat-Parast, M., Adams, S. G., and Jones, E. C. (2007a). "An empirical study of quality management practices in the petroleum industry." *Production Planning and Control*, 18(8): 693–702.

Mellat-Parast, M., Jones, E. C., and Adams, S. G. (2007b). "A relationship between Six Sigma and Malcolm Baldrige Quality award." *Quality Progress*, 40(9): 45–51.

SaferPak. (2012). *Deming Prize*. http://www.saferpak.com/deming_prize_art1.htm.

Singels, J., Ruel, G., and Van de Water, H. (2000). "ISO 9000 series certification and performance." *International Journal of Quality and Reliability Management*, 18(1): 62–75.

Summers, D. C. S. (2005). *Quality Management Creating and Sustaining Organizational Effectiveness*. Upper Saddle River, NJ: Pearson Prentice Hall.

Tummala, V. M. R. and Tang, C. L. (1996). "Strategic quality management, Malcolm Baldrige awards and ISO 9000 certification: Core concepts and comparative analysis." *International Journal of Quality and Reliability Management*, 13(4): 8–38.

Further Reading

Mellat-Parast, M. and Jones, E. C. (2006). "A relationship between Six Sigma and Malcolm Baldrige Quality award." *Proceedings of the 2006 Institute of Industrial Engineers Annual Conference*, May 20–24, Orlando, FL.

Review Questions

Question 1: (Overview) A set of active practices that an organization uses to create, assimilate, disseminate, and apply knowledge is

(a) Knowledge management
(b) Organizational memory
(c) Organizational learning
(d) Tacit knowledge

Question 2: The value obtained from web-enabled knowledge management applications depends on (or is the result of) (A) ease of software usage, (B) providing customers and suppliers with service data, (C) legal and regulatory affairs, (D) providing internal users with useful data

(a) A C
(b) A B D
(c) B C D
(d) A B C D

Question 3: (Research paper) The Baldrige criteria usually address system or enterprise-wide improvement whereas Six Sigma attempts to produce micro-level or specific improvement

(a) True
(b) False

Appendix 3A: A relationship between Six Sigma and Malcolm Baldrige Quality Award

3A.1 Introduction

The Malcolm Baldrige National Quality Award (MBNQA) has evolved from a means of recognizing best quality management practices to a comprehensive framework for world-class performance, where it is widely used as a model for process improvement (Flynn and Saladin 2001). Companies strive to achieve higher quality of products and/or services and enhance their profitability through implementing the Baldrige criteria. Interestingly, these criteria are well established by the Six Sigma process improvement method (SSPIM).

While the Baldrige criteria and the SSPIM have followed different development paths, they have the same roots. The US Congress established the MBNQA in 1987 to enhance US competitiveness by promoting quality awareness, recognizing the quality and business excellence of US companies, and publicizing the successful performance of these companies. On the other hand, the SSPIM was developed by Motorola in the late 1980s as an improvement initiative focused on quality (Feld and Stone 2002). However, both originated from quality management philosophies, and attempt to enhance performance, increase profitability, and improve quality. Accordingly, it is worth investigating how the Baldrige criteria and the SSPIM might be related to each other.

The purpose of this article is to investigate the relationship between the SSPIM and the Baldrige model. We use the Baldrige framework as a generic model for process improvement within an organization. We refer to the SSPIM from a management point of view—as a number of independent projects within a firm—in order to link it to the implementation of the Baldrige model (Lupan et al. 2005). We argue that it is possible to implement Six Sigma philosophies within the Baldrige framework and integrate it with the overall quality system of the organization.

3A.2 The Baldrige model for quality

Despite being described as a "badge of honor" (Dow et al. 1999), the MBNQA is much more than quality award for an organization. Garvin (1991) described it as "the most important catalyst for transforming American business." Further, Juran (1994) argues that the MBNQA is a helpful model for getting into world-class quality. MBNQA has been primarily used as a framework for business improvement rather than as an award for quality. The Baldrige model consists of seven categories, as follows:

1. Leadership
2. Strategic planning
3. Customer and market focus

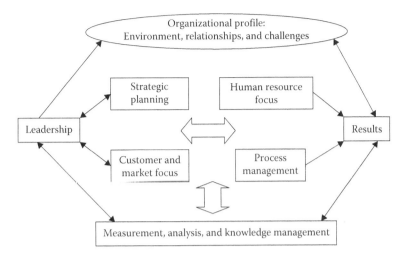

Figure 3A.1 Baldrige criteria for performance excellence framework: A systems perspective. (From www.quality.nist.gov. With permission.)

4. Measurement, analysis, and knowledge management
5. Human resource focus
6. Process management
7. Business results

MBNQA is not only used as a model for quality management implementation, but can also be helpful as a self-assessment tool, providing a framework for continuous business process improvement. Figure 3A.1 shows the framework for the MBNQA. The framework has evolved over time (since 1988) to address the challenges of the dynamic business environment, and has progressed as a comprehensive, integrated system for overall organizational performance management. It has also been used as a guideline for organizations to achieve higher levels of quality (National Institute of Standards and Technology 2005).

3A.3   Six Sigma methodology

According to Feld and Stone (2002), Six Sigma is a data-driven philosophy used to drive management decisions and actions across an organization. Caulcutt (2001) indicates that Six Sigma reduces waste, increases customer satisfaction, and improves processes with a considerable focus on financially measurable results. For the purpose of this chapter, Six Sigma is defined as the relentless pursuit of process variation reduction and breakthrough improvements that increase customer satisfaction and impact the bottom line (Rasis et al. 2002a).

A popular framework for implementing a Six Sigma methodology is the DMAIC process. DMAIC is the key process of a standard framework for the SSPIM approach and is shown in Figure 3A.2.

Figure 3A.2 DMIAC framework.

According to Jing and Li (2004), the psychology of this approach is that key process input variables are narrowed down to a vital few with the idea that having control of the vital few will allow for good control of the whole picture. DMAIC is widely used when a product or process is already in existence but performing inadequately. It is a management strategy that seeks to make an organization more effective and efficient. DMAIC focuses on eliminating unproductive steps, developing and applying new metrics, and using technology to drive improvement (De Feo and Barnard 2004). Six Sigma has a strategic component aimed at not only developing commitment to it, but also active involvement of higher management. That strategic component is the responsibility of management to identify the key processes of their organization, measure their effectiveness and efficiency, and initiate improvements of the worst performing processes. We need to recognize that while the Baldrige model tends to address the system or the enterprise-wide improvement, Six Sigma attempts to produce microlevel or specific improvement. We believe that firms can achieve higher levels of performance in the Baldrige criteria through implementing Six Sigma projects within the company that focus on the key criteria addressed within the Baldrige model.

3A.4   The Baldrige model and Six Sigma

We found that the goal of the SSPIM is to reduce variation in the product/service/process in dimensions (characteristics) that are essential for customer satisfaction. To do so, the firms need to analyze their products, processes, and services and determine the variables that affect customer satisfaction. The experience of Motorola with Six Sigma helped the company to win the Baldrige Award in 1988. According to Sumberg (2005), the SSPIM quality laid the foundation for Motorola to be the first company to win the Baldrige Award. We hypothesize that the linkage between these two initiatives may provide a strong indicator to business success as evidenced by the success of Motorola.

The advantage of the Baldrige model to other quality improvement systems (e.g., TQM and PDCA) revolves around the systems approach within the Baldrige model. The Baldrige model specifically addresses the areas companies need to concentrate on to improve their performance. In addition, it should be noted that the Baldrige model is a self-assessment model. Companies can use the Baldrige model to assess their performance over time, identify their strengths and weaknesses, and focus on the areas that need improvement. This valuable feedback mechanism in the Baldrige model makes it attractive to the majority of companies to use it as a tool to improve their performance over time. In sum, we argue that the Baldrige model is a better choice for companies in their never-ending journey for excellence in quality.

We can incorporate Six Sigma methodology to the Baldrige model. Byrne and Norris (2003) explain how Six Sigma methodology and the Baldrige model are related. They state that "Baldrige provides the framework, Six Sigma the methodology." Six Sigma proponents argue that winning quality awards, improving quality, and increasing customer satisfaction is achieved through the Six Sigma methodology (Douglas and Erwin 2000). We need to clarify that while the Baldrige model is typically applied at the enterprise level, Six Sigma is deployed as a number of independent projects (throughout the enterprise). Accordingly, by implementing Six Sigma methodology (as independent projects) within the Baldrige model, firms can achieve higher operational performance and efficiency (through Six Sigma projects) while operating under the guidelines and the framework of the Baldrige criteria. It that regard, the two approaches are complementary. The DMAIC cycle of the SSPIM can be employed as a practical tool for addressing the requirement in each category within the Baldrige framework. We can correlate that the Baldrige model assists in defining the goals and objectives and Six Sigma provides the techniques and tactics to achieve these objectives.

The integration of the SSPIM with quality management systems is getting more attention in recent years. Lupan et al. (2005) addressed the relationship between Six Sigma and the ISO 9001:2000 quality system. They proposed a strategy for the implementation of the SSPIM as an improvement solution for ISO 9001:2000. Their approach was focused on the integration of the DMAIC cycle with the PDCA process approach in ISO 9001:2000. Similar to ISO 9001:2000, we suggest that SSPIM can be effectively integrated into the Baldrige model. We also want to suggest that the benefits of integrating Six Sigma to the Baldrige model is that it provides more verifiable cost saving to companies and has the ability to gain greater acceptance due to the popularity among companies for bottom-line cost reductions. We suggest that integration of traditionally Six Sigma standards such as projects should demonstrate a minimum of $100,000 earnings before interest and taxes (EBIT) provides further validation that Baldrige-modeled companies are successful organizations.

In comparison with the other popularized quality certification initiative ISO 9001:2000 quality system, we suggest that the Baldrige model has more specific criteria which helps in defining quality-related projects in more effective way. The criteria for the Baldrige model have been clearly stated and described. Further, in the Baldrige model firms need to achieve business performance as well as customer satisfaction. While ISO 2004:2000 (a guideline document to ISO 9001:2000 for performance improvement) emphasizes the effectiveness of a quality management system on organizational performance, it does not provide any practical procedure to improve the overall performance of the organization. Rather, it mainly focuses on the overall quality of the firm with respect to customer satisfaction. The Baldrige model offers a broader view of quality, in which it relates quality to customer satisfaction and business results (performance). The ISO 9001:2000 "aims to enhance [customer] satisfaction through the effective application of the [quality] system" (section 1.1b of ISO 9001:2000) whereas the focus of the Baldrige model is on performance excellence for the entire organization in an overall management framework—which includes both quality and business performance (National Institute of Standards and Technology 2002). This could lead one to conclude that ISO 9001:2000 was designed to document organizational procedures in contrast to Baldrige which is designed to evaluate, measure, and critique an organizational performance. We suggest that Six Sigma is a better theoretical and real-world fit with the Baldrige Award than other quality initiative such as the ISO 9001:2000.

3A.4.1 Implementing Six Sigma within the Baldrige model

Our approach for implementing the SSPIM within the Baldrige model is consistent with the approach Lupan et al. (2005) have taken to relate Six Sigma with ISO 9001:2000. According to this approach, the implementation of the Baldrige model would mean:

- A focus on performance excellence for the entire organization in an overall management framework.
- Identifying and tracking all-important organizational results: customer satisfaction, product/service, financial objectives, human resource, and organizational effectiveness.

We need to recognize that whereas the Baldrige model does not provide any specific methodology, tool, or technique that can be incorporated, it does offer general guidelines.

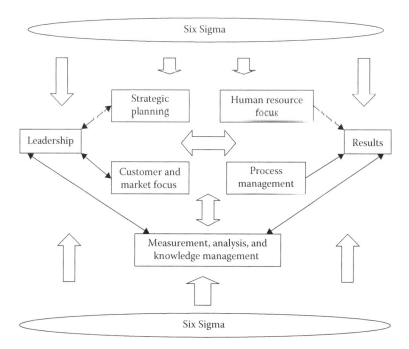

Figure 3A.3 An integrative approach to Six Sigma implementation in the Baldrige model.

As such, we believe that the SSPIM is a practical, appropriate, and useful tool that can be used to make the Baldrige model goals into more operational measures.

Accordingly, we may apply Six Sigma methodology on each of the seven Baldrige categories to evaluate it after Six Sigma is implemented, compare that with the level of the category before, and assess the improvement we have made in the Baldrige model. In this respect, whereas companies are using the Baldrige criteria as a self-assessment tool for evaluating the overall quality of the firm (with respect to the requirement of the Baldrige model), improving enterprise-wide quality level could be achieved through different Six Sigma projects.

The major difference is that Six Sigma projects may not be independent; rather they are all focusing on achieving higher levels of quality set by the Baldrige criteria. Figure 3A.3 shows the integration of Six Sigma into the Baldrige model. In this integrative approach, SSPIM is used to improve each category of the Baldrige model. Depending on the constraints or requirement of the organization, SSPIM can be implemented separately on each category or simultaneously.

3A.5   Integrating Six Sigma with the Baldrige model

The proposed integrative Six Sigma and Baldrige model is based on the following principles:

- The Six Sigma methodology is linked with the Baldrige model. In fact, it becomes part of the Baldrige model (and not separate from it) for achieving performance excellence. Such a performance requirement is set by top management. After the establishment of such goals (which need to be aligned with the requirements of the Baldrige model) SSPIM is used to improve processes, and meet quality objectives.

- SSPIM can be applied to all types of the projects, processes, and products. The selection, administration, and control mechanisms are directed by the top management.
- The Baldrige self-assessment includes both the assessment of each of the seven categories, in the Baldrige model, as well as the efficiency and effectiveness of the SSPIM. Such an approach toward quality ensures that the company is gaining benefit from implementing the SSPIM.
- The proposed model for the integrative Six Sigma–Baldrige framework is capable of addressing the core values of the Baldrige model. Through the focus of the SSPIM on management by fact (data analysis and statistics) and goal setting (focus on results) the Baldrige core values: visionary leadership, customer-driven excellence, organizational and personal learning, valuing employees and partners, agility, focus on the future, managing for innovation, social responsibility, and systems perspective can be addressed.

3A.5.1   *The case for Motorola*

To show how the principles described in this chapter can be applied in a real-world setting, we provide an example from a company that has effectively implemented Six Sigma and has been awarded the Baldrige Award twice: Motorola.

Motorola's success rests on integrating quality management initiatives with statistical quality control tools and techniques. Having been influenced by Japanese management practices at one of the previously owned Motorola's plants: Quasar, Motorola's management learned how to improve the quality of their products. Under Motorola's leadership, Quasar was losing market share to foreign competitors that sold products of better quality and lower cost. In 1974 Motorola sold Quasar to a Japanese consumer electronic company, Matsushima. Under Japanese management, the factory made drastic improvements in the quality of its products, producing TV sets with 1/20th number of defects that were made under Motorola's management. It was interesting to note that Matsushima was using the same workforce, technology, and design as Motorola. After a visit to the factory, Motorola's management realized that such surprising results could be achieved when an organization is focused on processes, people, and quality (ICFAI Center for Management Research: http://icmr.icfai.org/casestudies.catalogue.Operatios/OPER050.htm). In other words, they found that it is the quality system that leads a company to produce higher quality products. They also realized that in order to improve the quality of their products, they need to change their focus of quality improvement from product attributes to operational procedures.

This shift in thinking about quality within Motorola resulted in an emphasis in the systems approach, the interactions between different processes within the organization and their overall impact on performance. By focusing on management (leadership), people (human resource management), process (process management), flow of information (information, analysis and knowledge management), voice of customers (customer and market focus), and commitment to higher product quality (strategic planning) Motorola was successful in reducing the defect rate in its process to the Six Sigma level (3.4 defects per million). Such a dramatic improvement could not be achieved by just focusing on process management using statistical quality control tools and techniques. Rather, as evidenced by the Quasar case it is the result of improvement in all aspects of organization, leadership, strategic planning, customer and market focus, measurement, analysis and knowledge management, human resource focus, and process management, that led to achievement of higher level of performance that coincidently are the criteria of the Baldrige model.

In its current stage, Six Sigma and the Baldrige model have commonalities. With reference to the Baldrige criteria, Six Sigma is directly related to customer focus and satisfaction, process management, information, analysis, and knowledge management. These have been partially integrated now. The challenge is how to implement Six Sigma in areas such as leadership, strategic planning, and human resource management to achieve a full integration. We have the recommendations for the full integration of Six Sigma within the Baldrige model.

3A.5.2   Leadership and Six Sigma

Six Sigma projects can be effectively defined to enhance the quality of the leadership in an organization. In that regard, the feedback from the Baldrige model can be used as a base to define Six Sigma projects with focus on leadership development. Companies can assess how their leaders lead and how they should lead—both come from the Baldrige model. Then, they can develop Six Sigma projects that aim at meeting the desired leadership goals. Developing leadership standards can be a useful tool. Motorola has its own "4e's plus always 1" leadership standard which addressees key leadership characteristics; (1) envision: developing vision, strategies, and a viable plan to achieve them; (2) energize: create an environment for employees to excel and innovate; (3) edge: make tough, effective, and timely decisions; (4) execute: achieve results in a timely manner; and (5) ethics and character: conduct business practices professionally and ethically. Companies that do not integrate their leadership within Six Sigma projects are subject to fail in achieving their desired outcome from Six Sigma project. Honeywell was astute enough to understand that it needed to integrate its SSPIM initiative with the organizational leadership (Carter et al. 2005).

3A.5.3   Human resource management and Six Sigma

The feedback from the Baldrige self-assessment can be used to identify potential Six Sigma projects focusing particularly on human resource development. In addition, the SSPIM enhances organizational knowledge-base through training and systematic learning. But the impact of the SSPIM on human resource development goes beyond that. "There are hundreds of human resource issues that should be addressed when you begin Six Sigma," says David Silverstein, CEO and president of Breakthrough Management Group, a Six Sigma consulting firm based in Longmont, Colorado. This includes developing a plan for delivery of Six Sigma training, preparing human resource teams for their role in Six Sigma, and creating and implementing a communication strategy that will keep the organization informed about Six Sigma achievements. "The human resources team controls the history and the culture of the company," he says. "They are a vital part of the system that supports Six Sigma," Silverstein says (Fister 2003). At Lockheed Martin, an interview with Six Sigma personnel identified a few behaviors that were critical to the success of Six Sigma. Most of these behaviors were crucial conversations—such as slow rate of response in employee needs, observing unproductive practices, and attention given by top management to the problems—that enable Six Sigma projects to progress or stall, depending on how these behaviors were handled (Carter et al. 2005).

3A.5.4   Strategic planning and Six Sigma

Honeywell was another company that successfully implemented the SSPIM within the entire organization. Honeywell developed a new generation of Six Sigma which was called

Six Sigma plus. This initiative was primarily a quality strategy which was developed through the merger of Allied Signal and Honeywell in 1999. In Six Sigma Plus, Honeywell focused on key strategic objectives: providing more value to its customer though empowering the employees, improving its processes, products, and services, and capitalizing on e-Business. In fact, Honeywell's Six Sigma Plus program was focusing on implementing high impact projects consistent with its strategic planning. It employed rigorous project selection process which was in line with the company's overall strategic plan.

The above are just a few examples of the link between Six Sigma and the Baldrige model. As noted, the link between the SSPIM and the Baldrige model has been already established in categories such as process management, customer and market focus, and information, analysis, and knowledge management. Companies can achieve significant improvement in their Six Sigma projects if they fully integrate it with the Baldrige model.

3A.6   Conclusion

Organizations try to implement the Baldrige model as a means of achieving business excellence. The seven categories within the Baldrige model are integrated and related with the purpose of addressing business challenges so that companies can be competitive in dynamic business environment. However, the Baldrige model does not provide any tool or technique that can be used by the firm for achieving the desired performance and/or quality objectives.

Six Sigma is a practical methodology for reducing cost, improving quality, and fostering continuous improvement in the product/process. Because Six Sigma originated through quality, it provides time-tested tools that have enabled continuous improvement for organizations such as GE and Motorola that lead to better performance. Also, it has been widely used by companies which are looking to achieve higher level of customer satisfaction and profitability. This research suggests that due to previous acceptance of Six Sigma by other companies, the ability to produce bottom-line savings, and its foundations in quality would be a strong addition to the Baldrige model. Furthermore, we need to note that one of the requirements of the Baldrige model in business results.

Finally, we need to note that one of the requirements of the Baldrige model is achieving better performance levels. We believe that integrating the SSPIM into the Baldrige model not only helps firms achieve higher levels of performance and customer satisfaction in each category of the Baldrige model but also provides them with a useful methodology for pursuing quality and performance level set by the Baldrige model.

References

Byrne, G. and Norris, B. (2003). "Drive Baldrige level performance." *Six Sigma Forum Magazine,* May, pp. 13–21.

Carter, L., Ulrich, D., and Goldsmith, M. (2005). *Best Practices in Leadership Development and Organization Change.* San Francisco, CA: Pfeiffer Publisher.

Caulcutt, R. (2001). "Why is Six Sigma so successful?" *Journal of Applied Statistics,* 28(3&4): 301–306.

De Feo, J. and Barnard, W. (2004). *Juran Institute's Six Sigma: Breakthrough and Beyond.* New York: McGraw-Hill.

Douglas, P. C. and Erwin, J. (2000). "Six Sigma focus on total customer satisfaction." *The Journal for Quality & Participation,* March/April, pp. 45–49.

Dow, D., Samson, D., and Ford, S. (1999). "Exploding the myth: Do all quality management practices contribute to superior quality performance?" *Production and Operations Management,* 8(1): 1–27.

Feld, K. and Stone, W. (2002). "Using Six Sigma to change and measure improvement." *Performance Improvement,* 41(9): 20–26.

Flynn, B. B. and Saladin, B. (2001). "Further evidence on the validity of the theoretical models under-lying the Baldrige criteria." *Journal of Operations Management*, 19(3): 617–652.

Garvin, D. A. (1991). "How the Baldrige award really works?" *Harvard Business Review*, 69(6): 80–93.

Jing, G. and Li, N. (2004). "Claiming Six Sigma." *Industrial Engineer*, February, pp. 37–39.

Juran, J. M. (1994). "The upcoming century for quality." *Quality Progress*, 27(8): 29–37.

Lupan, R. et al. (2005). "A relationship between Six Sigma and ISO 9000:2000." *Quality Engineering*, 17: 1–7.

National Institute of Standards and Technology (NIST). (2002). Baldrige Criteria Quality Program, "Baldrige, Six Sigma, & ISO: Understanding your options." CEO issue sheet, www.quality.nist.gov/.

National Institute of Standards and Technology (NIST). (2005). Baldrige Criteria Quality Program, www.quality.nist.gov/Business-Criteria.htm

Rasis, D., Gitlow, H., and Popovich, E. (2002a). "Paper organizers international: A fictitious Six Sigma green belt case study 1." *Quality Engineering*, 15(1): 127–145.

Rasis, D., Gitlow, H., and Popovich, E. (2002b). "Paper organizers international: A fictitious Six Sigma green belt case study 2." *Quality Engineering*, 15(2): 259–274.

Sumberg, B. B. (2005). "Linking business needs and lessons learned to education." *Training & Development*, February, p. 70.

Further Readings

Does, R. J., van den Heuvel, E. R., de Mast, J., Bisgaard, S. (2002). "Comparing non-manufacturing with traditional applications of six sigma." *Quality Engineering*, 15(1): 177–182.

Gale, S. F. (2003). "Building frameworks for Six Sigma success." *Workforce*, 82(5): 64–69.

Appendix 3B: Comparing quality management practices between the United States and Mexico*

3B.1   Introduction

Globalization and international trade, along with advances in information technology, have dramatically increased competition worldwide. To compete in a global market, firms must be equipped with the latest technology, up-to-date information, skilled employees, and advanced managerial skills.

The concept of total quality management (TQM) developed as a result of intense global competition. Companies with international trade and global competition have paid consider-able attention to TQM philosophies, procedures, tools, and techniques. The importance of international quality management research first appeared in the *Journal of Decision Sciences* (Kim and Chang 1995). However, little empirical research has been conducted on interna-tional quality management (Rao et al. 1999), especially in the cross-cultural arena (Sila and Ebrahimpour 2003). While most quality management research has focused on comparing practices in the United States and Japan in early 1990, recent quality management research has been extended to other countries and regions around the world (Rao et al. 1999). Most of these studies, however, use different instruments and constructs for comparing quality manage-ment practices across countries. Accordingly, the need to develop instruments for comparing quality management practices across countries based on a generic and acceptable framework for quality management is clear. The development of the Malcom Baldrige National Quality Award (MBNQA) model as a basis for quality management practices and research has gained attention in recent years in the United States (Bell and Elkins 2004; Curkovic et al. 2000; Garvin 1991; Stephens et al. 2005). The application of the MBNQA is not limited to the United States.

* By Mahour Mellat Parast, Stephanie G. Adams, Erick C. Jones, S. Subba Rao, and T. S. Raghu-nathan.

The universality of the MBNQA and its relationship with many quality management constructs has made it a useful framework for studying quality management practices (Samson and Terziovski 1999). In fact, most national and international quality awards have been influenced by the MBNQA criteria (DeBaylo 1999; Ettore 1996), including the European Quality Award, the Mexican National Quality Award, the Brazilian National Quality Award, the Egyptian Quality Award, and the Japanese Quality Award. Accordingly, cross-cultural comparison of quality management based on the MBNQA criteria could be helpful in understanding how countries view quality management practices.

The purpose of this appendix is to compare quality management practices in the United States and Mexico, based on the MBNQA. More specifically, this research focuses on understanding the similarities and differences of quality management practices in the United States and Mexico, as well as the critical success factors.

3B.2   Literature review

The importance of TQM in manufacturing and service organizations has increased significantly over the past 20 years. Firms with international trade devote special attention to the tools and techniques recommended by quality management experts. Early studies on international quality practices were mainly focused on developed countries such as the United States and Japan (Benson et al. 1991; Ebrahimpour and Johnson 1992; Flynn 1992; Garvin 1986; Khoo and Tan 2003; Knotts and Tomlin 1994; Reitsperger and Daniel 1991; Richardson 1993; Rogers 1993). Moreover, the empirical literature has extended its scope by studying and comparing quality management practices in other developed and developing countries, such as the United States and Taiwan (Madu et al. 1995; Solis et al. 1998); Asia and the South Pacific (Corbett et al. 1998); Japan, Korea, and Denmark (Dahlgaard et al. 1990); the East and West (Russia, Taiwan, Japan, Korea, Finland, Estonia, Denmark, India, Sweden, England, and New Zealand) (Dahlgaard et al. 1998); Costa Rica and the United States (Tata et al. 2000); the United States, Mexico, China, and India (Quazi et al. 2002); the United States and Mexico (Burgos Ochoategui 1998); Canada and Mexico (Galperin and Lituchy 1999); the UK and Ireland (McAdam and Jackson 2002); and the West, East, and Russia (Selivanova and Eklof 2001). Most of these studies attempt to understand the critical success factors of TQM. Because of lack of development of valid instruments, the results of such studies cannot be generalized, yet they provide insights about quality management in the international context. As Sila and Ebrahimpour (2003) indicate, the question regarding the universality of quality management practices has not been answered and more empirical, cross-country, and industry-specific research is needed in quality management.

The universality of the MBNQA and its relationship to quality management constructs has made it a useful framework for studying quality management practices. Samson and Terziovski (1999) state:

> Although there are always going to be debates about how to categorize elements of a holistic process and framework like TQM, it is necessary to decompose it in some way to facilitate analysis. Since the most pervasive and universal method has been awards criteria such as the MBNQA, we have chosen to follow that framework.

Despite the application of the MBNQA criteria in practice, few theoretical or empirical studies have been reported in the literature, especially in the international context. Initial application of the MBNQA model as a framework for identifying quality management constructs was described by Steeples (1992). Previous research on quality management in the international context has proved the universality and applicability of MBNQA in comparative quality management studies (Prajogo 2005; Rao et al. 1997, 1999). Rao et al. (1998, 1997, 1999) compared quality management practices in different countries. They found that top management support is a significant factor influencing strategic quality planning, human resource development, supplier quality (SQ), quality results (QR), and customer orientation practices. They found significant relationships between top management support and all of the quality management practices and results: information and analysis, strategic quality planning, human resource development, quality assurance, supplier relationship, customer focus and orientation, and QR.

In summary, previous research supports the applicability of the MBNQA criteria in comparative analysis of quality management practices in the international context. The models developed by Rao et al. (1997, 1998, 1999) have resulted in the development of a set of reliable and valid instruments that can be used in comparative quality management practices.

3B.3   Theoretical framework: Comparative management

Traditionally, research in comparative management fits within three schools of thought (Corbett et al. 1998): culture free (Haire et al. 1966), convergence (Form 1979), and culture specific (Hofstede 1980). In this appendix, a culture-free perspective is employed, indicating that, based on the MBNQA criteria, there is no difference between quality management practices in the United States and Mexico.

Despite the development of different TQM models, little conceptualization has been done on quality management in the international and global context. The international quality management model, developed by Rao et al. (1999), is the only quality management model that has been developed, verified, and tested in the international context. The instrument consists of nine dimensions for quality management:

1. *Quality leadership (QL)*: Addresses the critical role of management in driving companywide quality management efforts (Flynn et al. 1994; Puffer and McCarthy 1996).
2. *Quality information and analysis (QIA)*: Refers to the availability of information systems in the company, and the procedures and systems that provide accurate and timely information for the managers to make quality decisions (Flynn et al. 1994).
3. *Strategic planning process of quality management (SPP)*: Incorporates the integration of quality and customer satisfaction issues into strategic and operational plans, which allows firms to set clear priorities, establish clear target goals, and allocate resources for the most important things (Godfrey 1993).
4. *Support for human resource development (SHRD)*: Emphasizes the essential role of employee training and their involvement in quality-related decisions in the company (Rao et al. 1997).
5. *Quality assurance of products and services (QAP)*: Refers to the systematic approach used by organizations for maintaining and improving the quality of their products and services (Rao et al. 1997).
6. *SQ*: Acknowledges the importance of suppliers in achieving higher levels of quality in an organization (Flynn et al. 1994).
7. *QR*: Indicates how much internal operations, customer satisfaction, and market and financial performances in the company have been improved by quality management practices (Juran 1993; Steeples 1992).

8. *Customer focus and satisfaction (CFS)*: Specifies how much the company evaluates the feedback from its customers in improving quality (Schonberger 1994).
9. *General matters* (social responsibility) *(GM)*: Stresses the practice of the company's responsibility and its social role in society, such as improvement of education, safety, and health care in the community (Florida 1996).

3B.4   Hypotheses

According to the culture-free approach in comparative management study, differences in cultural practices do not affect the practice of quality in organizations. Therefore, the following hypotheses have been developed:

- Hl: There is no difference in QL for quality management between the United States and Mexico.
- H2: There is no difference in QIA for quality management between the United States and Mexico.
- H3: There is no difference in the SPP between the United States and Mexico.
- H4: There is no difference in support of human resource development and management for quality management between the United States and Mexico.
- H5: There is no difference in quality assurance for products and services for quality management between the United States and Mexico.
- H6: There is no difference in SQ for quality management between the United States and Mexico.
- H7: There is no difference in QR for quality management between the United States and Mexico.
- H8: There is no difference in CFS for quality management between the United States and Mexico.
- H9: There is no difference in GM (social responsibility) for quality management between the United States and Mexico.

3B.5   Research methodology

3B.5.1   Survey instrument

Survey research is the methodology adapted for this appendix. Surveys provide an important source of basic scientific knowledge, and surveys are powerful methods for obtaining information about many aspects of a problem. The type of survey used in this research was a written mail questionnaire that reduces bias, and the researchers' opinions will not influence the respondents' answers. This is especially true for studies involving large sample sizes and large geographic areas.

The following steps were used to develop and validate the measurement instrument (Rao et al. 1999). First, the theoretical dimensions underlying quality practices were conceptualized. While the 1992 version of the MBNQA model with seven constructs was used as a starting point for developing the survey, the instrument has been tested, modified, and refined in different stages to reflect all aspects of quality management. The final instrument, which was used in data collection, has nine constructs. A questionnaire was developed to measure these constructs. The same questionnaire was used to collect data from the United States and Mexico, and a five-point Likert scale was used to measure the items (Tables 3B.1 and 3B.2).

Table 3B.1 Group Statistics

Group	N	Mean	Std. deviation	Std. error mean
QL	113	2.82035	0.787986	0.74127
	258	2.67946	0.693048	0.43147
QIA	113	2.46637	0.733218	0.68975
	258	2.51512	0.648947	0.40402
SPP	113	2.50000	0.638217	0.60038
	258	2.41705	0.590604	0.36769
SHRD	113	3.10265	0.935172	0.87974
	258	3.25581	0.847694	0.52775
QAP	113	3.67345	0.767070	0.72160
	258	3.58992	0.705437	0.43919
SQ	113	2.98496	0.780250	0.73400
	258	3.02481	0.722354	0.44972
QR	113	3.55498	0.904062	0.85047
	258	3.44845	0.748031	0.46570
CFS	113	2.86549	0.70978	0.66771
	258	2.84410	0.69315	0.37551
GM	113	2.54159	0.722313	0.67950
	258	2.72093	0.607947	0.37849

Table 3B.2 Correlation Matrix

	QL	QIA	SPP
QL			
Pearson correlation	1		
Significance (two-tailed)	–		
N	371		
QIA			
Pearson correlation	.679	1	
Significance (two-tailed)	.001	–	
N	371	371	
SPP			
Pearson correlation	.831	.693	1
Significance (two-tailed)	.001	.001	–
N	371	371	371
SHRD			
Pearson correlation	.704	.689	.714
Significance (two-tailed)	.001	.001	.001
N	371	371	371
QAP			
Pearson correlation	.639	.655	.623
Significance (two-tailed)	.001	.001	.001
N	371	371	371

(*Continued*)

Table 3B.2 Correlation Matrix (Continued)

	QL	QIA	SPP
SQ			
Pearson correlation	.852	.656	.576
Significance (two-tailed)	.001	.001	.001
N	371	371	371
QR			
Pearson correlation	.651	.605	.618
Significance (two-tailed)	.001	.001	.001
N	371	371	371
CFS			
Pearson correlation	.721	.655	.749
Significance (two-tailed)	.001	.001	.001
N	371	371	371
GM			
Pearson correlation	.641	.592	.631
Significance (two-tailed)	.001	.001	.001
N	371	371	371

Note: All correlations are significant at the .01 level (two-tailed).

3B.5.2   Data collection

Data were collected from the United States and Mexico to study the relationship between the quality practices across the two countries (Rao et al. 1999). The population of the study in the United States consisted of a random sample of quality managers and professionals who are members of the American Society for Quality (ASQ). A total of 1500 surveys were mailed, and 258 completed surveys were received (a response rate of 17%). In Mexico, a list of manufacturing companies was obtained from the Monterrey Institute of Technology from which a random sample of 200 firms was taken. Out of 200 surveys, 113 completed surveys were received (a response rate of 56%). The results from the survey were analyzed using SPSS software.

SHRD	QAP	SQ	QR	CFS	GM
1					
–					
371					
.595	1				
.001	–				
371	371				
.610	.671	1			
.001	.001	–			
371	371	371			

(Continued)

(Continued)

SHRD	QAP	SQ	QR	CFS	GM
.630	.655	.685	1		
.001	.001	.001	–		
371	371	371	371		
.687	.709	.677	.723	1	
.001	.001	.001	.001	–	
371	371	371	371	371	
.726	.588	.620	.619	.765	1
.001	.001	.001	.001	.001	–
371	371	371	371	371	371

References

Bell, R. R. and Elkins, S. (2004). "A balanced scorecard for leaders: Implications of the Malcolm Baldrige National Quality Award criteria." *SAM Advanced Management Journal,* 69(1): 12–17.

Benson, P. G., Saraph, J. V., and Schroeder, R. G. (1991). "The effects of organizational context on quality management: An empirical investigation." *Management Science,* 17: 1107–1124.

Burgos Ochoategui, F. R. (1998). "The effect of cultural differences in the adoption TQM practices in Mexico and the United States." PhD dissertation. Texas A&M University, Arlington.

Corbett, L. M., Adam, E. E. Jr., Harrison, N. J., Lee, T. S., Rho, B.-H., and Samson, D. (1998). "A study of quality management practices and performance in Asia and the South Pacific." *International Journal of Production Research,* 36(9): 2597–2607.

Curkovic, S. et al. (2000). "Validating the Malcolm Baldrige national quality award framework through structural equation modeling." *International Journal of Production Research,* 38(4): 765–791.

Dahlgaard, J. J. et al. (1998). "Quality management practices: A competitive study between East and West." *International Journal of Quality & Reliability Management,* 15(8/9): 812–826.

Dahlgaard, J. J., Kanji, G., and Kristensen, K. (1990). "A comparative study of quality control methods and principles in Japan, Korea, and Denmark." *Total Quality Management,* 1: 115–132.

DeBaylo, P. W. (1999). "Ten reasons why the Baldrige model works." *The Journal for Quality and Participation,* 22(1): 24–28.

Ebrahimpour, M. and Johnson, J. L. (1992). "Quality, vendor evaluation and organizational performance: A comparison of US and Japanese firms." *Journal of Business Research,* 25: 129–142.

Ettore, B. (1996). "Is the Baldrige still meaningful?" *Management Review,* 85(3): 28–31.

Florida, R. (1996). "Lean and green: The move to environmentally conscious manufacturing." *California Management Review,* 39: 80–105.

Flynn, B. B. (1992). "Managing for quality in the US and in Japan." *Interfaces,* 22(5): 69–80.

Flynn, B. B., Schroeder, R. G., and Sakakibara, S. (1994). "A framework for quality management research and an associated instrument." *Journal of Operations Management,* 11: 339–366.

Form, W. (1979). "Comparative industrial sociology and the convergence hypotheses." *Annual Review of Sociology,* 5: 1–25.

Galperin, B. L. and Lituchy, T. R. (1999). "The implementation of total quality management in Canada and Mexico: A case study." *International Business Review,* 8: 323–349.

Garvin, D. A. (1986). "Quality problems, policies, and attitudes in the United States and Japan: An exploratory study." *The Academy of Management Journal,* 29(4): 653–673.

Garvin, D. A. (1991). "How the Baldrige Award really works?" *Harvard Business Review,* 69(6): 80–93.

Godfrey, A. B. (1993). "Ten areas for future research in total quality management." *Quality Management Journal,* 1: 47–70.

Haire, M., Ghiselli, E. E., and Porter, L. W. (1966). *Managerial Thinking: An International Study.* New York: John Wiley & Sons.

Hofstede, G. (1980). *Culture's Consequences: International Differences in Work-Related Values.* London: Sage.

Juran, J. M. (1993). "Why quality initiatives fail?" *Journal of Business Strategy,* 14(4): 35–38.

Khoo, H. H. and Tan, K. C. (2003). "Managing for quality in the USA and Japan: Differences between the MBNQA, DP, and JQA." *The TQM Magazine,* 15(1): 14–24.

Kim, K. Y. and Chang, D. R. (1995). "Global quality management: A research focus." *Decision Sciences,* 26(5): 561–568.

Knotts, R. and Tomlin, S. (1994). "A comparison of TQM practices in the US and Mexican companies." *Production and Inventory Management Journal,* 35(1): 53–58.

Madu, C. N., Kuei, C., and Lin, C. (1995). "A comparative analysis of quality practice in manufacturing firms in the US and Taiwan." *Decision Sciences,* 26: 621–636.

McAdam, R. and Jackson, N. (2002). "A sectoral study of ISO 9000 and TQM transitions: The US and Irish brewing sector." *Integrated Manufacturing Systems,* 13(4): 255–263.

Prajogo, D. (2005). "The comparative analysis of TQM practices and quality performance between manufacturing and service firms." *International Journal of Service Industry Management,* 16(3): 217–228.

Puffer, S. M. and McCarthy, D. J. (1996). "A framework for leadership in a TQM context." *Journal of Quality Management,* 1: 109–130.

Quazi, H. A., Hing, C. H., and Meng, C. T. (2002). "Impact of ISO 9000 certification on quality management practices: A comparative study." *Total Quality Management,* 13(1): 53–67.

Reitsperger, W. D. and Daniel, S. J. (1991). "A Comparison of quality attitudes in the USA and Japan: Empirical evidence." *Journal of Management Studies,* 28(6): 585–599.

Richardson, J. (1993). "Restructuring supplier relationships in US manufacturing for improved quality." *Management International Review,* 33: 53–67.

Rogers, R. E. (1993). "Managing for quality: Current differences between Japanese and American approaches." *National Productivity Review,* 12: 503–517.

Rao, S. S., Raghunathan, T. S., and Solis, L. E. (1997). "A comparative study of quality practices and results in India, China, and Mexico." *Journal of Quality Management,* 2(2): 235–250.

Rao, S. S., Raghunathan, T. S., and Solis, L. E. (1998). "The best commonly followed practices in the human resource dimension of quality management in the new industrialized countries (NIC): The case of India, China, and Mexico." *International Journal of Quality & Reliability Management,* 16(3): 215–225.

Rao, S. S., Solis, L. E., and Raghunathan, T. S. (1999). "A framework for international quality management research: Development and validation of a measurement instrument." *Total Quality Management,* 10(7): 1047–1075.

Samson, D. and Terziovski, M. (1999). "The relationship between total quality management practices and operational performance." *Journal of Operations Management,* 17: 393–409.

Schonberger, R. J. (1994). "Human resource management lessons from a decode or total quality management and reengineering." *California Management Review,* 36: 109–123.

Selivanova, I. and Eklof, J. (2001). "Total quality management in the West, East and Russia: Are we different?" *Total Quality Management,* 12(7/8): 1003–1009.

Sila, I. and Ebrahimpour, M. (2003). "Examination and comparison of the critical factors of total quality management (TQM) across countries." *International Journal of Production Research,* 41(2): 235–268.

Solis, L. E. et al. (1998). "Quality management practices and quality results: A comparison of manufacturing and service sectors in Taiwan." *Managing Service Quality,* 8(1): 46–54.

Steeples, M. M. (1992). *The Corporate Guide to the Malcolm Baldrige National Quality Award.* Homewood, IL: Business One Irwin.

Stephens, P. R., Evans, J. R., and Matthews, C. H. (2005). "Importance and implementation of Baldrige practices for small business." *Quality Management Journal,* 12(3): 21–38.

Tata, J., Parasad, S., and Matwan, J. (2000). "Benchmarking quality management practices: US versus Costa Rica." *Multinational Business Review,* 8(2): 37–42.

Further Readings

Agus, A. (2004). "TQM as a focus on improving overall service performance and customer satisfaction: An empirical study on police service sector in Malaysia." *Total Quality Management,* 15(5/6): 615–628.

Anderson, E. W., Farnell, C., and Lehmann, D. R. (1994). "Customer satisfaction, market share, and profitability: Findings from Sweden." *Journal of Marketing,* 58: 53–66.

Dean, J. W. and Bowen, D. E. (1994). "Management theory and total quality: Improving research and practice through theory development." *The Academy of Management Journal,* 19(3): 392–418.

Flynn, B. B. and Saladin, B. (2000). "Relevance of Baldrige constructs in an international context: A study of national culture." *Academy of Management Proceedings,* OM: C1.

Hinkle, D. E., Wiersma, W., and Jurs, S. G. (1994). *Applied Statistics for the Behavioral Sciences.* 3rd edn. Boston, MA: Houghton Mifflin.

Lee, M. C. and Hwan, I. S. (2005). "Relationships among service quality, customer satisfaction, and profitability in the Taiwanese Banking Industry." *International Journal of Management,* 22(4): 635–648.

Metzler, K. et al. (2005). "The relationship between customer satisfaction and shareholder value." *Total Quality Management and Business Excellence,* 16(5): 671–680.

Noh, S. J., Park, H., and Park, M. (2004). "Multidimensional quality assessment of multimedia telecommunication systems for enhancing customer satisfaction." *Total Quality Management,* 15(7): 899–908.

Pelled, L. H. and Xin, K. R. (2000). "Relational demography and relationship quality in two cultures." *Organization Studies,* 216(6): 1077–1094.

Powel, T. C. (1995). "Total quality management as competitive advantage: A review and empirical study." *Strategic Management Journal,* 16: 15–37.

Robinson, C. J. and Malhotra, M. K. (2005). "Defining the concept of supply chain quality management and its relevance to academic and industrial practice." *International Journal of Production Economics,* 96: 315–337.

Sousa, R. and Voss, C. A. (2002). "Quality management revisited: A reflective review and agenda for future research." *Journal of Operation Management,* 20(1): 91–109.

Spencer, B. A. (1994). "Models of organization and total quality management: A comparison and critical evaluation." *The Academy of Management Review,* 19(3): 446–471.

Teagarden, M. B., Butler, M. C., and Van Ginlow, M. A. (1992). "Mexico's moquila industry: Where strategic human resource management makes a difference." *Organizational Dynamics,* 20: 34–48.

Westlund, A. H. et al. (2005). "On customer satisfaction and financial results in the Swedish real estate market." *Total Quality Management & Business Excellence,* 16(10): 1149–1159.

Wilson, D. D. and Collier, D. A. (2000). "An empirical investigation of the Malcolm Baldrige National Quality Award causal model." *Decision Sciences,* 31: 361–383.

chapter four

Six Sigma without Lean

> What has been will be again, what has been done will be done again,
> there is nothing new under the sun.
>
> **—Ecclesiastes 1:9**

4.1   Introduction

The Six Sigma process that many philosophers feel is an extension and modernization of total quality management (TQM) into a systemized approach has been studied extensively. This text extends the author's belief that Six Sigma is an extension of quality theory, industrial engineering science, statistics, and firm-wide economics. Although other researchers focus on how the history of quality evolved into Six Sigma, we believe that business dynamics moved business to "just in time" quality. Economically justifiable quality that moved the company's bottom line was the initial intent of Six Sigma.

The author of this text as an initial practitioner was a consultant for quality initiatives and later an academic researcher of the philosophies and description of the growth and emergence of the field both from a personal and a historical perspective. For example, some initial training of Six Sigma I experienced (Juran's Six Sigma) did not initially include statistics except for defect determination at the initial sigma-level but had an underpinning that would easily allow for statistical tools such as multi-variable analysis tools [ANOVA, linear regression, and statistical process control (SPC)] to be easily integrated.

4.1.1   The Six Sigma thinking in this text

The main ideas of Six Sigma thinking are important. The author suggests that the mindset of Six Sigma should be that Six Sigma takes a business problem, converts it to engineering problem that uses statistics, then develops an engineering solution, and finally converts that to a business solution.

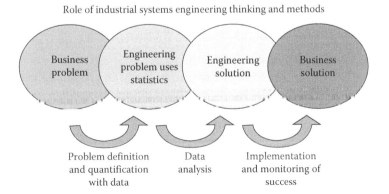

The integration of Deming's quality-based approaches with the Frederich Taylor Scientific Management-based approach has driven the Six Sigma processes. The scientific management philosophy that focuses on data, specifically on measures that focus on financial performance, to make business decisions has driven American business thought since the early 1900s. The long-running myopic focus only on bottom line resulted in businesses not appreciating the long term customer service perception in the form of the customer's perception of their firm's quality.

The axiom that cheaper products are easier to make in the short run but may put the organization out of business in the long term if another company is perceived to have a better product became true in the 1980s. The reaction to this situation was that firms moved completely away from bottom line focus to the total quality focus at the expense of short- and long-term profits to save their business perception and brand. Due to the intrinsic and hard to measure components of TQM and related processes, many companies associated lack of business performance to quality initiatives. This was further complicated by the movement of large firms to implement large-scale expensive computer system enterprise resource planning (ERP) systems. The implementation of ERP systems into firms was associated with a quality philosophy business process re-engineering (BPR). BPR initiatives that often led to layoffs were negatively associated with improving the firms' quality. Though the philosophy was sound, the integrators of these systems utilized it to justify the costs of the ERPs through labor reduction. Companies struggled with the need for positive customer perception of quality, the measurable improvement of the firm's products, and staying profitable as a company. This led to the emergence of a nonBPR, cost-focused way of improving a firm's quality, the Six Sigma.

This chapter provides many examples on using Six Sigma methods and tools and seeks to provide a background to the philosophy from its beginnings at Motorola; description of the personnel; approaches developed; management, scientific, statistical, and engineering tools; and shortcomings. The purpose of separating Six Sigma from Lean Six Sigma is that the use of Lean Six Sigma may not be relevant to some organizations. Understanding the distinction between Six Sigma, Lean, and Lean Six Sigma is relevant in that most companies are not one-size-fits-all cases. Though many trainings and consultants combined the philosophies for business reasons, these need to be reviewed to make sure they will work individually. This text chooses to separate them out and distinguish the parts. These then can be integrated if desired.

4.2   Background

In the 1980s, there was a perceived lack of quality associated with American firms and a push toward the Japanese view of quality which was based ironically on the theories of an American, William E. Deming. The increased market share realized by Japanese firms, mainly by Toyota, motivated organizations to study and romanticize the ideas of the Toyota Production Systems (TPSs) as the benchmark for firm quality. The TPS approach supports modern quality and the Lean concepts of today. The Six Sigma foundations were laid by quality management initiatives by some of the noted quality gurus. Some well-known quality gurus whose technical propositions contributed to Six Sigma are Philip B. Cosby, W. Edwards Deming, Armand Feigenbaum, Kaoru Ishikawa, Joseph M. Juran, Walter A. Shewart, and Genichi Taguchi. Most consider that Deming's foundational contributions in conjunction with those of the aforementioned gurus led to the most comprehensive theory of TQM.

TQM was popular initially with organizations except that its various components by many gurus caused confusion and made it difficult to associate the business results with the initiative. Researchers associate Six Sigma as the next step to quality theory seeking to improve the problems associated with TQM and maintain and improve the benefits.

4.2.1 *Which Six Sigma points are retained from TQM?*

The benefits associated with TQM include

1. Marketing in which everyone in an organization is responsible for the quality of goods and services produced by the firm.
2. Significant investment in education and training; training includes problem solving using quality tools. The traditional seven tools of quality include: control charts, histograms, check sheets, scatter plots, cause-and-effect diagrams, flowcharts, and Pareto charts. These are described further in the text.
3. Decision made by management that focuses on customer satisfaction. The traditional seven quality management tools include affinity diagrams, interrelationship digraphs, tree diagrams, matrix diagrams, prioritization matrices, process decision program charts, and activity network diagrams (Sower et al. 1999). The quality functional deployment (QFD) chart and the house of quality (HOQ) diagrams combine some of the management technique tools. These are described in the text.

4.2.2 *Which Six Sigma point changed from TQM?*

TQM was not enough or maybe too much in that it was highly philosophical and provided few "hard" facts or ideas for practitioners to follow. It was an all-inclusive philosophy that presented potential to transform the way in which all types of businesses could be managed. The philosophy was too broad and cross-cutting. Six Sigma extended mainly to the Deming concepts of TQM and provided a specific metric that allowed organizations to have a clear direction and goal (Sigma level) to achieve. It provided a structured concept and philosophical ideas into a methodology that can be followed to obtain process improvements.

4.2.3 *Six Sigma roots*

One manner of describing Six Sigma is that it is an approach that builds on the philosophies and tools of TQM following the direction provided by statistical performance goals. Another way of stating this is that the roots of Six Sigma can be traced to two primary sources: TQM and the statistical metrics originating at Motorola Corporation (Arnheiter and Maleyeff 2005).

4.2.3.1 *From product quality program to firm-wide decision-making strategy*

Currently, many see Six Sigma as an organizational strategy to meet business goals. Initially, Six Sigma was not a broad, long-term decision-making business strategy than a narrowly focused quality management program. Originally, it was a reliability performance goal that needed to be achieved for Motorola products. Pepper and Spedding (2010) mention that Six Sigma was developed in the 1980s at

Motorola by Bill Smith, a reliability engineer (Brady and Allen 2006). As it was recognized that most of the product lines needed to hit this type of performance goal to minimize defects, the company's overall performance would improve while seeking this goal.

Forming a doctrine for Six Sigma was initiated on the findings through internal reviews that different implementation approaches were used by employees due to difference in interpretations (Business Balls 2012). In the American Society for Quality Conference, the concept of Six Sigma was introduced as a performance improvement methodology that identifies defects in a scale of parts per thousand, later changed to parts per million (ppm) by engineers such as Mikel J. Harry. Noted pioneers of this concept in Motorola are Bob Galvin, the chairman, and Robert Galvin, the then CEO of Motorola. Using Bill Smith's yield theory, business processes were also brought under the area of application of Six Sigma (Quality Council of Indiana 2007).

4.3   Corporates support Six Sigma popularity

Six Sigma's popularity and acceptance came through the work of Jack Welch, the then CEO of General Electric (GE) in 1995. Welch had observed the success experienced through Bill Smith's approach and intensely championed and led the Six Sigma methodology in GE (Black and Revere 2006). He not only championed the approach but also utilized the personnel-driven belt systems as a means of identifying high-level directors and managers at GE in the future. Consequently, companies such as Allied Signal (now merged with Honeywell) and GE adopted this methodology with huge bottom-line results. This further popularized the concept and many business strategies were reframed that led to considerable financial success.

Extensive publicity by 1998, including emphasis in annual reports on Six Sigma, is associated with Bill Smith and Motorola in the late 1980s, and subsequently popularized by GE and Jack Welch in the early 1990s. Other implementations that popularized Six Sigma include companies such as Honeywell, Dow, and Goodyear inviting key customers and supply chain partners to participate in their Black Belt and other education programs. DuPont uses Six Sigma resources to reduce costs; GE utilizes Six Sigma as consulting services to customers. In late 1999, Ford Motor Company became the first major automaker to adopt a Six Sigma strategy. At Ford, each car has approximately 20,000 off fixture displays (OFDs). Therefore, if Ford were to attain Six Sigma quality, approximately one car in every 15 produced would contain a defect (Truby 2000). The results were real financial savings, verified by financial representatives, returned to the business.

4.3.1   Industries other than manufacturing

The application of Six Sigma in a variety of industries is well documented in the literature. Examples in the manufacturing sector include Motorola and GE (Pande et al. 2000) as the most famous, whereas it has also had success in the construction industry (Stewart and Spencer 2006) and accounting practices (Brewer and Bagranoff 2004). A current shift in literature is focused on the application of Six Sigma principles in the service sector (Antony 2006; Chakrabarty and Tan 2007; Sehwail and Deyong 2003). The importance of identifying key performance metrics is a recurring theme in the literature. Antony (2006) emphasizes the importance of aligning projects to business objectives, and in agreement with Sehwail and Deyong (2003), reflects that the definition of Six Sigma as a quality measure must be taken in the context of service industries. For example, "a defect may be defined as anything which does not meet customer needs or expectations. It would be illogical to assume that all defects are equally good when we calculate the sigma capability of a process" (Antony 2006, p. 246). In other words, there is so much possible variation in the customer response that it is difficult to fit them within the constraints of whether they are merely defects or not. Factors such as management commitment and open communication are essential for successful implementation as with any attempt at continuous improvement.

4.4 Challenges of Six Sigma

The challenges associated with Six Sigma (Pepper and Spedding 2010) initiatives include

1. Corporate-wide metrics are hard to measure
   The Six Sigma metrics are hard to develop into firm-wide performance because they have to be expanded to include not just manufacturing operations but other functions as well. The challenge for most companies is firm-wide benchmarking and measurement, and how and when to convert the Six Sigma metric or corporate Z score. With Six Sigma, the value of an organization's output includes not just quality, but availability, reliability, delivery performance, and after-market service. Most companies seek Six Sigma performance within each major functional output that contributes to the end customer value. The Six Sigma corporate metric is applied in a broad fashion, striving for Six Sigma level performance at the lowest level of activity or at the lowest invested cost.

2. The excessive training expenses for an "in-house" champion belt system
   Champion belt infrastructure. In answer to this, it can be said that Six Sigma provides a clear focus on measurable financial returns through a sequential and disciplined manner, and establishes an "infrastructure of champions" with its training style of introducing "belt" qualifications (green, black, master black belts, etc.) within the organization to lead the way in data-driven decision making for improvement efforts (Antony 2004).

3. The "in-house" belt system becomes more focused on internal prestige and promotion, not improvement. The champions may also become more consultant based and over-promise success. The training for and solutions put forward by Six Sigma can be prohibitively expensive for many businesses, and the correct selection of improvement projects is critical (Senapati 2004). Antony (2004) discusses the nonstandardization of training efforts (in terms of belt rankings, etc.), and how this accreditation system can easily evolve into a bureaucratic menace, in which time and resources are misspent focusing on the number of "belts" within the organization, and not on the performance issues at hand.

4. Consulting, not improvement, of the belt system is an attempt to develop "in-house" expertise. The danger is that this strategy may face the dangers that most consultancy practices face, which tend to overpromise and underdeliver. Consultants tend to recommend but do not have the power or funding to implement. Sponsors are given the improvement process that is to be implemented, then the "in-house" consultant or black or green belt leaves without ensuring that the implementation has taken place. This is a big challenge to the philosophy.

5. Six Sigma has a statistics-heavy, technical approach to process control. In order to prevent it from becoming "watered down" or "ignored" great care should be taken to focus on the wider philosophy behind the structured approach to Six Sigma. The statistics can be supplemented with specialists and engineers.

4.5 Six Sigma and engineers

4.5.1 Six Sigma and engineering education

The fact that Six Sigma was discovered by an engineer seems to escape many people. Bill Smith, the originator of the philosophy was a reliability engineer. The idea of using a Six Sigma metric was developed by Bill Smith at Motorola in response to sub-standard product quality traced in many cases to decisions made by engineers when designing component parts.

Traditionally, reliability engineering comes from a curriculum related to industrial and manufacturing engineers who have been trained in advanced statistics, reliability, and defects that are related to manufacturing in their base undergraduate accredited curriculum. They are also provided foundational knowledge in scientific management, human factors, and health systems engineering that address the employee (Deming) side of the Six Sigma. Other engineering undergraduate curriculums, including some mechanical engineering programs, also focus in these areas.

4.5.2 Controversy between engineers education and green/black belts

Some employers and academics question the need for "specialized" training of belts when these engineers have been trained to complete these tasks in a more robust manner. The Six Sigma training programs are seen as a means to teach advanced engineering skills to nonengineers. Some academics believe that the lack of robust training, especially in statistics, may lead to the reduction of effectiveness of the belts. In practice, the engineering students have adapted and seek Six Sigma certification prior to graduation and many companies hire both engineers and green and black belts.

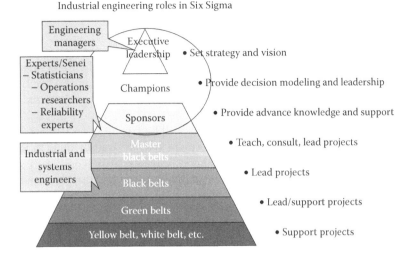

Industrial engineering roles in Six Sigma

4.6   *The statistics of Six Sigma*

The term "Six Sigma" refers to a statistical measure of defect rate within a system. The term "Sigma" is a Greek letter used to describe variability in mathematics. Its classical use was to describe defects per unit. The higher the Sigma level the lower the number of defects. The process is so well understood and controlled that those six standard deviations will fit between the average output and spec limit.

It could be said that Abraham de Moivre (1773) declared that "sigma" is measured or observed as standard deviations from the mean. These are on the order of a +3 and −3 and standard deviations from the mean represent a normal distribution. This leads to the logic that Six Sigma would be ±6 standard deviations from the mean. Thus, it can be said that the framework for applying Six Sigma is a set of variables that have a normal distribution.

We can further play on the term "sigma" which refers to variance squared in mathematics. Variance is the difference between the average and the actual performance achieved. It has been commonly used in statistics as a measure of variance due to the fact that by squaring variance you do not have to manipulate when variance is negative. Traditional product specifications were based on "Three-Sigma" or 99.7% performance reliability for binomial distributions. Historically, this had been acceptable in business, but as many products became based on sub-components, sub-parts, and sub-assemblies the number of opportunities for failure or to incur a defect became higher. The opportunities went from 100s, 1000s, to millions. If you look at traditional defective rates at 99.7%, then .03 defects per 100 does not seem too bad. If we look at this for 3000 ppm then it is not acceptable.

Historically, product design engineers use the "three-sigma" rule when evaluating whether or not an acceptable proportion of manufactured components would be expected to meet tolerances.

When a component's tolerances were consistent with a spread of six standard deviation units of process variation, about 99.7% of the components for a centered process would be expected to conform to tolerances. That is, only 0.3% of parts would be nonconforming to tolerances, which translates to about 3000 nonconforming parts per million (NCPPM).

As products designs became more complex and defective products were becoming more prevalent, customers started demanding higher quality. Three-sigma quality for each OFD was no longer acceptable.

For example, consider a product that contains 1000 OFDs. If, for each OFD, three-sigma quality levels are achieved, only about 5% of the products would be defect free.

The calculation used to obtain this probability requires raising the fraction conforming (0.997) to the power of 1000, and is based on the binomial probability distribution (Devore 2000).

Assuming 1000 OFDs, only 37% of products will be free of defects if the quality level at each OFD averaged 99.9%, and 90% of products will be free of defects if the quality level at each OFD averaged 99.99%.

Other industries face similar challenges in achieving superior quality. In addition to the consumer electronics industry, other products with a large number of OFDs include automobiles, engines, airframes, and health care. Manufacturers of medical devices and other products where defects in the field may cause harm must achieve almost perfect quality. Companies that manufacture less-complex products but sell them in very large volumes also need to focus on achieving superior quality.

4.7   The 3.4 defects per million or long-term shift confusion

Six Sigma is defined as using underpinned by statistical technique that presents a structured and systematic approach to process improvement, aiming for a reduced defect rate of 3.4 in a million, or Six Sigma (Brady and Allen 2006). At Motorola, as Six Sigma was being developed a relationship between component quality and final product quality was observed, it was discovered that, from lot-to-lot, a process tended to shift a maximum of 1.5 sigma units (McFadden 1993). For example, if the company sets a goal for final product quality of 99.7% and the products include about 1000 OFDs, then the 3.4 NCPPM corresponding to the Six Sigma metric would become the standard against which all decisions were made. In some popular explanations of Six Sigma, the "sigma score" is converted to a short-term measure. To make the conversion from the long-term values, 1.5 is added to the standard normal score. With that approach, the "0 Sigma" shown on the slide would be reported as a "1.5 sigma" level, 3 Sigma as 4.5, and 4.5 as 6.0. That is the reason we have 3.4 defects per million instead of 0.002 defects per million. The 1.5 sigma shift is given as process shift in short-term measurement as some practitioners explain. Many academics have chosen not to use this arbitrary addition. The sigma levels are simply the standard normal values. Depending on the constraints, the environment in which processes thrive, and the culture of the organization the level of sigma becomes significant for the required level of performance. It can be identified by a ±1.5 shift in sigma levels. The relevant rate of failure allowed for the level is given in Table 4.1.

Table 4.1 Defects and Sigma Levels

Sigma level	Defects per million opportunities (DPMOs) (ppm)
6	3.4
5	233
4	6210
3	66810
2	308770
1	697672

Source: Quality Council of Indiana (2007).

4.8   Methodologies

Some of the original training associated with Six Sigma was based on training by Joseph Juran. The methodology of the training that was provided to Motorola as the Six Sigma Methodology was being further developed. The author worked with this original version and noted several interesting observations. The foundational principle used in Juran's six-step method "is that the output (Y) of a process is dependent the inputs (X) to the process. What comes out of a process is a result of what goes in. We can say that an output (Y) is a function of inputs (X's). Stated mathematically reads $Y = f(X)$."

4.8.1   The Juran Six Sigma approach

The Juran Six Sigma approach consists of six consecutive phases, each one a prerequisite for performing the next:

1. In the *define* phase, a serious problem is identified and a project team is formed that is given the responsibility and resources for solving the problem.
2. In the *measure* phase, data are gathered and analyzed that describe with precision and accuracy: What is happening (Y's)—what is the current, or baseline, level of performance of the process that creates the problem. The *measure* phase also produces some preliminary ideas of the possible causes (X's) of the problem.
3. In the *analyze* phase, theories are generated as to what the problem is, and by means of testing theories, root causes (X's) are identified.
4. In the *improve* phase, root causes (X's) are removed by designing and implementing changes to the process that had been causing the problem.
5. In the *control* phase, new controls are designed and implemented, which prevent the original problem from returning and which also hold the gains made by improvement.
6. In the *replicate* phase, the knowledge, insights, and know-how acquired by the team are used to correct other quality problems and identify new quality improvement projects.

Dr. J. M. Duran believed that quality drives an organization's revenues and reduces lower cost of deficiencies. Both improve the overall financial performance. Organizations that have initiated breakthrough improvement efforts report significant results. The approach was derived through the Juran trilogy of

1. Quality planning designs and/or redesigns new goods and services. The right set of service features is delivered to the appropriate customers and service deficiencies are kept to a minimum.
2. Quality improvement (breakthrough improvement) reduces deficiencies in existing goods, services, or processes. Reaching this new level of performance is frequently referred to as breakthrough.
3. Quality control ensures the results of the first two processes are sustained over time.

4.8.2   Different methodologies

Six Sigma provides two methodologies to help resolve problems and meet critical customer needs. The DMAIC methodology is the most known version or methodology. This DMAIC process is associated with operational process improvement. It was originally associated with Deming's, plan–do–check–act steps and was associated with the MAIC steps. There was early recognition that Six Sigma should not only be about process improvement but should also include process development. To expand the methodology initially for management planning, the D step or *define* was added and this process was labeled Design for Six Sigma or DFSS. Later it was recognized that the D step was necessary for operational Six Sigma to be a successful firm-wide approach. Multiple DFSS steps were developed by several companies for product development. This book will provide more detail in this area.

Six Sigma brings structure to process improvement by providing the user with a more detailed outline of Deming's plan–do–check–act cycle by guiding the initiative through a five stage cycle of define–measure–analyze–improve–control (DMAIC) (Andersson et al. 2006; Pande et al. 2000).

4.8.3 Six Sigma statistical tools

As Six Sigma matured, and through the integration of Bill Smith yield defect theory, more statistical analysis were included. Eventually, each stage has a number of corresponding tools and techniques such as SPC, design of experiments, and response surface methodology, providing the user with an extensive tool box of techniques, so as to measure, analyze, and improve critical processes in order to bring the system under control (Keller 2005).

4.9 Conclusion

It is essential that Six Sigma should be understood to be a philosophy as well as a scientific approach and this has been gaining acceptance (Porter and Parker 1993). The idea of continuous improvement that makes business sense is the foundation of Six Sigma.

References

Andersson, R., Eriksson, H., and Torstensson, H. (2006). "Similarities and differences between TQM, Six Sigma and Lean." *The TQM Magazine*, 18(3): 282–296.

Antony, J. (2004), "Some pros and cons of Six Sigma: An academic perspective." *The TQM Magazine*, 16(4): 303–306.

Antony, J. (2006). "Six Sigma for service processes." *Business Process Management*, 12(2): 234–248.

Arnheiter, E. D. and Maleyeff, J. (2005). "The integration of Lean management and Six Sigma." *The TQM Magazine*, 17(1): 5–18.

Black, K. and Revere, L. (2006). "Six Sigma arises from the ashes of TQM with a twist." *International Journal of Health Care Quality Assurance*, 19(3): 259–266.

Brady, J. E. and Allen, T. T. (2006). "Six Sigma literature: A review and agenda for future research." *Quality and Reliability Engineering International*, 22: 335–367.

Brewer, P. C. and Bagranoff, N. A. (2004). "Near zero-defect accounting with Six Sigma." *The Journal of Corporate Accounting & Finance*, 15(2): 67–72.

BusinessBalls.(2012).http://www.businessballs.com/sixsigma.htm#six_sigma_quality_observations.

Chakrabarty, A. and Tan, K. C. (2007). "The current state of Six Sigma application in services." *Managing Service Quality*, 17(2): 194–208.

Devore, J. L. (2000). *Probability and Statistics for Engineering and the Sciences*, 5th edn. Pacific Grove, CA: Duxbury Press, pp. 119–126.

Keller, P. A. (2005). *Six Sigma Demystified*. New York: McGraw-Hill.

McFadden, F. R. (1993). "Six-Sigma quality programs." *Quality Progress*, 26(6): 37–42.

Pande, P. S., Neuman, R. P., and Cavanagh, R. R. (2000). *The Six Sigma Way*. New York: McGraw-Hill.

Pepper, M. P. J. and Spedding, T. A. (2010). "The evolution of lean Six Sigma." *International Journal of Quality & Reliability Management*, 27(2): 138–155.

Sehwail, L. and Deyong, C. (2003). "Six Sigma in health care." *Leadership in Health Services*, 16(4): 1–5.

Senapati, N. R. (2004). "Six Sigma: Myths and realities." *International Journal of Quality & Reliability Management*, 21(6): 683–690.

Sower, V. E., Savoie, M. J., and Renick, S. (1999). *An Introduction to Quality Management and Engineering.* Upper Saddle River, NJ: Prentice-Hall, pp. 33–45.

Stewart, R. A. and Spencer, C. A. (2006). "Six-sigma as a strategy for process improvement on construction projects: A case study." *Construction Management and Economics*, 24: 339–348.

Truby, M. (2000). "Nasser, Ford embrace data-driven quality plan." *Detroit News*, January 26, p. F1.

Further Readings

Abdulmalek, F. A. and Rajgopal, J. (2007). "Analyzing the benefits of Lean manufacturing and value stream mapping via simulation: A process sector case study." *International Journal of Production Economics*, 107: 223–236.

Anne', D. C. (2000). "Modern mobile laboratories." *Pollution Engineering*, 32(8): 37–39.

Antony, J., Escamilla, J. L., and Caine, P. (2003). "Lean Sigma." *Manufacturing Engineer*, April.

Bamber, C. J., Sharp, J. M., and Hides, M. T. (1999). "Factors affecting successful implementation of total productive maintenance: A UK manufacturing case study perspective." *Journal of Quality in Maintenance Engineering*, 5(3): 162–181.

Bamber, L. and Dale, B. G. (2000). "Lean production: A study of application in a traditional manufacturing environment." *Production Planning & Control*, 11(3): 291–298.

Bartezzaghi, E. (1999). "The evolution of production models: Is a new paradigm emerging?" *International Journal of Operations & Production Management*, 19(2): 229–250.

Bendell, T. (2005). "Structuring business process improvement methodologies." *Total Quality Management*, 16(8/9): 969–978.

Bendell, T. (2006), "A review and comparison of Six Sigma and the Lean organizations." *The TQM Magazine*, 18(3): 255–262.

Boughton, N. and Arokiam, I. (2000). "The application of cellular manufacturing: A regional small to medium enterprise perspective." *Proceedings of the Institution of Mechanical Engineers Part B—Journal of Engineering Manufacture*, 214(8): 751–754.

Burke, R. J. et al. (1995). "The effect of inspector errors on the true fraction nonconforming: An industrial experiment." *Quality Engineering*, 7(3): 543–550.

Chan, P. S. and Wong, A. (1994). "Global strategic alliances and organizational learning." *Leadership & Organization Development Journal*, 15(4): 31–36.

Cusumano, M. A. (1994). "The limits of 'Lean'." *Sloan Management Review*, 35(4): 27–45.

Devane, T. (2004). *Integrating Lean Six Sigma and High Performance Organizations Leading the Charge toward Dramatic, Rapid and Sustainable Improvement.* San Francisco, CA: Pfeiffer, an imprint of Wiley.

Emiliani, M. L. (2001). "Redefining the focus of investment analysts." *The TQM Magazine*, 13(1): 34–50.

Emiliani, M. L. (2003), *Better Thinking Better Results.* Kensington, CT: The Center for Lean Business Management.

Fitzgerald, K. R. (1997). "Chrysler training helps suppliers trim the fat."

Friedman, S. (2000). "Where materials and minds meet?" *Package Printing and Converting*, 47(2): 24–25.

George, M. L. (2002). *Lean Six Sigma: Combining Six Sigma Quality with Lean Speed.* New York: McGraw-Hill.

Gunter, B. (1987). "A perspective on the Taguchi methods." *Quality Progress*, 20(6): 44–52.

Hall, R. (1992). "Shall we all hang separately?" *Industry Week*, 241(17): 65.

Hancock, W. M. and Zayko, M. J. (1998). "Lean production: Implementation problems." *IIE Solutions*, 30(6): 38–42.

Harrison, J. (2006). "Six Sigma vs. Lean manufacturing: Which is right for your company?" *Foundry Management & Technology*, 134(7): 32.

Harry, M. and Schroeder, R. (2000). *Six Sigma.* New York: Doubleday, p. 65.

Hendry, L. (1998). "Applying world class manufacturing to make-to-order companies: Problems and solutions." *International Journal of Operations & Production Management*, 18(11): 1086–1100.

Higgins, K. T. (2005). "Lean builds steam." *Food Engineering: The Magazine for Operations and Manufacturing Management*, http://www.foodengineeringmag.com/Articles/Feature_Article/1e1b90115c2f8010VgnVCM100000f932a8c0____

Hines, P., Holweg, M., and Rich, N. (2004). "Learning to evolve: A review of contemporary Lean thinking." *International Journal of Operations & Production Management*, 24(10): 994–1011.

Holweg, M. (2007). "The genealogy of Lean production." *Journal of Operations Management*, 25: 420–437.

Hopp, W. J. and Spearman, M. L. (2001). *Factory Physics.* 2nd edn. New York: Irwin/McGraw-Hill, p. 25.

HubPages. http://tamarawilhite.hubpages.com/hub/History-of-Lean-Six-Sigma

Inman, R. R. (1999). "Are you implementing a pull system by putting the cart before the horse?" *Production and Inventory Management Journal*, 40(2): 67–71.

Irani, S. A. (2001). *Value Stream Mapping in Custom Manufacturing and Assembly Facilities.* Columbus, OH: Department of Industrial, Welding and Systems Engineering, The Ohio State University.

Ireland, F. and Dale, B. G. (2001). "A study of total productive maintenance implementation." *Journal of Quality in Maintenance Engineering*, 7(3): 183–191.

James-Moore, S. M. and Gibbons, A. (1997). "Is lean manufacture universally relevant? An investigative methodology." *International Journal of Operations & Production Management*, 17(9): 899–911.

Jina, J., Bhattacharya, A. K., and Walton, A. D. W. (1997). "Applying Lean principles for high product variety and low volumes: Some issues and propositions." *Logistics Information Management*, 10(1): 5–13.

Joiner, B. L. (1994). *Fourth Generation Management.* New York: McGraw-Hill.

Katayama, H. and Bennett, D. (1996). "Lean production in a changing competitive world: A Japanese perspective." *International Journal of Operations & Production Management*, 16(2): 8–23.

Kotter, J. P. (1995). "Leading change: Why transformation efforts fail?" *Harvard Business Review*, 73(2): 59–67.

Kumar, M. et al. (2006). "Implementing the lean sigma framework in an Indian SME: A case study." *Production Planning & Control*, 17(4): 407–423.

Lester, M. A. (2000). "Quick drying enables single-wafer cleans." *Semiconductor International*, 23(12): 54.

Lian, Y.-H. and Landeghem, H. V. (2007). "Analysing the effects of Lean manufacturing using a value stream mapping-based simulation generator." *International Journal of Production Research*, 45(13): 3037–3058.

Lynds, C. (2002). "Common sense evolution: TI automotive's company-wide Lean strategy brings lower costs by slashing waste." *Plant*, 61(7): 12.

Maleyeff, J. and Lewis, D. A. (1993). "Pre-control or X-bar charts: An economic evaluation using alternative cost models." *International Journal of Production Research*, 31(2): 471–482.

McAdam, R. and Lafferty, B. (2004). "A multilevel case study critique of Six Sigma: Statistical control or strategic change?" *International Journal of Operations & Production Management*, 24(5): 530–549.

McDonald, T., Van Aken, E. M., and Rentes, A. F. (2002). "Utilising simulation to enhance value stream mapping: A manufacturing case application." *International Journal of Logistics*, 5(2): 213–232.

Mika, G. (2006). "Six Sigma isn't Lean." *Manufacturing Engineering*, 137(1): 18.

Mileham, A. R. et al. (1999). "Rapid changeover—A pre-requisite for responsive manufacture." *International Journal of Operations & Production Management*, 19(8): 785–796.

Nelson, R. D. (2004). "How Delphi went Lean?" *Supply Chain Management Review*, 8(8): 32–37.

Ott, E. R. and Schilling, E. G. (1990). *Process Quality Control.* Ch. 7. New York: McGraw-Hill.

Parker, M. and Slaughter, J. (1994). "Lean production is mean production: TQM equals management by stress." *Canadian Dimension*, 28(1): 21.

Pepper, M. (2007). "A supply chain improvement methodology for the process industries." PhD thesis, School of Management and Marketing, University of Wollongong, Wollongong.

Porter, L. J. and Parker, A. J. (1993). "Total quality management—the critical success factors." *Total Quality Management*, 4(1): 13–22.

Pyzdek, T. (2000). "Six Sigma and Lean production." *Quality Digest*, January.

Quality. (2004). "Mitsubishi goes beyond MES to incorporate Lean." *Quality*, 43(10): 48–49.

Quality Council of Indiana. (2007). *Certified Six Sigma Black Belt Primer*. West Terre Haute, IN: Quality Council of Indiana.

Raisinghani, M. S. (2005). "Six Sigma: Concepts, tools, and applications." *Industrial Management and Data Systems*, 105(4): 491–505.

Rinehart, J., Huxley, C., and Robertson, D. (1997). *Just Another Car Factory? Lean Production and Its Discontents*. Ithaca, NY: ILR Press.

Rother, M. and Shook, J. (1999). *Learning to See: Value Stream Mapping to Add Value and Eliminate Muda*. Cambridge, MA: Lean Enterprise Institute.

Russell, R. S. and Taylor, B. W. (2000). *Operations Management*. Englewood Cliffs, NJ: Prentice-Hall.

Salvia, A. A. (1988). "Stoplight control." *Quality Progress*, 21(9): 39–42.

Sharma, U. (2003). "Implementing lean principles with the Six Sigma advantage: How a Battery company realized significant improvements." *Journal of Organizational Excellence*, 22(3): 43–52.

Sheridan, J. H. (1999). "Focused on flow." *Industry Week*, 248(19): 46–48.

Sheridan, J. H. (2000). " 'Lean Sigma' synergy." *Industry Week*, 249(17): 81–82.

Shingo, S. (1986). *Zero Quality Control—Source Inspection and the Poka-yoke System*. Cambridge, MA: Productivity Press.

Smith, B. (2003). "Lean and Six Sigma—A one-two punch." *Quality Progress*, 36(4): 37–41.

Spear, S. J. (2004). "Learning to lead at Toyota." *Harvard Business Review*, 82(5): 78.

Spector, R. and West, M. (2006). "The art of Lean program management." *Supply Chain Management Review*, www.scmr.com/article/329570-The_Art_of_Lean_Program_Management.php

Teresko, J. (2005). "It came from Japan!" *Industry Week*, February 1.

Warwood, S. J. and Knowles, G. (2004). "An investigation into Japanese 5-S practice in UK industry." *The TQM Magazine*, 16(5): 347–353.

Waurzyniak, P. (2005). "Lean at NUMMI." *Manufacturing Engineering*, 135(3): 73–81.

Wheat, B., Mills, C., and Carnell, M. (2003). *Leaning into Six Sigma: A Parable of the Journey to Six Sigma and a Lean Enterprise*. New York: McGraw-Hill.

White, R. E. and Prybutok, V. (2001). "The relationship between JIT practices and type of production system." *Omega: The International Journal of Management Science*, 29: 113–124.

Winter, D. (1983). "Job shops dig in their heels." *Production*, 92(1): 62–66.

Womack, J. P. and Jones, D. T. (1996). *Lean Thinking: Banish Waste and Create Wealth in Your Corporation*. London: Simon and Schuster.

Womack, J. P., Jones, D. T., and Roos, D. (1990). *The Machine that Changed the World*. New York: Rawson Associates.

Wong, A. and Hammond, J. H. (1991). *Dore'-Dore'*. Cambridge, MA: Harvard Business School Publishing.

Review Questions

Question 1: Identify the quality guru who believed the best approach in understanding the purpose of a quality system would be the four absolutes of quality management?

(a) Juran
(b) Crosby
(c) Deming
(d) Feigenbaum

Question 2: Which of the quality gurus would be most clearly identified as a proponent of improvement and breakthrough projects?

(a) Juran
(b) Ishikawa
(c) Deming
(d) Crosby

Question 3: Select the quality guru who endorsed spreading the quality improvement message by creating a core of quality specialists within a company.

(a) Ishikawa
(b) Juran
(c) Crosby
(d) Deming

Question 4: Juran often describes a parallel between HIS quality sequence and which of the following management processes?

(a) Production
(b) Financial
(c) Design engineering
(d) Auditing

Question 5: A commonly reported problem with Six Sigma projects deal with

(a) A desire to complete projects on time
(b) A failure to complete any project charter documentation
(c) A lack of business impact for the company
(d) A requirement that projects must be at least $100,000 in value

Question 6: What are the meanings of the following statements with respect to Lean Six Sigma and quality?

(a) Quality should be built in.
(b) Companies use to many inspectors.
(c) Manual inspections are not accurate.
(d) People rely on others to audit their work.

chapter five

Lean

> Life: a cycle. A series of events, meetings, and departures. Friends discovered, others lost, precious time, wastes away. Big droplet tears are shed for yesterday, but are dried in time for tomorrow, until all that remain are foggy, broken memories of a happy yesteryear.
>
> —Daniella Gallo

5.1   Introduction

Lean or Lean thinking is based on the Japanese quality movement specifically at the Toyota Production System (TPS) and was popularized by Womack and Jones (1996). This approach to manufacturing began shortly after the World War II, and was created by Taiichi Ohno while he was employed by the Toyota motor company. Many Japanese companies such as Toyota were hit by shortages after World War II in both capital and resources. Lean principles were established by Sakichi Toyoda around the same time (1920s) as quality principles were being accepted and implemented. It gained popularity in the American industrial setup where companies chose to exploit the economies of scale and increase production without concern for variety in products which was much needed in the Japanese setup. This encouraged copious amounts of scrap and inventory as the objective was to achieve maximum throughput at a given time. In other words much waste was created which gave rise to more problems such as loss of time and money in rework due to defects and holding costs that had to be dealt with.

All Toyota's employees were instructed to eliminate waste. Waste was defined as "anything other than the minimum amount of equipment, materials, parts, space, and time which are absolutely essential to add value to the product" (Russell and Taylor 2000, p. 737). Through a process of trial and error, Ohno would create a new manufacturing paradigm— the TPS (White and Prybutok 2001). The TPS became the dominant production model to emerge from a number of concepts around at the time (Bartezzaghi 1999; Katayama and Bennett 1996). As a result of the International Motor Vehicle Program (IMVP) benchmarking study, and the work of Womack et al. (1990), US and European companies began adapting the TPS under the title of just-in-time (JIT) to remain competitive with Japanese industry.

5.2   Lean enterprise

The management skills involved to deal with this issue is termed as "Lean thinking." It was popularized through the Toyota production method initiated by Taiichi Ohno. "A Lean process is one in which the value-added time in the process is more than 25% of the total lead time of that process." Lean practice can be used in all processes in any type of industry or organization to improve their efficiency. It is applicable to all stages in a supply chain and is thus collectively called the "Lean Enterprise."

It is known that a far-reaching performance is achieved by the use of Six Sigma and Lean to reduce waste. This in turn reduces cost, an important factor in determining the survival of an enterprise. Manufacturing with Lean in mind has certain features such as reduced lead times of processes, faster delivery, improved quality of end products, reduction in inventory such as the work in process (WIP), low customer service activities, core materials, and products at all customer interfaces. This gives room for incorporating changes.

The standardization of processes and improved service and productivity is now possible. Implementation of Lean depends on the area of application, the culture of that company, the industrial sector it belongs to and its internal customer setup (Bell 2006; Wortman et al. 2007).

5.3   Ohno seven sources of waste

It should be noted that for Lean to come into play all sources that lead to unproductive results or "wastes" must be identified in a systematic fashion. Waste was termed as "muda," the Japanese equivalent. According to the Ohno's established fundamentals there are seven sources. These are given in detail further:

1. Overproduction: This means that producing more than the directed level of product required to meet customer demand can lead to many complications. There is an increase in costs incurred due to extra handling and holding costs. Also, it may force the sale of these products at a discounted price.
2. Delay due to waiting: There are several instances during normal production hours that output might be delayed by waiting for a system to recover from downtime and due to lag created in response by an authorizing personnel or the machine.
3. Unutilized resource: An unutilized resource is equivalent to excess inventory in terms of investment cost and holding charges if any. Improper usage of skill set and experience of employees to bring about improved system models is also a kind of waste.
4. Excessive inventory: Cash flow is disrupted as floor space is taken up for inventory that is not needed currently. Sometimes these might be a result of change in requirements as specified by the customer.
5. Defective products: Erroneous processing can lead to defects. This will always lead to increase in investment for the purpose of inspection, changes in processes, design of product. Valuable time is lost in engineering these faults to an acceptable level and for maintenance downtime of equipment.
6. Transportation of inventory: Unnecessary movements between work centers derive excess risk in terms of operator safety and also cost for the transportation. A change in layout might help eliminate these negative impacts.
7. Complexity in processing: Excess WIP inventory can lead to addition in steps involved to achieve required throughput level. Time might be lost in reworking, redesigning, or due to ill-maintained equipment with respect to the total cycle time.

5.4   Lean elements of manufacturing

These wastes have a negative effect over some elements of manufacturing focused on by Lean. These are the following:

- Flow: Production must be smooth and optimized to the schedule. Line stations must be balanced by allotting station times according to customer needs termed as "takt" time.
- Cycle time: It must be low to meet demand on time.

- Value: The deliverance by the manufacturer to the customer for the investment.
- Pull planning: Process governs movement of product instead of raw material.
- Perfection: Profits of the organization must increase. For a low cost incurred enough value must be attributed to the customer.
- Management's decisions: Accomplishment is incomplete without support of management that can be gained by choosing projects with a higher return on investment (ROI).

5.5   Lean tools

Lean goals can be achieved via tools such as 5-S, total productive maintenance (TPM), mistake proofing, plan–do–check–act (PDCA), value stream mapping (VSM), and implements such as cellular manufacturing, "kaizen," "kanban," SMED for setup reduction. The measurement of profit returns of Lean practice has some options and these are: Using a feedback loop; industrial testing; recording and analyzing process performance results; and mathematical representation of process parameters.

Lean has been mostly viewed as a manufacturing philosophy. But one must be aware that—"Lean does not mean manufacturing, Lean means speed" (George 2002).

This is an insight that originates from Henry Ford's findings in his time when inventory was encouraged due to focus on large amount of units as output. To address the Japanese market wherein variety was "demanded" a tradeoff between high quality with low cost and high speed was introduced as the concept of Lean. The principle impact is on wait time that makes every process flexible. So, if we increase velocity by a factor of five the inventory levels reduce by the same factor for the same throughput. Make-to-order products are thus possible, that is, only that amount is produced such that demand is just met with reduced cost. It is known that by using a reduction factor of 80% the overhead charges, charges for keeping up quality and other hidden costs can decrease by 20% thus enabling higher returns. To measure this profitability that comes with the practice of Lean ideology we use the Lean Metric. It is used to estimate cycle efficiency denoted by the expression given below (George 2002).

$$\text{Process cycle efficiency} = \text{value-added time/total lead time}$$

An average of 25% cycle efficiency is deemed to be "world class" which implies that more than 25% of lead time must be devoted to value creation in the end product. It must be noted that personnel experienced with managing these sources of waste are best employed into these processes selected for Lean application.

This speed initiative can be applied to any and all processes including rework. It helps enhance the effect of Six Sigma tools such as design of experiments (DOEs). It aids in approaching the goals of Six Sigma much faster by eliminating the factors that create lag and do not add value (sources of waste) and by use of "rapid speed"/Lean tools. The number of process steps is effectively reduced, creating a compressed time environment accelerating targeted action.

5.6   Value stream mapping

The main tool to starting Lean is the VSM process. VSM provides the scope of the project by defining the current state and desired future state of the system. This future state map is then used to develop Lean improvement strategies, for example, parallel working and flexibility through employment of multi-skilled employees (requiring minimal expenditure). The benefits of VSM include

1. A common language when considering manufacturing processes.
2. It also brings together all of the Lean techniques which help practitioners avoid the temptation to cherry-pick one or two of the "easier" to implement. In fact, no other tool depicts the linkages between information and material flow such as VSM (Rother and Shook 1999). Their book *Learning to See* has become the definitive text for organizations starting a Lean journey.

The limitations of VSM as stated by Sheridan (2000) suggested that the practical nature limits the amount of detail collected and also detracts from the actual system workings (the action of using pencil and paper to draw the map may remove focus from the actual system being analyzed). This dynamic view looks beyond VSM as giving a quick, succinct overview of where "muda" is present, and develops the idea of the mapping process itself becoming a continuous tool, constantly being updated via software (Sheridan 2000).

5.7 Lean success

Originally, Lean philosophies were applied to large manufacturing operations in high-volume, low-variety facilities. Not surprisingly, following its inception at Toyota, some of the first Western companies to consider the transition to a Lean culture were US automotive manufacturers. Between the years of 1968 and 1978, US productivity increased by 23.6%, but the Japanese experienced an impressive 89.1% increase (Teresko 2005). If you cannot beat them benchmark them. Successful initiatives can be found in the automotive and aerospace sectors. Chrysler used resources to extend in-house training of Lean philosophy to its major suppliers, emphasizing the commitment needed from all parties in order to establish Lean, and realize the full potential for everyone involved (Fitzgerald 1997). Delphi took a multi-pronged approach, looking at supplier development, cost management, strategic sourcing, and quality problems, led by top management, again emphasizing the long-term commitment needed, and highlighting the importance of knowledge management to provide clear examples for the automotive sector (Nelson 2004).

Lean criticisms as described by researchers (Parker and Slaughter 1994; Rinehart et al. 1997) include that Lean does stem understanding, direction, and/or commitment from management (Hancock and Zayko 1998), and is difficult to implement in a heavily unionized culture of the industry (Hall 1992). Due to the fact that management and their budgets provide the backbone of any continuous improvement effort, the philosophy also requires that employees have transparency from management and be provided education and empowerment in the change process. Often management will not provide these tools to support the initiative. Given these challenges Lean implementation has not been as successful in the United States as other quality initiatives.

The counter points of the main reasons include the following:

1. Large product portfolios mean that each "job" is likely to be different and therefore production approaches cannot be standardized.
2. Products' characteristics create production constraints.

3. Job-shops or smaller firms simply cannot match the dominance or resources that the larger firms enjoy, allowing them to be inflexible along their supply chains (Bamber and Dale 2000).

4. The view that Lean is pro-company, not pro-employee, has some validity, and cannot be dismissed. For example, when employees are "empowered" employees may feel a sense of insecurity, perceiving Lean as a no-win situation and an employment threat. The opinion is that managements avoid accountability when problems arise, letting it filter downwards onto the lower levels of hierarchy (Parker and Slaughter 1994). This is to miss the fundamental underpinning of empowerment and cultural change, resulting from a failure by management to approach Lean with the correct goals.

5.8 Conclusion

Lean requires and relies on support of organizational values and true empowerment, which in itself is key to the sustainability of Lean. Without this we see an adverse effect on morale that leads to increasing levels of worker unhappiness and withdrawal, ultimately leading to failure.

Many employees see Lean not as an enabler for strategic development but as a tool for downsizing. Managements tend to concentrate on tools and practices, rather than viewing Lean as a philosophy, aiming to teach new improvement tools to employees, rather than immersing them in the practical side of solving opportunities for improvement with a Lean approach (Spear 2004).

References

Bamber, L. and Dale, B. G. (2000). "Lean production: A study of application in a traditional manufacturing environment." *Production Planning & Control*, 11(3): 291–298.

Bartezzaghi, E. (1999). "The evolution of production models: Is a new paradigm emerging?" *International Journal of Operations & Production Management*, 19(2): 229–250.

Bell, S. (2006). Lean Enterprise Systems. Northwest Lean Networks. http://www.nwlean.net.

Fitzgerald, K. R. (1997). "Chrysler training helps suppliers trim the fat."

George, M. (2002). *Lean Six Sigma: Combining Six Sigma Quality with Lean Speed*. New York: McGraw-Hill.

Hall, R. (1992). "Shall we all hang separately?" *Industry Week*, 241(17): 65.

Hancock, W. M. and Zayko, M. J. (1998). "Lean production: Implementation problems." *IIE Solutions*, 30(6): 38–42.

Katayama, H. and Bennett, D. (1996). "Lean production in a changing competitive world: A Japanese perspective." *International Journal of Operations & Production Management*, 16(2): 8–23.

Nelson, R. D. (2004). "How Delphi went Lean?" *Supply Chain Management Review*, 8(8): 32–37.

Parker, M. and Slaughter, J. (1994). "Lean production is mean production: TQM equals management by stress." *Canadian Dimension*, 28(1): 21.

Rinehart, J., Huxley, C., and Robertson, D. (1997). *Just Another Car Factory? Lean Production and Its Discontents*. Ithaca, NY: ILR Press.

Russell, R. S. and Taylor, B. W. (2000). *Operations Management*. Englewood Cliffs, NJ: Prentice-Hall.

Sheridan, J. H. (2000). "'Lean Sigma' synergy." *Industry Week*, 249(17): 81–82.

Spear, S. J. (2004). "Learning to lead at Toyota." *Harvard Business Review*, 82(5): 78.

Teresko, J. (2005). "It came from Japan!" *Industry Week*, February 1.

White, R. E. and Prybutok, V. (2001). "The relationship between JIT practices and type of production system." *Omega: The International Journal of Management Science*, 29: 113–124.

Womack, J. and Jones, D. T. (1996). *Lean Thinking: Banish Waste and Create Wealth in Your Corporation*. London: Simon and Schuster.

Womack, J., Jones, D. T., and Roos, D. (1990). *The Machine that Changed the World*. New York: Rawson Associates.

Wortman, B. et al. (2007). CSSBB Primer. Terre Haute, IN: Quality Council of Indiana.

Further Readings

Abdulmalek, F. A. and Rajgopal, J. (2007). "Analyzing the benefits of lean manufacturing and value stream mapping via simulation: A process sector case study." *International Journal of Production Economics*, 107: 223–236.

Andersson, R., Eriksson, H., and Torstensson, H. (2006). "Similarities and differences between TQM, Six Sigma and Lean." *The TQM Magazine*, 18(3): 282–296.

Antony, J. (2004). "Some pros and cons of Six Sigma: An academic perspective." *The TQM Magazine*, 16(4): 303–306.

Antony, J. (2006). "Six Sigma for service processes." *Business Process Management*, 12(2): 234–248.

Antony, J., Escamilla, J. L., and Caine, P. (2003). "Lean sigma." *Manufacturing Engineer*, April.

Arnheiter, E. D. and Maleyeff, J. (2005). "The integration of Lean management and Six Sigma." *The TQM Magazine*, 17(1): 5–18.

Bamber, C. J., Sharp, J. M., and Hides, M. T. (1999). "Factors affecting successful implementation of total productive maintenance: A UK manufacturing case study perspective." *Journal of Quality in Maintenance Engineering*, 5(3): 162–181.

Bendell, T. (2005). "Structuring business process improvement methodologies." *Total Quality Management*, 16(8/9): 969–978.

Bendell, T. (2006). "A review and comparison of six sigma and the lean organizations." *The TQM Magazine*, 18(3): 255–262.

Black, K. and Revere, L. (2006). "Six Sigma arises from the ashes of TQM with a twist." *International Journal of Health Care Quality Assurance*, 19(3): 259–266.

Boughton, N. and Arokiam, I. (2000). "The application of cellular manufacturing: A regional small to medium enterprise perspective." *Proceedings of the Institution of Mechanical Engineers Part B—Journal of Engineering Manufacture*, 214(8): 751–754.

Brady, J. E. and Allen, T. T. (2006). "Six Sigma literature: a review and agenda for future research." *Quality and Reliability Engineering International*, 22: 335–367.

Brewer, P. C. and Bagranoff, N. A. (2004). "Near zero-defect accounting with Six Sigma." *The Journal of Corporate Accounting & Finance*, 15(2): 67–72.

Chakrabarty, A. and Tan, K. C. (2007). "The current state of Six Sigma application in services." *Managing Service Quality*, 17(2): 194–208.

Chan, P. S. and Wong, A. (1994). "Global strategic alliances and organizational learning." *Leadership & Organization Development Journal*, 15(4): 31–36.

Cusumano, M. A. (1994). "The limits of 'Lean'." *Sloan Management Review*, 35(4): 27–45.

Harrison, J. (2006). "Six Sigma vs. Lean manufacturing: Which is right for your company?" *Foundry Management & Technology*, 134(7): 32.

Hendry, L. (1998). "Applying world class manufacturing to make-to-order companies: Problems and solutions." *International Journal of Operations & Production Management*, 18(11): 1086–1100.

Higgins, K. T. (2005). "Lean builds steam." *Food Engineering: The Magazine for Operations and Manufacturing Management*, http://www.foodengineeringmag.com/Articles/Feature_Article/1e1b90115c2f8010VgnVCM100000f932a8c0____

Hines, P., Holweg, M., and Rich, N. (2004). "Learning to evolve: A review of contemporary Lean thinking." *International Journal of Operations & Production Management*, 24(10): 994–1011.

Holweg, M. (2007). "The genealogy of Lean production." *Journal of Operations Management*, 25: 420–437.

Irani, S. A. (2001). *Value Stream Mapping in Custom Manufacturing and Assembly Facilities*. Columbus, OH: Department of Industrial, Welding and Systems Engineering, The Ohio State University.

Ireland, F. and Dale, B. G. (2001). "A study of total productive maintenance implementation." *Journal of Quality in Maintenance Engineering*, 7(3): 183–191.

James-Moore, S. M. and Gibbons, A. (1997). "Is Lean manufacture universally relevant? An investigative methodology." *International Journal of Operations & Production Management*, 17(9): 899–911.

Jina, J., Bhattacharya, A. K., and Walton, A. D. W. (1997). "Applying Lean principles for high product variety and low volumes: Some issues and propositions." *Logistics Information Management*, 10(1): 5–13.

Joiner, B. L. (1994). *Fourth Generation Management*. New York: McGraw-Hill.

Keller, P. A. (2001). "Recent trends in Six Sigma." *Annual Quality Congress Proceedings*, May 1, Tuscon, AZ. ASQ, Milwaukee, WI, pp. 98–102.

Keller, P. A. (2005). *Six Sigma Demystified*. New York: McGraw-Hill.

Kotter, J. P. (1995). "Leading change: Why transformation efforts fail?" *Harvard Business Review*, 73(2): 59–67.

Kumar, M. et al. (2006). "Implementing the Lean Sigma framework in an Indian SME: A case study." *Production Planning & Control*, 17(4): 407–423.

Lian, Y.-H. and Landeghem, H. V. (2007). "Analysing the effects of Lean manufacturing using a value stream mapping-based simulation generator." *International Journal of Production Research*, 45(13): 3037–3058.

Lynds, C. (2002). "Common sense evolution: TI automotive's company-wide lean strategy brings lower costs by slashing waste." *Plant*, 61(7): 12.

McAdam, R. and Lafferty, B. (2004). "A multilevel case study critique of Six Sigma: Statistical control or strategic change?" *International Journal of Operations & Production Management*, 24(5): 530–549.

McDonald, T., Van Aken, E. M., and Rentes, A. F. (2002). "Utilising simulation to enhance value stream mapping: A manufacturing case application." *International Journal of Logistics*, 5(2): 213–232.

Mika, G. (2006). "Six Sigma isn't Lean." *Manufacturing Engineering*, 137(1): 18.

Mileham, A. R. et al. (1999). "Rapid changeover—A pre-requisite for responsive manufacture." *International Journal of Operations & Production Management*, 19(8): 785–796.

Pande, P. S., Neuman, R. P., and Cavanagh, R. R. (2000). *The Six Sigma Way*. New York: McGraw-Hill.

Pepper, M. (2007). "A supply chain improvement methodology for the process industries." PhD thesis, School of Management and Marketing, University of Wollongong, Wollongong.

Pepper, M. P. J. and Spedding, T. A. (2010). "The evolution of Lean Six Sigma." *International Journal of Quality & Reliability Management*, 27(2): 138–155.

Pyzdek, T. (2000). "Six Sigma and lean production." *Quality Digest*, January.

Quality. (2004). "Mitsubishi goes beyond MES to incorporate lean." *Quality*, 43(10): 48–49.

Raisinghani, M. S. (2005). "Six Sigma: Concepts, tools, and applications." *Industrial Management and Data Systems*, 105(4): 491–505.

Rother, M. and Shook, J. (1999). *Learning to See: Value Stream Mapping to Add Value and Eliminate Muda*. Cambridge, MA: Lean Enterprise Institute.

Sehwail, L. and Deyong, C. (2003). "Six Sigma in health care." *Leadership in Health Services*, 16(4): 1–5.

Senapati, N. R. (2004). "Six Sigma: Myths and realities." *International Journal of Quality & Reliability Management*, 21(6): 683–690.

Sharma, U. (2003). "Implementing Lean principles with the Six Sigma advantage: How a Battery company realized significant improvements." *Journal of Organizational Excellence*, 22(3): 43–52.

Smith, B. (2003). "Lean and Six Sigma—A one-two punch." *Quality Progress*, 36(4): 37–41.

Spector, R. and West, M. (2006). "The art of Lean program management." *Supply Chain Management Review*, www.scmr.com/article/329570-The_Art_of_Lean_Program_Management.php

Stewart, R. A. and Spencer, C. A. (2006). "Six Sigma as a strategy for process improvement on construction projects: A case study." *Construction Management and Economics*, 24: 339–348.

Warwood, S. J. and Knowles, G. (2004). "An investigation into Japanese 5-S practice in UK industry." *The TQM Magazine*, 16(5): 347–353.

Waurzyniak, P. (2005). "Lean at NUMMI." *Manufacturing Engineering*, 135(3): 73–81.

Wheat, B., Mills, C., and Carnell, M. (2003). *Leaning into Six Sigma: A Parable of the Journey to Six Sigma and a Lean Enterprise*. New York: McGraw-Hill.

Winter, D. (1983). "Job shops dig in their heels." *Production*, 92(1): 62–66.

Review Questions

Question 1: When the term "pull" is used in Lean thinking, who is doing the pulling?

(a) Downstream operations
(b) The "takt" time
(c) The customer
(d) The cycle time

Question 2: Listed below are seven of the most widely recognized forms of "muda."

A. Overproduction
B. Inventory
C. Repair/reject
D. Motion
E. Processing
F. Waiting
G. Transportation

Which of the above are least likely to result in poor product quality?

(a) F G
(b) A B
(c) C E
(d) D F

chapter six

Lean Six Sigma

They are reinventing; attempting to reinterpret results of the election and the intentions of voters by subjective, not objective means.

—Karen Hughes

6.1 Introduction

In this chapter, we discuss the combining of two philosophies: Lean, which focuses on employee empowerment and standardization, and Six Sigma, which provides a structured approach and tools to facilitate financially justified process improvement. Though these philosophies appear to be originating from opposite directions, consultants and academics have chosen to use their common beginnings and integrate the philosophies. The Lean philosophy focuses on empowerment mindset or bottom-up approach and Six Sigma philosophy, which is management driven, has positive and negative points. This chapter describes how the combined approach supports and strengthens individual weaknesses.

6.1.1 History of Lean Six Sigma

Lean is a methodology derived from the Toyota Production System (TPS) created by Taiichi Ohno. The main cause for concern in it is elimination of waste and increase of speed that has been proven to affect quality beneficially. We must look at the historical development of continuous improvement methodologies of Six Sigma and Lean that have been used in sync as they have a shared history.

The concept of Lean Six Sigma originated in the book published in 2002 by Michael George, called "Lean Six Sigma: Combining Six Sigma with Lean Speed." Lean Six Sigma practitioners have similar roles and responsibilities according to the belt levels of Six Sigma (Wilhite 2012).

With popularization of Lean, it was realized that meeting customer requirements was the common goal and quality initiatives were initiated for the overall purpose. It was encouraged to avoid inspection and have intrinsic quality during their initial production. It is evident that this was not developed independently and was a consequence of the contributions of both ideologies.

6.1.2 What is the real difference?

Using the Lean approach, processes are not controlled with the basis of statistical measures. These are measures that account for errors in measurement due to variability factors and to relate quality with mathematical problem solving and root cause analysis after elimination of waste.

However, the Six Sigma approach lacks tools for improving speed, reducing costs, extended timeline, and slower achievement of savings. By combining the two methodologies, the shortcomings are mitigated and organizations must strive to formally integrate the two methodologies and tool sets (Devane 2004).

They are both positioned to drive toward the similar goals but Lean thinking bases sources of waste as unnecessary procedural steps whereas Six Sigma ideology suggests that variability in processes is the main cause of concern. The prioritizing of project objectives differ for both Lean and Six Sigma typically takes a longer time to be implemented.

- *Philip B. Crosby* stressed the involvement of senior management and brought into picture four absolutes of quality and measurement of quality costs.
- *W. Edwards Deming* introduced the plan–do–study–act (PDSA) for system improvement.
- *Armand Feigenbaumin* introduced total quality control/management and encouraged the involvement of the top management.
- *Kaoru Ishikawa* introduced 4M or cause–effect diagram, encouraged efforts for companywide quality control and established that the next operator is also a customer.
- *Joseph M. Juran* introduced the quality trilogy and more methods for quality cost measurement and pareto analysis.
- *Walter A. Shewart* brought about more developments in plan–do–check–act (PDCA). He used statistics for improvement and established the use of control charts.
- *Genichi Taguchi* introduced the loss function concepts, use of signal-to-noise ratio for quality purposes, and many experimental design methods.

The concept of Lean Six Sigma as an approach to process improvement is yet to be fully recognized as an individual area of academic research (Bendell 2006).

Most often the integration of Lean and Six Sigma is difficult because the two approaches have often been implemented in isolation (Smith 2003), creating Lean and Six Sigma subcultures to emerge within the organizations, which can cause conflict of interest and drain on resources (Bendell 2006).

The Six Sigma process complements the Lean philosophy because it provides tools and strategies to solve specific problems that are identified along a company's Lean journey: Lean philosophies reduce waste and seek to support standardization. Six Sigma focuses on project work on the identified variation from the proposed standard. Often it is seen to not entirely focus on the customer requirements; instead, it is envisioned as more of a cost-reduction exercise (Bendell 2005).

Similarities can again be drawn between Lean and Six Sigma, and the need for a culture of continuous improvement operating at all levels within an organization. Researchers see that Six Sigma provides a framework for Lean and more of a scientific approach to quality, so that through the use of control charts, processes can be kept on target, effectively reducing waste incurred through faulty measurement.

The Lean Six Sigma methodology uses a comprehensive methodology incorporating the key elements of both, as each stage can gain from the respective techniques. The Six Sigma road map of define, measure, analyze, improve, and control (DMAIC) includes Lean concepts of empowerment and education of everyone in the organization to identify and eliminate nonvalue adding activities (Higgins 2005). The integration of the two methodologies attempts to provide empowerment even at the higher level process analysis stages, so that employees have true ownership of the process.

Spector and West (2006) take the view of the practitioner, pointing out that when adopting Lean/Six Sigma, practitioners can find themselves commencing a large number of projects that yield insufficient results for the amount of time needed to complete them. In stark contrast, Mika (2006) takes the stance that the two approaches are completely incompatible with one another because Six Sigma cannot be embraced by the "average worker on the floor" (Mika 2006, p. 1). He argues that Lean is accessible by these workers, and encourages effective teamwork collaboration and participation through cross-functional teams.

Lean techniques allow significant changes to be made to an organization without a thorough understanding of the system. This may provide immediate results but if performed too often can lead to instability of the system. The risk of only using Lean techniques is that it would take too long to develop the necessary depth of understanding for a sustainable improvement initiative. One of the main criticisms of Lean is that the lack of understanding creates the unsustainable nature of many Lean initiatives. Using Lean tools and techniques for specific key areas identified by Six Sigma techniques appears to be more advantageous.

Sharma (2003) also describes the benefits of using Six Sigma techniques in conjunction with Lean, whereby strategic improvement goals are established by the company's leaders, and then a process of quality function deployment (QFD) is used to prioritize the project work. Although effective in this implementation, there is no comprehensive framework present that specifically integrates Lean and Six Sigma concepts through an implementation roadmap. The QFD approach can also be viewed as a more complicated approach to the selection of continuous improvement tools. The work of George (2002) can be seen to lead the exploration of Lean and Six Sigma techniques, providing benchmark work for future researchers. Lean philosophy underpins the framework, providing strategic direction and a foundation for improvement, orientating the general dynamics of the system by informing the current state of operations. From this, Lean thinking identifies key areas for improvement (hot spots). Once these hot spots have been identified, Six Sigma provides a focused, project-based improvement methodology to target these hot spots and ultimately drive the system toward the desired future state (Pepper and Spedding 2010).

This textbook approaches Lean Six Sigma as more of a philosophy based on Six Sigma. The Lean tools are used as an addition to the Six Sigma philosophy and used when appropriate. In the science and engineering chapter where the tools are extensively described the Lean tools are added to the control phases, because other alternative Six Sigma techniques that utilize different tools view Lean tools as one of the alternative Six Sigma techniques. Other techniques presented in the Applications section, which includes case studies and examples, include

1. Lean Six Sigma
2. Six Sigma logistics
3. Six Sigma for health-care IT

This researcher believes that by utilizing the text the correct fit for the type of Six Sigma, Lean, or other approaches can be derived.

References

Bendell, T. (2005). "Structuring business process improvement methodologies." *Total Quality Management*, 16(8/9): 969–978.

Bendell, T. (2006). "A review and comparison of Six Sigma and the Lean organizations." *The TQM Magazine*, 18(3): 255–262.

Devane, T. (2004). *Integrating Lean Six Sigma and High-Performance Organizations: Leading the Charge toward Dramatic, Rapid, and Sustainable Improvement.* Vol. 4. Hoboken, NJ: Wiley.

George, M. (2002). *Lean Six Sigma: Combining Six Sigma Quality with Lean Speed.* New York: McGraw-Hill.

Higgins, K. T. (2005). "Lean builds steam." *Food Engineering: The Magazine for Operations and Manufacturing Management,* http://www.foodengineeringmag.com/Articles/Feature_Article/1e1b90115c2f8010VgnVCM100000f932a8c0____

Mika, G. (2006). "Six Sigma isn't Lean." *Manufacturing Engineering,* 137(1): 18.

Pepper, M. P. J. and Spedding, T. A. (2010). "The evolution of Lean Six Sigma." *International Journal of Quality & Reliability Management,* 27(2): 138–155.

Sharma, U. (2003). "Implementing Lean principles with the Six Sigma advantage: How a battery company realized significant improvements." *Journal of Organizational Excellence,* 22(3): 43–52.

Smith, B. (2003). "Lean and Six Sigma—A one-two punch." *Quality Progress,* 36(4): 37–41.

Spector, R. and West, M. (2006). "The art of Lean program management." *Supply Chain Management Review,* www.scmr.com/article/329570-The_Art_of_Lean_Program_Management.php

Further Readings

Abdulmalek, F. A. and Rajgopal, J. (2007). "Analyzing the benefits of Lean manufacturing and value stream mapping via simulation: A process sector case study." *International Journal of Production Economics,* 107: 223–236.

Andersson, R., Eriksson, H., and Torstensson, H. (2006). "Similarities and differences between TQM, Six Sigma and Lean." *The TQM Magazine,* 18(3): 282–296.

Antony, J. (2004). "Some pros and cons of Six Sigma: An academic perspective." *The TQM Magazine,* 16(4): 303–306.

Antony, J. (2006). "Six Sigma for service processes." *Business Process Management,* 12(2): 234–248.

Antony, J., Escamilla, J. L., and Caine, P. (2003). "Lean Sigma." *Manufacturing Engineer,* April.

Arnheiter, E. D. and Maleyeff, J. (2005). "The integration of Lean management and Six Sigma." *The TQM Magazine,* 17(1): 5–18.

Bamber, C. J., Sharp, J. M., and Hides, M. T. (1999). "Factors affecting successful implementation of total productive maintenance: A UK manufacturing case study perspective." *Journal of Quality in Maintenance Engineering,* 5(3): 162–181.

Bamber, L. and Dale, B. G. (2000). "Lean production: A study of application in a traditional manufacturing environment." *Production Planning & Control,* 11(3): 291–298.

Bartezzaghi, E. (1999). "The evolution of production models: Is a new paradigm emerging?" *International Journal of Operations & Production Management,* 19(2): 229–250.

Black, K. and Revere, L. (2006). "Six Sigma arises from the ashes of TQM with a twist." *International Journal of Health Care Quality Assurance,* 19(3): 259–266.

Boughton, N. and Arokiam, I. (2000). "The application of cellular manufacturing: A regional small to medium enterprise perspective." *Proceedings of the Institution of Mechanical Engineers Part B— Journal of Engineering Manufacture,* 214(8): 751–754.

Brady, J. E. and Allen, T. T. (2006). "Six Sigma literature: A review and agenda for future research." *Quality and Reliability Engineering International,* 22: 335–367.

Brewer, P. C. and Bagranoff, N. A. (2004). "Near zero-defect accounting with Six Sigma." *The Journal of Corporate Accounting & Finance,* 15(2): 67–72.

Chakrabarty, A. and Tan, K. C. (2007). "The current state of Six Sigma application in services." *Managing Service Quality,* 17(2): 194–208.

Chan, P. S. and Wong, A. (1994). "Global strategic alliances and organizational learning." *Leadership & Organization Development Journal,* 15(4): 31–36.

Cusumano, M. A. (1994). "The limits of 'Lean'." *Sloan Management Review,* 35(4): 27–45.

Fitzgerald, K. R. (1997). "Chrysler training helps suppliers trim the fat."

Hall, R. (1992). "Shall we all hang separately?" *Industry Week,* 241(17): 65.

Hancock, W. M. and Zayko, M. J. (1998). "Lean production: Implementation problems." *IIE Solutions,* 30(6): 38–42.

Harrison, J. (2006). "Six Sigma vs. Lean manufacturing: Which is right for your company?" *Foundry Management & Technology,* 134(7): 32.

Hendry, L. (1998). "Applying world class manufacturing to make-to-order companies: Problems and solutions." *International Journal of Operations & Production Management*, 18(11): 1086–1100.

Hines, P., Holweg, M., and Rich, N. (2004). "Learning to evolve: A review of contemporary Lean thinking." *International Journal of Operations & Production Management*, 24(10): 994–1011.

Holweg, M. (2007). "The genealogy of Lean production." *Journal of Operations Management*, 25: 420–437.

Irani, S. A. (2001). *Value Stream Mapping in Custom Manufacturing and Assembly Facilities.* Columbus, OH: Department of Industrial, Welding and Systems Engineering, The Ohio State University.

Ireland, F. and Dale, B. G. (2001). "A study of total productive maintenance Implementation." *Journal of Quality in Maintenance Engineering*, 7(3): 183–191.

James-Moore, S. M. and Gibbons, A. (1997). "Is Lean manufacture universally relevant? An investigative methodology." *International Journal of Operations & Production Management*, 17(9): 899–911.

Jina, J., Bhattacharya, A. K., and Walton, A. D. W. (1997). "Applying Lean principles for high product variety and low volumes: Some issues and propositions." *Logistics Information Management*, 10(1): 5–13.

Joiner, B. L. (1994). *Fourth Generation Management.* New York: McGraw-Hill.

Katayama, H. and Bennett, D. (1996). "Lean production in a changing competitive world: A Japanese perspective." *International Journal of Operations & Production Management*, 16(2): 8–23.

Keller, P. A. (2001). "Recent trends in Six Sigma." *Annual Quality Congress Proceedings*, May 1, Charlotte, NC. ASQ, Milwaukee, WI, pp. 98–102.

Keller, P. A. (2005). *Six Sigma Demystified.* New York: McGraw-Hill.

Kotter, J. P. (1995). "Leading change: Why transformation efforts fail?" *Harvard Business Review*, 73(2): 59–67.

Kumar, M. et al. (2006). "Implementing the Lean Sigma framework in an Indian SME: A case study." *Production Planning & Control*, 17(4): 407–423.

Lian, Y.-H. and Landeghem, H. V. (2007). "Analysing the effects of Lean manufacturing using a value stream mapping-based simulation generator." *International Journal of Production Research*, 45(13): 3037–3058.

Lynds, C. (2002). "Common sense evolution: TI automotive's company-wide Lean strategy brings lower costs by slashing waste." *Plant*, 61(7): 12.

McAdam, R. and Lafferty, B. (2004). "A multilevel case study critique of Six Sigma: Statistical control or strategic change?" *International Journal of Operations & Production Management*, 24(5): 530–549.

McDonald, T., Van Aken, E. M., and Rentes, A. F. (2002). "Utilizing simulation to enhance value Stream mapping: A manufacturing case application." *International Journal of Logistics*, 5(2): 213–232.

Mileham, A. R. et al. (1999). "Rapid changeover—A pre-requisite for responsive manufacture." *International Journal of Operations & Production Management*, 19(8): 785–796.

Nelson, R. D. (2004). "How Delphi went Lean?" *Supply Chain Management Review*, 8(8): 32–37.

Pande, P. S., Neuman, R. P., and Cavanagh, R. R. (2000). *The Six Sigma Way.* New York: McGraw-Hill.

Parker, M. and Slaughter, J. (1994). "Lean production is mean production: TQM equals management by stress." *Canadian Dimension*, 28(1): 21.

Pepper, M. (2007). "A supply chain improvement methodology for the process industries." PhD thesis, School of Management and Marketing, University of Wollongong, Wollongong.

Pyzdek, T. (2000). "Six Sigma and Lean production." *Quality Digest*, January, p. 14.

Quality. (2004). "Mitsubishi goes beyond MES to incorporate Lean." *Quality*, 43(10): 48–49.

Raisinghani, M. S. (2005). "Six Sigma: Concepts, tools, and applications." *Industrial Management and Data Systems*, 105(4): 491–505.

Rinehart, J., Huxley, C., and Robertson, D. (1997). *Just Another Car Factory? Lean Production and Its Discontents.* Ithaca, NY: ILR Press.

Rother, M. and Shook, J. (1999). *Learning to See: Value Stream Mapping to Add Value and Eliminate Muda.* Cambridge, MA: Lean Enterprise Institute.

Russell, R. S. and Taylor, B. W. (2000). *Operations Management.* Englewood Cliffs, NJ: Prentice-Hall.

Sehwail, L. and Deyong, C. (2003). "Six Sigma in health care." *Leadership in Health Services*, 16(4): 1–5.

Senapati, N. R. (2004). "Six Sigma: Myths and realities." *International Journal of Quality & Reliability Management*, 21(6): 683–690.

Sheridan, J. H. (2000). "'Lean Sigma' synergy." *Industry Week*, 249(17): 81–82.

Spear, S. J. (2004). "Learning to lead at Toyota." *Harvard Business Review*, 82(5): 78.

Stewart, R. A. and Spencer, C. A. (2006). "Six Sigma as a strategy for process improvement on construction projects: A case study." *Construction Management and Economics*, 24: 339–348.

Teresko, J. (2005). "It came from Japan!" *Industry Week*, February 1.

Warwood, S. J. and Knowles, G. (2004). "An investigation into Japanese 5-S practice in UK industry." *The TQM Magazine*, 16(5): 347–353.

Waurzyniak, P. (2005). "Lean at NUMMI." *Manufacturing Engineering*, 135(3): 73–81.

Wheat, B., Mills, C., and Carnell, M. (2003). *Leaning into Six Sigma: A Parable of the Journey to Six Sigma and a Lean Enterprise*. New York: McGraw-Hill.

White, R. E. and Prybutok, V. (2001). "The relationship between JIT practices and type of production system." *Omega: The International Journal of Management Science*, 29: 113–124.

Wilhite, T. (2012). "Advantages and disadvantages of Six Sigma." HubPages—Business and Employment.

Winter, D. (1983). "Job shops dig in their heels." *Production*, 92(1): 62–66.

Womack, J. and Jones, D. T. (1996). *Lean Thinking: Banish Waste and Create Wealth in Your Corporation*. London: Simon and Schuster.

Womack, J., Jones, D. T., and Roos, D. (1990). *The Machine that Changed the World*. New York: Rawson Associates.

Deciding on firm qualities and Lean Six Sigma initiatives

chapter seven

The cost approaches and engineering economics of Six Sigma projects

When your pills get down to four order more.

—**Anonymous**

7.1   Introduction

Six Sigma is a popular management strategy that many companies have adopted. It was originally developed by Motorola in 1987 with the aim of reducing their defects to 3.4 ppm. The term "sigma" (σ) refers to the variation in the average of a process. Six Sigma refers to 3.4 defects per million opportunities (DPMOs) (Antony and Coronado 2002). The Six Sigma methodology encompasses both management and technical components in a standardized framework of define–measure–analyze–improve–control (DMAIC) for process and product improvement (Snee 2000). The concepts of DMAIC will be explained in further chapters. With more than two decades of successful implementation in major corporations, the success and benefit of Six Sigma are well documented (Kumar et al. 2008). The success of Six Sigma lies in its explicit objective of enhancing the "sigma level" of performance measures that reflect customer needs and requirements (Harry and Schroeder 2000). The ultimate outcome of Six Sigma is reduction of process variation through sustained effort (Hahn et al. 1999). Studies were also conducted to examine the effect of other quality improvement activities such as quality awards, quality certification, and other quality initiatives on the stock prices (Goh et al. 2003; Przasnyski and Tai 1999). The response of stock price returns for total quality management (TQM) companies was also studied (Adams et al. 1999; Hendricks and Singhal 1996; Jarrell and Easton 1997).

The implementation of Six Sigma in a company has not necessarily resulted in success. Numerous researches have been conducted to study the critical factors that affect the successful implementation of Six Sigma or the factors resulting in Six Sigma failure. For a project that is undertaken, it is very important to establish a link between the project and the business strategy. It should demonstrate the benefit to the project in monetary terms and the way in which it will help business strategy (Antony and Coronado 2002). In their study, Jiju Antony and Ricardo Banuelas Coronado (2002) very briefly explain various factors that are critical to a successful Six Sigma methodology. Some of these critical success factors are involvement and commitment of the management, cultural change communication, organizational infrastructure, and training.

Gary Hamel, in a *Harvard Business Review* article, described how suicides were committed in General Motors, which led to the company's failure (Hamel 2009). In his article, Forrest Breyfogle mentions the reasons for it as poor quality, unhealthy labor environment, lack of coherent brand identities, and poor power-trains cause adverse supplier relations (Breyfogle 2003). He states that the management values the financial "playing-games-with-numbers" more highly than inspired engineering. Gnanam, in his presentation on Lean Six Sigma (LSS) companies, lists nine factors as responsible for the failure of Six Sigma implementation, stating that almost all successful LSS companies eliminate the following failure modes: (a) failure to align Lean Six Sigma implementation with strategic direction/plan; (b) failure to create a sense of urgency and need for change; (c) failure to acquire the right knowledgeable leaders; (d) failure to acquire leadership knowledge; (e) failure to create quick wins; (f) failure to employ effective communication; (g) failure to link results to the bottom-line; (h) failure to enforce enterprise wide implementation; and (i) failure to provide recognition, awards, bonuses, etc. to active members (Gnanam 2011).

7.2   Cost approach

The goal of any business is to "make money" (Goldratt 1986). A company's success in making money is measured by its net profit or net income. The global market has forced companies to produce products or services of higher quality to survive the competition (Kumar et al. 2008). Quality management techniques have been implemented to achieve higher levels of profitability and organizational performance (Jarrell and Easton 1997). There are many quality management techniques that have been implemented by different companies, such as zero defects, quality circles, TQM, and business process re-engineering (BPR) (Marsh 2008).

Apart from many success stories, there are some companies which have failed to yield positive results even after Six Sigma implementation. An article from *The Wall Street Journal* states that according to a recent study, 60% of all corporate Six Sigma companies were unsuccessful in its implementation, and also discusses the reasons for the failure (Chakravorty 2010). One way to examine how well a company is performing is to analyze its stock performance (Brown and Warner 1985; Hendricks and Singhal 1996). Several research studies illustrate how Six Sigma led to financial gains at Motorola, General Electric, and other Fortune 200 companies (Goh et al. 2003; Hammer 2002; Harry and Schroeder 2000; Lucier and Seshadri 2001).

Some of the important metrics for evaluating a company's success in response to shareholder value are return on equity (ROE), earnings per share (EPS), price-earnings (PE) ratio, total return price (TRP), etc. Companies can be classified based on their present worth, future worth, annual worth, internal rate of return/minimum acceptable rate of return (MARR), etc. (White et al. 2010). While considering any method we need to be aware of inflation. When analyzing stock market performance over many years, it is important that the company consider inflation-adjusted performance. Comparing makes no sense if we do not include inflation (Al 2012).

7.3   Value approach

Both LSS and Six Sigma are expected to add value in terms of improved quality of product and service by applying them singly as well as in combination. Let us look at some principles of LSS and Six Sigma that show prospects of increasing value (Devane 2004).

Some of the LSS principles that are beneficial are

- Improving and proper sequencing of flow
- Reducing setup time
- Destroying waste
- Contingency measure
- Avoiding waste

But by using the LSS approach, processes are not controlled on the basis of statistical measures. These measures account for errors in measurement due to variability factors and relate quality with mathematical problem solving and root cause analysis after the elimination of waste.

Some of the Six Sigma principles that are beneficial are

- Focus on customer requirements
- Management of processes for their improvement
- Use of statistical tools
- Approval of projects with high returns

However, the Six Sigma approach lacks tools to enhance speed, reduce costs, extend timelines, and reduce the slow attainment of savings.

By combining the two methodologies, shortcomings are mitigated and organizations must strive to formally integrate the two methodologies and tool sets.

7.4   Application approach

Six Sigma processes focus on developing and producing products and services with the highest possible quality consistently. A product that has followed Six Sigma practices for its production has process capability, $C_{pk} > 1.5$. Six Sigma is a management strategy that strives to minimize error and the causes for it by concentrating on the throughput that is expected by the customer (Snee 1999). It uses statistical tools to increase the profit and quality of the organization. Six Sigma represents the standard deviation of the process and is a statistical term. It is usually a normally distributed process. ± 3 Sigma has 99.73% measurement for the process and ± 4 Sigma has 99.9996% measurement for the process. A probability of 0.00135 and 0.99865 is observed for 99.73% of the process, respectively (Quality Council of Indiana 2007).

In order to produce complex assemblies from various assemblies, Motorola focused on a process having normal distribution and normal process variation of the mean with a specification limit of ± 6 to produce less than 3.4 defects per million (Figure 7.1).

Table 7.1 demonstrates various defect quantifications for different sigma levels. Defect levels are quantified by parts per million (ppm).

Six Sigma processes are defined as DMAIC for many organizations, and are described clearly as follows:

- Define: Choose the appropriate practice for the process to be improved.
- Measure: Collect the data to measure the variable that is chosen to be improved.
- Analyze: Analyze the primary cause for the error and defect or the measured variation from the original practice.
- Improve: Minimize variation of the cause.
- Control: Check the process constantly to sustain improvements (Hahn et al. 1999).

The above process is altered to the following eight steps:

1. R—Recognize the state of the business.
2. D—Choose the appropriate practice in the process to be improved and frame plans that will support improvement of the current process.
3. M—Measure the system by collecting data.

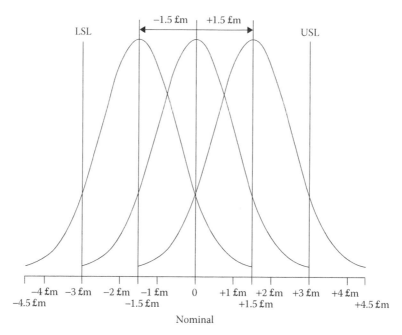

Figure 7.1 The ±1.5 Sigma shift. (From Gitlow, H., http://www.processexcellencenetwork.com/six-sigma-quality/articles/an-introduction-to-six-sigma-management/, 2012. With permission.)

Table 7.1 Defect Levels

Sigma level	ppm
6 Sigma	3.4
5 Sigma	233
4 Sigma	6210
3 Sigma	66810
2 Sigma	308,770
1 Sigma	697,672

Source: Terry, K., http://www.isixsigma.com/new-to-six-sigma/sigma-level/sigma-performance-levels-one-six-sigma/, 2010. With permission.

4. A—Analyze the primary cause of the error and defect or the measure variation from the original practices. Compare the measured data with the benchmark.
5. I—Minimize variation of the cause to reach the anticipated performance standard.
6. C—Check the process constantly to sustain improvements.
7. S—Standardize the process that is suitable and perform necessary action to reach the desired level.
8. I—Integrate to the system and constantly monitor the process (Harry and Schroeder 2000).

The advantages of implementing Six Sigma are listed as follows:

1. Improvement in productivity
2. Reduction of defects
3. Reduction of cost

4. Growth in market share
5. Improvement in customer relations
6. Improvement in product and service
7. Reduction of cycle time
8. Change in culture (Pande et al. 2000)

Six Sigma initiatives implemented by Motorola helped save them $940 million in three years. GE earned around $2.5 billion in the year 1999 by implementing Six Sigma. Honeywell followed the concepts of Six Sigma and saved $1.5 billion in the year 1997 (Hahn 1999).

According to Snee (1999), Six Sigma implementation is successful due to the following reasons:

- Top-level management is involved
- Well-practiced approach (DMAIC) is used
- Specified measures of success are used
- Well-defined statistical tools are followed
- Processes and customers are the main concentration in the process
- Short duration of projects

An organization achieves greater process quality than Six Sigma quality results if it follows Six Sigma concepts for many years. Defects become occasional in that they do not occur when the organization continuously follows Six Sigma concepts.

7.5   LSS/Quality decision matrix

Timely decisions can give bottom-line results as expected from LSS implementation. A Pugh matrix is used to weigh solutions against customer requirements.

The voice of the customer demonstrates these customer specifications. They also give valuable information about the performance metrics that play a key role in meeting the level of quality of service desired through critical-to-success factors. One must make sure that all of these are measurable, including attribute data. Then we determine the tools that are relevant to these results and the phases in which they can be utilized. Then these parameters are ranked and weight allotted to them appropriately.

Symbols are used to indicate increase, decrease, or no change from the expected performance measure. Every field in the matrix is checked for increase in levels. The highest level makes it the best solution. The matrix with the required tools for each success factor also makes implementation easier. A risk analysis can be conducted for refined solutions (Pereira 2007).

7.6   Determining economic justification of engineering projects

Organizations are continuously adapting to quality management techniques and tools to survive competition in the market. The underlying goal is to improve the revenue of the company. The Six Sigma methodology is an approach that has been utilized to improve the quality of operations. However, the main focus of Six Sigma for business is to have a direct impact on the company's earnings. Many organizations have attempted to use this methodology as an approach to improve their business and failed. Often, engineering economics calculations and the decision process of performing the recommended actions for Six Sigma projects are not performed and reviewed. In this research, we investigate the critical factors that are required for successful Six Sigma implementation using hurdle rate. Additionally, we evaluate different economic tools with respect to static earnings before interest and taxes (EBIT). The findings from this research will provide an insight into Six Sigma implementation and a framework to pick the correct economic tool for financial success.

The goal of any business is to "make money" (Goldratt 1986). A company's success in making money is measured by its net profit or net income. The global market has forced companies to produce products or services of higher quality to survive the competition (Kumar et al. 2008). Quality management techniques have been implemented to achieve higher levels of profitability and organizational performance (Jarrell and Easton 1997). There are many quality management techniques that have been implemented by different companies, such as zero defects, quality circles, TQM, and BPR (Marsh 2008). Due to significant differences in terms of construct development, research methodology, operationalizing quality management as single or multiple constructs, measuring performance at one level or multiple levels, and differences in data analysis techniques, all have contributed to producing mixed results in the relationship between quality management and firm performance (Ittner and Larcker 1997; Kaynak 2003; Molina et al. 2007).

Six Sigma methodology is often used to develop and or manage a business, department, organization, or product efficiently. In order to improve the performance of the company, organizations use Six Sigma tools to better serve their customers (Pande and Holpp 2002). However, numerous facts have shown that the implementation of Six Sigma in a company has not necessarily resulted in success. In this chapter, we discuss how these unsuccessful Six Sigma implementations in fact relate to the failure in establishing the relationship between Six Sigma project and business strategy with regard to the critical factors that exist in Six Sigma implementation. We investigate this relationship in financial terms by comparing the common EBIT practice in a Six Sigma company with financial analysis using engineering economics tools.

7.7 Previous economic studies

Previous studies also examined the effect of other quality improvement activities such as quality awards, quality certification, and other quality initiatives on stock prices (Goh et al. 2003; Przasnyski and Tai 1999). The response of stock price returns for TQM companies was also studied (Adams et al. 1999; Chakravorty 2010; Hendrics and Singhal 1996).

7.7.1 Impact of Six Sigma on performance of a company

In spite of its success, there are some companies that failed to yield positive results even after Six Sigma implementation. In his article in *The Wall Street Journal*, Dr. Chakravorty mentions that according to a recent study, 60% of all corporate Six Sigma companies were unsuccessful in the implementation of Six Sigma, and he also discusses the reasons for the failure (Brown and Warner 1985). One way to examine how well the company is doing is to look at its stock performances (Hendrics and Singhal 1996; Lucier and Seshadri 2001). Previous research studies illustrate how Six Sigma led to financial gains in Motorola, General Electric, and other Fortune 200 companies (Goh et al. 2003; Hammer 2002; Harry 1998; White et al. 2010).

Some of the important metrics for evaluating a company's success in response to shareholder value are ROE, EPS, PE ratio, TRP, etc. The companies can be classified based on present worth, future worth, annual worth, internal rate of return/MARR, etc. (Al 2012). While considering any method we need to be aware of inflation. Dr. Chakravorty in his blog states that, when analyzing stock market performance over multiple years, it is important that we consider *inflation-adjusted* performance. Comparing makes no sense if we do not include inflation (Hamel 2009).

7.8 Critical factors in Six Sigma implementation

Studies have investigated the critical factors that led to successful implementation of Six Sigma against the factors that caused Six Sigma failure. It is important for a project to be undertaken to demonstrate its financial benefit and its contribution toward competitiveness of the business in the market. This demonstration is aimed at establishing a relation between Six Sigma project and business strategy.

The critical success factors, which include management involvement and commitment, cultural change communication, organizational infrastructure, and training, contribute toward the successful implementation of Six Sigma methodology (Antony and Coronado 2002). Lack of understanding of the critical success factors could lead to business failure of the organization as demonstrated in the General Motors case (Breyfogle 2003). The additional contributing factors for the failure of General Motors were poor quality, unhealthy labor environment, lack of coherent brand identities, poor power-trains cause adverse supplier relations, and financial values rather than inspired engineering (Gnanam 2011). The following failures must be eliminated for successful Six Sigma implementation: (a) failure to align LSS implementation with strategic direction/plan, (b) failure to create a sense of urgency and need for change, (c) failure to acquire right knowledgeable leaders, (d) failure to acquire leadership knowledge, (e) failure to create quick wins, (f) failure to employ effective communication, (g) failure to link results to the bottom-line, (h) failure to enforce enterprise-wide implementation, and (i) failure to provide recognition, awards, bonuses, etc. to active members (Hahn et al. 1999).

The Six Sigma implementation failures as discussed earlier lead to the conclusion that in the selection and decision-making process, managers are faced with the hard-to-predict, stochastic nature of a project's outcome. Regarding this stochastic nature of the outcome, cost and benefit analysis can be employed as a way for managers to take into account this uncertainty, as stated by Geoff Riley from Eton College:

> In a world of finite public and private resources, we need a standard for evaluating trade-offs, setting priorities, and finally making choices about how to allocate scarce resources among competing uses. Cost benefit analysis provides a way of doing this. (Riley 2006)

Six Sigma generally distinguishes itself from other continuous improvement and quality initiatives in that it focuses on financial results or bottom-line business results. It can be found in many "best-selling" Six Sigma books that the most common economic performance metric is EBIT. Given that there are critical factors that are required for successful Six Sigma implementation, we suggest that the engineering economics "hurdle-rate" analysis may be more appropriate. Additionally, we evaluate the different engineering economics tools with respect to the static EBIT. The findings from this research will provide an insight into Six Sigma implementation and a framework to pick the correct economics tool for financial success.

7.9 Methodology

For this research, a five-step comparative approach framework, as suggested by Jones et al. (2013), is utilized for comparing the different engineering economics tools that are required in a Six Sigma project. The following sections detail the methodology that is used in this research.

7.9.1 Baseline

The first step of the five-step comparative framework is the baseline step. In this step, we describe when a traditional Six Sigma DMAIC project utilizes EBIT and engineering economics tools for financial and other benefits as shown in Table 7.2.

7.9.2 Identifying engineering economics metric

The baseline step is followed by identifying the engineering economics metric, which is the second step. In this step, we show the use of common engineering economics metric as a flowchart in Figure 7.2. The flowchart was adapted from Stevens (1994). The questions to determine Six Sigma viability must be custom defined based on the nature of the project in the "Define" stage of Six Sigma DMAIC methodology.

7.9.3 Comparison

The third step of the proposed methodology is comparison. In this step, the value obtained from using EBIT and a minimum of two engineering economics metrics identified from the flowchart in the previous step is compared. A sample table is provided in Table 7.3.

Table 7.2 EBIT and Engineering Economics in Six Sigma DMAIC Project

	Sub steps	Traditional Six Sigma financial metrics	Engineering economics (planned)
Define	Quality projects (sponsorships)	EBIT	ROA, ROI, NPV, IRR, PBP
	Define specific project (black belt)	None	ROA, ROI, NPV, IRR, PBP
Measure	Identify critical characteristics	None	ROA, ROI, NPV, IRR, PBP
	Clarify specifics and targets	None	ROA, ROI, NPV, IRR, PBP
	Measurement system validation	None	ROA, ROI, NPV, IRR, PBP
Analyze	Determine baseline	None	ROA, ROI, NPV, IRR, PBP
	Identify improvement objectives	None	ROA, ROI, NPV, IRR, PBP
	Evaluate process inputs	None	ROA, ROI, NPV, IRR, PBP
Improve	Formulate implementation plan	None	ROA, ROI, NPV, IRR, PBP
	Perform work measurement	None	ROA, ROI, NPV, IRR, PBP
	Lean execute the proposed plan	None	ROA, ROI, NPV, IRR, PBP
Control	Reconfirm plan	None	ROA, ROI, NPV, IRR, PBP
	Set up statistical process control	None	ROA, ROI, NPV, IRR, PBP
	Operational process transition	EBIT	ROA, ROI, NPV, IRR, PBP

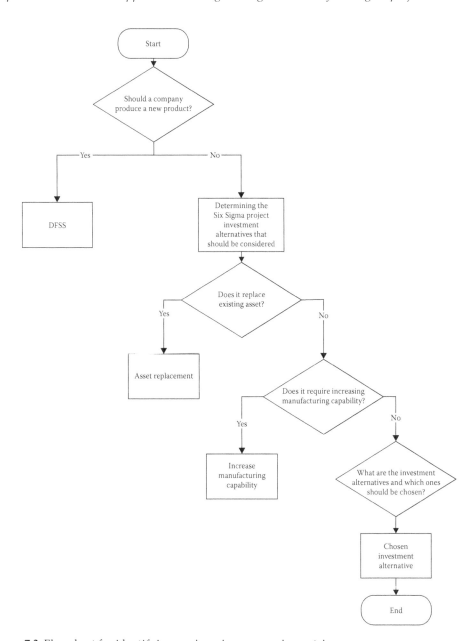

Figure 7.2 Flowchart for identifying engineering economics metric.

Table 7.3 Sample Comparison of EBIT, IRR, and NPV Metrics

	EBIT growth rate	IRR	NPV	Decision (based on EBIT)	Decision (based on IRR)
Project	*EBIT growth ratio	$0 = \sum_{j=0}^{n} \dfrac{X_j}{\left(1+i\right)^j}$	$\mathbf{NPV} = \sum_{j=0}^{n} \dfrac{X_j}{\left(1+k\right)^j}$	If EBIT growth ratio > X%, accept project	If IRR > MARR, accept project

*EBIT growth ratio (refer to Equation 7.1).

Table 7.4 Sample Sensitivity Analysis of EBIT, NPV, and IRR Metrics

Project	EBIT growth rate	NPV	IRR
Scenario 1			
Scenario 2			

7.9.4　Evaluation

Evaluation is the fourth step in which we identify the rate of EBIT growth that we define as the ratio of the average EBIT and the average revenue within the past four years of the project life time.

$$\text{EBIT Growth Rate} = \text{Average EBIT} / \text{Average Revenue.} \qquad (7.1)$$

For the purpose of evaluation in the study case section, we determine that if the EBIT growth rate $\geq 5\%$, then the project is accepted.

　　We also evaluate the internal rate of return (IRR) of the project, and take a decision whether to reject or accept the project based on it.

7.9.5　Sensitivity

Sensitivity is the fifth step that allows further investigation into the challenges and benefits of using EBIT and the identified engineering economics metrics. Table 7.4 provides a sample comparison of the sensitivity analysis.

7.10　Case study

In this case study, we will compare two scenarios of four-year-life-time projects with different revenues and cash flows. The expenses are expected to be 40% of the revenue and the working capital required in each year is expected to be 20% of the revenue in the following year. The product requires an immediate investment of $58,000 in plant and equipment. Depreciation is $12,500 each year. The cost of capital is 10%. Management defines the rate of EBIT growth as the ratio between the average EBIT and the average revenue within the four years of the project life time. Management has also determined that if the rate of EBIT growth within the four years of the project life time is greater than or equal to 5%, then they will accept the project, otherwise they will reject it. Tax is determined to be 40% of the EBIT in each year. We will see how the two scenarios show the difference in the decisions that are made on the basis of EBIT and the net present value (NPV) and IRR. Detailed calculations are provided in Table 7.5 for scenario 1 and in Table 7.6 for scenario 2.

7.10.1　Sample calculation

The revenue in the first year is $19,000. The EBIT for the first year is simply 60% times the revenue minus the depreciation, which is 60% ($19,000)–$12,500 = –$1,100. The net income for the first year is simply the EBIT for that year minus the tax. In year 1 since the EBIT is negative, we do not consider the tax rate because negative loss does not include

Table 7.5 Project Scenario 1

Year	Revenue	EBIT	Rate of EBIT growth	Net income	Cash flow	NPV	IRR	Decision
0					−58,000			
1	19,000	−1,100	Rate > 0.05	−1100	15,200	($1604.40)	11.52%	IRR > 0.10
2	21,000	100	Accept the	60	16,760			Accept the
3	41,000	12,100	project	7260	27,960			project
4	20,000	−500		−500	16,000			
Average	25,250	2,650 Rate	10.495%					

Table 7.6 Project Scenario 2

Year	Revenue	EBIT	Rate of EBIT growth	Net income	Cash flow	NPV	IRR	Decision
0			Rate > 0.05		−58,000			
1	10,000	−6,500	Accept the	−6500	8,000	($4804.45)	6.52%	IRR < 0.10
2	21,000	100	project	60	16,760			Reject the
3	41,000	12,100		7260	27,960			project
4	20,000	−500		−500	16,000			
Average	23,000	1,300 Rate	5.6522%					

any tax and hence the net income is −$1,100. In the second year, the EBIT is 100 and so the net income would be 100−40% (100) = 600. The cash flow for year 1 equals the net income for year 1 plus the working capital recoup for year 1 plus the depreciation per year, which is (−1,100) + 20% (19,000) + 12,500 = 15,200. Next, we can compute the average revenue and average EBIT from year 1 to year 4, and based on these averages, we can compute the rate of EBIT growth within four years of the project lifetime, which comes to 2,650/25,250 = 10.495%. Finally, we can compute the NPV and IRR of the project, which are 17,920 and 11.216%, respectively. Note that the MARR for the project is 10%, which is same as the cost of the capital.

The sensitivity analysis for the case study given earlier is based on some key assumptions. The most critical assumptions for the EBIT growth rate, cash flow, and IRR are the revenue growth rate and the estimated gross margin. We can see that by simply changing the inputs for these assumptions, EBIT growth rate and IRR are impacted (Table 7.7).

Table 7.7 Sample Sensitivity Analysis of EBIT, NPV, and IRR Metrics

Project	EBIT growth rate (%)	NPV	IRR (%)
Scenario #1	10.495	**$1604.40**	11.52
Scenario #2	5.6522	**$4804.45**	6.52

Note: The bold terms are the calculated net present values (NPVs) of each scenario. NPV is considered as one of the most important decision metrics for accepting or rejecting a project.

Table 7.8 Decision Comparison Based on EBIT Growth Rate, IRR, and NPV Metrics

	EBIT growth rate (%)	IRR (%)	NPV	Decision (based on EBIT)	Decision (based on IRR)
Scenario #1	10.495	**11.52**	**$1604.40**	Accept the project	Accept the project
Scenario #1	5.6522	6.52	**$4804.45**	Accept the project	Reject the project

Note: The bold terms are the calculated net present values (NPVs) of each scenario. NPV is considered as one of the most important decision metrics for accepting or rejecting a project.

7.11   Results and discussions

7.11.1   Decision comparison based on EBIT growth rate and IRR

Table 7.8 gives the decision comparison on whether to accept or reject the project both for scenarios 1 and 2.

We can see that in scenario 2, the decision based on EBIT growth rate is different from the decision based on IRR. This difference in decision is simply because the decision based on EBIT growth does not take into account the time value of money.

7.12   Conclusion

This chapter describes the approaches for evaluating the benefits of Six Sigma methodology from a firm perspective. We presented a research method for determining the economics for Six Sigma projects. The research provides the initial formulation of a methodology and the engineering economics tools that account for the stochastic nature of the market for successful Six Sigma implementation in an organization. The findings have ramifications on decision making, which is an important characteristic of management. Additional results have been published elsewhere.

References

Adams, G., Grant, M., and Kristie, S. (1999). "Revisiting the stock price impact of quality awards." *Omega*, 27(6): 595–604.

Al. (2012). "100-years of inflation-adjusted stock market history." *Observations*, March 21. Retrieved from http://observationsandnotes.blogspot.com/2011/03/stock-market-100-year-inflation-history.html.

Antony, J. and Coronado, R. B. (2002). "Critical success factors for successful implementation of Six Sigma projects in organizations." *TQM Magazine*, 14(2): 92–99.

Breyfogle, F. W. (2003). *Implementing Six Sigma: Smarter Solutions Using Statistical Methods*. 2nd edn. New York: John Wiley & Sons.

Brown, S. J. and Warner, J. B. (1985). "Using daily stock returns: The case of event studies." *Journal of Financial Economics*, 14(1): 3–31.

Chakravorty, S. S. (2010). "Where process-improvement projects go wrong?" *The Wall Street Journal*. Retrieved January 25 from http://online.wsj.com/article/SB1000142405274870329800457445747131393 8130.html.

Devane, T. (2004). *Integrating Lean Six Sigma and High Performance Organizations: Leading the Charge toward Dramatic, Rapid and Sustainable Improvement*. San Francisco: Pfeiffer, an imprint of Wiley.

Gitlow, H. (2012). "An introduction to Six Sigma management." Process excellence network. December 16, 2012. November 16, Boulder, CO, ASQ Boulder Section 1313 Monthly Meeting. Retrieved from http://www.processexcellencenetwork.com/six-sigma-quality/articles/an-introduction-to-six-sigma-management/.

Gnanam, J. (2011). "Why Lean Six Sigma implementations fail?" ASQ Boulder Meeting, Boulder, CO.

Goh, T. N., Low, P. C., Tsui, K. L., and Xie, M. (2003). "Impact of Six Sigma implementation on stock price performance." *Total Quality Management & Business Excellence*, 14(7): 753–763.

Goldratt, E. M. (1986). *The Goal: A Process of Ongoing Improvement*. New York: North River Press.

Hahn, G. J. et al. (1999). "The impact of Six Sigma improvement—A glimpse into the future of statisticians." *The American Statistician*, 53(2): 208–215.

Hamel, G. (2009). "25 stretch goals for management." *Harvard Business Review*, Retrieved December 31, 2012, from http://blogs.hbr.org/hbr/hamel/2009/02/25_stretch_goals_for_managemen.html

Hammer, M. (2002). "Process management and the future of Six Sigma." *MIT Sloan Management Review*, 42(3): 26–32.

Harry, M. (1998). "Six Sigma: A breakthrough strategy for profitability." *Quality Progress*, 31(5): 60–64.

Hendricks, K. B. and Singhal, V. R. (1996). "Quality awards and the market value of the firm: An empirical investigation." *Management Science*, 42(3): 415–436.

Ittner, C. D. and Larcker, D. F. (1997). "The performance effects of process management techniques." *Management Science*, 43(4): 522–534.

Jarrell, S. L. and Easton, G. S. (1997). "An exploratory empirical investigation of the effects of total quality management on corporate performance." in Lederer, P. J. and Karmarkar, U. S. (eds.) *The Practice of Quality Management*. Boston, MA: Kluwer Academic Publishers, pp. 11–53.

Jones, E. C. (2013). "Modern quality using six sigma techniques: A practical guide." Book manuscript, University of Texas, Arlington.

Kaynak, H. (2003). "The relationship between total quality management practices and their effects on firm performance." *Journal of Operations Management*, 21(4): 405–435.

Kumar, M. et al. (2008). "Common myths of Six Sigma demystified." *International Journal of Quality & Reliability Management*, 25(8): 878–895.

Lucier, G. T. and Seshadri, S. (2001). "GE takes Six Sigma beyond the bottom line." *Strategic Finance*, 82(11): 40–46.

Marsh, S. A. (2008). "Six Sigma: A passing fad or a sign of things to come?".

Molina, L., Llorensmontes, J., and Ruizmoreno, A. (2007). "Relationship between quality management practices and knowledge transfer." *Journal of Operations Management*, 25(3): 682–701.

Pande, P. and Holpp, J. (2002). *What Is Six Sigma?* New York: McGraw-Hill.

Pande, P., Neuman, R. P., and Cavanagh, R. R. (2000). *The Six Sigma Way: How Ge, Motora and other Top Companies are Honing their Performance*. New York: McGraw-Hill.

Pereira, R. (2007). "The Pugh Matrix." Lean Six Sigma Academy. Retrieved from http://lssacademy.com/2007/06/19/the-pugh-matrix/.

Przasnyski, Z. H. and Tai, L. S. (1999). "Stock market reaction to Malcolm Baldrige National Quality Award announcements: Does quality pay?" *Total Quality Management*, 10(3): 391–400.

Quality Council of Indiana (2007) *Certified Six Sigma Black Belt Primer*. West Terre Haute, IN: Quality Council of Indiana.

Riley, G. (2006). "Cost benefit analysis." Retrieved December 30, 2012, from http://www.tutor2u.net/economics/revision-notes/a2-micro-cost-benefit-analysis.html

Snee, R. D. (1999). "Discussion: Development and use of statistical thinking. A new era." *International Statistical Review*, 67(3): 255–258.

Snee, R. D. (2000). "Impact of Six Sigma implementation on quality engineering." *Quality Engineering*, 12(3): ix–xiv.

Stevens, G. T., Jr. (1994). *The Economic Analysis of Capital Expenditures for Managers and Engineers*. Boston, MA: Pearson Custom Publishing.

Terry, K. (2010). "Six Sigma performance levels." iSixSigma. Retrieved October 2, 2013, from http://www.isixsigma.com/new-to-six-sigma/sigma-level/sigma-performance-levels-one-six-sigma/.

White, J., Case, K., and Pratt, D. (2010). *Principles of Engineering Economic Analysis*. 5th edn. New York: John Wiley & Sons, 237, 278–295.

Review Questions

Question 1: Variances from budget for a project:

    A. Are used to focus on corrective action efforts
    B. Are the difference between planned and actual
    C. Indicate if the project manager did a poor job of controlling costs
    D. Are usually expressed in standard deviation units from the norm
      (a)   A B
      (b)   B C D
      (c)   C D
      (d)   A B C D

Question 2: An organized, disciplined approach to problem solving in most Six Sigma organizations is called

  (a) SIPOC
  (b) DPMO
  (c) DMAIC
  (d) PDCA

chapter eight

Lean Six Sigma and firm performance: A managerial perspective[*]

> When your pills get down to four order more.
>
> **—Anonymous**

In this chapter, the effect of Six Sigma on firm performance is discussed. The aim is to provide a holistic and systematic review of the existing body of knowledge in both academic and practitioner literatures. The chapter aims to answer the following question: Does Six Sigma improve firm performance?

8.1 Introduction

It is a common practice that firms look for new tools, methodologies, and techniques to improve their performance and enhance their competiveness in the marketplace. One of the popular management thinking and practices is Six Sigma, where through its structured approach to process improvement it enables organizations to achieve superior performance (Choo et al. 2007a, 2007b; Linderman et al. 2003, 2006).

In order to better understand Six Sigma and examine its relationship to firm performance, we need to look at its origin, what it intends to do, and how it achieves its goals. We also need to look at the scope and domain of Six Sigma, its promises, and its limitation. We also need to have a definition for Six Sigma. This helps us to understand what Six Sigma is really about. Then, we can argue whether Six Sigma improves performance or not.

Six Sigma has its roots in quality control and quality management (Goeke and Offodile 2005). The proponents of Six Sigma argue that it improves organizational process and leads to improvement in organizational performance (Dasgupta 2003; Linderman et al. 2003; Pantano et al. 2006). Six Sigma is defined as a business process improvement approach that seeks to find and eliminate causes of defects and errors, reduce cycle times and cost of operations, improve productivity, meet customer expectations better, and achieve higher asset utilization and returns (Evans and Lindsay 2005). Another definition for Six Sigma is provided by Hammer (2002) who argues that Six Sigma employs a project-based methodology to solve a specific performance problem recognized by an organization. Douglas and Erwin (2000) argue that the focus of Six Sigma is on the customer rather than on the product.

Although scholars and practitioners cite numerous examples of the positive effect of Six Sigma on firm performance (e.g., Foster 2007; Hoerl 1998; Johnson 2005; Roberts 2004; Rucker 2000; Shafer and Moeller 2012; Swink and Jacobs 2012), there are concerns and criticisms about the effectiveness and impact of Six Sigma projects. Companies such as Bank of America and Citigroup have reported significant improvement in their performance as the result of implementing Six Sigma programs (Roberts 2004; Rucker 2000). Alternatively,

---

[*] By Mahour Parast.

some argue that Six Sigma is simply a repackaging of traditional quality management which is subject to the limitations and criticism of the quality programs, so there is nothing new here (Dahlgaard and Dahlgaard-Park 2006). The proponents of Six Sigma argue that it helps organizations to be more ambidextrous by focusing on both efficiency and innovation (Schroeder et al. 2008). The answer to the relationship between Six Sigma and firm performance cannot be easily addressed because most organization have other improvement programs in place so it would be difficult to argue that which program gets the credit for what type of improvement and how.

The purpose of this chapter is to provide different perspectives to the effect of Six Sigma programs on firm performance. To do so, the foundation of Six Sigma and its underlying assumptions is discussed. Later, the effect of Six Sig Sigma on enhancing innovation is explained. Finally, the role of contextual variables and factors in the relationship between Six Sigma and performance is discussed.

8.2   What is special about Six Sigma?

Whereas some argue that Six Sigma is simply the repackaging of the quality control and quality management tools and techniques, a review of the literature and existing body of knowledge shows that Six Sigma has different strategies to process improvement than traditional quality management programs (Mellat-Parast 2011).

The fundamental difference between Six Sigma and other process improvement programs such as TQM, Lean, and the Baldrige model is related to the ability of Six Sigma of providing an organizational context that facilitates problem solving and exploration across the organization. Whereas Six Sigma programs have their roots in the quality movement, they are different from other quality programs due to their limited time-frame, measurable and quantifiable goals, and the project structure (Andersson et al. 2006; Dahlgaard and Dahlgaard-Park 2006). These characteristics of Six Sigma program gives them a more measurable, objective, and quantifiable capability which is less emphasized in other quality management programs. So, the answer to the critics of Six Sigma and those who argue that Six Sigma does not provide anything new to quality management and process improvement domain is that it has a different perspective to process improvement. Please note that at this stage, we are not claiming that Six Sigma has a significant impact on firm performance; rather, we just argue that it is different (to a certain degree) than what is already known in quality control and quality management.

Having said that, the proponents of Six Sigma claim that Six Sigma enables organizations to become more ambidextrous by switching structure, act organically when being challenged by new ideas, and operate mechanically in focusing on efficiency (Schroeder et al. 2008). Ambidextrous organizations manage trade-off between conflicting goals (alignment and adaptation) through utilizing and implementing "dual structures." In these forms of organizations whereas some of the business units are focused on efficiency, other business units emphasize innovation and change (Duncan 1976; Gibson and Birkinshaw 2004). This dual structure enables organizations to focus on both exploitation and exploration, addressing both efficiency and innovation (March 1991). Whereas these are promising statements about Six Sigma, we need to look deeper and examine whether Six Sigma can really achieve the above objectives.

First, the review of both practitioner and academic literature suggests that Six Sigma is characterized under the family of process management programs (Benner and Tushman 2003; Evans and Lindsay 2005; Hammer 2002; Mellat-Parast 2011). Process improvement programs, by their nature, favor incremental (exploitative) innovation at the expenses of neglecting radical (explorative) innovation. Popular and well-known programs such as TQM, business process reengineering, Lean and Six Sigma all focus on improving, rationalizing, and enhancing organizational processes (Hammer and Champy 1993; Harry and Schroeder 2000; Powell 1995). Because process improvement programs are technically designed to reduce variance in the system, they may impede practices related to product innovation and radical change. In other words, they are not designed to deal with "high variability" processes because their focus is to reduce variability.

Second, understanding and identifying critical characteristics of existing customers is one of the principles of Six Sigma (Dasgupta 2003; Evans and Lindsay 2005; Harry 1998; Linderman et al. 2003; Mellat-Parast 2011). Overemphasis on process improvement efforts along with capturing of the needs and preferences of existing customers may weaken the ability of the firm to identify new customers and introduce new products and/or services. We should realize that incremental process innovation such as Six Sigma programs are fundamentally designed to meet the needs of existing customers (Benner and Tushman 2003; Mellat-Parast 2011). As such, the ability of Six Sigma program to enhance the innovation capability of a firm in design and development of new products/services for new customers is questionable.

Third, Six Sigma programs differentiate themselves from other quality improvement programs through maintaining a strong emphasis on setting specific goals (Linderman et al. 2003). Setting a clear goal is central to Six Sigma where customer requirements are translated into the development of Six Sigma project goals (Pande et al. 2000; Schroeder et al. 2008). However, the notion of "setting specific goals" is in contrast with the viewpoints of the founders of quality management (Deming 1986; Linderman et al. 2003). As such, it could be argued that Six Sigma programs cannot initiate, develop, and maintain sustainable quality systems, and cannot address the core principles of quality management such as a culture of learning, continuous improvement of processes, and a system view of the organization (Dean and Bowen 1994; Mellat-Parast 2011).

Back to our initial question, we need to look deeper into the scope, boundary, and premises of Six Sigma programs in order to understand their impact on firm performance. Because Six Sigma has been categorized under process improvement programs, it is reasonable to continue the discussion on this domain. Therefore, a review of the literature on process management is necessary to answer these questions.

8.3   Process management

Process management programs can influence firm performance at both operational and strategic levels (Benner and Tushman 2003). At the strategic level, process management programs such as TQM, the Baldrige model, business process reengineering, Lean and Six Sigma appear to have a positive impact on business results and are able to improve profitability (Das et al. 2000; Douglas and Judge 2001; Easton and Jarrell 1998; Hendricks and Singhal 1996, 2001a, 2001b; Kaynak 2003; Powell 1995). At the operational level, process management programs are used in order to increase the transformation of input (e.g., raw material, labor) to the products and/or services (Silver 2004).

An important consideration in effective implementation of process management programs is how to maintain the trade-off between both strategic and operational objectives, and evaluate both short-term (operational) and long-term (strategic) impacts of process management programs (Klassen and Menor 2007). For example, inventory management requires taking into account the cost of holding the inventory (operations) whereas at the same time considering the safety stock to maintain an acceptable level of customer satisfaction (strategic). Effective inventory management programs are those which address both these objectives simultaneously.

At the same line, the stability of the environment should be considered in the development and implementation of process management programs. The ability of the firm to respond to changes in a highly dynamic and evolving market will be at risk if process management maintains a narrow and tight scope on operational measures. For example, it is not a wise decision to focus entirely on inventory reduction and cost reduction in material handling and logistics in a highly devolving, innovative and dynamic environment. This approach to process management may weaken the ability of the firm to respond to change in the market and customer preferences (Bower and Christensen 1995; Klassen and Menor 2007). Whereas in stable markets emphasis on efficiency and variance reduction is valued, evolving markets require flexibility and adaptability. Accordingly, the management of organizational processes and their relationship to firm performance changes as well.

Following the above discussion, we realize that organizational, contextual, and environmental variables have a significant impact on the effectiveness of process management programs. In fact, research shows that too much emphasis on process management may impede the firm's ability of innovation and responsiveness to new customers (Sterman et al. 1997). Back to our discussion on Six Sigma programs and keeping in mind that Six Sigma is best defined under the umbrella of process management programs, organizational, contextual, and environmental factors have a significant impact on the effectiveness of Six Sigma programs.

Until here, we established two things: (1) Six Sigma is part of the broader process management programs, and (2) the effect of process management programs on firm performance should be examined with reference to the organizational, contextual, and environmental conditions. Following the above discussion, the next step is to understand organizational processes.

8.3.1 Understanding organizational processes

Generally speaking, organizational processes can be defined under three different categories: Work processes, behavioral processes, and change processes (Garvin 1998). Starting with work processes, these are primarily concerned with accomplishing tasks. They can be divided into processes that produce goods/services (operational processes) and those that support them (administrative processes). We expect to see improvement in operations when work processes are redesigned and restructured. This means that both work processes and administrative processes should be improved.

Then we have the behavioral processes; they deal with behavioral patterns across the organization. These include decision making, communication, and learning processes. Finally, change processes are those that deal with the sequence of activities over time. They describe how individuals, groups, and organizations act, develop, and grow over time. Table 8.1 presents a review of organizational processes.

Whereas one can develop certain programs to improve each type of these processes, significant improvement in firm performance does not happen unless all of these processes are improved (Garvin 1998). In addition, we need to carefully understand the interaction and interconnectedness among these three types of the processes in order to achieve higher level of firm performance. For example, a significant improvement in production efficiency cannot happen without improving its supporting administrative processes, the way decisions are made in the organization (behavioral process), and understanding the nature and pattern of organizational change over time (change processes).

Table 8.1 Organizational Processes

Processes	Work processes	Behavioral processes	Change processes
Definition	Sequences of activities that transform inputs into outputs	Widely shared patterns of behavior and ways of acting/interacting	Sequences of events over time
Role	Accomplish the work of the organization	Infuse and shape the way work is conducted by influencing how individuals and groups behave	Alter the scale, character, and identity of the organization
Major categories	Operational and administrative	Individual and interpersonal	Autonomous and induced, incremental and revolutionary
Examples	New product development, order fulfillment, strategic planning	Decision making, communication, organizational learning	Creation, growth, transformation, decline

Source: Garvin, D. A., *MIT Sloan Management Review*, 39, 34–50, 1998.

Now by looking back to Six Sigma program as a process management initiative and by reviewing different types of organizational processes, we realize that Six Sigma programs are primarily concerned with design, development, and improving on improving work processes (mostly task processes). Six Sigma programs are not able to improve behavioral and change processes for two reasons: (1) Six Sigma programs maintain a strong focus on specifying measurable and quantifiable goals (Linderman et al. 2003). Whereas in the case of work processes (especially task processes) that might be achievable, we are not able to set specific and measurable goals for improving and/or restructuring behavioral and change processes, and (2) Six Sigma programs are developed through translating the voice of customers (primarily the external customers) to specific improvement projects. This is easily achievable in task and operational processes because the customer is easily defined in this context. What is the voice of customers in behavioral and change processes? The answer to this question is not clear. In fact, such a link between the voice of external customers and behavioral and change processes has not been established.

The above discussion underscores the limitation of Six Sigma in improving organizational processes where it is directly linked to only one type of organizational process. Therefore, Six Sigma programs cannot develop, maintain, and establish sustainable process improvement programs because these programs are not designed to integrate all processes in their process improvement efforts. Six Sigma program only improves operational processes, and to the extent that a firm performance is directly linked to improvement in its operational processes, Six Sigma program can be directly linked to firm performance.

With this being said, we realize that in order to get the best out of Six Sigma programs, we need to understand how these programs can improve organizational processes. Next, we review how process management can improve firm performance.

8.3.2 Process management and firm performance

Whereas the proponents of process management cite numerous benefits from process improvement programs, the effect of process management on firm performance has been mixed, and has failed to yield promising results (Benner and Tushman 2003). Researchers have failed to find any significant relationship between process management programs and firm performance (Powell 1995; Samson and Terzivoski 1999). There is not sufficient evidence to suggest that process improvement programs increase firm performance. For example, whereas process management programs have increased performance in the auto industry, they appear to decrease performance in the computer industry (Ittner and Larcker 1997). One could argue that process management programs are best implemented in the stable markets (e.g., auto industry) whereas their applicability in the evolving markets is limited (e.g., computer industry).

We already acknowledged that process improvement programs such as quality management can present firms with a trade-off between the short-term and long-term benefits (Sterman et al. 1997). A fundamental question on the effectiveness of process improvement programs such as Six Sigma is to determine whether firms emphasizing Six Sigma programs can achieve the dual goal of short-term (efficiency) and long-term (innovation) performance. Furthermore, to the extent that firms execute process improvement programs (such as Six Sigma), how are the dual goals of control (efficiency) and learning (innovation) addressed?

Previous studies on Six Sigma address the role of Six Sigma as a highly structured and disciplined approach to process improvement. Whereas there is agreement on the ability of Six Sigma on enhancing operational performance, there is little understanding on the effect of Six Sigma to improving firm's performance over time (Foster 2007). One compelling argument is that Six Sigma programs do not improve the innovation capability of the firms so that they cannot provide a sustainable quality performance over time. In order to address this issue, we need to examine whether Six Sigma enhances innovation capability of the firm.

8.3.3   Process management and innovation

Because process management programs are technically designed to reduce variance in organizational processes, overemphasis on these programs may have a negative impact on the ability of the firms to undertake radical change (Benner and Tushman 2003). Too much focus on process management may negatively affect the long-term performance of the firm because these programs do not address radical change (Garvin 1991; Hill 1993). To find the impact of process management on firm performance in both the short and long terms, we need to take a closer look on the effect of process management on innovation.

How process management interacts with the innovation capability of the firms? We need to understand that process management influences innovation of the firm in several ways. First, by using process management programs we are inherently trying to balance the allocation of resources to activities across the firm (Christensen and Bower 1996; Klassen and Menor 2007). Therefore, we may more or less emphasize specific activities and processes that are directly related to the firm's operations and business strategy. Second, process management deals with minimizing sources of variability in internal and external activities (Pannirselvam et al. 1999; Silver 2004). This may result in focusing on specific types of innovations that are consistent with reducing variability in the processes (Henderson and Clark 1990).

Whereas proponents of Six Sigma argue that Six Sigma can enhance the innovation capability of the firm (Byrne et al. 2007), there are different forms of innovation. Does Six Sigma enhance all types of innovation?

There are two dimensions for innovation: (1) the degree to which the innovation is close to the current technological path of the firm, and (2) the degree to which the innovation is serving existing markets/customers (Abernathy and Clark 1985). Therefore, to determine the impact of process management programs, we need to consider the following organizational and contextual factors:

1. Incremental versus radical innovation
2. The processes, procedures, and routines within the firm (work, decision making, and change processes)
3. New customers/markets or existing customer/markets

We should realize that process improvement programs are fundamentally designed and built on the current technological base of the firm (Benner and Tushman 2003; Green et al. 1995). Six Sigma programs assume that the current organizational processes are sound but they need minor (incremental) improvement to be more efficient (Hammer 2002).

Table 8.2 provides a summary of the relationship between Six Sigma and innovation. As we can see, Six Sigma programs are best effective when they are designed and developed to improve the current technological path of the firm in order to address the needs of existing customers.

The relationship among Six Sigma, innovation, and firm performance is presented in Figure 8.1. Six Sigma has a positive effect on incremental innovation of the firm. Customer satisfaction for existing customers is increased as organizations invest more on their Six Sigma programs. As the result of improvement in customer satisfaction, along with incremental innovation, firm performance is improved (Mellat-Parast 2011).

Table 8.2 Six Sigma and Technological Innovation of the Firm

Technological innovation	Existing customers	New customers
Current technological path	Focus on structure and control, focus on Six Sigma is very effective	A balance between structure and innovation, Six Sigma may be effective
New technological path	A balance between structure and innovation, Six Sigma may be effective	Focus on innovation, emphasis on Six Sigma may not be very effective

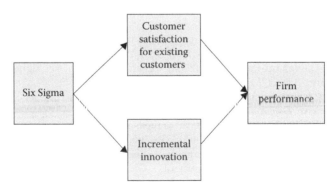

Figure 8.1 The effect of Six Sigma on firm performance. (Adapted from Mellat-Parast, M., *International Journal of Project Management*, 29, 45–55, 2011.)

8.4 Final notes

Our primary question in this chapter was to investigate whether Six Sigma enhances firm performance. We began our discussion by looking at Six Sigma from a process management perspective. This was a very valid and reliable conceptualization of Six Sigma.

We next discussed different types of organizational processes and realized that Six Sigma programs are fundamentally designed to improve work processes. In industries where task processes are the major determinates of performance, we expect to see a strong link between Six Sigma and firm performance. As the relationship between process improvement and firm performance diminishes, Six Sigma and similar programs are less effective in improving firm's performance over time.

Firms would be able to benefit from Six Sigma if two conditions are met: (1) the firm (or the business unit) is focused on efficiency (process improvement), and (2) the rate of change in the industry is low (stable industry).

How can organizations maximize their return from investment in Six Sigma? To get the best out of Six Sigma programs, organizations need to carefully address the needs of their current customers while monitoring the formation of new markets and/or customers. In its current form, Six Sigma programs do not guarantee a sustainable competitive advantage for the firms. Managers could expect to observe improvement in their processes and firm performance due to Six Sigma implementation and its focus on existing processes, products, and customers; however, such benefits may or may not be sustainable over time.

References

Abernathy, W. J. and Clark, K. (1985). "Mapping the winds of creative destruction." *Research Policy*, 14(1): 3–22.

Andersson, R., Eriksson, H., and Torstensson, H. (2006). "Similarities and differences between TQM, Six Sigma and Lean." *The TQM Magazine*, 18(3): 282–296.

Benner, M. J. and Tushman, M. L. (2003). "Exploitation, exploration, and process management: The productivity dilemma revisited." *Academy of Management Review*, 28(2): 238–256.

Byrne, G., Lubowe, D., and Blitz, A. (2007). "Using a Lean Six Sigma approach to drive innovation." *Strategy & Leadership*, 35(2): 5–10.

Choo, A. S., Linderman, K. W., and Schroeder, R. G. (2007a). "Method and context perspectives on learning and knowledge creation in quality management." *Journal of Operations Management*, 25(4): 918–931.

Choo, A. S., Linderman, K. W., and Schroeder, R. G. (2007b). "Method and psychological effects on learning behaviors and knowledge creation in quality improvement projects." *Management Science*, 53(3): 437–450.

Christensen, C. M. and Bower, J. L. (1996). "Customer power, strategic investment, and the failure of leading firms." *Strategic Management Journal*, 17(3): 197–218.

Dahlgaard, J. J. and Dahlgaard-Park, S. M. (2006). "Lean production, Six Sigma quality, TQM and company culture." *The TQM Magazine*, 18(3): 263–281.

Das, A. et al. (2000). "A contingent view of quality management—The impact of international competition on quality." *Decision Sciences*, 31: 649–690.

Dasgupta, T. (2003). "Using the Six Sigma metric to measure and improve the performance of a supply chain." *Total Quality Management*, 14(3): 355–366.

Dean, J. W. and Bowen, D. E. (1994). "Management theory and total quality: Improving research and practice through theory development." *The Academy of Management Journal*, 19(3): 392–418.

Deming, W. E. (1986). *Out of Crisis*. Cambridge, MA: MIT Center for Advanced Engineering Study.

Douglas, P. C. and Erwin, J. (2000). "Six Sigma's focus on total customer satisfaction." *Journal for Quality and Participation*, 23(2): 45–49.

Douglas, T. J. and Judge Jr., W. Q. (2001). "Total quality management implementation and competitive advantage: The role of structural control and exploration." *Academy of Management Journal*, 44: 158–169.

Duncan, R. B. (1976). "The ambidextrous organization: Designing dual structures for innovation." In Kilmann, R. H., Pondy, L. R., and Slevin, D. (eds.) *The Management of Organization*, vol. 1. New York: North-Holland. pp. 167–188.

Easton, G. S. and Jarrell, S. L. (1998). "The effects of total quality management on corporate performance: An empirical investigation." *Journal of Business*, 71(2): 253–307.

Evans, J. R. and Lindsay, W. M. (2005). *The Management and Control of Quality*, 6th edn. Mason, OH: South-Western.

Foster, S. T., Jr. (2007). "Does Six Sigma improve performance?" *Quality Management Journal*, 14(4): 7–20.

Garvin, D. A. (1991). "How the Baldrige Award really works?" *Harvard Business Review*, 66(6): 80–93.

Garvin, D. A. (1998). "The processes of organization and management." *MIT Sloan Management Review*, 39(4): 34–50.

Gibson, C. B. and Birkinshaw, J. (2004). "The antecedents, consequences, and role of organizational ambidexterity." *Academy of Management Journal*, 47(2): 209–226.

Goeke, R. J. and Offodile, F. (2005). "Forecasting management philosophy life cycle: A comparative study of Six Sigma and TQM." *Quality Management Journal*, 12(2): 34–46.

Green, S., Garvin, M., and Smith, L. (1995). "Assessing a multidimensional measure of radical innovation." *IEEE Transactions of Engineering Management*, 42(3): 203–214.

Hammer, M. (2002). "Process management and the future of Six Sigma." *MIT Sloan Management Review*, 43(2): 26–32.

Hammer, M. and Champy, J. (1993). *Reengineering the Corporation: A Manifesto for Business Revolution*. New York: Harper Business.

Harry, M. (1998). "Six Sigma: A breakthrough strategy for profitability." *Quality Progress*, 31(5): 60–64.

Harry, M. J. and Schroeder, R. (2000). *Six Sigma: The Breakthrough Management Strategy Revolutionizing the World's Top Corporations*. New York: Currency.

Hendricks, K. B. and Singhal, V. R. (1996). "Does implementing an effective TQM program actually improve operating performance? Empirical evidence from firms that have won quality awards." *Management Science*, 43(9): 1258–1274.

Hendricks, K. B. and Singhal, V. R. (2001a). "Firms characteristics, total quality management, and financial performance." *Journal of Operations Management*, 19(3): 269–285.

Hendricks, K. B. and Singhal, V. R. (2001b). "The long-term stock price performance of firm with effective TQM programs." *Management Science*, 47(3): 359–368.

Henderson, R. M. and Clark, K. B. (1990). "Architectural innovation: The reconfiguration of existing product technologies and the failure of established firms." *Administrative Science Quarterly*, 35(1): 9–30.

Hill, R. (1993). "When the going gets tough: A Baldrige Award winner on the line?" *The Executive*, 7(3): 75–79.

Hoerl, R. W. (1998). "Six Sigma and the future of quality profession." *Quality Progress*, 31(6): 35–42.

Ittner, C. D. and Larcker, D. (1997). "The performance effects of process management techniques." *Management Science*, 43(4): 522–534.

Johnson, K. (2005). "Six Sigma delivers on-time service." *Quality Progress*, 38(12): 57–59.

Kaynak, H. (2003). "The relationship between total quality management practices and their effects of firm performance." *Journal of Operations Management*, 21: 405–435.

Klassen, R. D. and Menor, L. J. (2007). "The process management triangle: An empirical investigation of process trade-offs." *Journal of Operations Management*, 25(5): 1015–1036.

Knowles, G., Whicker, L., Femat, J. H., and Canales, F. D. C. (2005). "A conceptual model for the application of Six Sigma methodologies to supply chain improvement." *International Journal of Logistics: Research and Applications*, 8(1): 51–65.

Kwak, Y. H. and Anbari, F. T. (2006). "Benefits, obstacles, and future of Six Sigma approach." *Technovation: The International Journal of Technological Innovation, Entrepreneurship, and Technology Management*, 26(5/6): 708–715.

Lai, K. and Cheng, T. C. E. (2003). "Initiatives and outcomes of quality management implementation across industries." *OMEGA*, 31(2): 141–154.

Laszlo, E. (1994). "The evolutionary project manager." In Cleland, D. I. and Gareis, R., *Global Project Management Handbook*. New York: McGraw-Hill. pp. 31–37.

Linderman, K. et al. (2003). "Six Sigma: A goal theoretic perspective." *Journal of Operations Management*, 21(2): 193–203.

Linderman, K. W., Schroeder, R. G., and Choo, A. S. (2006). "Six Sigma: The role of goals in improvement teams." *Journal of Operations Management*, 24(6): 779–790.

March, J. G. (1991). "Exploration and exploitation in organizational learning." *Organization Science*, 2(1): 71–87.

Mellat-Parast, M. (2011). "The effect of Six Sigma projects on innovation and firm performance." *International Journal of Project Management*, 29(1): 45–55.

Pande, P. S., Neuman, R. P. and Cavanagh, R. R. (2000). *The Six Sigma Way: How GE, Motorola, and Other Top Companies Are Honing Their Performance*. New York: McGraw-Hill.

Pannirselvam, G. P. et al. (1999). "Operations management research: An update for the 1990s." *Journal of Operations Management*, 18(1): 95–112.

Pantano, V., O'Kane, P., and Smith, K. (2006). "Cluster-based Six Sigma development in small and medium sized enterprises." *Proceedings of 2006 IEEE International Conference on Management of Innovation and Technology*, June 21–23, Singapore.

Powell, T. (1995). "Total quality management as competitive advantage: A review and empirical study." *Strategic Management Journal*, 16(1): 15–37.

Roberts, C. M. (2004). "Six Sigma signals." *Credit Union Magazine*, 70(1): 40–43.

Rucker, R. (2000). "Citibank increases customer loyalty with defect-free processes." *Journal for Quality and Participation*, 23(4): 32–36.

Samson, D. and Terziovski, M. (1999). "The relationship between total quality management practices and operational performance." *Journal of Operations Management*, 17(4): 393–409.

Schroeder, R. G. et al. (2008). "Six Sigma: Definition and underlying theory." *Journal of Operations Management*, 26(4): 536–554.

Shafer, S. M. and Moeller, S. B. (2012). "The effects of Six Sigma on corporate performance: An empirical investigation." *Journal of Operations Management*, 30(7–8): 521–532.

Silver, E. A. (2004). "Process management instead of operations management." *Manufacturing and Service Operations Management*, 6(4): 273–279.

Sterman, J. D., Repenning, N. P., and Kofman, F. (1997). "Unanticipated side effects of successful quality programs: Exploring a paradox of organizational improvement." *Management Science*, 43(4): 503–521.

Swink, M. and Jacobs, B. W. (2012). "Six Sigma adoption: Operating performance impacts and contextual drivers of success." *Journal of Operations Management*, 30(6): 437–453.

Further Readings

Anderson, S. D. (1999). "Project quality and project managers." *International Journal of Project Management*, 10(3): 138–144.

Antilla, J. (1992). "Standardization of quality management and quality assurance: A project viewpoint." *International Journal of Project Management*, 10(4): 208–212.

Beer, M. (2003). "Why total quality management programs do not persist? The role of management quality and implication for leading a TQM transformation." *Decision Sciences*, 34(4): 623–642.

Biedry, J. (2001). "Linking Six Sigma analysis with human creativity." *Journal of Quality & Participation*, 24(4): 36–38.

Bower, J. L. and Christensen, C. M. (1995). "Disruptive technologies: Catching the wave." *Harvard Business Review*, 73(1): 43–53.

Brown, S. L. and Eisehardt, K. M. (1997). "The art of continuous change: Linking complexity theory and time-paced evolution in relentlessly shifting organizations." *Administrative Science Quarterly*, 42(1): 1–34.

Buch, K. and Tolentino, A. (2006). "Employee perceptions of the rewards associated with Six Sigma." *Journal of Organizational Change Management*, 19(3): 356–364.

Cameron, K. and Barnett, C. (2000). "Organizational quality as a cultural variable: An empirical investigation of quality culture, processes, and outcomes." In Cole, R. E. and Scott, W. (eds.) *The Quality Movement and Organizational Theory.* Thousand Oaks, CA: Sage. pp. 271–294.

Christensen, C. M. (1998). *The Innovator's Dilemma: When New Technologies Cause Great Firms to Fail.* Boston, MA: Harvard Business School Press.

Cicmil, S. (2000). "Quality in project environments: A non-conventional agenda." *International Journal of Quality & Reliability Management*, 17(4/5): 554–570.

Cicmil, S. and Terziowski, M. (1999). "Total quality in project management—Towards an integrated concept." In Sumanth, D., Werther, W. and Takala, J. (eds.) *Productivity and Quality Management Frontiers VIII, The Proceedings of 8th International Conference for Productivity and Quality Research.* June, Vaasa, Sweden.

Fleming, J. H., Coffman, C., and Harter, J. (2005). "Manage your human sigma." *Harvard Business Review,* 83(7/8): 107–114.

Hahn, G. J., Doganakosy, N., and Hoerl, R. (2000). "The evolution of Six Sigma." *Quality Engineering*, 12(3): 317–326.

Herzog, N. V., Polajnar, A., and Tonchia, S. (2007). "Development and validation of business process reengineering (BPR) variables: A survey research in Slovenia companies." *International Journal of Production Research*, 45(24): 5811–5834.

Khang, D. B. and Myint, Y. M. (1999). "Time, cost and quality trade-off in project management: A case study." *International Journal of Project Management*, 17(4): 249–256.

McAdam, R. and Lafferty, B. (2004). "A multilevel case study of Six Sigma: Statistical control or strategic change?" *International Journal of Operations and Production Management*, 24(5): 530–549.

Raisinghani, M. S. et al. (2005). "Six Sigma: Concepts, tools and applications." *Industrial Management & Data Systems*, 105(4): 491–505.

Sodhi, S. M. and Sodhi, N. S. (2005). "Six Sigma pricing." *Harvard Business Review*, 83(5): 135–142.

Somasundaram, S. and Badiru, B. A. (1992). "Project management for successful implementation of continuous quality improvement." *International Journal of Project Management*, 10(2): 89–101.

Turner, J. R. and Cochrane, R. A. (1993). "The goals and methods matrix: Coping with projects for which the goals and/or methods of achieving them are ill-defined." *International Journal of Project Management*, 11(2): 93–101.

Tushman, M. L. and Murmann, J. P. (1998). "Dominant designs, technology cycles, and organizational outcomes." In Staw, B. and Cummings, L. (eds.) *Research in Organizational Behavior*, vol. 20. Greenwich, CT: JAI Press, pp. 231–266.

Westphal, J. D., Gulati, R., and Shortell, S. M. (1997). "Customization or conformity? An institutional perspective on the content and consequences of TQM adoption." *Administrative Science Quarterly*, 42(2): 366–394.

Wilklund, H. and Wilklund, P. S. (2002). "Widening the Six Sigma. An approach to improve organizational learning." *Total Quality Management*, 13(2): 233–239.

York, K. M. and Miree, C. E. (2004). "Causation or covariation: An empirical re-examination of the link between TQM and financial performance." *Journal of Operations Management*, 22(3): 291–311.

Zbaracki, M. J. (1998). "The rhetoric and reality of total quality management." *Administrative Science Quarterly*, 43(3): 602–636.

Zhao, X., Yeun, A. C. L., and Lee, T. S. (2004). "Quality management and organizational context in selected service industries of China." *Journal of Operations Management*, 22(6): 575–587.

Review Questions

1. How can we define Six Sigma?
2. What is the relationship between Six Sigma and process improvement?
3. Some argue that Six Sigma is different than other process improvement programs. Can you provide an argument to support it?

4. Some argue that Six Sigma is the same as other process improvement programs. Can you provide an argument to support it?
5. Explain different types of processes in a firm. Can Six Sigma improve all of them? Why or why not?
6. Does Six Sigma improve firm performance? Why or why not?
7. Does Six Sigma improve innovation? Why or why not?
8. If you want to get the best out of Six Sigma what do you do?
9. In which industries do you think you can implement Six Sigma most effectively? In which industries Six Sigma is least effective?
10. In your opinion, what are the limitations of Six Sigma?

chapter nine

Benchmarking high-performing Six Sigma companies

> Experience is the worst teacher. It always gives the test first and the instruction afterward.
>
> —Anonymous

9.1   Introduction

The aim of any company is to improve its bottom line. Several attempts have been made to improve the company's bottom line by introducing various methods and technologies. One of the most popular methods among all is Six Sigma methodology. Many companies have successfully implemented this methodology whereas others have failed. Various studies have been carried out in an attempt to look for possible causes of failure of the Six Sigma methodology. We first identified a set of Six Sigma companies and categorized them into high performing and low performing based on their stock performances 5 years before Six Sigma implementation, 5 years after implementation, and 10 years after implementation. This study also took into account future worth analysis using the minimum acceptable rate of return. The influence of inflation rate over the rate of return was also considered. The next objective was to investigate and analyze the factors responsible for successful Six Sigma implementation. Based on previous studies, we identify a set of success factors for Six Sigma implementation. We hypothesize that if the company meets this factor, the factor is responsible for the success of the Six Sigma implementation in the company. The proposed results define a model which describes what the high-performing companies do different from the low-performing companies.

9.1.1   Lean Six Sigma criteria testing model

Primary objective of the testing model is to identify a financial variable. This financial variable will be considered as the basis for sorting Six Sigma companies. The types of Six Sigma companies will be defined as high performance companies (HPCs) and low performance companies (LPCs). We seek to answer the question-if we invest $1000 in all the companies considered for the study, which company will provide us the maximum amount of return over a specific time period? The time period will be decided upon when the company implemented Six Sigma.

Once the companies are classified, the next target is to investigate the performance of each company throughout the Six Sigma implementation phase. This process is done by analyzing a set of failure factors. It will be helpful to conclude if that particular factor leads to the failure of a Six Sigma implementation in a company. The hypothesis statement (for each individual failure factor) is that if the company meets this factor, the factor is responsible for the failure of the Six Sigma implementation in the company. An alternate hypothesis is that if the company does not meet that particular factor then there is no negative impact on the company. The assumptions are listed below.

$$A_i = \text{failure factors } A_1, A_2, A_3, \ldots, A_n.$$

$$B_j = \text{companies } B_1, B_2, B_3, \ldots, B_m.$$

where:
  n is the total number of failure factors considered for the study
  M is the total number of companies studied

The null and the alternate hypothesis statement:

  H_0: If the company B_j meets Factor A_i, then A_i has a negative impact on the company B_j
  H_a: If the company B_j does not meet Factor A_i, then A_i has no negative impact on the company B_j

This section gives the details of the steps that need to be followed to obtain the objective model that differentiates the practices performed for high- and low-performing companies. The steps are detailed into phases to provide better clarity of methodology. The steps are listed as follows:

- Obtain a list of companies that have implemented Six Sigma.
- Data availability and finalization of companies for the study
- Data collection

9.1.1.1   Identification of Six Sigma companies

9.1.1.1.1   Obtain a list of companies implementing Six Sigma.   A total of 62 Six Sigma companies were obtained by conducting a complete literature search through journals, research/conference papers, and web blogs and other web sources in addition to the Fortune 500 list. Simultaneously, information was gathered regarding the implementation dates for Six Sigma in these companies (Table 9.1).

9.1.1.1.2   Data availability and finalization of companies for the study.   The data that are needed for this purpose is the stock performance for a particular company extended from 5 years before the date of Six Sigma implementation to 10 years after implementation. This gave a total of 15 years of time span for the study for every company. Data were obtained from an online source (Y charts). Information was unavailable for a few companies during the time span which was required, reducing the number of companies to 43. After carefully looking at our finalized companies, we decided that it would be more sensible to compare company performances sector-wise. The most ideal comparison would be comparison based on industries. But due to the limitation of data, we subjected our comparison sector-wise.

9.1.1.1.3   Data collection.   To evaluate a company's performance, information regarding the stock prices was gathered over a 15-year time span. These values were required to compute necessary variables in further analysis.

9.1.1.1.4   High performance organization.   High performance organization (HPO) concentrates on altering the organizational culture, structure, and work capability in such a

Table 9.1 List of Six Sigma Companies Obtained for the Studies

S. no.	Six Sigma companies	Implementation dates
1	Tenneco Automotive	2001
2	Allied Signal	Late 1993
3	Texas Instruments	Early 1993
4	Bank of America	2001
5	Raytheon	January 1999
6	Merrill Lynch	April 2001
7	Motorola	1986
8	Corning	1994
9	Whirlpool	1997
10	Boeing	1999
11	Dow Chemical	1999
12	Johnson Controls	1999
13	Macy's	2001
14	Terex	2001
15	Circuit City	2001
16	Aon	2002
17	General Cable	2002
18	Plexus	2002
19	Cigna	2003
20	DuPont	1999
21	Caterpillar	2000
22	Coventry Health Care	2003
23	IBM	1990s
24	Kodak	2002
25	General Electric	1995
26	Home Depot	July 1, 2001
27	Honeywell	January 2000
28	3M	December 2003
29	Lockheed Martin	September 2007
30	Xerox	2003
31	Starwood Hotels	2001
32	Westinghouse	Early 1980s
33	Walmart	August 2006
34	Exxon Mobil	May 2009
35	Chevron	2000
36	Conoco Philips	2000
37	Cardinal Health Care	Late 1990s–2004
38	Berkshire Hathaway	June 1998
39	HP	1994
40	JPMorgan CHASE	1998
41	Citigroup	1997
42	McKesson	1999

(*Continued*)

Table 9.1 List of Six Sigma Companies Obtained for the Studies (Continued)

S. no.	Six Sigma companies	Implementation dates
43	P&G	2008
44	Target	2003–2004
45	Johnson & Johnson	1997
46	Medco Health Solution	2005
47	Dell	Early 2000
48	Best Buy	2003–2004
49	Intel	2008
50	Fedex	2006
51	UPS	2003
52	Microsoft	2004
53	American Express	Late 1998
54	Commonwealth Health Corp.	March 1998
55	Asea Brown Bovari	1993
56	EMC	2000
57	Quest Diagnostics	2001
58	Textron	2002
59	Covance	2002

way that the people accounts for more ownership and responsibility. HPO contains a high performance team (HPT) within the organization which aids everyone to participate in it. It mainly focuses in achieving the goals and strategy framed by the organization. HPT is also responsible for continuous improvement or Kaizen. The best part of HPT is that it offers equal chance to each and every person in the team. There is a famous quote of Benjamin Franklin to support this notion, "If we don't all hang together, we shall surely hang separately" (Devane 2004).

HPO can be defined as the structure which consists of many HPTs composed of individual persons who work independently on challenging tasks (Devane 2004). The team members present in the HPTs will be the same throughout. They will be responsible for the outcome for any process, product or services from the company. HPO also consists of the Kaizen team that focuses on continuous improvement.

In the case of HPO strategy, reward, information, and other decisions are downstream in the process or services in such a way that mistakes, error detection, and planning are close to the source. It gives credit to the human resources involved in the HPTs by rewarding them when they achieve specific goals and they also provide pay for the level of work they are performing. Information systems team informs the HPTs about the hierarchical structure and other key details about the team.

9.1.1.1.4.1 Outcomes of HPO. The outcomes achieved by HPO are listed below.

- HPO consists of distinctive HPTs that will be responsible for pushing down strategy and other ideas to the structure.
- Responsibilities are distributed equally to the human resources present in the HPT which enables the level of accountability and expectations of deliverables from each person to increase.

- High focus on the measurement system and performance are the outputs for setting up a high level of expected goal.
- The structure of the organization can rapidly adapt to change in external conditions.
- HPO consists of lesser hierarchical structure than the traditional or low performance organization.
- The total work force understands the concepts and business process strategy better due to the local team goal setting.

9.1.1.1.4.2 Criteria for productive work. To obtain productive work, there are six criteria that need to be followed for achieving it. The employer following the six criteria is said to be more productive. The six criteria are listed below.

1. Decision making is self-governed.
2. Productive work requires continuous study that needs specific goal making and regular feedback about the process or services.
3. There should be high morale within the team. Each and every person in the team should respect everyone else in the team.
4. Proper vision has to be set which will drive the organization to work toward it.
5. There should be difference in the content of work and there should be constant pace for the work.
6. There should be meaning to what the team is performing. Value is vital information for the task performed by the organization.

To analyze the productivity of the work, the above-mentioned questions can be put up and the feedback can be used for corrective performance. HPO allows having high score for each human resource based on the above criteria. These criteria can be handled in four possible ways that are listed below.

1. It is used by top-level management to analyze the usefulness of the HPO.
2. It is used to record the purpose of benchmarking for progress evaluation later.
3. It is used as the diagnostic tool for redesign purposes.
4. It is used when the organization undergoes restructuring during HPT transition (Devane 2004).

9.2   Lean Six Sigma priority evaluation models

The objective is to calculate the performance of a company by obtaining the future value of the investment. From the previous discussion let us assume that we invest an initial amount of $1000 in every company 5 years before Six Sigma implementation. The second investment of $1000 is said to be done at the year during which Six Sigma is implemented. We seek to obtain the future value of the first investment (5 years before Six Sigma implementation) at the year of Six Sigma implementation and the future value of second investment after 5 and 10 years from the year of Six Sigma implementation. The future values are to be calculated based on the expected minimum attractive rate of return (MARR) or the hurdle rate (White et al. 2010). Effects of inflation are also taken into consideration over the study period. Corresponding inflation rates over 5 years before implementation and over 5 and 10 years after implementation are computed. The difference between the future value due to the expected actual rate of return and future value due to inflation would provide us an approximate profit or loss incurred by the company due to Six Sigma implementation.

9.2.1   Estimation of MARR

Principles of Engineering Economic Analysis states that one of the ways to establish MARR is using the rate of return actually achieved over past particular number of years (White et al. 2010, pp. 237, 278–295). The rate of return can be calculated by using a log ratio: ("faculty. washington.edu"). MARR equation is provided in the following equation:

$$R = \ln \frac{P_1}{P_0},$$

(9.1)

where:
  R is the rate of return per year
  P_1 is the closing price of stock on the last day of the year
  P_0 is the closing price of the stock on the first day of the year

The rate of return for every year was calculated using Equation 9.1. The average rate of return was calculated over: (1) 5 years before Six Sigma implementation, (2) 5 years after Six Sigma implementation, and (3) 10 years after Six Sigma implementation.

9.2.2   Discount for inflation

As stated earlier, the effect of inflation on the returns was also taken into consideration. We obtained the annual inflation rates over the 15-year span and calculated the expected inflation rates over: (1) 5 years before Six Sigma implementation, (2) 5 years after Six Sigma implementation, and (3) 10 years after Six Sigma implementation, as corresponding to the MARR values calculated in the previous step. Inflation data is obtained from the web source, Inflationdata.Com (1996–2012).

9.2.3   Calculate performance of each company

The performance of a company can be computed from its future values. We calculate two future values: (1) based on the expected MARR and (2) based on the average inflation rates (Hartman 2007).

$$F = P * e^{(r*N)}.$$

(9.2)

The future values are calculated every 5 years as mentioned earlier: (1) 5 years before Six Sigma implementation, (2) 5 years after Six Sigma implementation, and (3) 10 years after Six Sigma implementation. The difference between the future values obtained from the MARR and the future values based on the inflation rate would provide us with the actual profit/loss incurred. This can be calculated by the following equation:

$$FV_{\text{real return}} - FV_{\text{only inflation}} = \text{actual profit}.$$

(9.3)

Percentage of profit/loss is then computed. It is provided in the following equation:

$$\% \text{ Profit or loss} = \frac{\text{Initial investment} - (\text{profit or loss incurred})}{\text{Initial investment}} * 100.$$

(9.4)

Initial investment in our case is equal to $1000. The profit or loss calculation is done for 5 years before Six Sigma implementation, 5 years after, and 10 years after Six Sigma implementation. The difference (Δ) between the profit/loss incurred 5 years before Six Sigma implementation and the profit/loss incurred after 10 years of Six Sigma implementation would illustrate a clear view to us of the company's performance over those 10 years during which Six Sigma was implemented. This would become the basis for classification of companies as high or low performing.

9.3   Non-LSS projects in the LSS organization—How do we manage non-LSS projects? Evaluation and comparison of practices between high- and low-performing companies

9.3.1   Study factors critical for Six Sigma implementation

There have been several attempts to study which factors affect the success of Six Sigma implementation in a company. These factors are elaborated in various research papers, journal articles, articles in the newspaper, magazines, etc. from which 16 sources were reviewed to investigate which factors affect the successful implementation of Six Sigma. Due to the repetition of factors among the sources, it narrowed the sources to nine articles. These nine articles provided 24 factors which are critical to the success of Six Sigma. To look for identifiable factors, a criterion that a factor should appear in at least 60% of the nine articles were set. With this criterion, six identifiable success factors were gathered.

After the data collection for every company, information regarding all the parameters necessary for the company's performance evaluation from rate of return, to inflation rate, future value calculations based on these rates to the percentage profit or loss are gathered. Table 9.2 illustrates the example of Dow Chemical for all of these performance parameters.

Table 9.2 List of Finalized Six Sigma Companies

S. no.	Finalized companies	Implementation date
Sector	**Financial**	
1	Bank of America	2001
2	American Express	Late 1998/1999
3	AON	2002
4	Citigroup	1997
5	J.P. Morgan Chase	1998
Sector	**Health care**	
6	Conventry Health Care	2003
7	Cigna	2003
8	Covance	2002
9	Quest Diagnostics	2001
Sector	**Conglomerates**	
10	3M	December 2003
11	Dow Chemical	1999

(Continued)

Table 9.2 List of Finalized Six Sigma Companies (Continued)

S. no.	Finalized companies	Implementation date
Sector	**Basic materials**	
12	Chevron	2000
13	Conoco Phillips	2000
Sector	**Consumer goods**	
14	Ford	1999
15	Johnson Controls	1999
16	Tenneco Automotive	2001
17	Whirlpool	1997
18	Xerox	2003
Sector	**Technology**	
19	Corning	1994
20	Dell	Early 2000
21	EMC Corp	2000
22	HP	1994
23	IBM	1990
24	Microsoft	2004
25	Plexus	2002
26	Texas Instruments	Early 1993
Sector	**Services**	
27	Best Buy	2003
28	Cardinal Health	1990s
29	FedEx	2006
30	Home Depot	July 1, 2001
31	Macy's	2001
32	Mckesson	1999
33	Starwood Hotels	2001
34	Target	2003
35	Ups	2003
36	Walmart	August 2006
Sector	**Industrial goods**	
37	Boeing	1999
38	General Cable	2002
39	General Electric	1995
40	Honeywell	2000
41	Raytheon	1999
42	Terex	2001
43	Textron	2002

Similarly, calculations were performed for all the other companies. The sector-wise division of the companies along with their percentage profit/loss and the delta (Δ) values is shown in Tables 9.3 through 9.5. Plots were plotted to clearly show the high-performing companies and the low-performing companies. Figure 9.1 illustrates the graphs of all the companies included in the financial sector. Tables 9.4 and 9.5 and Figure 9.2 help us in classifying the Six Sigma companies into high and low performing.

Table 9.3 Table Illustrating the Rate of Returns, Inflation Rates, Future Values, and % Profit or Loss for DOW Chemical

	Year	Real R%	% Inflation	FV (actual rate)	FV (inflation)	Profit due to SS	% Profit/Loss
			Rate of returns, FV for Dow Chemical				
	1994	0.174155348	0.0261				
	1995	0.05606468	0.0281				
	1996	0.097096122	0.0293				
	1997	0.2557499	0.0234				
	1998	-0.107445142	0.0155				
Average	**1999**	**0.095832539**	**0.02448**	1,614.721824	1130.20609	484.51573	-51.5484
	1999	0.39570868	0.0219				
	2000	-1.28767541	0.0338				
	2001	-0.03003579	0.0283				
	2002	3.531933204	0.0159				
	2003	0.31096406	0.0227				
Average 5 Years +	**2004**	**0.584351455**	**0.02452**	18,573.89828	1130.43216	17,443.4661	1644.3466
	2004	0.185436887	0.0268				
	2005	-0.121073928	0.0339				
	2006	-0.09571401	0.0324				
	2007	-0.012103022	0.0285				
	2008	-0.943098483	0.0385				
10 Years + Average	**2008**	**0.193720472**	**0.02827**	6,939.326478	1326.70709	5,612.6193	461.2619

Table 9.4 List of High-Performing Companies

S. no.	High-performing companies	5 years before	5 years after	10 years after	Delta (Δ)
1	General Cable	1352.3686	1399.872	39,094.43	37,742.0654
2	Tenneco Automotive	−207.4446	344.7258	760.2807	967.7253
3	Cardinal Health Care		190.3228	780.463	590.1402
4	Dow Chemicals	−51.5484	1644.347	461.2619	512.8103
5	Texas Instruments	−85.6872	245.3067	301.6357	387.3229
6	J.P. Morgan Chase	72.7650	34.8286	194.0062	121.2412
7	Quest Diagnostics	−136.37	172.0623	42.0723	178.4424
8	Conoco Phillips	−66.2698	−25.4848	−12.64	53.6298

Table 9.5 List of Low-Performing Companies

S. no.	Low-performing companies	5 years before implementation	5 years after implementation	10 years after implementation	Delta (Δ)
1	Honeywell	10,510.3	−148.43	−158.90	−10,669.20
2	Dell	7,772.413	−135.59	−203.70	−7,976.11
3	Home Depot	4,282.066	3790.92	3217.82	−1,064.25
4	Citigroup	778.6114	−62.37	−95.09	−873.70
5	CIGNA	742.8669	43.75	−66.23	−809.09
6	3M	21.4058	−147.28	−126.90	−148.31
7	Chevron	−22.6943	−91.77	−51.47	−28.78
8	Ford	65.0819	−171.94	−226.73	−291.81

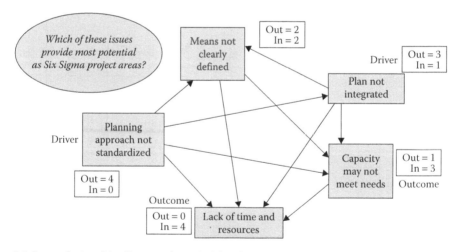

Figure 9.1 Interrelationship diagram for prioritization.

9.3.2   *Examine and compare failure factors for companies under our study*

Two pairs of industries each consisting of a high-performing and a low-performing company were analyzed. Factors that were practiced by the high-performing companies to succeed and factors that the low-performing companies failed to implement which led to the failure of Six Sigma initiatives were examined.

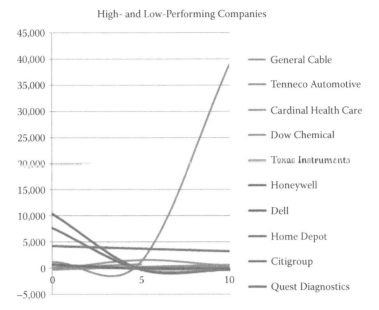

High- and Low-Performing Companies

— General Cable
— Tenneco Automotive
— Cardinal Health Care
— Dow Chemical
— Texas Instruments
— Honeywell
— Dell
— Home Depot
— Citigroup
— Quest Diagnostics

Figure 9.2 Trends in company's performances.

Sources were obtained to study the Six Sigma implementation in these companies. The first pair of companies was classified under Conglomerate industry with Dow Chemical as the high-performing company and 3M as the low-performing company. Five articles supported the implementation of Six Sigma at Dow Chemical and six articles to critique Six Sigma implementation at 3M. The other pair of companies belonged to the service industry. Quest Diagnostics was identified as a high-performing company and Cigna was identified as a low-performing company. Three articles were identified for each of these companies to support the success and failure in implementing Six Sigma.

All the articles relating to the high-performing companies were carefully examined to check if the companies met any of the six success factors. Similarly, a study was conducted for the low-performing companies. Analysis was made whether the low-performing company followed the success factors. On the basis of this study, companies can be distinguished by which factors make them closer to successfully implement Six Sigma initiatives.

The 16 sources which were referred were narrowed down to nine due to repetition of success factors and sources. These nine articles provide the 19 success factors for Six Sigma implementation. These factors were further reduced to six identifiable factors with a criterion that these factors should appear in at least 60% of the articles.

Table 9.6 provides six success factors for Six Sigma implementation. To explain the success and failure of Six Sigma implementation on the basis of these critical factors, we illustrate with an example of high-performing and low-performing companies.

The sources that we found provide information regarding the positive propensity of Dow Chemical toward the critical success factors. All six factors that were listed in Table 9.6 were observed to be fulfilled by Dow Chemical. 3M, which was categorized as a low-performing company, also followed these factors on the basis of the references. Similarly, comparison is made for the implementation of Six Sigma in the service industry on the basis of these factors. The two companies under the service industry identified

Table 9.6 Top Six Success Factors for Six Sigma Implementation

1	Active involvement from the management
2	Acceptance to change/culture
3	Effective communication
4	Effective training/right people
5	Linking Six Sigma to business goals/objectives
6	Developing leadership knowledge

as high- and low-performing companies were Quest Diagnostics and Cigna, respectively. Comparing these factors with the implementation of Six Sigma in Quest Diagnostics and Cigna, it is observed that Quest Diagnostics fulfilled all the six factors while Cigna failed to observe four out of the six. Considering from the first study for the conglomerate industry, one can conclude that except for the factors which were used to shortlist from the sources, there are other factors which contribute to the success of Six Sigma initiative. While failing to observe the success factors for Six Sigma implementation clearly lead to failure of Six Sigma implementation at Cigna.

On observing the sources closely, we found other factors that were critical to the success or failure of Six Sigma implementation in the companies. Table 9.7 shows a list of these factors. Considering that 60% criterion of these factors appear in at least 60% of the sources, we reduced the findings to five factors critical to a successful Six Sigma implementation. These included: (1) no change in CEO during the Six Sigma initiative, (2) identifying the right areas for implementing Six Sigma, (3) appointing a separate infrastructure/body to support quality programs in the company, (4) encouraging team work, and (5) most importantly, continued commitment to achieve a successful Six Sigma initiative.

From the five factors critical to the Six Sigma implementation, failure to follow two factors leads to unsuccessful Six Sigma initiatives in the low-performing companies. The failure factors were: (1) change in CEO during the Six Sigma initiative, and (2) identifying wrong areas of application for Six Sigma implementation.

With respect to Dow Chemical's approach to Six Sigma implementation, Michael Marx in his article on Six Sigma mentions the customer-centric Six Sigma philosophy of Dow chemical (Marx 2005). From the *Supply Chain Management Review*, Tom Gurd's statement on Dow's Six Sigma initiative suggests that the company made an effort to thoroughly train its employees in Six Sigma philosophy. He also mentioned that almost 60% of Dow's employees were exposed to Lean/Six Sigma concepts (Schlegel and Smith 2005).

Table 9.7 List of Other Factors Critical to Six Sigma Implementation

1	No change in CEO during the Six Sigma initiative
2	Areas of application
3	Avoiding aggressive application of Six Sigma
4	Appointing a separate infrastructure and body to support quality in the company
5	Encouraging team work
6	Global documentation
7	Continued commitment

This highlights the efforts of Dow Chemical to expand the Six Sigma campaign enterprise-wide, which is one of the most important factors for the success of Six Sigma. Case studies at Dow Chemical explain to us how the implementation of Six Sigma was carried out for the projects undertaken. We could clearly see that the team followed a measure, analyze, improve, and control (MAIC) methodology. The second case study describes Dow's vision toward Six Sigma, importance of effective communication right from the corporate level to the employees, to the people, customer-driven attitude and acceptance of cultural changes and organization structure (Tannenbaum 2003).

Implementation of Six Sigma at 3M had never been smooth. The company is known for its innovation and creativity linked to its products. When McNerney was announced as the CEO in December 2000, the company's stock jumped nearly 20%. He used the principles of Six Sigma in the company with the aim of lowering costs and increasing efficiency. The Six Sigma methodology was very well implemented at 3M. Thousands of employees were trained in Six Sigma, the company was focused on customer satisfaction, and leadership development was also encouraged. Due to the perfect environment for Six Sigma, the company posted a high rise in stock performance. But this did not last for long. When McNerney left for Boeing, the company began to experience a downfall. McNerney never saw the base of the company which is nothing but innovation. He could not manage the balance between efficiency and innovation. Too much discipline had ceased the creativity in the employees. When the new CEO came in at 3M, he had to go back to focus on the core of the company, that is, encourage innovation in the research laboratories. He made an official company policy to allow employees to use 15% of their time in pursuing independent projects. Defenders of Six Sigma at 3M claim that a more systematic new-product introduction process allows innovations to get to market faster. Other innovators such as Fry, the Post-it note inventor, disagreed with the philosophy of the Six Sigma defenders at 3M. He placed the blame for 3M's recent lack of innovative sizzle squarely on Six Sigma application in 3M research labs (Hindo 2007).

The success of Quest Diagnostics was in identifying the use of Six Sigma in health care. Being one of the early pioneers of Six Sigma in health care, they already had a well-documented progress over a long haul. The sources provided us with information which confirmed that Quest Diagnostics found the right areas/projects in the company where Six Sigma implementation was necessary and thus proved beneficial. They proceeded with customers as their main focus (Quest Diagnostics). The documents support the factor of not implementing Six Sigma as a standalone tool. At Quest they integrated the Six Sigma implementation with Lean tools and Kaizen (Quest Diagnostics).

An employee at Cigna in his article explains how Six Sigma was implemented in the organization, and how it did succeed, but failed within a short time. He brings to light the problems which were first invisible because everything seemed normal as running smoothly in the company. But it was soon when the company realized it needed improvement. It redesigned the Six Sigma initiative with an aim to implement the initiative enterprise-wide (Javinett 2010).

9.3.3   Difference between traditional and high performance organization

High performance organization does not follow the same hierarchical tree structure pattern. The difference between a HPO and a traditional organization is shown in Figure 9.3. The first structure consists of a top level manager who is responsible for managing next level managers. The sub-level managers will be heading separate teams. Sub-level managers will be responsible for reporting the information to the top

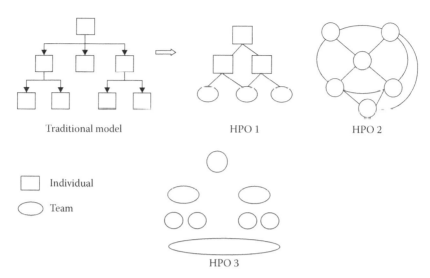

Traditional model   HPO 1   HPO 2

Individual

Team

HPO 3

Figure 9.3 Traditional structure versus high performance organization structure. (Devane, T.: *Integrating Lean Six Sigma and High Performance Organizations Leading the Charge toward Dramatic, Rapid and Sustainable Improvement.* 2004. Copyright Wiley-VCH Verlag GmbH & Co. KGaA. Reproduced with permission.)

level managers. Next type of HPO consists of a network of teams which has a centralized team that is responsible for coordinating with the other teams. Last type of HPO consists of a team at the top position and it also has two management teams which are represented by a medium-sized ellipse. There are various HPTs beneath the medium-sized ellipse. Finally, there is a service team which provides service to all the other teams. It is present at the bottom of the structure and is represented by a long ellipse.

HPT consists of managers who fix goals and pull down information regarding goal setting, budgeting, planning, and scheduling. In the case of a traditional organization, individual performance units are measured by the supervisor. Traditional companies have several layers of management whereas in the case of an HPO it will be 10 to 12 people who are responsible for reporting to the top manager. Let us consider an example in which an organization has a hundred employees and the top manager will not be evaluating the whole 100 reports. Instead, there will be 10 to 12 people who are responsible for reporting the progress of those 100 employees. The policy of HPO is the redundancy to observe business fluctuation as a function. That requires multiple skills in one person than the skills usually he/she possesses. In the case of quality inspection in HPOs, an individual or group will be responsible for the quality inspection whereas in the case of a traditional organization the supervisor or the quality department is responsible for quality inspection.

Groups of tasks are assigned to each team and each team is responsible for the particular task assigned in HPO. In the case of a traditional organization, tasks are assigned to an individual. The individual in the team is responsible or accountable for the task. When the HPO is prevented from multiskilling then group coordination follows. In the case of traditional industry, people are often uncoordinated. Productive work is taken into the consideration and the work structure is designed in such a manner in HPO. Productive work is not accounted for day-to-day interaction in traditional industry. Knowledge of design principles in HPO is known to the people in HPT. In the case of a traditional industry, the principles of design are known only to the few people. Responsibility is one level more than the actual work in traditional organizations. In the case of HPO, it is at the same level (Devane 2004) (Table 9.8).

Table 9.8 Comparison of HPO and Traditional Organization

	High performance organization	Traditional organization
Basic unit of performance	HPT consists of managers who fixes goal and pulls down the information regarding goal setting, budgeting, planning, and scheduling	In traditional organization, the individual performance units is measured by the supervisor
Layers of management	Layer of management will be less. In case of high performance organization it will be 10 to 12 people who are responsible for reporting to the top manager. Let us consider an example that an organization has 100 employees and the top manager will not be evaluating the whole 100 reports instead there will be 10 to 12 people who are responsible for reporting the progress of those 100 employees	Traditional companies have several layer of management
Redundancy to observe business fluctuation	The policy of the high performance organization is that redundancy to observe the business fluctuation is as a function. That is it requires multiple skills in one person than the skills usually he/she possess	Redundancy of parts
Work product development	Groups of tasks are assigned to each teams, each team is responsible for the particular task assigned in HPO	Tasks are assigned to individual. Individual in the team is responsible or accountable for the task
Work quality	In case of quality inspection in HPO individual or group will be responsible for the quality inspection	In case of traditional organization supervisor or the quality department is responsible for quality inspection
Information on productive work	Productive work is taken into the consideration and the work structure is designed in such a manner in HPO	Productive work is not accounted for day to day interaction in traditional industry
Knowledge of high performance organization	Knowledge of design principles in HPO is known to the people in the HPT	In the case of traditional industry, the principles of design are known only to a few people
Coordination and control of work	In case of HPO, it is at the same level	Responsibility is one level more than the actual work

9.4   Lean Six Sigma/no Lean Six Sigma management integration strategy

Some common themes within all established Six Sigma strategies are well analyzed by the true voice of customer (VOC). A majority of the activities of Six Sigma project are carried out by a team assembled for a specific purpose as defined in the project charter.

DMAIC as mentioned earlier is an operational methodology of implementing Six Sigma. But when the decisions at the executive level are required, strategic planning becomes an integral part of executive level management. Initiative is more than simply training everyone on how to use the technical tools. Particularly in executive level managements of health care industry, strategic planning is bound by guidelines and regulations that have been taken into account without altering the present structure of public health and safety. Decisions have to be taken by the executive management with the help of strategic planning. The best option(s) for achieving the objective is selected rather than taking to improving a process. In DMAIC methodology, operational level managements develop a plan for improving the process. The need for a new methodology that can assist the executive management in identifying and prioritizing the goals that are aligned with the mission and vision of the organization can be fulfilled by DEAPS.

It is the opinion of the researchers that the tools of Six Sigma and Lean techniques provide near limitless opportunities for organizations to improve the way they conduct business. The evaluation of applied Six Sigma methodologies apart from DMAIC, new methodologies from DFSS like DEAPS demonstrate the necessity of management initiatives to adapt to meet the dynamic needs of the organization. The strength of the DEAPS approach is its focus on seeking valid information when solid expertise is lacking.

References

Devane, T. (2004). *Integrating Lean Six Sigma and High Performance Organizations Leading the Charge toward Dramatic, Rapid and Sustainable Improvement*. San Francisco, CA: Pfeiffer, an imprint of Wiley.

Hartman, J. C. (ed.) (2007). *Engineering Economy and the Decision Making Process*. New Jersey: Pearson Education Inc.

Hindo, B. (2007). "At 3m, a struggle between efficiency and creativity." *Bloomberg Business Week*. http://www.businessweek.com/magazine/content/07_24/b4038406.htm

Inflationdata.Com. (1996–2012). Capital professional services, LLC. December 16, 2012. http://www.inflationdata.com/.

Javinett, L. (2010). "Cigna's quality system evolution: Ready for the next step." *Process Excellence Network*. July. http://www.processexcellencenetwork.com/process-improvement-case-studies/articles/cigna-s-quality-system-evolution-ready-for-the-nex/.

Marx, M. (2005). *Dow Chemical Company—Six Sigma*. Retrieved from i Six Sigma: http://www.isixsigma.com/industries/chemicals/dow-chemical-company-six-sigma.

Schlegel, G. and Smith, R. (2005). "The next stage of supply chain excellence." *Supply Chain Management Review*. March.

Tannenbaum, K. (2003). "Applying Six Sigma methodology to energy saving projects: Case studies." US Department of Energy: Energy Efficiency and Renewable Energy. http://texasiof.ces.utexas.edu/texasshowcase/pdfs/casestudies/cs_dow_sixsigma.pdf

White, J., Case, K., and Pratt, D. (eds.) (2010). *Principles of Engineering Economic Analysis*. 5th ed. New York: John Wiley & Sons.

Review Questions

Question 1: Which of the following is an acceptable justification for initial Lean Six Sigma projects?

A. Cost savings
B. Customer satisfaction
C. Internal organizational challenges
D. Design improvements
  (a) A B
  (b) B C
  (c) A C D
  (d) A B C D

Appendix 9A: Developing an instrument for measuring Six Sigma implementation

9A.1 Introduction

The emergence of Six Sigma as a robust process improvement methodology has gained momentum in recent years. Six Sigma has been identified as a powerful tool that dramatically improves performance, enhances process capability, and produces bottom-line results for organizations (Dasgupta 2003; Linderman et al. 2003; Pantano et al. 2006). Having been developed from quality management philosophy, Six Sigma has attracted academic research attention in the recent years. According to Hammer (2002), Six Sigma is a project-based methodology that attempts to solve a specific performance problem recognized by an organization.

The influence and effect of Six Sigma initiatives on organizational performance and profitability are numerous. According to Hoerl (1998), GE's operating margins has increased from 13.8% to 14.5%, which is worth about $600 million, stemmed from Six Sigma quality initiatives. In 2002, at least 25% of Fortune 200 companies claimed they have Six Sigma program (Hammer 2002). Organizations are looking toward Six Sigma not only as a way to make themselves more competitive and to save money wherever possible, but also to keep their customer base returning for additional business opportunities. With regard to customer satisfaction, Kondo (2002) found that eliminating flaws, defects, rework, and deficiencies results in greater customer satisfaction. While Six Sigma has been extensively used in industry, the academic research in Six Sigma lags behind its practice in industry (Linderman et al. 2003).

As organizations are looking toward Six Sigma programs, there is a need to address the issue of effective implementation of Six Sigma projects in firms. In an attempt to do so, this chapter focuses on the development of an instrument to determine Six Sigma implementation in firms. We operationalize Six Sigma implementation using the Plan–Do–Check–Act (PDCA) approach. Through careful review of relevant literature we found a need for developing instrument for understanding Six Sigma implementation in firms (Blakeslee 1999; Goh et al. 2003; Harry 1998). However, little empirical study has been conducted on operationalizing Six Sigma implementation in firms.

9A.2   Background and literature review

9A.2.1   Six Sigma methodology

According to Feld and Stone (2002), Six Sigma is a data-driven philosophy used to drive management decisions and actions across an organization. Caulcutt (2001) indicates that Six Sigma reduces waste, increases customer satisfaction, and improves processes with a considerable focus on financially measurable results. For the purpose of this work, Six Sigma is defined as a set of methodologies and techniques used to improve quality and reduce cost, utilizing a structured and disciplined methodology for solving business problems.

A popular framework for implementing a Six Sigma methodology is the define, measure, analyze, improve, and control (DMAIC) process. DMAIC is the key process of a standard framework for a Six Sigma approach and is shown below in Figure 9A.1.

According to Wei-min (2006), the psychology of this approach is that key process input variables (KPIVs) are narrowed down to a vital few with the idea that having control of the vital few will allow for good control of the whole picture. DMAIC is widely used when a product or process is already in existence but performing inadequately. DMAIC focuses on eliminating unproductive steps, developing and applying new metrics, and using technology to drive improvement (De Feo and Barnard 2004). Another popular approach associated with Six Sigma projects, define, measure, analyze, design, and verify (DMADV) is used for developing new products/services (Figure 9A.2). While the focus of DMAIC is on eliminating waste and improving an existing process, DMADV is primarily utilized to develop new products/services for addressing customer needs. While the effectiveness of DMADV as a tool for process design is under question (Hammer 2002), we intended to use both approaches since we believe that provides a better understanding of Six Sigma projects in firms.

Six Sigma has a strategic component aimed at not only developing commitment to it, but also active involvement of higher management. That strategic component is the responsibility of management to identify the key processes of their organization, measure their effectiveness and efficiency, and initiate improvements of the worst-performing processes.

9A.2.2   Literature review

As evidenced by Linderman et al. (2003), academic research in Six Sigma has lagged behind its practice in the industry. While rigorous academic research is needed to fill the gap between theory and practice of Six Sigma little academic research has been carried out to understand the effective implementation of Six Sigma projects in firms.

Most research on Six Sigma practices have been primarily focused on anecdotal evidence and case studies. Academic research on Six Sigma has been accelerated in recent years (Choo et al. 2007a, 2007b; Linderman et al. 2003, 2006; McAdam and Lafferty 2004).

Figure 9A.1 DMIAC framework.

Figure 9A.2 DMADV framework.

While early research on Six Sigma has been focused on the technical side of Six Sigma in terms of tools, techniques, and methodologies, recent studies have paid attention to the psychological, contextual, and human side of Six Sigma.

With reference to Six Sigma implementation as a structured and systematic process improvement methodology, little had been done on the design and structure of Six Sigma as a uniformed methodology. While companies implement Six Sigma in a variety of ways, it is important to determine how much of those efforts are done in a systematic and structured way. Standardization of methodologies reduces variability in the processes so it is crucial for companies to implement Six Sigma in a structured and systematic methodology so that they can benefit the most from their efforts.

To achieve that, we have employed the PDCA framework for Six Sigma implantation. The PDCA cycle is a well-established framework for process improvement.

9A.3   Construct development

Eight constructs were evaluated. These constructs come from the literature of Breyfogle (2003) and Pyzdek (2003) along with other constructs developed by the researchers after observing several Six Sigma organizations. These constructs investigate variations in implementing Six Sigma. However, this is not intended to imply that these cover the only differences in implementations. The recent study on improvement projects (including Six Sigma) shows that the quality of such projects is affected by both methods and psychological variables (Choo et al. 2007a). In another study, the effect of contextual variables on Six Sigma implementation has been addressed (Choo et al. 2007b). Our instrument addresses both the methods aspect of Six Sigma (PDCA) as well as the psychological and organizational aspect (e.g., role of black belts, financial responsibility, and executive commitment). The following is a list with a brief description of the eight constructs evaluated.

Black Belt Roles: Black belts are used differently across organizations. The purpose of this construct is to measure the degree to which the roles of black belts are dedicated to Six Sigma or if they have to split their time between their Six Sigma responsibilities and other day-to-day management activities. The role of black belts in effective implementation of Six Sigma projects is addressed in previous studies (Hammer 2002; Linderman et al. 2003).

DMAIC versus DMADV: These two Six Sigma processes are designed to be used for specific types of projects. The purpose of this construct is to measure if they are used according to their intended use as defined by Breyfogle (2003).

Plan: Deming describes how projects are managed in the following process; PDCA. It is sometimes referred to as a PDCA cycle (Averson 1998). For both DMAIC and DMAVD, this is the "define" step. This construct contains the first steps to start a project such as project selection, project planning, and project scope and metrics. The purpose of this construct is to compare how organizations actually start projects and how DMAIC and DMADV recommend starting projects (Breyfogle 2003; Pyzdek 2003). A higher score indicated a better alignment with Pyzdek and Breyfogle's recommendations.

Do: The purpose of this construct is to measure how companies perform the second step in the PDCA cycle. For both DMAIC and DMAVD, this is the "measure" stage in the Six Sigma process. Experiments and tests are performed to evaluate current performance and improvements on a project. Examples of these tools are design of experiments (DOEs) and failure modes and effect analysis (FMEA) (Breyfogle 2003).

Check: The purpose of this construct is to measure the third step in the PDCA cycle to see if the new ideas will perform as expected. Both DMAIC and DMADV have "analyze" as the third step. The purpose of this construct is to determine if the organizations check the data with statistical tools to determine which sources of variation are critical to the process. Another purpose is to determine if judgments are made on the data before the statistical analyses are performed. A higher score indicates that statistical tests are used to determine critical variations instead of assumptions.

Act. The purpose of this construct is to measure the last step in the PDCA cycle which encompasses the last two steps in the Six Sigma processes. Although they are different in nature, the last two steps are used to set and implement plans to make sure that the changes or new ideas are successful.

Financial responsibility: The purpose of this construct is to measure how leaders of the project are held accountable for the reported benefits as well as the benefits to the leaders when the project goes well. The more accountability and reward for projects, the higher the score.

Executive support: The purpose of this construct is to measure how involved the executives are with the projects. Questions are asked about how often executives are in meetings and the ease of communication. The more often executives are in meetings and the easier the communication is with executives, the higher the score.

9A.4   Methodology

The first major step was to identify a group of organizations for the purpose of conducting an in-depth study of their Six Sigma practices. It was also important to have good access to financial data. Another consideration was the size of the company. The following describes how the companies were selected for this research.

Fortune 500® provides a list of the top 500 companies sorted by gross income (*USA Today* 2005). We initially decided to utilize this list for several reasons. The first is that most of the companies have been around prior to the advent of Six Sigma. This is beneficial for comparing aspects before and after a Six Sigma implementation. Second, larger corporations have a broad range of businesses and so corporate initiatives are better publicized and have a large impact on company performance whereas smaller companies may be more impacted by specific customer demand for a specific product or business unit.

The next step was to compile a list of Fortune 500® companies that have implemented Six Sigma. The website www.sixsigmacompanies.com annually reports a list of these companies (Marx 2005). In 2005 the site reported 180 companies. From this list of 180 companies, 20 companies were chosen randomly for an evaluation of their stock performance and further investigation of their Six Sigma implementation. From the above list, three companies have been selected. In each company, contacts with managers and Six Sigma project manager have been made to identify potential Six Sigma projects. A total number of 29 people have been identified. These are managers who specifically work on Six Sigma projects and have great knowledge and understanding of Six Sigma implementation in their company.

9A.5   Analysis

9A.5.1   Black belt roles

Despite the standardization of Six Sigma implementations, there are different ways to deploy black belts in an organization. Breyfogle (2003) recommends a black belt's only job is to be a black belt. Part of his definition of a black belt is that a black belt's time should not be split between black belt roles and managing a production line, troubleshooting, design work, and other duties that are associated with management or ownership of the

Table 9A.1 Cronbach's Alpha for Black Belt Roles

Cronbach's Alpha	No. of items
0.391	3

product or service the organization provides (e.g., line management at a factory or sales manager). However, some companies do not use this strict definition. Using this definition of black belt roles, six questions were developed by the researchers to determine if a company forces black belts to split time between "regular duties" and black belt duties such as coaching and guiding Six Sigma projects. Pyzdek (2003) refers to black belts not as project managers but as change agents. His vision of a black belt is someone who has extensive training in statistics and Six Sigma methodologies and helps guide multiple teams through Six Sigma projects.

The initial Cronbach's Alpha was 0.146, which is considerably below the target of 0.700. Therefore, a stepwise reduction of questions was used to increase the internal reliability of the construct. The reductions removed one question per step from the construct if eliminating a question raised the Cronbach's Alpha and no more than half of the original questions have been removed.

Table 9A.1 shows the final Cronbach's Alpha. However, a final Cronbach's Alpha of 0.391 indicates the questions need to be reconsidered for use in this or any future study. After reviewing the questions there were several factors that may have lead to the low internal reliability. The first factor could have been the use of the word "only" when describing the roles of a Black Belt in questions 1 and 3. In fact, Black Belts have many responsibilities and therefore the word "only" may have caused confusion in the respondent's answers. The word "only" could have been implied in question 5. Questions 2 and 5 referred to process owners and area managers. These questions could have different meanings between organizations (or divisions within an organization). Question 6 was the only one that had consistency between respondents within organizations. Last, the construct started with only six questions which does not allow for very many questions to be eliminated.

9A.5.2   DMAIC versus DMADV

The purpose of the DMAIC versus DMADV construct was intended to measure the usage of these two common Six Sigma processes. Questions were asked to determine if companies were using DMAIC and DMADV according to Simon's (2000) definitions. DMAIC is used to improve current processes or product whereas DMADV is used to develop new processes or products. The same method of question reduction was used but only a summary of the reductions will be shown. Table 9A.2 shows the reduction summary.

Table 9A.2 DMAIC versus DMADV Summary of Reductions

Step	Initial Alpha	Question to eliminate	New Alpha
1	0.324	13	0.377
2	0.377	15	0.43
3	0.43	12	0.523
4	0.523	14	0.587
5	0.587	16	0.676
Final Alpha	0.676		

The final Cronbach's Alpha was 0.676 which is considered marginally acceptable. After reviewing the questions it was noticed that all of the questions asked negatively were removed. Some of the questions asked if current processes or products were improved using DMADV steps and the others asked if new processes or products were improved using DMAIC steps, which is the opposite of how Simon (2000) recommends using DMAIC and DMADV. The responses tended to show inconsistent responses within companies. This could indicate that some practitioner noticed the difference in the use of DMAIC and DMADV and others did not.

9A.5.3   Plan

The next set of questions focused on the planning stage of projects. Pyzdek (2003) and Breyfogle (2003) both recommend that meetings are held with executives, accountants guide financial reasoning, executive direction takes precedence over manager's interests, metrics are clearly defined, and the scope of the project is narrowed for both DMAIC and DMADV. These concepts were tested using the questions listed in Table 9A.3 shows Cronbach's Alpha reduction.

The final alpha calculated was 0.640. This indicated marginal acceptability for this construct. Again, the reductions eliminated most of the questions asked negatively. After reviewing the questions, it was apparent that some of the same confusing wording was used again. Questions 19 and 24 used the word "mostly" when describing the reason for making a decision. This term indicated that the majority of a decision was based on one specific factor. The intent of the question was to determine if the factor was used and therefore the word "mostly" could have been removed from the questions. Question 22 specifies that managers decide on the projects to work on which could be interpreted in many ways. There are managers at all levels and the intent of the question was to see if managers specific to the process or product decide on the projects (e.g., the sales manager decides on the projects that are in the sales department). Question 23 asked about when accountants are used in the project selection and uses the word "only." This question may have been better without it. Question 25 could be confusing because it is difficult to measure current performance without having a predetermined metric.

9A.5.4   Do

The "do" step of the PDCA cycle is considered as the Measure stage in both DMAIC and DMADV terminology. This is where KPIVs, key process output variables (KPOVs), current performance, and other aspects of a product or process are measured (Breyfogle 2003). Table 9A.4 shows the final Cronbach's Alpha.

Table 9A.3 Final Cronbach's Alpha for Plan

Step	Initial Alpha	Question to eliminate	New Alpha
1	0.296	23	0.408
2	0.408	25	0.459
3	0.459	24	0.524
4	0.524	22	0.588
5	0.588	19	0.64
Final Alpha	0.64		

Table 9A.4 Final Cronbach's Alpha for Do

Step	Initial Alpha	Question to eliminate	New Alpha
1	0.178	35	0.399
2	0.399	34	0.464
3	0.464	32	0.478
4	0.478	30	0.548
5	0.548	33	0.610
Final Alpha	0.610		

This construct tested to be marginally acceptable with a final Cronbach's Alpha of 0.610. After reviewing the questions, some of the same confusing wording was used again. Questions 30 and 35 concern the precision of measurement instruments being 10 and 2 times the specification tolerance, respectively. These questions were not mutually exclusive because if a company required 10 times the specification tolerance for measurement devices it would not disqualify the statement of Question 35 (at least 2 times the specification tolerance). Question 32 asked if input variables are known in the define stage. This question is intended to be false because they are supposed to be found during the "measure" phase but sometimes the inputs are known before this stage. Question 33 uses the word "assumed" to be the reason for believing the accuracy of a testing device. This could be interpreted as none of the measuring devices are checked before use. But there are instances where devices are not easily checked. Question 34 asked if key input and output variables are determined in the Analyze phase. According to Breyfogle (2003), they are supposed to be found in the Measure phase.

9A.5.5   Check

The "check" stage of the PDCA cycle is seen in both DMAIC and DMADV as Analyze. This is the step where all the data taken from the measure phase is analyzed and inferences are made from the data. The questions used in this construct are to measure whether this stage is performed properly by the organization or not. Cronbach's Alpha reduction summary is shown in Table 9A.5.

Check's internal reliability tested to be marginally acceptable with a final Cronbach's Alpha of 0.681. After reviewing the questions, there were some that had confusing wording. Question 38 asked if root causes "or critical customer requirements" are determined in the *analysis* phase. This could have caused consistency issues because critical customer requirements should have already been found during the *define* stage. The "or critical customer requirements" statement should be removed from the question. The answer to

Table 9A.5 Cronbach's Alpha for Check

Step	Initial Alpha	Question to eliminate	New Alpha
1	0.532	44	0.586
2	0.586	43	0.603
3	0.603	45	0.633
4	0.633	38	0.638
5	0.658	41	0.681
Final Alpha	0.681		

Question 41 could change depending on the software package used to perform the ANOVA and therefore could cause variation in the responses within a company. Question 43 states, "or critical customer requirements" which should have already been determined in a prior step which could have led to variations in the responses. Question 44 asked about "gaps and opportunities" which is performed in the *analyze* phase. However, sometimes it is performed by the selection committee and not a well-known step to all involved in a project. So green belts might not be aware of this step and thus could have caused variation in the responses. Question 45 states, "all sources of variation need to be lowered." This could be interpreted in two unintended ways. First, the variation is typically what is lowered, not the source of variation. Second, generally sources of variation that are apparent have statistical significance. It could be perceived that "all" sources of variation need to be fixed.

9A.5.6   Act

The last step of the PDCA cycle is "Act." The implementation or deployment of a project occurs in this phase. This construct was intended to measure what companies do in this phase. Table 9A.6 shows the reduction of questions. The reductions eliminated the negatively asked questions and the final Alpha is marginally acceptable.

This construct tested to be marginally acceptable with a final Cronbach's Alpha of 0.640. After reviewing the questions there was confusing language used that could have led to lower than needed internal reliability. Questions 52 and 53 asked if it is necessary to track "its properties" after a project is completed. This could be interpreted as the project's properties or the source of the variation's properties. Question 54 asked if the project leader decides how to transfer the ownership of the project back to the process owner. There should be a standard procedure for the transfer but the project leader typically has the responsibility to make the transfer happen. Question 55 asked if changes are kept until the root causes of the problems have disappeared. This could cause variation because there are other reasons for the changes to be stopped. Question 56 was also eliminated; however, the cause for variation in this problem is unknown as a response plan in a standard step in Six Sigma projects.

9A.5.7   Financial responsibility

Six Sigma has many aspects outside of its process methodologies. One aspect not mentioned often in literature is how the financial impacts of projects are reported and audited. Some companies have internal audits to determine if the project actually met its predicted savings while other companies do not. This construct is not specific to Six Sigma and can be used on nonSix Sigma companies as well. Table 9A.7 shows the reductions and additional reductions only showed marginal improvement.

Table 9A.6 Cronbach's Alpha for Act

Step	Initial Alpha	Question to eliminate	New Alpha
1	−0.011	56	0.421
2	0.421	55	0.53
3	0.53	54	0.567
4	0.567	53	0.61
5	0.61	52	0.64
Final Alpha	0.64		

Table 9A.7 Cronbach's Alpha for Financial Responsibility

Step	Initial Alpha	Question to eliminate	New Alpha
1	0.700	59	0.783
2	0.783	63	0.811
3	0.811	66	0.832
4	0.832	64	0.843
Final Alpha	0.843		

Financial responsibility was considered internally reliable with a final Cronbach's Alpha of 0.843. Its initial Cronbach's Alpha was considered acceptable at 0.700 but Questions 59, 63, 64, and 66 were removed to shorten the survey. Any further reduction would have lowered the final Cronbach's Alpha.

9A.5.8   Executive support

The last construct is one that textbooks and consultant groups emphasize. Executive support is deemed critical to the success of any program regardless if it is Six Sigma related or not. Table 9A.8 shows the reductions.

Executive support was considered internally reliable with a final Cronbach's Alpha of 0.851. Its initial Cronbach's Alpha was 0.544. Only question 76 needed to be deleted to make the construct internally reliable. The question asked if executives question the "true" benefits of the projects. This question would have been more appropriate without the word "true." Questions 70, 72, 74, and 75 were removed to shorten the survey. In addition, they increased Cronbach's Alpha. Any further reduction would have lowered the final Cronbach's Alpha.

9A.5.9   Final reduced questionnaire

By using Cronbach's Alpha to validate each construct, a master list of questions that can be used for future research has been developed. Table 9A.9 shows the remaining questions and the final Cronbach's Alpha for each construct.

9A.6   Summary and conclusions

The eight constructs tested were intended to evaluate different aspects of a Six Sigma implementation. After the iterative reductions, all but one of the constructs tested to be at least marginally reliable. With further refinement, the questionnaire developed could be used to determine characteristic differences between Six Sigma implementations of companies.

Table 9A.8 Cronbach's Alpha for Executive Support

Step	Initial Alpha	Question to eliminate	New Alpha
1	0.544	76	0.751
2	0.751	70	0.795
3	0.795	72	0.838
4	0.838	74	0.851
Final Alpha	0.851		

Table 9A.9 Final Questionnaire Summary

Construct	Questions	Cronbach's Alpha
Black Belt Roles	2, 3, 6	0.391
DMAIC vs. DMADV	7, 8, 9, 10, 11	0.676
Plan	17, 18, 20, 21, 26	0.640
Do	27, 20, 29, 31, 36	0.610
Check	37, 39, 40, 42, 46	0.681
Act	47, 48, 49, 50, 51	0.640
Financial Responsibility	57, 58, 60, 61, 62, 65	0.843
Executive Support	67, 68, 69, 71, 73, 75	0.851

The process used to calculate the differences in the constructs is given by Nunnaly (1978). He describes the mean discriminal process (MDP), which is basically the arithmetic mean, and how it is used to analyze the differences between scaled data. Even though the individual ranges for responses overlap, Nunnaly recommends the MDP of the construct's responses to calculate a response to the stimuli. Table 9A.10 shows that there were differences between the two companies. A Walsh test was then conducted on the differences between the companies to determine if the companies responded in the same manner to the questionnaire (Siegel 1956). The Walsh test showed the companies responded significantly different at an alpha of 0.086. This is the highest certainty the test could show given the number of constructs and sample size. This rejects the null hypothesis of *H2* that there is no difference in Six Sigma implementations between the two companies.

The construct with the biggest difference was black belt roles. The results show that company B splits up black belts' time between black belt roles and other responsibilities. Another construct that showed a difference was financial responsibility. Company A appears to hold its employees more accountable for their financial reports than company B. Finally, it appears that company A uses DMAIC and DMADV much closer to how it is recommended by Pyzdek (2003) and Breyfogle (2003) than company B.

Even though the black belt role construct did not have a high internal reliability, it is clear that there are major differences in how the two studied companies studied the use of black belts within their organizations. In discussions with company A, they stated employees are pulled from management or engineering and given a 2-year rotation in a black belt position. During this time they are removed from their previous roles and given the responsibility to coach and lead Six Sigma projects within their division. Question 6 showed the questionnaire tested for this by asking, "A black belt must split

Table 9A.10 MDP Comparison of Companies A and B

	Company A	Company B	Difference
BB Roles	3.52	2.47	1.06
Financial	3.67	2.70	0.97
IC vs. DV	4.37	3.68	0.69
Executive	3.48	2.97	0.50
Act	4.27	3.88	0.39
Plan	3.53	3.32	0.21
Do	3.68	3.58	0.10
Check	4.14	4.44	−0.30

their time between black belt roles and department or area management." In discussions with company B, they stated black belts are trained and then encouraged to implement Six Sigma into their day-to-day management responsibilities.

The financial responsibility construct had the next largest difference between companies. This could indicate that company A holds their leaders more responsible for their financial reporting. Company A's responses indicated that there are audits that check the financial reports and that leaders are held responsible if a project did not realize its financial benefits.

DMAIC versus DMADV also had differences between the companies' responses. After looking at the raw data, it could be interpreted that company A uses the two processes more closely to Simon's (2000) recommendations.

The last construct that showed a difference between the companies is executive support. The primary questions that showed differences involved the speed in which messages were returned by executives and the active involvement with executive management.

This questionnaire, however internally reliable it is, showed the differences in the companies' respective Six Sigma programs. This was the original purpose of the questionnaire. Therefore, the questionnaire may need to be re-worded before it is used again, but the premise of the constructs and questions did show measurable differences between the two companies.

9A.6.1   General findings

The final part of this research aimed to relate how the company performances relate to the responses given by Six Sigma practitioners. Unfortunately, only two companies volunteered for this study, therefore a complete analysis cannot be performed on the data relating the questionnaire responses to the success metric. But there were noticeable differences in the responses from the two companies.

Without more companies' participation it is not possible to mathematically show relationships between these differences in the constructs and the success metric. However, the data suggested there are differences in how Six Sigma is practiced and there were differences in success after their Six Sigma implementations. This could lead to finding relationships between implementation techniques and performance.

In conclusion, companies A and B had differences in their success and in how they implemented Six Sigma. The differences found in their Six Sigma programs could have been responsible for the differences in performance. Assuming that the differences in the constructs are responsible for the performance differences, companies should have dedicated black belts, hold project leaders responsible for their financial reporting, and use DMAIC and DMADV according to their intended use.

Assuming the implementations of Six Sigma caused the changes in performances and that the differences in implementations caused the differences in performance, several recommendations can be made.

1. Black belts should be dedicated to the Six Sigma program and not forced to split their time between their Six Sigma responsibilities and other management responsibilities.
2. Employees should to be held responsible for their financial impacts by either by positive or negative reinforcement.
3. Companies should use DMAIC and DMADV, according to Simon (2000), who indicated that DMAIC should be used for improving and DMADV should be used for developing.
4. Executives should be actively involved with projects.

9A.6.2 Limitations

There are some critical assumptions that need to be mentioned. The assumption of validity when comparing the responses to the questionnaire may have been violated. Only three of the constructs had a final Cronbach's Alpha higher than the necessary 0.700 to be considered valid. Four of the constructs are considered marginally valid because their final Cronbach's Alpha was above 0.600. One construct was not considered valid because it was well below 0.600.

In addition to marginal validity, most of the questions eliminated were asked negatively. This indicates that the questions asked in the negative either did not have the intended opposite meaning, were too confusing, or otherwise constructed in a way that invalidated the questions.

This research has many useful concepts however it does have several limitations. In addition, the lower than needed internal reliability of the constructs indicate the questionnaire needs further refinement. However, on face value, three of the constructs have differences between the two companies and two of the constructs show similarities. This indicates that there are variations between the characteristics of the two programs and they could be related to the success metric. With the addition of several other companies and the elimination of less meaningful or unreliable constructs, the methodology provided in this research could show which areas can the companies improve in their Six Sigma program.

9A.6.3 Future research

This research has built the foundation for future study of Six Sigma within companies. Future research should include refinement of the questionnaire, particularly with the roles of black belts within companies. In addition, with the participation of more companies there is the possibility of finding relationships between the performance of the companies and how they implement Six Sigma.

9A.6.4 Implications for managers

The implications of this research are that it is possible to measure the inputs and outputs of a Six Sigma implementation. For researchers, this methodology of measuring the implementation techniques and relating them to company performance could be applied to other types of management methodologies such as Lean, Kaizen, or total quality management (TQM). If future results show relationships between implementation patterns and performance, it could help practitioners implement Six Sigma with a higher likelihood of company success.

References

Blakeslee, J. (1999). "Implementing the Six Sigma solution: How to achieve quantum leaps in quality and competitiveness?" *Quality Progress*, July, pp. 77–85.

Breyfogle, F. W. III. (2003). *Implementing Six Sigma*. 2nd edn. New Jersey: John Wiley & Sons.

Choo, A. S., Linderman, K. W., and Schroeder, R. G. (2007a). "Method and context perspectives on learning and knowledge creation in quality management." *Journal of Operations Management*, 25: 918–931.

Choo, A. S., Linderman, K. W., and Schroeder, R. G. (2007b). "Method and psychological effects on learning behaviors and knowledge creation in quality improvement projects." *Management Science*, 53(3): 437–450.

Dasgupta, T. (2003). "Using the Six Sigma metric to measure and improve the performance of a supply chain." *Total Quality Management*, 14(3): 355–366.

Goh, T. N. et al. (2003). "Impact of Six Sigma implementation on stock price performance." *Total Quality Management & Business Excellence*, September, pp. 753–763.

Hammer, M. (2002). "Process management and the future of Six Sigma." *MIT Sloan Management Review*, 43(2): 26–32.

Harry, M. (1998). "Six Sigma: A breakthrough strategy for profitability." *Quality Progress*, May, pp. 60–64.

Hoerl, R. W. (1998). "Six Sigma and the future of quality profession." *Quality Progress*, June, pp. 35–42.

Kondo, Y. (2002). "Customer satisfaction: How can I measure it?" *Total Quality Management*, 12(7/8): 867–872.

Linderman, K. et al. (2003). "Six Sigma: A goal theoretic perspective." *Journal of Operations Management*, 21: 193–203.

Linderman, K. W., Schroeder, R. G., and Choo, A. S. (2006). "Six Sigma: The role of goals in improvement teams." *Journal of Operations Management*, 24: 779–790.

McAdam, R. and Lafferty, B. (2004). "A multilevel case study critique of Six Sigma: Statistical control or strategic change?" *International Journal of Operations and Production Management*, 24(5): 530–549.

Pantano, V., O'Kane, P., and Smith, K. (2006). "Cluster-based Six Sigma development in small and medium sized enterprises." *Proceedings of 2006 IEEE International Conference on Management of Innovation and Technology*, June 21–23, Singapore.

Pyzdek, T. (2003). *The Six Sigma Handbook: A Complete Guide for Green Belts, Black Belts, and Managers at All Levels*. New York: McGraw-Hill.

Further Readings

Buch, K. and Tolentino, A. (2006). "Employee perceptions of the rewards associated with Six Sigma." *Journal of Organizational Change Management*, 19(3): 356–364.

Byrne, G., Lubowe, D., and Blitz, A. (2007). "Using a Lean Six Sigma approach to drive innovation." *Strategy & Leadership*, 35(2): 5–10.

Fleming, J. H., Coffman, C., and Harter, J. (2005). "Manage your human sigma." *Harvard Business Review*, 83(7/8): 107–114.

Gottfredson, M. and Aspinall, K. (2005). "Innovation versus complexity: What is too much of a good thing?" *Harvard Business Review*, 83(11): 62–71.

Raisinghani, M. S. (2005). "Six Sigma: Concepts, tools and applications." *Industrial Management & Data Systems*, 105(4): 491–505.

Sodhi, S. M. and Sodhi, N. S. (2005). "Six Sigma pricing." *Harvard Business Review*, 83(5): 135–142.

Wilklund, H. and Wilklund, P. I. (2002). "Widening the Six Sigma: An approach to improve organizational learning." *Total Quality Management*, 13(2): 233–239.

Yang, H. M. et al. (2007). "Supply chain management Six Sigma: A management innovation methodology at the Samsung Group." *Supply Chain Management: An International Journal*, 12(2): 88–95.

*Appendix 9B: Six Sigma deployment success according to shareholder value**

9B.1 Introduction

Companies often look for a "recipe for success" with the latest management initiatives such as Lean, total quality management (TQM), and business process reengineering. With these initiatives, companies hope to achieve success in their industry as well as success when the stock market is down. There are several ways by which companies measure success, such as profit, price of stock, and other various financial metrics. The impact on financials is a critical evaluation of the accomplishment of the management initiative. Six Sigma is the most popular

---

* By Erick C. Jones and Angela Garza.

management strategy that affects the quality of a company's product. Although companies usually take on these management strategies to benefit themselves financially, shareholders are also affected by these management initiatives. Shareholder value is commonly known as wealth the company brings to the shareholder. A way to measure this wealth is to evaluate the return on equity (ROE), which measures a "corporation's profitability by revealing how much profit a company generates with the money the shareholders have invested" (Return on Equity 2009). Shareholders are interested to know what the profit of their money invested is. In a recent article in the *Wall Street Journal* titled "Where process—Improvement project go wrong," Dr. Chakravorty explores the reasons behind Six Sigma program failures (Chakravorty 2010). An explanation is needed for why some companies are successful in their Six Sigma implementation while others are not. A similar study of the evaluation of a company's success was conducted by Jim Collins in his management book *Good to Great*.

9B.1.1   Good to Great

In *Good to Great*, several companies were researched to identify the characteristics that help a company transition from short-lived success to long-term results. Jim Collins identifies seven characteristics demonstrated by the companies that achieve long-term results: Level 5 leadership, first who … then what, confront the brutal facts, the hedgehog concept, a culture of discipline, technology accelerations, and the flywheel and the doom loop (Collins 2001). These characteristics are common attributes that Collins found in the companies that shifted from good companies to great companies. Each quality was determined through extensive research and interviews from key executives of the companies.

9B.1.2   Six Sigma

Six Sigma has become a popular management strategy that has provided success for some companies and failure for others. According to Roger Schroeder, who compared field data to literature, Six Sigma is defined as "an organized, parallel—mesostructure to reduce variation in organizational processes by using improvement specialists, a structured method, and performance metrics with the aim of achieving strategic objectives" (Schroeder et al. 2008). The Six Sigma initiative is a customer-focused strategy that centers on improving quality while lowering costs. The methodology followed varies depending on the undertaken project, but the standard operational Six Sigma methodology represents the five phases of Six Sigma: define, measure, analyze, improve, and control (DMAIC). Six Sigma takes a process that is operating a three sigma level and increases a process to function at a much higher level by reducing defects to 3.4 defects per million opportunities, which in turn reduces the cost of poor quality. The most important benefit of this strategy is the impact Six Sigma has on financial services. A typical Six Sigma project saves a minimum of $100,000. These projects last no more than 120 days; however, several of these projects can be carried out for a greater length of time. By improving the cost of quality to a Six Sigma level, revenue is no longer wasted on nonvalue-added projects, which drives profits to increase (George 2002). Researchers suggest that generally companies that incorporate Six Sigma in their management strategy typically see benefits within the first year of implementation. The financial success of Six Sigma does not begin to materialize until after 18 months have elapsed in continual improvement implementation. While an aggressive deployment sees benefits in 6 to 12 months, most companies realize benefits somewhere between 6 and 18 months (Snicker 2004).

9B.1.3   Lean Six Sigma

Lean methodologies have been around much longer that Six Sigma. This approach was heavily used by Henry Ford and has roots in the Toyota Production System (Holweg 2007). Lean concepts focus on eliminating waste and have shown significant success in manufacturing. The Lean philosophy consists of many concepts such as 5s (sorting, straighten, sweeping, standardizing and sustaining) and just-in-time (JIT) production. As Six Sigma has evolved, many companies have added a Lean aspect to their management strategy in this Lean Six Sigma. The combination of Lean and Six Sigma leads to a common principle "the activities that cause the customer's critical-to-quality issues and create the longest time in delays in any process offer the greatest opportunity for improvement in cost, quality, capital and lead time" (George 2002). The methodology of Lean Six Sigma is the same as Operational Six Sigma, but it incorporates Lean tools as well.

9B.1.4   Use of Lean Six Sigma for company growth and compliance

Six Sigma's focus is very similar to the criteria of top quality awards and ISO standards. One such quality award is known as the Malcolm Baldrige National Quality Award (MBNQA). The MBNQA is "given by the president of the United States to businesses [...] that apply and are judged to be outstanding in several areas: leadership; strategic planning; customer and market focus; measurement analysis; and knowledge management; workforce focus; process management; and results" (Frequently asked questions about Malcolm Baldrige International Award 2009). The award focuses on the different perspectives of quality that make a business successful. To qualify for this award is almost as prestigious as winning the award. Using the Baldrige model coupled with Six Sigma "helps firms achieve higher levels of performance and customer satisfaction in each Baldrige category" (Mellat–Parast et al. 2007).

Companies utilize Six Sigma as a strategy to explore business and meet future required standards such as ISO 9000, specifically ISO 9000:2008, and ISO 14000, which are two important quality standards. These standards are not required by governments, but adhering to these standards and registering for certification demonstrate a company's need for excellent quality. Moreover, certification of these standards is recognized internationally. The ISO 9000 series consists of five specifications "designed to ensure all aspects of a business affecting product or service quality are addressed in the organization's quality program, and they are formally structured and documented in auditable form" (Cater and Pasqualone 1995).

The ISO 14000 standard focuses on regulating as well as improving the environmental impact of the processes of a company and setting environmental goals that can be achieved by approaching the objectives systematically (ISO 14000 essentials). Similar to quality management techniques, it is important to strategically plan how processes can be environmentally friendly. The benefits of ISO 14000 include "reduced cost of waste management, savings in consumption of energy and materials, lower distribution costs, improved corporate image among regulators, customers and the public, and framework for continual improvement and environmental performance" (Business benefits of ISO 14000, 2009). Not only will registering for the standard assist in the improvement of quality management, but it will show other companies, as well as customers, that the quality of a product is affected by environmental factors as well.

9B.1.5 Previous study

An initial study was conducted previously with the team at the University of Nebraska–Lincoln. This study sought to determine some of the underlying factors that affect a company's deployment of Six Sigma and determine long-term success. The approximate implementation dates were determined through articles and book references. The study showed that Six Sigma affects stock market price value as opposed to sharp declines seen by the general market during the period studied. Conclusions can be drawn that Six Sigma companies withstand financial stresses on the market. Six Sigma companies seemed to have fared much better than their counterpart companies that do not deploy this initiative; these comparison companies saw a much smaller increase or even larger decreases in stock prices during the same period of study due to lack of management strategies to aid their financial situations (Garza and Jones 2010).

For companies that implemented Six Sigma, there existed a financial difference between companies that have been successful or unsuccessful in their deployment. Sixty-four percent of companies deemed "successful" in our study, saw profits increase over a 100% after the implementation of Six Sigma. Many of these companies were experiencing declining profits prior to implementation, and they completely reversed the decline with Six Sigma deployment. Thirty-six percent of unsuccessful Six Sigma companies showed a significant decrease in profits, and 64% of unsuccessful companies showed only a minor decrease in profits. Additionally 81% of these unsuccessful companies had demonstrated an increase in profits prior to the addition of Six Sigma. The research determined that three main characteristics impacted performance: quality awards and standards (MBNQA criteria, ISO 9000 and ISO 14000 certifications), mission statement, inclusion of Six Sigma stated core values, and improving quality. In this study, we build upon previous research by evaluating the financial impacts of the aforementioned successful and unsuccessful companies (Garza and Jones 2010).

9B.2 Study methodology

The current study evaluates different industries to determine if ROE and earnings per share (EPS) are a good metric for evaluating a company's success with Six Sigma in response to shareholder value.

9B.2.1 Population sample (representative companies)

To evaluate the effects of Six Sigma deployment, 75 companies are to be chosen from Fortune 500, an annual list compiled and published by *Fortune* of the top 500 companies in America. To contrast companies that are successful, unsuccessful, and a control group, this study identified three companies from each industry: two companies that implemented Six Sigma and one that did not. The 10 industries currently being utilized are shown in Table 9B.1.

The implementation dates were found in order to determine a period for studying the effects of Six Sigma deployment. These dates are the approximate dates of implementation found in different articles discussing their implementation. Because the financial benefits of Six Sigma are usually realized after 3 years, a period of 5 years after implementation will be studied to analyze the long-term financial benefits. The implementation dates of the Six Sigma companies that are currently in the study are in Table 9B.2.

Table 9B.1 Companies

Industry	Six Sigma company 1	Six Sigma company 2	Comparison companies
Diversified financials	General Electric	Aon	Fannie Mae
Health care: insurance and managed care	CIGNA	Coventry Health Care	Wellpoint
Aerospace and defense	Boeing	Honeywell	United Technologies
Chemicals	Dow Chemical	Dupont	PPG Industries
Construction and farm machinery	Caterpillar	AGCO	Terex
Motor vehicles and parts	Johnson Controls	Ford	Navistar Intl.
General merchandisers	Macy's	Target	J.C. Penny Corp.
Network and other communications equipment	Motorola	Corning	Harris Corp.
Specialty retailers	Home Depot	Circuit City	Costco Wholesale
Electronics, electrical equipment	General Cable	Whirlpool	Harman Intl. Industries

Table 9B.2 Implementation Dates of Current Six Sigma Companies in Study in Ascending Order

Company	Implementation date	Company	Implementation date
Motorola	1986	Circuit City	2001
Corning	1994	AON	2002
General Electric	1995	General Cable	2002
Whirlpool	1997	Plexus	2002
Boeing	1999	CIGNA	2003
Dow Chemical	1999	DuPont	1999
Johnson Controls	1999	Caterpillar	2000
Macy's	2001	Ford	2000
Home Depot	2001	Coventry Health Care	2003
Terex	2001	Target	2004

Implementation dates range from 1986 (date of creation of Six Sigma) to 2004. No companies which implemented Six Sigma after 2004 were studied because a 5-year period of study after implementation would be possible.

9B.2.2 Measurement criteria

To determine the financial impact, we will analyze the differences in the financial data of the companies in the study. We utilized the metric ROE, which is measured by using earnings after taxes divided by the shareholder equity. We chose this metric because it represents shareholder value and can easily be found for public companies. We utilize three different ROE values for each company in our investigation: ROE at implementation, ROE 5 years prior to implementation, and ROE 5 years after implementation. We will use two study periods to demonstrate pre- and postSix Sigma implementation. We determine the inflection points by evaluating the ROE values. An example of a two-study period can be found in Figure 9B.1.

Figure 9B.1 Two study periods if implementation date is 2000.

In order to determine whether a difference exists between the financial results of companies that implemented Six Sigma and companies that did not, a one-sided, paired two sample for means t-test is constructed at an alpha level of 0.05. We have determined an initial set of successful, unsuccessful, and comparison companies, which is shown in Table 9B.3.

The difference between the rates of change over study period 1 and over study period 2 of each Six Sigma company chosen and the difference between the rates of change over study period 1 and over study period 2 of the comparison company will be found in order to construct the one-sided paired two sample for means t-test. A one-sided t-test was chosen to show that the average difference of ROE rates of change of Six Sigma companies is significantly greater than those of companies that do not implement Six Sigma. Results of this t-test should demonstrate a greater impact on Six Sigma companies' ROE. This statistical test has the following hypothesis:

$$H_0 : u_s \leq u_n,$$

$$H_a : u_s > u_n,$$

where:

u_s is the average difference of the rate of change of ROE over study period 1 and the rate of change of ROE over study period 2 of Six Sigma companies

u_n is the average difference of the rate of change of ROE over study period 1 and the rate of change of ROE over study period 2 of comparison companies

Table 9B.3 Six Sigma Companies with Comparison Companies and Range of Study

Industry	Six Sigma company	Comparison company	Range of study
Diversified financials	General Electric	Fannie Mae	1990–2000
Health care: insurance and managed care	CIGNA	Wellpoint	1998–2008
Aerospace and defense	Boeing	United Technologies	1994–2004
Chemicals	Dow Chemical	PPG Industries	1994–2004
Construction and farm machinery	Caterpillar	Terex	1995–2005
Computer peripherals	EMC	Western Digital	1995–2005
Motor vehicles and parts	Johnson Controls	Navistar Intl.	1995–2005
General merchandisers	Macy's	J.C. Penny Corp.	1996–2006
Network and other communications equipment	Motorola	Harris Corp.	1981–1991
Specialty retailers	Home Depot	Costco Wholesale	1996–2006
Electronics, electrical equipment	Whirlpool	Harman Intl. Industries	1992–2002

9B.2.3   Criteria for success

After determining Six Sigma is a financial impact, an analysis will be conducted to evaluate whether one company is more successful in its Six Sigma deployment than the other. To uncover whether there is a difference between successful Six Sigma companies and unsuccessful Six Sigma companies, a standard for success must be established. A company is successful if it satisfies two sets of conditions:

1. The company sees an increase in ROE per period in study period 2.
2. The percent increase in ROE per study period 2 exceeds ROE per study period 1.

Condition 2: (Future)

1. The company sees an increase in EPS in study period 2.
2. The percent increase in EPS in study period 2 exceeds EPS in study period 1.

9B.2.4   Identify characteristics

Once successful and unsuccessful Six Sigma companies are established, the study will explore the underlying factors that lead to successful Six Sigma deployment. Three characteristics will be studied: quality awards and standards, number of employees, and Lean philosophy. An ANOVA test will be constructed to see which characteristics significantly affect the success of Six Sigma deployment. To determine how much these characteristics affect Six Sigma deployment, interviews will be conducted with the Six Sigma companies. The first characteristic evaluated will involve awards and ISO standards that demonstrate quality and continual improvement. To establish a score for each company, four aspects of awards and ISO standards that demonstrate quality and continual improvement. To establish a score for each company, four aspects of awards and ISO standards will be studied: won a MBNQA, apply and follow the MBNQA criteria, are certified to follow ISO 9000 standard, and are certified or follow the ISO 14000 standard. A score will be calculated by determining the weights for these aspects, assigning one point if an aspect is met, and then summing the weighted values. The second characteristic involves the number of employees. Any management strategy's success depends on the involvement of the employees, so the number of employees may affect the success of deployment because with a large number of employees, employee involvement is harder to achieve. The final characteristic to be explored is whether the company implemented a Lean philosophy along with Six Sigma. Each company will be give a point if they deployed Six Sigma coupled with Lean thinking.

9B.3   Conclusion

Six Sigma is a management strategy that many companies utilize in order to impact their financial results. This impact in important to companies, but shareholders may want to know how this management strategy affects their investment. Whereas many companies embrace Six Sigma culture, not all truly reap the benefits of this initiative. Similar to the study conducted by Dr. Chakravorty, this study will explore the reasons Six Sigma fails with respect to shareholders. These people invest their money trusting that companies will provide high returns. This study will evaluate how Six Sigma impacts ROE and EPS as well as identify the factors that lead some Six Sigma companies to bring their shareholders wealth while others are unsuccessful in their attempts.

References

Business benefits of ISO 14000. (2009). International organization for standardization.

Cater, D. J. and Pasqualone, R. G. (1995). "ISO 9000—A perspective on a global quality standard." *IEEE Transactions on Industry Applications*, 31: 61–67.

Caulcutt, R. (2001). "Why is Six Sigma so successful?" *Journal of Applied Statistics*, 28(3/4): 301–306.

Chakravorty, S. S. (2010). "Where process improvement projects go wrong?" *The Wall Street Journal*. http://online.wsj.com/article/SD10001424052748703298004574457171313938130.html

Collins, J. (2001). *Good to Great*. New York: HarperCollins Publishers Inc.

De Feo, J. A. and Barnard, W. W. (2004). *Juran's Six Sigma: Breakthrough and Beyond*. New York: McGraw-Hill.

Feld, K. G. and Stone, W. K. (2002). "Using Six Sigma to change and measure improvement. *Performance Improvement*, 41(9): 20–26.

Frequently asked questions about Malcolm Bridge International Award. (2009). National Institute of Standards Technology.

Garza, A. and Jones, E. (2010). "Six Sigma deployment success according to profit." White Paper University of Nebraska–Lincoln.

George, M. L. (2002). *Lean Six Sigma: Combining Six Sigma Quality with Lean Speed*. New York: McGraw-Hill Companies, Inc. pp. 4–31.

Holweg, M. (2007). "The genealogy of lean production." *Journal of Operations Management*, 25(2): 420–437.

ISO 14000 essentials. (2009). International Organization for Standardization. http://www.iso.org/iso/iso_catalogue/management_standards/iso_9000_iso_14000/iso_14000_essentials.htm

Mellat–Parast, M., Jones, E., and Adams, S. (2007). *Six Sigma and Baldrige: A Quality Alliance, Quality Progress*, 40(9): 45.

Nunnaly, J. (1978). *Psychometric Methods*. New York: McGraw-Hill.

Return on Equity (ROE). (2009). Investopedia, http://www.investopedia.com/terms/r/returnonequity.asp

Schroeder, R. G. et al. (2008). "Six Sigma: Definition and underlying theory." *Journal of Operations Management*, 4: 536–554.

Siegel, S. (1956). *Nonparametric Statistics for the Behavioral Sciences*. New York: McGraw-Hill.

Simon, K. (2000). "DMAIC Versus DMADV." http://www.isixsigma.com/library/content/c001211a.asp

Snicker, R. (2004). *Implementing Six Sigma: A Planning Guide for Executive Teams*. Madison, WI: Oriel Inc.

USA Today. (2005). List of the Fortune 500. http://usatoday30.usatoday.com/money/companies/2004-03-22-fortune-500-list_x.htm

Wei-min, L. I. (2006). "A DMAIC approach to optimize the process of replacing brushes of start-generator." *Mathematics in Practice and Theory*, 6: 133–141.

Six Sigma and HR: Getting the right talent and industrial engineers in key Six Sigma roles

> There is one quality which one must possess to win, and that is definiteness of purpose, the knowledge of what one wants, and a burning desire to possess it.
>
> —Napoleon Hill

10.1 Introduction

The challenges associated with Six Sigma (Pepper 2010) initiatives include

1. Corporate-wide metrics are hard to measure.

   The Six Sigma metrics are hard to develop into firm-wide performance because they have to expand to include not just manufacturing operations but other functions. The challenge for most companies is firm-wide benchmarking and measurement, and how and when to convert the Six Sigma metric or corporate Z score. With Six Sigma, the value of an organization's output includes not just quality, but availability, reliability, delivery performance, and after-market service. Most companies seek Six Sigma performance within each major functional output that contributes to the end-customer value. The Six Sigma corporate metric is applied in a broad fashion, striving for Six Sigma level performance at the lowest level of activity or at the lowest invested cost.

2. The excessive training expenses for an "in house" champion belt system.

   Champion belt infrastructure. In answer to this, it can be said that Six Sigma provides a clear focus on measurable financial returns through a sequential and disciplined manner, and establishes an "infrastructure of champions" with its training style of introducing "belt" qualifications (green, black, master black belts, etc.) within the organization to lead the way in data-driven decision making for improvement efforts (Antony 2004).

3. The "in house" belt system becomes more focused on internal prestige and promotion not improvement. The champions may also become more consultant based and overpromise success. The training for and solutions put forward by Six Sigma can be prohibitively expensive for many businesses, and the correct selection of improvement projects is critical (Senapati 2004). Antony (2004) discusses the nonstandardization of training efforts (in terms of belt rankings, etc.), and how this accreditation system can easily evolve into a bureaucratic menace, where time and resources are misspent focusing on the number of "belts" within the organization, and not the performance issues at hand.

4. Consulting not improvement. The belt system is an attempt to develop "in-house" expertise. The danger is that this strategy may face the dangers of most consultancy practice, which tend to overpromise and under-deliver. Consultants tend to recommend but not have the power or funding to implement. Sponsors are given the improvement process that is to be implemented then "in-house" consultant or black or green belt leaves without ensuring that the implementation has taken place. This is a big challenge in the philosophy.

5. Statistics heavy analysis. Six Sigma has is a statistics-heavy, technical approach to process control. In order to prevent it becoming "watered down" or "ignored" improvement approaches great care should be taken to focus on the wider philosophy behind the structured approach to Six Sigma. The statistics can be supplemented with specialists and engineers.

These challenges are mostly associated with human resources. The need for specialized personnel who can handle the statistic-based calculations is key to Six Sigma. We begin by looking at the engineering education that Lean Six Sigma (LSS) is based upon.

10.2   LSS in education (graduate and undergraduate)

10.2.1   Six Sigma and engineers

10.2.1.1   Six Sigma and engineering education

The fact that Six Sigma was discovered by an engineer seems to escape many people. Bill Smith, the originator of the philosophy was a reliability engineer. The idea of using a Six Sigma metric was developed by Bill Smith at Motorola in response to sub-standard product quality traced in many cases to decisions made by engineers when designing component parts.

Traditionally, reliability engineering come from a curriculum related to industrial and manufacturing engineers who have, in their base undergraduate accredited curriculum, been trained in advanced statistics, reliability, and defects that are related to manufacturing. They also provide foundational knowledge in scientific management, human factors, and health systems engineering that address the employee (Deming) side of the Six Sigma. Other engineering undergraduate curriculum including some mechanical engineering programs have a focus in these areas.

10.2.1.2   Controversy between engineers education and green/black belts

Some employers and academics question the need for "specialized" training of belts when these engineers have been trained to complete these tasks in a more robust manner. The Six Sigma training programs are seen as a means to teach advanced engineering skills to nonengineers. Some academics believe that the lack of robust training, especially in statistics, may lead to the reduction of effectiveness of the belts. In practice, the engineering students have adapted and seek Six Sigma certification prior to graduation and many companies hire both engineers and green and black belts.

10.2.2   Aspects of Lean Six Sigma

The courses offered in the industrial engineering (IE) discipline enable the engineers to handle three aspects of LSS. First, IEs are prepared for data collection by the education in time study and work measurement. Further, the procedure to setup the experiments is aided by the study of design of experiments (DOEs). Second, IEs are educated on the statistical evaluation tools through multivariate statistics and probability courses. Third, senior capstone courses prepare the IE to make recommendations based on project assessment.

10.2.3   Roles of Lean Six Sigma

The traditional roles (project manager, data analyst, continuous improvement engineer, etc.) of an industrial engineer are the same roles that Six Sigma organizations assign to the black belt, master black belt, or champions. The academic education of an industrial engineer provides a sufficient background to play the roles as champions, statistical evaluator, and economic justifier (Figure 10.1).

10.2.3.1   Industrial roles

Industrial engineers are also trained in basic engineering and additional courses which enable them to act as a change agent. The various tools used in LSS are part of the IE curriculum. Many IEs suggest that LSS is nothing more than using IE tools to focus on financial performance instead of efficiency improvement. The versatile nature of the industrial engineers helps them to perform in any part of the LSS projects. Also, many tools used in the Six Sigma like flowcharts or cause-and-effect diagrams are included in most IE academic curriculum.

10.2.3.2   Six Sigma–IE interface framework

Six Sigma–IE interface framework (SS–IE IF). The framework describes three components that relate Six Sigma initiatives with IE. The components include

- Six Sigma–industry interface
- Sig Sigma–IE academic interface
- Six Sigma–IE specialized knowledge interface

The Six Sigma interface shows the DMAIC process and the specific tools that are utilized in those steps. These tools have been envisioned to support the successful implementation of Six Sigma projects. Most of the tools described are taught in an IE undergraduate program and most IE and engineering management (EM) curriculums. Lean tools include 5-S, TPM, mistake proofing, visual systems, and value stream mapping.

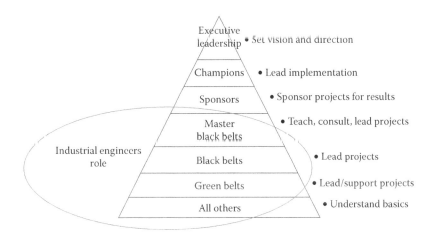

Figure 10.1 Role of Six Sigma in industrial engineering projects.

The Six Sigma–IE academic interface describes the general academic classes that cover the topics and theory that support the correct use of Six Sigma tools. The premise is to correctly use tools such as regression, multi-variable analysis, and statistical process control (SPC). It is important to have foundational knowledge of calculus-based statistics. Classes such as engineering statistics (descriptive, hypothesis testing, inferential), applied statistics (ANOVA, Regression, DOEs), and statistical quality control allow mastery of these tools. Industrial quality control focuses on capability ratios whereas applied statistics and quality control focuses more on control charts. Oftentimes, many pundits who criticize Six Sigma identify that most statistical components are not correctly used or understood, minimizing the effectiveness of these tools. Utilizing IEs in the execution of these tools would provide a higher probability of using these tools correctly and better results for Six Sigma projects and initiatives. Table 10.1 shows the interface between the IE academic curricula and the Six Sigma process. The course names used in this study are typical course names and are used to represent courses that are typically available in IE undergraduate and graduate programs. Additionally, some courses such as project management and advanced classes in engineering economics may be provided as a graduate class in IE departments. These courses would teach the student how to perform a stakeholder analysis which is commonly used in the Define phase.

The Six Sigma IE specialized knowledge interference as shown in the Table 10.1 highlights the special knowledge of the industrial engineer. IEs can also participate in Six Sigma in specialized roles that can support its initiatives. Some Six Sigma organizations create specialized positions that assist them in successfully implementing Six Sigma. Some of these positions include statistics specialist, Lean specialist, and leadership liaison. Separating these roles from traditional Six Sigma roles provides an organization the ability to supplement their Six Sigma process. An additional role not shown is the economist, which is a role in which the specialist evaluates and verifies the earnings or savings for the project. This role would typically work in close partnership with accounting (Table 10.2).

The integration of the job functions and opportunities for IEs in Six Sigma are not included in our framework, but it is an important component for IE visibility in Six Sigma organizations. This integration is shown in the table below. Six Sigma is based upon the plan–do–check–act (PDCA) premise. We show how different job functions are identified under both PDCA and Six Sigma steps to show the flexibility of the frameworks use for nonSix Sigma organizations. Overtaking this table with the other tables describes how the IE who seeks these positions can identify the skills they learned and can utilize them in those job functions. For example, if an IE works as a CEO, he/she will use skills learned in EM. This table also suggests that those skills will allow one to effectively perform the stakeholder analysis that is aligned in the Define phase when executing Six Sigma events (Table 10.3).

Table 10.1 Six Sigma Industrial Engineering Academic Interference

Define/academic discipline	Measure/academic discipline	Analyze/academic discipline	Improve/academic discipline	Control/academic discipline
Project management, engineering management, industrial quality, engineering economy	Applied statistics and quality control, engineering statistics, and data analysis	Industrial quality control, engineering statistics and data analysis, applied statistics, and quality control	Quality engineering use of experimental design and other techniques, engineering statistics, and data analysis	Industrial quality control, applied statistics, and quality control

Table 10.2 Six Sigma Industrial Engineering Specialized Knowledge

Special knowledge	Industrial tools	Academic courses
Statistics	Multivariate chart, chi square, regression, factorial design of experiments, measurement system analysis	Applied statistics and quality control, engineering statistics, and data analysis
Lean	5-s, TPM, mistake proofing	Industry quality control, applied statistics and quality control
Leadership	Value stream mapping, through process map, stakeholder analysis, system map, interview process, prioritization matrix	Project management, engineering management, industry quality control, engineering economy

Table 10.3 Plan–Do–Check–Act

Plan	Do/check	Act
Define	Measure/analyze/improve	Control
CEO, COO, plant manager, project manager, and operations manager	Project manager, engineer, consultant, and programmer	IE consultant, safety engineer, and ergonomist

10.3   LSS integration into the Malcolm Baldrige Award

The Baldrige criteria usually address system- or enterprise-wide improvement, while Six Sigma attempts to produce micro level or specific improvement. The Six Sigma process improvement method is a useful tool that can transform the goals of the Baldrige model into reality. The two methods are complementary.

Organizations strive to produce higher quality products and services and enhance profitability by implementing the Baldrige criteria. These criteria are well represented in the Six Sigma process improvement method. While the Baldrige criteria and SSPIM have followed different development paths, they have similar roots. Both quality activities originated from management philosophies that attempted to enhance performance, increase profitability, and improve quality. The Baldrige framework can be used as a generic model for process improvement in an organization. SSPIM is referred from a management point of view as independent projects in a firm that focus on improving operational and business performance. It is possible to implement the Six Sigma philosophies within the Baldrige framework and integrate them with overall quality system of the organization.

10.3.1   Baldrige model for quality and Six Sigma method

The Baldrige model consists of seven categories:

1. Leadership
2. Strategic planning
3. Customer/patient and market focus
4. Measurement, analysis, and knowledge management
5. Work force focus
6. Process management
7. Results

Six Sigma is defined as the relentless pursuit of process variation reduction and break-through improvements that increase customer satisfaction and impact the bottom line. While the Baldrige model tends to address system or enterprise-wide improvements, firms can achieve higher levels of performance in the Baldrige criteria by implementing Six Sigma projects that focus on the key Baldrige criteria. Companies can benefit from integrating Six Sigma and Baldrige. Others have explained how both are interrelated and how "Baldrige model provides the framework, Six Sigma the methodology" (De Feo and Barnard 2004). Six Sigma proponents say that winning quality awards, improving quality, and increasing customer satisfaction is achieved through Six Sigma (Douglas and Erwin 2000). The Baldrige model is typically applied at enterprise level; Six Sigma is deployed as a number of independent projects throughout the enterprise. Accordingly, by implementing Six Sigma as independent projects within the Baldrige model, firms can achieve higher operational performance and efficiency through Six Sigma projects while operating under these guidelines and the framework of Baldrige criteria. In that regard, the two approaches are complementary. The DMAIC cycle of SSPIM can be employed as a practical tool for addressing the requirement in each category within the Baldrige framework.

10.3.2   Six Sigma within Baldrige model

According to one approach, the implementation of the Baldrige model would mean

- A focus on performance excellence for the entire organization in an overall management framework.
- Identifying and tracking all-important organizational results: customer satisfaction, product/service, financial and human resources organizational effectiveness.

While the Baldrige model does not provide any specific method or tool that can be incorporated, it does offer general guidelines. As such, SSPIM is a practical, appropriate, and useful tool that can be used to transform the goals of Baldrige model into reality. Accordingly, the Six Sigma concepts can be applied to improve the performance of each of seven categories of Baldrige model. In other words, Six Sigma projects facilitate process improvements in a firm through focus on Baldrige criteria.

The major difference is that Six Sigma projects might not be independent; rather, they are all focused on achieving higher levels of quality set by the Baldrige criteria (Figures 10.2 and 10.3).

The integration of Six Sigma with Malcolm Baldrige is based on the following principles:

- Six Sigma becomes part of Baldrige model for achieving performance excellence. The top management sets such requirements. After establishing such goals which need to be aligned with Baldrige requirements, SSPIM can be used to improve processes and meet the quality objectives.
- SSPIM can be applied to all types of projects, processes, and products. Top management directs the selection, administration, and control mechanisms.
- The Baldrige self-assessment includes both assessment of seven categories as well as the efficiency and effectiveness of SSPIM.
- The proposed model for the integrative Six Sigma–Baldrige framework is capable of addressing the core values of the Baldrige model. Through the focus of SSPIM on management by fact (data analysis and statistics) and goal setting, the Baldrige core values can be addressed.

Figure 10.2 Malcolm Baldrige Award process.

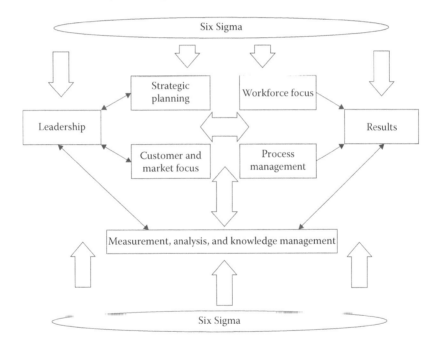

Figure 10.3 Malcolm Baldrige integration into Six Sigma.

10.3.3   Full integration

Currently, Six Sigma and Baldrige model are linked but fully integrated. The challenges are implementing Six Sigma in areas such as leadership, human resources/work force management, and strategic planning to achieve full integration.

10.3.4   Leadership

Six Sigma projects can be effectively defined to enhance the quality of the leadership in an organization. In that regard, the feedback from the Baldrige model can be used as a basis to define Six Sigma projects that focus on leadership management. Motorola has its own "4E's + Always 1" leadership standard that addresses key leadership characteristics: envision, energize, edge, execute, ethics, and character.

10.3.5 Human resource/work force management

Feedback from Baldrige self-assessment model can be used to identify potential Six Sigma projects focusing specially on human resource development. In addition, SSPIM enhances the organizational knowledge base through training and systematic learning. But the impact of SSPIM on human resource development goes beyond that.

10.3.6 Strategic planning

Honeywell developed special approach called Six Sigma plus, which focused on strategic objectives, providing more values to its customers by empowering the employees, improving its products, processes, and services.

Thus, integrating Six Sigma into Baldrige not only helps firms achieve higher levels of performance and customer satisfaction in each Baldrige category, it also provides firms with a useful method for pursuing quality and performance set by the Baldrige model.

10.4 LSS integration into ISO 9000

ISO 9001 is an internationally acclaimed standardization methodology to facilitate interrelationships globally between suppliers and purchasers. It has been popularized by large enterprises when they orient themselves with this body of literature for quality and processes using LSS ideologies (Chinvigai et al. 2010).

However, use of LSS has not caught on with some of the smaller businesses due to budget and resource constraints. ISO 9001 has been quicker to achieve and has become some sort of a prerequisite for popularity with customers. It does remain important though to integrate ISO 9000 processes with LSS methodologies. It has been propagated by various researchers (Wessel and Burcher 2004). It eliminates the requirement to update with ever-improving quality literature the ISO standard the organization is associated with. The emphasis on accurate records assists the LSS processes and makes process improvements attainable.

10.5 LSS integration into logistics

A manager's responsibility is to judiciously allocate resources that would automatically affect costs of holding inventory, labor, transportation, and process. The overall profits of the sequence of operations in a supply chain can be dramatically increased. Thus, designing of the flow of material and equipment in the workspace is important to optimize costs invested in improved logistics.

A network modeled by instilling the LSS thinking and its methodologies was proposed (Jones, and Dejing).

We incorporate the designing of a strategic logistics network in the Analyze phase of the DMAIC process.

1. Define: They must decide and tally with the sponsor's goal of ROI within the organizations strategic mission boundaries.
2. Measure: It is more accurate due to encouragement in collecting accurate and sufficient data by Six Sigma ideology.
3. Analyze: Data is the only base of reliability in making decisions on modeling. A robust framework can be created by using tools such as DOEs.
   In general, the following steps can be used for a project:
   a. Collate data within limits
   b. Optimization constraints are to be applied
   c. Compare with alternatives for minimized cost function with improved service
   d. Decide among the alternatives
4. Improve. The aim now must be to eliminate waste using Lean thinking. Only the best practices must be incorporated for transformation of structure in the supply chain.
5. Control: The variations must now be kept in check by recording the performance in an improved framework and how long these results last. Using LSS tools key performance parameters will be selected to be tracked during the testing phase.

References

Antony, J. (2004). "Some pros and cons of Six Sigma: An academic perspective." *The TQM Magazine*, 16(4): 303–306.

Chinvigai, Ch., Dafaoui, E., and El Mahmedi, A. (2010). "ISO 9001: 2000/2008 and Lean-Six Sigma integration toward to CMMI-Dev for performance process improvement. Evaluation and optimization of innovative production systems of goods and services." *2010 International Conference of Modeling and Simulation*, May 10, Hammamet, Tunisia.

De Feo, J. A. and Barnard, W. (eds.) (2004). *Juran Institute's Six Sigma: Breakthrough and Beyond*. New York: McGraw-Hill.

Douglas, P. C. and Erwin, J. (2000). "Six Sigma focus on total customer satisfaction." *The Journal for Quality &Participation*, March/April, pp. 45–49.

Jones, E. C. and Dejing, K. (2010). "Evaluation and optimization of innovative production systems of goods and services." *Proceedings of the 2011 Industrial Engineering Research Conference*, May 11, Reno, Nevada.

Pepper, M. P. J. and Spedding, T. A. (2010). "The evolution of lean Six Sigma." *International Journal of Quality & Reliability Management*, 27(2): 138–155.

Senapati, N. R. (2004). "Six Sigma: Myths and realities." *International Journal of Quality & Reliability Management*, 21(6): 683–690.

Wessel, G. and Burcher, P. (2004). "Six Sigma for small and medium-sized enterprises." *The TQM Magazine*, 16(4): 264–272.

Appendix 10A: Quality control measurement of an engineering productivity index

10A.1   Introduction

In today's society, there appears to be a growing trend among individuals who believe they should give themselves a raise by cutting back the hours they work in order to spend more time on side businesses or hobbies that bring them more satisfaction. This contradicts some of the older Taylor-oriented philosophies of a fair day's work for a fair day's pay. Organizations should expect some form of commitment to productivity by their knowledge workers, in order to get the return on investment they expect from the higher salaries paid to knowledge workers. Because of economic pressure and the use of business process re-engineering, corporate right-sizing and corporate re-structuring, employees are not as likely to have one job throughout their lifetime as was the case in previous generations. Nevertheless, companies should expect good work from their personnel while they are on the job (Amabile et al. 2002).

Knowledge work requires a special set of skills related to an area of expertise, such as those of a project manager, an engineer, a salesperson, a consultant, a manager, or a healthcare professional. But it requires much more than a technical competence to be successful as a knowledge worker. Savvy knowledge workers understand that these additional skills include the ability to acquire and transfer knowledge effectively. Knowledge is the stock-in-trade of knowledge workers—it is both the process and the product of their work (Fisher 1998).

Knowledge workers resist any attempt to have their productivity "measured" in any traditional sense. Given the complex nature of their work and the multiple relationships between various factors to be considered by these workers, traditional performance measurement techniques such as work measurement have not been very effective in identifying best practices. Most see themselves more as "artists" or "brain workers" than production workers, and most believe that there are no fair and equitable measures that could be used to establish how well they perform. However, management maintains that "what you cannot measure, you cannot control."

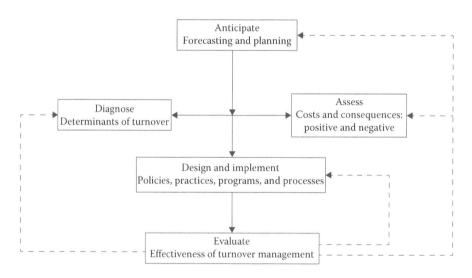

Many of the fundamental quality control (QC) principles have been well proven in academic studies and in practice from industry. Collection of meaningful data and proper statistical analysis has led to improved quality through process control actions. This study aims to demonstrate how QC concepts can be used to efficiently measure knowledge worker productivity.

This research provides a new approach for measuring cognitive, or knowledge productivity. This approach is called the statistical evaluation of cognitive turnover control system, or SECtCS methodology.

This study presents performance analysis of an engineering productivity index using statistical process control (SPC) charts. The data obtained on working engineers leads to improved measurement and control which includes analysis of a powerful method for establishing control limits, identifying run patterns, and finding statistical outliers of performance.

10A.2   Background

This research focuses on the productivity of the engineer as the knowledge worker. In order to measure the statistical components of the engineering productivity, a cognitive productivity index for engineers was created. A questionnaire was created to test a representative group of working engineers for cognitive turnover. Cognitive turnover (CT) is a condition in which knowledge workers are judged to have lower productivity because of lack of motivation related to thoughts of quitting the job and to extreme burnout. A linear regression model was utilized to validate the independent variables for the questionnaire which included variables for turnover, burnout, and job satisfaction against a self-scored dependent variable for CT.

The CT model was developed based upon a total of 51 engineers and related service providers. Tables 10A.1a and 10A.1b contain demographics of the respondents.

The age of respondents ranged from 20 to 64 with a mean age of 25. The length of service for respondents ranged from 1 month to 480 months with a mean of 35 months. In the study there were 15 females, which represented 29% of the test population. Also, there were 34 respondents who, while working, were looking for a better job, which represents 67% of our population. Finally, there were only eight respondents working a secondary job or activity.

The variables used to measure CT were challenges, depersonalization, personal achievement, and promotion. These variables were selected based on the research by the researcher on job turnover, burnout and work motivation. The questionnaire was tested for reliability and validity using Cronbach's alpha and R-square, respectively.

Multiple linear regression was used to develop a model shown in the following equation:

$$CT = 1.1986 \text{ (mean challenges)} + 1.5749 \text{ (mean depersonalization)}$$
$$- 1.7123 \text{ (mean personal achievement)} - 0.9351 \text{ (mean promotion)} + 5.1224. \tag{10A.1}$$

CT is the dependent variable, which represents the degree of CT ranging from 1 to 10. Scores that are 1–4 represent low cognitions to leave and generally low burnout indications. Scores 5–8 represent moderate burnout and leaving cognitions, with expected lower

Table 10A.1a Demographics of Respondents by Age

Category	Mean	Std deviation	Minimum	Maximum	Number of participants
Age (Years)	25.5957	7.496	20	64	51

Table 10A.1b Demographics of Respondents by Sex

Category	Number	Percent to total	Number of participants
Male	36	71	51
Female	15	29	51

Table 10A.2 CT Index Score Description Table

Score	CT	Considering leaving	Description
1–2	No	No	Not burned out, high productivity
3–4	No	Occasionally	Light burnout, good productivity
5–7	Yes	Open for other jobs	Medium/high burnout low productivity
8 10	Yes	Strongly considering	High burnout low productivity, possible sabotage

productivity. Scores 9 and above may represent detrimental burnout and possible sabotage if departure is not eminent; lower productivity is expected (Table 10A.2).

10A.3 Application of quality control principles

The idea that statistics may be important in assuring the quality of products and processes goes back to the advent of mass production. Statistical quality assurance includes mainly three techniques, quality control, establishment of tolerance limits, and acceptance sampling. The word *quality*, when used technically, refers to some measurable property of a product, process, or in our case a condition.

In order to assure quality, a process has to have variability that does not fluctuate greatly between products or processes. When the variability present in a process is confined to chance variation, the process is in a state of statistical control. This state is usually attained by finding and eliminating the problem that is creating the assignable variation. Most processes have this variation, and the most common method for detecting serious deviations is by using quality control charts. There is a differentiation between control charts for measurements and control charts for attributes, depending on whether the observations are measurements that count data or monitor attributes. This research is a control chart for measurement (Johnson 2000).

The most common working statistical tools in quality control are control charts. Control charts are characterized by a process mean and upper and lower control limits. As data are gathered about the performance measure, control charts portray the data as a function of time and change of performance (mean and variation) can be monitored. Trends can be seen and data outside the control limits indicate that the process is no longer in control. The use of control charts is frequently called "statistical process control" (SPC).

10A.4 Use of control charts for cognitive turnover

After the questionnaire and regression model development, turnover participants participated in a 12-week charting of responses for the control chart development of CT.

The control chart used in this study, called the SECtCS CT, is very similar to a typical $\bar{x}$ or R chart. They are similar with respect to the central line, the UCL, and LCL lines, zone calculations, axes normalization, and the pattern-analysis rules.

In the development SECtCS CT, we first use historical data on the different CT mean scores for similar professionals. This approach is related to the chart of individuals for manufacturing processes (Montgomery and Runger 1999).

Second, the CT scores from a questionnaire are plotted on the y-axis. SPC charts plot actual measurement values. Each point on the chart is intended to reflect the current state of the system. This approach allows for instantaneously detecting of changes in the normally stable process. It has been noted in our research that the length of time that an individual stays at a company may keep them from leaving due to benefits and pension investment. This propensity to stay may cause CT to increase over time and may distort the findings.

CT scores were charted on an individual event as opposed to the averages of the CT scores for the individual. With this type of chart, the sample size for process control is 1 as opposed to the traditional sample size of 5 to 6. Because the sample size is not 5 or 6, the assumption of normality is challenged. In this study, the respondent's scores were tested for a normal distribution using a K-S test and the null hypothesis could not be rejected. The main differences between the control charts for individuals and the traditional x control charts are that the control chart for individuals tracks independent variables.

The traditional $\bar{x}$ chart plots the average of some defined variable like a process. This distinction is focused because for this research the individuals CT score is tracked over time, not their average CT score.

10A.4.1   Chart run rules

With typical SPC charts, six generally accepted patterns indicate that the process is out of statistical control. The rules are as follows: SPC rule 1, 1 point beyond zone A; SPC rule 2, 9 points in a row in Zone C or beyond; SPC rule 3, 6 points in a row steadily increasing or decreasing; SPC rule 4, Fourteen points in a row alternating up and down; SPC rule 5, 2 out of 3 points in a row in Zone A or beyond; SPC rule 6, 4 out of 5 points in a row in Zone B or beyond. If any of these patterns were observed, a typical response would be to investigate the process for potential problems. Once the source of variation is corrected, the process should return to under control. More details are available in literature (Bauch and Chung 2001).

10A.5   Chart development

The test population and sample group were the respondents who identified themselves as nonCT, because we are testing to identify respondents who have excessive CT trauma, we utilized a chart that was derived from a normal group. This allows the identification of excessive CT by the SECtCS control chart.

Previously, 51 working engineers' CT index scores were predicted using the linear regression model. After taking the questionnaire, the valid independent variables were placed in linear regression equation (Equation 10A.1) and their CT score was predicted. This predicted value was called the CT index. Through self-assessment 28 of the 51 respondents identified themselves as nonCT respondents. The normalized questionnaire variable data for the respondents who identified themselves as nonCTs was used to calculate the central line, UCL, and LCL lines, and the zone boundaries. A chart of individuals approach was used; the central line will be defined as the mean of the CT indexes (Figure 10A.1).

For our research the following control chart components were calculated from our 28, nonCT, respondents.

Table 10A.3 depicts the information from nonCT respondents with respect to the SECtCS CT. The information gives $\bar{x}$ as 4.3571 which represents the mean for people who were not CT. The standard deviation is given as 2.0040. Then utilizing the traditional UCL and LCL calculation, the upper and lower limits for the given zones were listed on the charts.

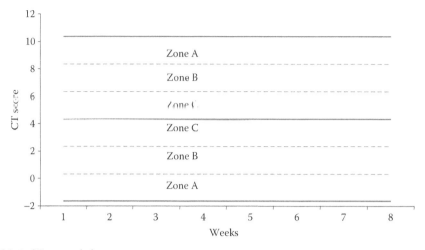

Figure 10A.1 CT control chart.

Table 10A.3 Construct Means and Standard Deviations

$\bar{x}$	Standard deviation	CL	Upper zone C	Lower zone C	Upper zone B	Lower zone B	Upper zone AUCL	Lower zone ALCL
4.3571	2.0040	4.3571	6.3611	2.3532	8.3651	0.3492	10.3690	−1.6555

10A.6   Results

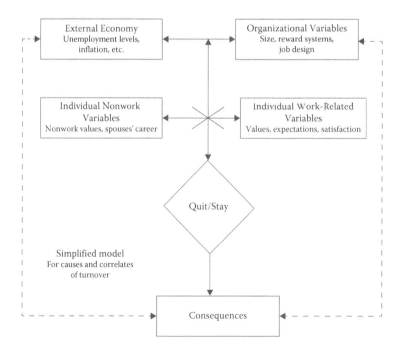

Because this research is more focused on the CT scores that are out of control, more emphasis was placed on CT scores that violate the run rules with respect to the upper control limits. In other words, values below the CL are not considered relevant because they would represent CT scores that would not indicate a problem for this research. The most relevant analysis would be those scores that are measured above the CL and these are the CT scores that are analyzed for violation of the run rules.

There were nine respondents who were tracked over a 12-week period. Of the nine respondents, four defined themselves as having CT whereas the others did not define themselves as CT. Over the 12-week period, the most significant results are listed in Figure 10A.2.

Respondent 1's control chart shows that the respondent is out of control by violating SPC rule 6, four out of five points in a row in zone B or beyond. Also, rule 3 is violated six points in a row steadily increasing. This respondent self-scored himself as a CT (Figure 10A.3).

Respondent 2 does not have an out-of-control SPC chart. Further, this respondent did not self-score himself or herself as CT and this is consistent with their chart (Figure 10A.4).

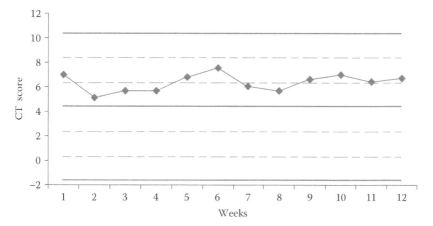

Figure 10A.2 SPC chart respondent 1.

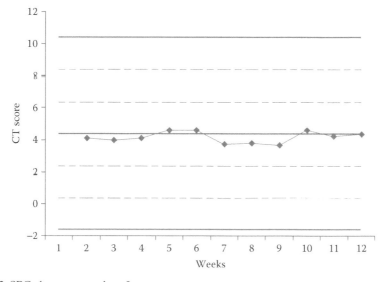

Figure 10A.3 SPC chart respondent 2.

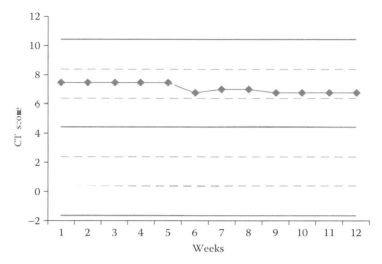

Figure 10A.4 SPC chart respondent 3.

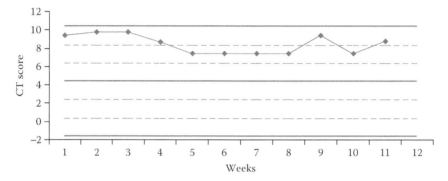

Figure 10A.5 SPC chart respondent 8.

Respondent 3 violates SPC rule 6, four out of five points in a row in zone B or beyond. This respondent also did not consider himself or herself a CT on the initial dichotomous self-scoring questionnaire. This respondent's score barely exceeded zone C into zone B in the latter part of the study, this may be why this person considered themselves a nonCT. They appear to be on the borderline (Figure 10A.5).

Respondent 8 violates SPC rule 5, two out of three points in a row in zone A and beyond, SPC rule 6, four out of five points in a row in zone B or beyond, and SPC rule 2, nine points in a row in zone C or beyond. This respondent also scored himself or herself as a CT on the self-scored CT dichotomous index. This respondent incurred quite a bit of added responsibility during this 12-week process and some of the questionnaires were taken quickly.

The results confirm that respondents, who initially scored themselves as CT, also have SPC charts that showed that they were out-of-control with respect to normal respondents. Also, it appears out of the four respondents who scored themselves as nonCT's, three were borderline CT's. This would follow along with the theory that many people who believe they are not CTs do actually measure as CT. This further explains why this problem needs to be identified and addressed from organizations as a whole.

Though all respondents show some form of CT there are some respondents who had more profound out of control patterns on the control charts than others. It appears that on a weekly basis there is not a large change from week to week for some of the respondents. It appears that if respondents' scores were tracked on a 4- to 5-week basis (monthly) there may be a larger trend. After informal discussions with respondents during the process events like performance reviews, company layoff news, and personal life events may have affected the score on a certain week. Further, the countries' current economy, an economy with recession like indicators, may have an impact on burnout. This affect on burnout can definitely impact respondents CT scores.

10A.7   Conclusions

In this research, we demonstrated a quantitative measurement technique for knowledge workers in several ways. First, we described a method for measuring an index that measures when knowledge workers have reduced mental focus which affects their work productivity. Next, we used quality control to truly identify that the problem is out of control. The benefits are that companies can use these techniques to identify low output engineers and knowledge workers. Further, companies can save money on nonproductive behaviors that may lead to company turnover and reduced work output.

There are other potential applications using the quality control component of this SECtCS methodology. One opportunity is to measure the group means of project teams or different engineering departments and compare it to other projects or the engineering population as a whole. This would require using a more traditional $\bar{x}$ chart, but could provide valuable company insight to their engineering team's production. Also, the SECtCS methodology provides for a re-charting of the indexes after a company intervention has taken place. This allows the company to evaluate the solutions they may have implemented to address the problems. The challenge that may occur with this research is that management and personnel may not be willing to address the problems that this methodology may identify and be willing to take action.

The SECtCS methodology can potentially go into any organization and indicate which common causes are relevant for that organization and then evaluate the knowledge workers who work there. The true benefit is that after personnel or teams are measured, the scores can be utilized to determine if a department, group, project, or organization as a whole has an out of control CT group. Upon identification, some type of intervention can take place for these engineers which will allow the company to become more profitable. Also, they can cut their losses on personnel or departments that are not providing the productivity that is necessary for the group, project, or organization to remain profitable.

References

Amabile, T. M., Hadley, C. N., and Kramer, S. J. (2002). "Creativity under the gun." *Harvard Business Review*, 80(8):52–61.

Bauch, G. T. and Chung, C. A. (2001). "A statistical project control tool for engineering managers." *Project Management Journal*, 32(2):37–44.

Fisher, K. and Fisher, M. D. (1998). *The Distributed Mind: Achieving High Performance through the Collective Intelligence of Knowledge Work Teams*. New York: Amacom.

Johnson, R. A. (2000). *Miller & Freunds's Probability and Statistics for Engineers*. New Jersey: Prentice Hall.

Montgomery, D. C. and Runger, G. C. (1999). *Applied Statistics and Probability for Engineers*, 2nd edn. New York: John Wiley & Sons.

Further Readings

Bassman, E. S. (1992). *Abuse in the Workplace, Management Remedies and Bottom Line Impact.* Westport, CT: Quorum Books.

Boyatzis, R., McKee, A., and Goleman, D. (2002). "Reawakening your passion for work." *Harvard Business Review*, 80(4):87–94.

Evans, J. R. and Lindsay, W. M. (1993). *The Management and Control of Quality.* St. Paul, MN: West Publishing Company.

Grant, E. L. and Leavenworth, R. S. (1988). *Statistical Quality Control.* New York: McGraw-Hill Company.

Mobley, W. H. (1982). *Employee Turnover: Causes, Consequences, and Control.* Reading, MA: Addison-Wesley.

Spector, P. E. and Fox, S. (2002). "An emotional-centered model of voluntary work behavior some parallels between counterproductive work behavior and organizational citizenship behavior." *Human Resource Management Journal Review*, 12(2): 269–292.

Appendix 10B: The value of industrial engineers in Lean Six Sigma organizations[*]

10B.1   Introduction

Lean Six Sigma (LSS) is a methodology that allows organizations to "maximize shareholder value by the fastest rate of improvement in customer satisfaction, cost, quality, process, speed, and invested capital" (George 2002). There are several approaches that can be utilized by LSS such as define, measure, analyze, improve, and control (DMAIC); define, measure, analyze, design, and verify (DMADV); or identify, design, optimize, and validate (IDOV). The DMAIC methodology utilized by Six Sigma is employed to reduce the resources wasted within existing processes (Yang and Haik 2009). To design a new product to be of Six Sigma quality, the design for Six Sigma (DFSS) approach is needed; this process operates on either the DMADV or IDOV methodology.

10B.2   Background

The standard improvement model of the LSS is DMAIC. The *define* phase establishes the project goal and defines the resources needed for the project. In the *measure* phase, the details of the process are measured. In the *analyze* phase, various root causes for the problem are identified and statistical data analyses are conducted to find the most important root causes of the problem. The root causes identified in the *analyze* phase are then improved upon in the *improve* phase. During the *improve* phase, it must be determined through the use of facts and data whether the solution found is viable. An organization utilizing Six Sigma makes changes in the process to attain the goal that was identified during the *define* phase. During the *improve* phase, the solutions are implemented to prevent the problems from reoccurring. During the various phases, a variety of tools including statistical evaluations are utilized (Pyzdek and Kellere 2009). These tools and how to use them are taught in various courses in industrial engineering (IE) curriculum. The knowledge and skills taught in the IE curriculum enables the Industrial Engineers to play a variety of roles in Six Sigma. The purpose of this appendix is to describe the relationship between the IE academic curriculum, the variety of roles played by the industrial engineers in Six Sigma organization, and the variety of engineering analysis tools used.

---

[*] By Erick C. Jones, Michael W. Riley, and Trevor Battieste.

10B.3   Aspects of Lean Six Sigma

There are three important aspects of an LSS project: data collection, statistical analysis and evaluations of data collected, and statistical evaluations. Data collection is an important aspect of any type of research study. There are two types of data collection techniques: quantitative and qualitative. Quantitative data collection utilizes numerical and statistical processes to answer specific questions (Pyzdek and Kellere 2009). Statistical data analysis can result in descriptive measures (mean, mode, median, range, and standard deviation) or inferential measures where numerous hypotheses can be tested. Statistics can be used in a variety of ways to support inquiries or program assessments/evaluations. Data can be collected in a variety of ways such as through experimentation. Qualitative data is obtained through methods such as interviews, on-site observations, and focus groups that are in narrative rather than numerical form (Maxwell 1996; Patton 1990, 2000; Steckler et al. 1992) and are then analyzed by searching for themes and patterns. The second feature of a LSS project is performing statistical evaluations on the data collected to predict the results. Some of the tools used in the statistical evaluations are regression analysis, control charts, scatter plots, and ANOVA. The final aspect of an LSS project is to provide results of the statistical evaluation. In this phase, the important causes of the problems, the economic impact of these issues, and the methods to reduce and control these issues are provided. The IE discipline trains engineers in all these aspects through the various courses throughout a typical undergraduate degree program.

10B.4   Roles in Lean Six Sigma

The different roles in a Six Sigma project are shown in the Figure 10B.1. The executive leadership role is fulfilled by a person who sets a vision and direction for the project. In most cases, this role is filled by a top level executive or manager. Champions are the leaders of the implementation of the project. This role can be filled by a district manager or senior manager. The financial backing required by the project is provided by the sponsors. The master black belt is the person who can teach, consult, or lead the technical aspects of a Six Sigma project. Black belts and green belts are typically the ones who lead or support the project.

Figure 10B.1 Role in Six Sigma projects.

10B.5   Academic discipline

10B.5.1   Aspects of the Lean Six Sigma

The courses offered in the IE discipline enables the engineers to handle the three aspects of LSS. First, IE's are prepared for data collection by the education in time study and work management. Further, the procedure to set up the experiments is aided by the study of design of experiments (DOEs). Second, IE's are educated on the statistical evaluation tools through multivariate statistics and probability courses. Third, senior capstone courses prepare the IE to make recommendations based on project assessment.

10B.5.2   Roles in Lean Six Sigma

The traditional roles (project manager, data analyst, continuous improvement engineer, etc.) of an industrial engineer are the same roles that the Six Sigma organizations assign to the black belt, master black belt, or champions. The academic education of an industrial engineer provides sufficient background to enable them to perform these roles. Some additional education may be helpful to prepare engineers to play the roles of champions, statistical evaluator, and economic justifier (Figure 10.1).

10B.6   Industrial roles

Industrial engineers are also trained in basic engineering and additional courses which enable them to act as a change agent. The various tools used in LSS are part of the IE curriculum. Many IEs suggest that LSS is nothing more than using IE tools to focus on financial performance instead of efficiency improvement. The versatile nature of the industrial engineers helps them to perform in any part of the LSS projects. Also many tools used in the Six Sigma like flowcharts or cause-and-effect diagrams are included in most IE academic curriculum.

10B.7   Six Sigma–IE interface framework

In this chapter, we introduce a Six Sigma–IE interface framework (SS–IE IF). The framework describes three components that relate Six Sigma initiatives with IE. The components include

1. Six Sigma–Industry interface
2. Six Sigma–IE academic interface
3. Six Sigma–IE specialized knowledge interface

We describe those components and demonstrate how the framework can reveal the value of using IEs in all of the Six Sigma roles. *The Six Sigma Interface* shows the DMAIC process and the specific tools that are utilized in those steps. These tools have been envisioned to support the successful implementation of Six Sigma projects. Most of the tools described are taught in an IE undergraduate program and most IE and engineering management (EM) curriculums. Table 10B.1 shows the tools typical taught to IEs that allow an industry to execute Six Sigma DMAIC. Typically, the Lean aspects of Six Sigma or the Lean in LSS are demonstrated in the *control* phase. Specifically, the Lean tools include 5-S, TPM, mistake proofing, visual systems, and value stream mapping.

Table 10B.1 Six Sigma–Industry Interface

Define	Measure	Analyze	Improve	Control
Interview process	Hypothesis testing	Process capability analysis	Factorial design of experiments	Control plans
Language processing	Analysis of variance	x–y map	Fractional factorials	Visual systems
Prioritization matrix	Quality function deployment	FMEA	Data mining	5-S
System map	Flow down	Multi-vari chart	Blocking	TPM
Stakeholder analysis	Measurement system analysis	Chi-square	Response surface methodology	Mistake proofing
Though process map	Graphical methods	Regression	Multiple response methodology	SPC/APC
Value stream mapping	Process behavior charts	Buffered tolerance limits	Theory of constraints	

Six Sigma–IE academic interface describes the general academic classes that cover the topics and theory that support the correct use of Six Sigma tools. The premise is that to correctly use tools such as regression, multi-variable analysis, and statistical process control (SPC), it is important to have foundational knowledge of calculus-based statistics. Classes such as engineering statistics (descriptive, hypothesis testing, inferential), applied statistics (ANOVA, regression, design of experiments), and statistical quality control allow for mastery of these tools. Industrial quality control focuses on capability ratios whereas applied statistics and quality control focuses more on control charts. Oftentimes, many pundits who criticize Six Sigma identify that most statistical components are not correctly used or understood, minimizing the effectiveness of these tools. Utilizing IEs in the execution of these tools would provide a higher probability of using these tools correctly and better results for Six Sigma projects and initiatives. Table 10B.2 shows the interface between the IE academic curricula and the Six Sigma process. The course names used in this study are typical course names and are used to represent courses that are typically available in IE undergraduate and graduate programs. Additionally, some courses such as project management and advanced classes in engineering economics may be provided

Table 10B.2 Six Sigma–Industrial Engineering Academic Interface

Define/academic discipline	Measure/ academic discipline	Analyze/ academic discipline	Improve/ academic discipline	Control/ academic discipline
Project management, engineering management, industrial quality control, engineering economy	Applied statistics and quality control, engineering statistics and data analysis	Industrial quality control, engineering statistics and data analysis, applied statistics and quality control	Quality engineering: Use of experimental design and other techniques, engineering statistics and data	Industrial quality control, applied statistics and quality control

Table 10B.3 Six Sigma IE Specialized Knowledge

Special knowledge	Industrial tools	Academic courses
Statistics	Multi-vari chart, chi-square, regression, factorial design of experiments, measurement system analysis (Pyzdek and Kellere 2009)	Applied statistics and quality control, engineering statistics and data analysis
Lean	5-S, TPM, mistake proofing (Pyzdek and Kellere 2009)	Industry quality control, applied statistics and quality control
Leadership	Value stream mapping, though process map, stakeholder analysis, system map, interview process, prioritization matrix (Pyzdek and Kellere 2009)	Project management, engineering management, industry quality control, engineering economy

Table 10B.4 Job Integration by Six Sigma Step

Define	Measure/analyze/improve	Control
CEO, COO, plant manager, project manager, operations manager	Project manager, engineer, consultant, programmer	IE consultant, safety engineer, ergonomist

as a graduate class in IE departments. These courses would teach the student how to perform a stakeholder analysis which is commonly used in the *define* phase.

Six Sigma–IE specialized knowledge interface as shown in Table 10B.3 highlights the special knowledge of the industrial engineer (IE). IEs can also participate in Six Sigma in specialized roles that can support Six Sigma initiatives. Some Six Sigma organizations create specialized positions that assist them in successfully implementing Six Sigma. Some of these positions include statistics specialist, Lean specialist, and leadership liaison (or change agent). Separating these roles from traditional Six Sigma roles provides an organization the ability to supplement their Six Sigma process. An additional role not shown is economist, which is a role in which the specialist evaluates and verifies the earnings or savings for the project. This role typically works in close partnership with accounting.

The *integration of the job functions and opportunities for IEs in Six Sigma* is not included in our framework, but it is an important component for IE visibility in Six Sigma organizations. This integration is shown in Table 10B.4. Six Sigma is based upon the plan–do–check–act (PDCA) premise. We show how different job functions are identified under both PDCA and Six Sigma steps to show the flexibility of the frameworks use for non-Six Sigma organizations. Overlaying this table with the other tables describes how the IE who seeks these positions can identify the skills they learned and can utilize them in those job functions. For example, if an IE works as a CEO, he/she will use skills learned in EM. This table also suggests that those skills will allow one to effectively perform the stakeholder analysis that is aligned in the *define* phase when executing Six Sigma events.

10B.8   Implementation of the framework

We demonstrate examples of how to use the framework. For example, a CEO (Table 10B.4) who represents the Six Sigma role of executive leader (Figure 10B.1) would use project management (Table 10B.2) knowledge in the *define* stage (Table 10B.3) as seen in Figure 10B.2. In addition, the leader may use an interview process (Table 10B.3) as special knowledge

CEO	Define	Project management	Leadership
Table 4 ⟹	Table 1 ⟹	Table 2 ⟹	Table 3

Figure 10B.2 Example 1 of how to use the framework.

IE consultant	Control	Industry quality control	Lean
Table 4 ⟹	Table 1 ⟹	Table 2 ⟹	Table 3

Figure 10B.3 Example 2 of how to use the framework.

for leadership. A specific example is the CEO leading an executive round table using interview process.

Another example is that of IE consultants (Table 10B.4) using their industrial quality control (Table 10B.3) knowledge to perform as Lean Trainer (Table 10B.1) in the *control* phase of the Six Sigma project as seen in Figure 10B.3.

10B.9   Proposed study

A test instrument will be utilized to test the validity of this framework. The questionnaire will be used to quantify what role an IE has at the Six Sigma company, which tools they utilize, and which job function or title they have at the company. The test population will include personnel who work at Six Sigma companies and participate in one of the active roles described in the chapter. Additionally, a second instrument will be utilized to verify the IE curriculum and body of knowledge outlined in this chapter. A sample set of IE programs will be evaluated by their posted curriculum and interviews from various faculty and department heads of IE departments. At the conclusion of the analysis, the framework will be updated so that it provides an accurate depiction of IE–SS interface.

10B.10   Conclusion

Many courses in the IE discipline provide the foundation for various roles in Six Sigma initiatives that are implemented in many companies. This chapter describes how the foundational knowledge that is important in Six Sigma is supported by the coursework provided in most IE curriculums. We also describe how typical role and or jobs that are associated with Six Sigma are related to IE curricula or IE body of knowledge. We provided an industrial engineering–Six Sigma interface framework that shows how the Six Sigma roles utilize the IE body of knowledge in order to use the standard Six Sigma tools. This chapter also provides a path forward for organizations that seek to extend Six Sigma with DFSS and move toward a new process and product development using the IE body of knowledge.

This chapter shows how the IE field has provided a critical foundation to Six Sigma. It can be implied that the IE body of knowledge has been reorganized and marketed to create Six Sigma given that Six Sigma foundation is based upon PDCA principles popularized in the IE field. All IEs should be able to utilize this document to demonstrate that IEs can play a critical part to successful Six Sigma initiatives at most companies. The framework shows that IEs can play all Six Sigma roles and additional roles such as statistician, Lean leader, economic auditor and consultant. This framework should be used to ensure that IEs can market and demonstrate the skill set during Six Sigma initiatives.

References

George, M. L. (2002). *Lean Six Sigma: Combining Six Sigma Quality with Lean Speed*. New York: McGraw-Hill.

Maxwell, J. (1996). *Qualitative Research Design: An Interactive Approach*. Thousand Oaks, CA: Sage.

Patton, M. Q. (1990). *Qualitative Evaluation and Research Methods*. 2nd edn. Newbury Park: Sage Publications.

Patton, M. Q. (2002). *Qualitative Research and Evaluation Methods*. 3rd edn. Thousand Oaks, CA: Sage.

Pyzdek, T. and Kellere, P. (2009). *The Six Sigma Handbook*. 3rd edn. New York: McGraw-Hill.

Steckler, A. et al. (1992). "Toward integrating qualitative and quantitative methods: An introduction." *Health Education Quarterly*, 19:1–8.

Yang, K. and Haik, B. E. (2009). *Design for Six Sigma: A Roadmap for Product Development*. 2nd edn. New York: McGraw-Hill.

The science and engineering of Lean Six Sigma

chapter eleven

LSS process, steps, tools, and statistics

For as laws are necessary that good manners may be preserved, so there is no need of good manners that laws may be maintained.

—Machiavelli

Overview	
Steps	Substeps
Define	• Quality projects (sponsorship)
	• Define specific project (black belt)
Measure	• Identify critical characteristics
	• Clarify specifics and targets
	• Measurement system validation
Analyze	• Determine baseline
	• Identify improvement objectives
	• Evaluate process inputs
Improve	• Formulate implementation plan
	• Perform work measurement
	• Lean execute the proposed plan
Control	• Reconfirm plan
	• Set up statistical process controls
	• Operational process transition

Source: PWD Group LLC.

Recommended Tools by Belt						
		Define	Measure	Analyze	Improve	Control
Stage 1	Yellow belt	Interview process Language processing System map (flowchart)		X–Y map/QFD/ house of quality Fishbone diagram	EVOP	Control plans Visual systems 5-S SPC/APC
Stage 2	Green belt	Prioritization matrix Thought process map Stakeholder analysis Thought process map	Hypothesis testing Flowdown Measurement system analysis Graphical methods Analysis of variance	FMEA Multi-vari chart Chi-square Regression Pareto analysis	Design of experiments Fractional factorials Data mining Blocking	TPM Mistake proofing

(Continued)

Recommended Tools by Belt (Continued)

		Define	Measure	Analyze	Improve	Control
Stage 3	Black belt	Value stream map Gantt chart development Project management	Process behavior charts QFD	Buttered tolerance limits Process capability analysis	Theory of constraints Multiple response optimization Response surface methodology	

Source: PWD Group LLC.

Notes: Black belts are expected to know the continued knowledge obtained throughout green belt and yellow belt certifications. The same goes for green belt and yellow belt. Empty cells represent that skill is not required for that belt stage; however, a general knowledge of the step is expected.

5-S, sort, set in order, shine, standardize, and sustain; APC, advanced process control; EVOP, evolutionary operations; EVOP, evolutionary operations; FMEA, failure modes and effect analysis; QFD, quality function deployment; SPC, statistical process control; TPM, total productive maintenance.

11.1   Introduction

No matter which approach is deployed for improvement teams in any organization, the team will need to know what is expected of them. That is where standard improvement processes like DMAIC and design for Six Sigma (DFSS) are of great help because it provides the team with a road map. Several sources provide knowledge about DMAIC. The purpose of this book is to focus on Lean Six Sigma (LSS) processes implementing DMAIC and DFSS methodologies using efficient tools that are helpful and also provide a clear picture of how to implement an improvement program in any organization. Finally, case studies are illustrated on how these tools are deployed effectively in Six Sigma projects.

One of the advantages of implementing LSS approach is that there cannot be any competition between Lean and Six Sigma approaches. Different certifications are provided to make it easier for organizations to hire professionals to implement quality programs. The various Six Sigma certifications are yellow, green, black, master black belt, and champion. By training the employees and giving the certifications, it makes them eligible to employ the tools that are associated with each belt and implement them in an organization. These trainings are necessary because wrong selection of tool or phase will ruin the entire project making Six Sigma projects worthless. This will make the sponsors even more reluctant in encouraging the Six Sigma projects. So, this book explains the importance of every tool and its appropriateness in usage.

In general, Six Sigma methodologies can be separated into operation Six Sigma (OSS) and DFSS. The define, measure, analyze, improve, control (DMAIC) roadmap is commonly referred to operational Six Sigma (Breyfogle 2003). DMAIC has been traditionally applied for improving existing manufacturing processes and products; however, it is also valuable for design or improving nonmanufacturing processes. DMAIC methodology should be used when a product or process is in existence at the company but is not meeting customer satisfaction or is not performing adequately (Figure 11.1).

The standard improvement model of the LSS is DMAIC. The *define* phase establishes the project goal and defines the resources needed for the project. In the *measure* phase the details of the process are measured. In the *analyze* phase, various root causes of the problem

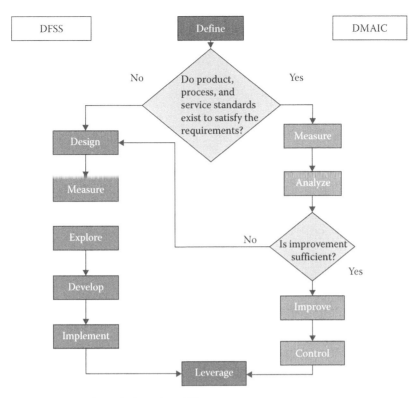

Figure 11.1 Comparison of DMAIC and DFSSR.

are identified and statistical data analyses are conducted to find the most important root causes of the problem. The root causes identified in the *analyze* phase are then improved upon in the *improve* phase. During the *improve* phase, it must be determined through the use of facts and data whether the solution found is viable. An organization utilizing Six Sigma makes changes in the process to attain the goal that was identified during the *define* phase. During the *improve* phase, the solutions are implemented to prevent the problems from recurring (Figure 11.2).

As mentioned earlier, DMAIC is related to the plan–do–check–act (PDCA) quality methodology. With reference to Six Sigma implementation as a structured and system atic process improvement methodology, little has been said on the design and structure of Six Sigma projects. While companies implement Six Sigma in a variety of ways, it is important to determine how much of those efforts are done in a systematic and structured way.

It is important to note that the success of any quality improvement project requires the coexistence of both attention to detail and innovation. While a structured method facilitates attention to detail, contextual variables such as the role of black belts and executive commitment are the key to provide a cultured innovation (Naveh and Erez 2004). The literature on Six Sigma has paid little attention to the role of black belts in Six Sigma projects and how they can facilitate successful implementation of Six Sigma.

PDCA framework can be related to Six Sigma implementation for process improvement. The steps in Six Sigma (DMAIC or DMADV) are similar to PDCA (Choo et al. 2007; Shewart 1931, 1939). The PDCA cycle is well established for process improvement where it focuses

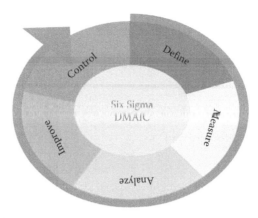

Figure 11.2 Quality circle for DMAIC.

on continuous learning and knowledge creation (Deming 1993). While companies may use different terminologies for their Six Sigma projects, by adhering to a united and well-established framework such as PDCA, attempt is made to integrate Six Sigma projects under the PDCA cycle. Thus, framing Six Sigma projects within a PDCA cycle provides a more comprehensive view of Six Sigma implementation in organizations. The basic constructs of Six Sigma project implementation given by Jones et al. (2011) are listed as follows:

- Black belt roles
- DMAIC
- Plan
- Do
- Check
- Act
- Financial responsibility
- Executive supports

There are differences among the companies while implementing Six Sigma projects. While companies such as GE reports millions of dollars in savings from Six Sigma projects, Six Sigma projects do not benefit all organizations. In other words, to benefit from Six Sigma projects, companies need to follow the methodological aspect of Six Sigma (e.g., PDCA), while addressing the contextual variables (e.g., leadership commitment). In that regard, we expect to see differences in the implementation of Six Sigma projects and hence there is a significant difference among organizations regarding Six Sigma implementation. Organizations that follow a structured methodology for Six Sigma outperform organizations that do not follow a structured methodology. Revised hypothetical statements are provided by Jones et al. (2011), which helps in making decisions for DMAIC and DFSS research (DFSS-R) models.

11.2   Hypothetical statements

- Executive support significantly affects the role of black belts in Six Sigma.
- Executive support significantly affects the financial responsibilities of team members in Six Sigma implementation.

- Financial responsibility of Six Sigma member positively affects the selection of appropriate methodology (DMAIC).
- The role of black belts positively affects the selection of appropriate methodology (DMAIC).
- The selection of the appropriate methodology (DMAIC) has a significant effect on the plan phase of the PDCA cycle.
- The selection of the appropriate methodology (DMAIC) has a significant effect on the do phase of the PDCA cycle.
- The selection of the appropriate methodology (DMAIC) has a significant effect on the check phase of the PDCA cycle.
- The selection of the appropriate methodology (DMAIC) has a significant effect on the act phase of the PDCA cycle.
- The plan phase in the PDCA cycle has a significant effect on the performance of Six Sigma projects.
- The do phase in the PDCA cycle has a significant effect on the performance of Six Sigma projects.
- The check phase in the PDCA cycle has a significant effect on the performance of Six Sigma projects.
- The act phase in the PDCA cycle has a significant effect on the performance of Six Sigma projects (Table 11.1).

11.3   Define

11.3.1   Chartering a Six Sigma project

The key point that drives the motivation to improve a process or product is to find a compelling and concrete reason to initiate an opportunity for a business investment. The task of investigating the root cause of a problem is herculean due to the complex nature of interrelated processes. The diagnosis of the case is done in a sequential order as shown in Figure 11.3.

- The issue is to be pulled out through the labyrinthine network of tasks that support the objective.
- The stakeholders involved in the supply chain: suppliers, customers are defined.
- The area of application or scope of the project is recognized.
- The effects are characterized through analysis using quantitative proof from the realm of statistics such as variability, capability performance measures, financial cash flow equations, etc.
- The problem statement needs to be formulated keeping in mind the SMART objective. SMART implies a specific, measureable, achievable, relevant, and timed objective that is oriented to the organization's overall strategic aim. It gives rise to the goal statement of defining the positive outcomes of undertaking this project.
- A team is formed to aid this process and performs activities for recognizing customer requirements.
- A charter is created which is a comprehensive documentation of all of the above details.

The problem statement must not be presumptive of the solutions and results. Its impact is stated through a measured variable for cost. Appropriate assumptions are taken into consideration before presenting the business case diagnosis that delegates business

Table 11.1 Tools for Each Phase in DMAIC

Define	Measure	Analyze	Improve	Control
Project selection tools	Operational definitions	Pareto charts	Brainstorming	Control plans
PIP management processes	Data collection plan	C & E matrix	Benchmarking	Standard operating procedures
Value stream map	Pareto chart	Fishbone diagrams	TPM	Training plan
Financial analysis	Histogram	Brainstorming	5S	Communication plan
Project charter	Box plot	Basic statistical tools	Line balancing	Implementation plan
Multi generational plan	Statistical sampling	Constraint identification	Process flow improvement	Visual systems
Stakeholder analysis	Measurement system analysis	Time type analysis	Replenishment pull	Mistake proofing
Communication plan	Control charts	Nonvalue-added analysis	Sales and operation planning	Process control plans
SIPOC map	Process cycle efficiency	Hypothesis testing	Setup reduction	Project commissioning
High-level process map	Process sizing	Confidence intervals	Generic pull	Project replicating
Nonvalue-added analysis	Process capability C_p & C_{pk}	FMEA	Kaizen	PDCA cycle
VOC and Kano analysis	QFD	Simple and multiple regression	Poka-yoke	5S
QFD	Flow down	ANOVA	FMEA	TPM
RACI and Quad charts	ANOVA	Queuing theory	Hypothesis testing	SPC/APC
Stakeholder analysis	Review with sponsor	Analytical batch sizing	Solution selection matrix	Review with sponsor
Prioritization map		Chi-square	"To-be" process maps	
Review with sponsor		x–y map	Piloting and simulation	
		Review with sponsor	Theory of constraints	
			Fractional factorials	
			Data mining	
			Blocking	
			Response surface methodology	
			Review with sponsor	

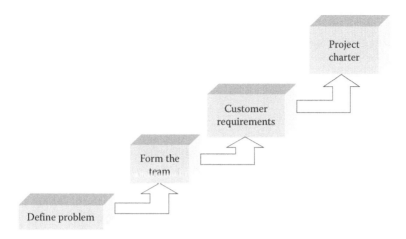

Figure 11.3 Define—plan.

implications and records the study undertaken. It is viewed as a "living" document. The responsibilities of team members are assigned on the basis of experience (belt) level.

11.3.2   The assignments of the team

The assignments of the team include the following:

- Assessing voice of customer
- Examining indexes and departmental guide books, blueprints, master schedules, etc.
- Auditing recorded data and examining within the constraints and boundaries laid out
- Interpreting the gap in performance in terms of costs incurred

A progress forecast for the target is framed according to the deadline placed and a plan is communicated in appropriate format such as an email. It is for the purpose of tracking the progress in the uprooting of the problem using tools such as Gantt charts (bar charts). A report is prepared finally to project the performance and cost–benefit ratios for the process changes and phase transitions in the project depicted against a timeline. Other tools such as project evaluation and review technique (PERT) and critical path method (CPM) also supplement the evaluation of timeline. Activity flow structure is established through work breakdown structure (WBS).

A high-level process map—suppliers or the provider of inputs; inputs or resources; process or set of activities to transform inputs to objective outputs; outputs or specified results; customers or recipient of output, identifies the customer.

The levels of customer categorization are decided by the Kano model as, given by

- Dissatisfiers (basic)
- Satisfiers (variable)
- Delighters (latent)

Voice of the customer is the first step to assigning weights to customer requirements. It helps build the decision matrix by considering cost, quality, features, and availability

Figure 11.4 Critical to quality.

(CQFA) factors to frame a value grid and matching with the relevant customer base for competency. The voice of customer usually implies use of quality function deployment (QFD) for providing technical characteristics to satisfy customer's needs as well as that which directly affects customer retention and service level.

11.3.3   Methods to collate information

- Surveys—multilevel, targeted
- Interviews—multilevel, targeted
- Market research
- QFD
- Data warehousing
- Data mining
- Customer audits, etc. are methods to collate information

Segregation of this information is done in accordance with the type of customer:

1. Internal
2. External

Analyzing the information through statistical tests, line graphs, control charts, Pareto analysis provides a criticism given the sampling errors. Customer requirements not completely met, inefficiencies, waste sources are some of the variation sources that need to be concentrated upon. A critical-to-quality (CTQ) tree is used to quantify these requirements from the described by VOC. The flow diagram for this step is given below. The numerical format is designed by validating requirements through needs for every customer level (Figure 11.4).

　　Project management is a necessary practice to maximize the benefits of a business by streamlining the working of the team with the goals of the management of the organization. The duties for the team managers also involve ensuring the application of appropriate functional elements through stages in phase with time and in a cost-effective manner. The effectiveness is measured by reduction in variance and obtaining results such as cycle time reduction, improved workflow, reduction of internal defects, reduced yield loss, customer returns, warranty claims, etc. The effectiveness of this phase lies in the selection of the right tools.

11.4   Measure

Measure phase paves way for feedback in the system. This phase initiates the necessity of data from the management. Once a project is defined, all the information is gathered from the existing process. Information includes key input variable and key output variable. The key input variable drives the key output variables in the system. This phase concentrates on evaluating each supply chain component. The main aim of the measure phase is to analyze the criticality and severity of the problem.

The key activities involved in the measure phase are listed as follows:

- Identify issue of measure and process error
- Gather data for the above purpose. As mentioned earlier, this triggers the result of triangulating to the issue which causes lowering of the stability of the system.
- The collected data are interpreted using different modes of communication such as charts, graphs, etc.
- Then the normal baseline for the process is calculated which enables comparing it with the obtained results.

Some of the quantified tools used in the measure phase are listed as follows:

- Data collection and accuracy
- Process description
- Focus and prioritization
- Generating and organizing ideas
- Quantifying and describing variation

11.4.1 Data collection and accuracy

In order to verify the accuracy of the data collection and data logging carried out by the team, it has to be checked for accuracy. Data collection and accuracy is used for accounting for the variation in the measurement and measuring system. Common tools used for accounting data collection and data accuracy are gage repeatability and reproducibility (GR&R) and check sheets. This is done to improve the reliability of the system. This tool is concerned with the process being measured for the same metric over a finite number of trials and obtaining nearly the same result every trial. It is also concerned with different people taking the same measurement on the same metric and arriving at an accurate result. Check sheet is the most effective data-recording technique. As the event occurs, a check mark is placed on the check sheet. As a number of events occur, the observer uses the same check mark to mark the occurrence of the event.

11.4.2 Process description

When a project is identified, information regarding the existing process is gathered. Important information on the key input variables that drive the key output variables are identified in this phase. Some of the tools that come under process description tools are listed as follows:

- Process complexity value stream map
- Process cycle efficiency
- Time value analysis

Complexity value stream map helps us to identify the value-added and nonvalue-added process in the process flow diagram. This also aids in identifying the targeted output and product flow. Process cycle efficiency gives the relationship between the amount of value-added time and the total cycle time in the entire process. It is calculated as the ratio of the amount of value-added time to the total process cycle time. Time value analysis separates value-added and nonvalue-added time in the process.

$$\text{Process cycle efficiency} = \frac{\text{Amount of value-added time}}{\text{Total process cycle time}} \times 100. \tag{11.1}$$

11.4.3   Focus/prioritization tools

This tool is generally used for identifying the problems faced in the system. Some of the tools used in focus/prioritization are given as follows:

- Pareto chart
- Failure modes and analysis
- Cause-and-effect matrix

This tool is used to rank the significance for the occurrence of the events that cause the problem. It is ranked right from the most significant problem to the least significant problem. Pareto chart is constructed by selecting the subject of interest to measure and gather data regarding the number of quantity and the cost associated with failure or quality problems. FMEA is a tool which is used for analytical approach to determine the failure modes and its potential causes. This phase works in identifying the failure mode based on the past data and products or processes using the shared failure mechanism logic. Cause-and-effect matrix tool is used to segregate the least important variables from the useful variables which affect the key output variable.

11.4.4   Quantifying and describing variation

Describing variation is the important aspect in the measure phase, because variation in quality can end up in rework and scrap. It further increases the overhead, lead time, etc. To reduce variation, it must be spotted and causes be identified. Different types of quantifying and describing variation tools are described further.

- Control charts
- Process capability charts

As discussed earlier, this tool is used to analyze the variation in data set. The variation in the data set can be normal or can be noticeable due to the variables which affect the key output variable. There are two types of variation; they are common-cause variation which is always present in the process to a greater or lesser extent. Another variation is special cause variation which occurs due to the change in the variable that greatly affects the key output variable. Process capability discusses about how well the process fits in the customer range of specifications. The main thing to ensure before assuring capability is to check for process stability. It is mandatory to have stable results from the control chart before accounting for process capability.

11.4.5 Generating and organizing ideas

Brainstorming and cause-and-effect diagram are the two tools used in generating and organizing ideas. Brainstorming is the technique that enables team members to generate different ideas that will suit certain process. The effective brainstorming involves the recording of all ideas developed during the brainstorming session. Cause-and-effect diagram is also known as fishbone diagram which comes under organizing tools. It helps in identifying the potential causes for the problem and also to investigate it thoroughly. It is an important tool for brainstorming.

11.5 Analyze

By the end of measure phase, the Six Sigma team should have enough amounts of data and information about the case under study. *Analyze* phase helps in drawing relationships between the cause and effects. It provides clear roots of the causes that make the variations in the process and so on. So, much of the *analyze* phase is carried out in order to link the relationships between the input and the output variables. To perform the *analyze* section effectively, there are certain tools to be used which are listed below. As mentioned earlier, there are belt certifications as yellow, green, and black followed by master black belt. The tool described does not comply for all the belts. The distinction of tools based on belts is explained in later chapters.

- Pareto charts
- Ishikawa diagram
- Scatter diagram
- Hypothesis testing
- Design of experiments (DOEs)
- Regression analysis
- Data sheets
- Run charts
- Histograms
- ANOVA

The main objective of the *analyze* phase is to verify the root cause of a problem using the above-mentioned tools and the output of the phase should contain process analysis, root cause analysis, and quantification of the performance gap. The detailed causes of the effects analyzed in this phase helps the *improve* phase use time effectively to correct the causes creating the underlying problems. The credibility of the *analyze* phase can be reviewed with following questions by Michael L. George (2002):

- What fatal causes are the team going to work on in the *improve* phase?
- What are the major reasons to focus on only specific causes?
- What are the other fatal and potential causes the team investigated and how is it concluded that they are not the actual causes?
- What indicates that addressing the cause will improve the selected improvement process under study?

The Ishikawa diagram is the most efficient tool to study the cause-and-effect relationships. Statistical tools such as regression analysis and hypothesis testing. Often people use their

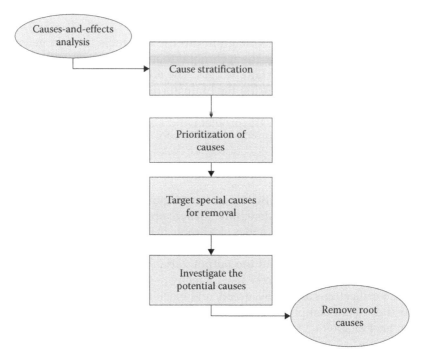

Figure 11.5 Pictorial representation of *analyze* phase.

gut feeling or intuition to understand the cause of the problem and try solving with little or no information. This might lead to jumping to causes and not exactly identify the root cause underlying the problem. In that case, the statistical tools are used to provide credible information on root causes (Figure 11.5).

11.6 Improve

The *improve* phase begins at the end of the *measure* phase. The objective is to pursue a logical link between the causes and actions. Throughout most of the *measure* and *analyze*, a team is challenged to bring out their creativity and inclusively in deciding the potential causes to be investigated, how the data is to be displayed, and how to interpret the messages. The team in this stage is required to think broadly in a more focused manner with a practical mindset, now that they know what the causes are, what specific changes can they make in the process to come over the causes, and which methods will achieve the desired affect?

The process moves on to *action* from the *analyze* step. The figure below shows the three-step process of the improve phase. The generation and implementing of viable solutions that will root out special causes is the purpose of the *improve* phase. The goal of these solutions is to perform corrective action on root causes and also should be financially feasible to implement. An effective implementation plan to install the solution must be designed, once the best solution is found (Figure 11.6).

Several tools and activities are required to obtain an effective improvement plan. The quality certification one holds determines the type of tools he/she can employ for the improvement plan. The tools for each belt are segregated in the future chapters. Of all the tools that are employed for the DMAIC, those most commonly used in the *improve* phase

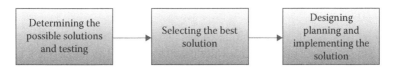

Figure 11.6 Three step solution for the *improve* phase.

involve the broadest mix of both Lean and Six Sigma tools. Some of the tools of Lean used in the *improve* phase are total productive maintenance, pull systems, setup reduction, and so on, and are used to eliminate work-in-process (WIP) and time delays; DOEs and process mapping are tools that represent approaches that are obtained from Six Sigma/quality improvement tradition.

Several tools that are most commonly used in the *improve* phase of LSS are as follows:

1. Brainstorming
2. Cost–benefit
3. Benchmarking
4. Failure mode and effects
5. Analysis
6. Simulations
7. Scatter diagrams
8. DOEs
9. Action plans
10. PDCA

The first four activities are used to obtain solutions that will bring out the root causes. The remaining activities are performed to compare the alternative solutions, its related cost–benefit analysis, a "should-be" process map, and any related data. Once the solution team identifies a solution, it is optimized based on testing and analysis. The removal of the root causes and the problem should be done before proceeding to the implementation. The same tools can be used to contrast the problem before and after the test customer feedback can also be employed. The results can be compared to the target of the performance to be achieved. The new process and the new inputs are to be designed by the team, once the testing is complete.

The next step in the improvement plan is the development of the implementation plan. The plan should be able to answer the basic who, when, where, and how questions. It should include a schedule for the plan to be carried out, the resource requirements, and a structure of the work breakdown. The plan should also contain the appropriate approvals from the executive sponsor and the top management. Once all of the above procedures are completed and the approvals are attained, the improvement plan should be fully implemented.

In this phase, there is a focus on implementation. Some of the most common questions one would ask are listed as follows:

- What measures to counteract the problems did the team develop?
- How did the decision among alternatives be decided for the improvement plan?
- How did the team realize that the measures taken by them would affect the causes confirmed in *measure*?
- What happened when the first step of implementation was done? What changes were required to be made to refine the improvements?

11.7   Control

Control phase ensures the future operational performance of the system. This phase enables to implement the solution obtained from the *improve* phase. Decisions for this process are made based on adopt to change, abandon it, or repeat the problem solving step. If it is practical, then the new procedure is made operational. The modified solution ensures the long-term performance. Once the new operating procedures are deployed, new measurement and control system are also developed. The team selects appropriate statistical process and control charts. This paves the way for favorable performance and cost saving for the system through the *control* phase.

Control phase is used to maintain the performance of the updated model unless there is a better way. This phase also provides inputs to monitor the process so that the problems can be resolved when there is any anomaly. The common tool used in *control* phase is statistical process control.

Statistical process control is a tool for the *analyze* phase using control charts so that processes can be monitored for changes due to nonrandom factors based on the assumption that all processes are variable, these variations can be natural, and assignable to a cause. Statistical process chart is used to achieve process control by implementing it to the key input variables. The main components of statistical process control are as follows:

- Create a control chart
- Isolating and removing special causes of variation
- Procedures for immediate detection and correction of future problem

Control chart is a tool which is used to analyze the variation in data set. The variation in the data set can be normal or can be noticeable due to the variables which affect the key output variable. Control chart describes the stability of the process. It is concerned with gathering specific set of data plotted against the sample size. Components present in the control charts are data points, center line, and control limits. These inputs are fed to the process capability analysis. The constraint applicable is that the process has to be stable for it to analyze for its capability. This information determines the natural variation of the process which is the control parameter that must lie within a specified tolerance range.

Control limits can be obtained using control chart. If there are any points outside the control limit then it is a problem that has occurred, which causes deviation in the process. These deviations can be analyzed by investigating the changes in the process. Specific tools such as cause-and-effect diagram can be used to isolate the source for the problem. The specific variation is reduced by the above process and the common variation is only left out. The common variation occurs due to machine used, activities followed, etc. This can be reduced by bringing in changes to the process.

The above-mentioned methods ensure that the process is in control and deliver outputs according to the customer's specifications. As mentioned above, control phase includes stability and performance of the process in future. This component of the step can be achieved using two methods, the first method being: to train personnel in usage of control chart and the second being: to formulate contingent procedures for special cases. These cases can be categorized based on type of action.

- Damage control
- Remedial action

Attributes and variable chart contribute to benefits of using this predictive tool that is effective for continuous change efforts too. These benefits can be translated to profits and it is found that the cost incurred for data collection personnel training investigation and correction duly compensate. The efficiency of control charting is determined by appropriate selection of process parameters that are to be controlled. These variables can be identified using DOEs and analysis of variance such as Pareto analysis and subgrouping.

These critical variables are thus named as key process input variables (KPIVs). Key process output variables are the outcomes that determine the process capability based on degree of effect of KPIV on a process.

Types of control chart for variables are as follows:

- $\bar{X}$-R charts
- Run charts (limited data-single point)
- MX-MR charts (limited data-moving average)
- $\bar{X}$R-MR chart (limited data)
- X's chart (Sigma known)
- Median chart
- Q sum chart (cumulative charts)
- Moving average
- Exponentially weighted moving average (EWMA)

These charts provide valuable insights because each variable requires collating of data. Control chart for attributes are as follows:

- p chart (fraction defective)
- np chart (number of defectives)
- c chart (number of defects)
- u chart (number of defects per unit)

These attributes characterize measurements of process time. The possible outcomes of the process are as given in Table 11.2.

For the process to be out of control, it must have a limit of three standard deviations which is equal to 99.73% out of range.

Control plan is a document that describes the critical parameter of the process that translates the quality characteristics. Types of control plan are as per ISO/TS 16949:2002 and the advanced product quality planning (APQP) (2000) for automotive sector.

- Prototype
- Prelaunch
- Production

Table 11.2 Process Outputs

Average	Range	Control
Out of control	In control	Process out of control
In control	Out of control	Process out of control
Out of control	Out of control	Process out of control
In control	In control	Process in control

Control plan documentation must include team member, data of issue, process and sub steps, specifications, key variable, process capability, reaction plan, etc.

11.8 DFSS-R introduction

DFSS-R is derived from Six Sigma methodology. It was developed by Dr. Erick C. Jones in 2006 (Jones et al. 2011). DFSSR is a research methodology focused on reducing variability, removing defects, and getting rid of wastes from processes, products, and transactions. It contains three main predominant phases: plan, predict, and perform. The methodology is then broken down into seven main phases: define, measure, analyze, design, identify, optimize, and verify (Figure 11.7).

This figure gives a visual representation of the phases and how they relate to each other. Each of these phases will be briefly described within the following text.

11.8.1 Define

Within the *define* phase, a problem statement is to be created and the basis of material from which the process will draw information from is to be defined. It is critical to correctly denote all the information and resources that will be used for future phases of the process. This allows for the writer to begin framing his/her thoughts through discerning relevant subjects and further investigating the subjects necessary to enrich the subsequent phases. The *define* stage also serves as a springboard to educate the reader and prospective client for this method. Prospective tools that can be utilized are thought process mapping, system mapping, and interview processes among other observational techniques. Furthermore, standards of performance are created for the project to follow. Doing so aids in ensuring the process remains on course throughout the experiment as well as for evaluation purposes of what the overarching process is trying to achieve. The result of a successful *define* phase is a clearly defined problem statement and foundation for the methodology to propel itself upon.

11.8.2 Measure

The *measure* phase is that in which the metrics are established for the methodology. It is in this phase that the validation of measurement systems occurs. Additionally, the capability to accurately quantify this data and criteria for success must also be determined. This can be attained through the use of hypothesis testing, flowcharting, analysis of variance,

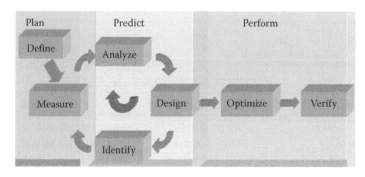

Figure 11.7 X DFSS-R.

and other graphical methods. It is crucial to establish a correct metric from the tools for it will determine if the solution will successfully address the problem. Without the correct metric, the rest of the methodology will be undesirable. Upon completion of this phase, the finalization of performance standards should be fortified.

11.8.3 Analyze

The *analyze* phase is that in which the identification of the sources of conflict occurs in addition to screening potential causes. It is here where tools such as fishbone diagram, Pareto analysis, multivariate charting, and other statistical observations transpire. Fishbone diagrams are useful for determining the causes of a problem and feeding them into one central issue. Pareto analysis can be applied in a multitude of ways to observe data and patterns from the figures. It focuses on the principle of the 80/20 rule in which 80% of the issue can be solved by addressing 20% of the problem. The results of this tool are then displayed in a Pareto chart with a graphical representation where the sources of the problem are displayed as well as a linear summation that details the contribution of each source. Other forms of regression and multivariate analysis are useful for interpretation and elimination methods. The result to take away from this phase is statistical information upon which empirical decisions can be made.

11.8.4 Design

Within the *design* phase, decision analysis and solution storming takes place. This is where possible solutions are created to address the problems obtained from the *analysis* phase. Some tools employed by the *design* phase are scenarios, multiple response optimization, and formulating theories of constraints. It is important to thoroughly deliberate a possible solution that does not create more problems. The goal of this phase is to come away with possible solutions which will be further investigated in the subsequent phases.

11.8.5 Identify

For the *identify* phase, the focus is on impact analysis. This section lays out what the return from the solutions will attain. Some mainstay tools for this section are cost–benefit analysis, return on investment (ROI) analysis, 5-S, and mistake proofing. These tools are used to justify the courses of actions being investigated by the DFSS-R. The ROI is a particularly beneficial tool to convey the financial aspects to potential investors. Cost–benefit analysis can have a general approach that lays out more than just financial returns which may produce a tipping factor when multiple financial avenues seem desirable. In addition, mistake proofing is pivotal to this phase where a solution is meant to fully address all potential issues. The take away is a direction for the methodology to pursue.

11.8.6 Optimize

The *optimize* phase is where it all comes together. It is here where the process establishes operational analysis and integration analysis of all the preceding phases. Fine tuning occurs of each phase to find the prime course of action. The scenarios are broken down and evaluated based upon how well they address the metric. New scenarios are compared to previously analyzed data and conducted as trials. These are then repeated, analyzed, and modified until a final methodology is definite.

11.8.7   Verify

Finally, it is the *verify* phase that proves the outcomes by results verification, redefining the new capabilities, and closing documentation. This includes implementation of what has been proposed by the methodology. To accomplish successful execution continuous improvement, control plans, and mistake proofing are focused upon. Mistake proofing is revisited as part of continuous improvement. Both of these tools are contained within a control plan that is to be conducted and evolved. From this phase, a final product and maintenance regiment is to be obtained.

References

Breyfogle III, F. W. (2003). *Implementing Six Sigma: Smarter Solutions Using Statistical Methods*. New York: John Wiley & Sons.

Choo, A. S., Linderman, K. W., and Schroeder, R. G. (2007). "Method and context perspectives on learning and knowledge creation in quality management." *Journal of Operations Management*, 25(4): 918–931.

Deming, W. E. (ed) (1993). *The New Economics for Industries, Government, Education*. Cambridge, MA: MIT Press.

George, M. L. (2002). *Lean Six Sigma: Combining Six Sigma Quality with Lean Speed*. New York: McGraw-Hill.

Jones, E. C., Mellat-Parast, M., and Adams, S. G. (2011). "Developing an instrument for measuring Six Sigma implementation." *International Journal of Services and Operations Management*, 9(4): 429–452.

Naveh, E. and Erez, M. (2007). "Innovation and attention to detail in quality improvement paradigm." *Management Science*, 50(11): 1576–1586.

Shewart, W. A. (1931). *Economic Control of Quality of Manufacturing Products*. New York: D. Van Nostrand.

Shewart, W. A. (1939). *Statistical Method from the Viewpoint of Quality Control*. Washington, DC: The Graduate School, Department of Agriculture.

chapter twelve

Fundamental statistics and basic quality tools

In God we trust
All others must use Data.

—W. Edwards Deming

12.1   Introduction

The goal of Six Sigma is to enhance quality by minimizing variability. In the process of variation reduction, several statistical methods are employed to verify the quality at different stages of the project. This verification involves planning, collecting, and analyzing the data on quality characteristics of the entire population from the sample observations. However, the variability in the population makes the verification challenging.

The science of statistics provides the means of defining the populations with variability and the methods for estimating the population quality from observed samples. The generalization of population quality is feasible for both products and services. The most commonly utilized statistical methods used are as given in Table 12.1 (adapted from Krishnamoorthi 2006).

To develop a good understanding of these methods and use them effectively, a good understanding of the basics of probability and statistics is essential. This chapter provides the basic discussion on the theory of probability and statistics, and the tools that will be utilized in the later chapters of the textbook. The chapter is presented in a different format than the rest of the chapters in this book with the intention to assist the reader in understanding the basics of statistics and fundamental quality tools.

12.2   Basic definitions

The basic definitions of population, parameter, sample, statistic, and data are as follows:

Population: The term *population* refers to the collection of items that are of interest. For example, an engineer might be interested in determining the defect rate in the M8 bolts manufactured in the third shift. In this case, the population or the item of interest is the M8 bolts manufactured in the third shift.

Parameter: It is the numerical summary measure used to describe a population characteristic. In most situations, we are interested in the mean (μ), variance (σ^2), and the count (N) within the population. For example, the average diameter of all the M8 bolts is 4.165 mm.

Sample: The term *sample* is a subset of the population. The process of selecting samples is known as sampling. There are many types of sampling: simple random sample, stratified sample, cluster sample, accidental sample, and others. In this textbook, we will limit our discussion to random sampling in which each item selected from the population has an equal chance of being included.

Table 12.1 Statistical Methods in Quality

Method	Application
Control charts	To control processes that will result in products of uniform quality
Sampling plans	To establish the adequacy of products from sample information
Designed experiments	Determining the best combination of levels of process parameters to obtain desired levels of quality characteristics
Regression analysis	Determining the variables that affect the quality characteristics and the extent of their effect
Reliability engineering	Understanding the factors that affect the life of parts and assemblies and improving their longevity
Tolerancing	Determining allowable variability in product and process variables so the products can be produced economically while meeting customer needs

Statistic: The term *statistic* is the numerical summary measure used to describe the sample characteristics. The equivalent sample characteristics: sample mean ($\bar{X}$), sample variance (s^2), and sample count (n).

Data: The term *data* refers to information that is being observed or collected. The data are broadly classified into two main categories: variable data and attribute data. The variable data come from measurement characteristics such as length, width, and weight. Examples of variable data include 5′11″, 230 lbs, and 500 cft. Attribute data come from inspection characteristics such as color, appearance, taste, and fit. The data will be in counts or proportions and are classified as good/bad, tight/loose, yes/no, and so on.

The reason to understand the data classification is that different types of statistical methods will be employed to analyze them as the mathematics is different.

12.3 Module 1—Empirical methods

12.3.1 Frequency distribution

Population variability is best understood using frequency distribution tool. This graphical tool provides a visual input on how the units in a population are distributed within the possible range of values. The challenge in employing this tool for a population is the enormous amount of time and resources it takes to develop it for the population. So, frequency distribution is developed using the data from samples. Such a sample frequency distribution is called histogram. The procedure to develop a histogram is explained next.

12.3.2 Histogram

The histogram is a very effective tool for understanding the variability distribution of a population from the sample data. The steps to develop a histogram are illustrated as follows:

Step 1: Record the observation of the randomly selected items from the population of interest.
Step 2: Find the largest and the smallest values and find the difference between them. This difference is called the range (R) of the data.
Step 3: Divide the range into a convenient number of bins (also called class intervals). A few rules are available in determining the number of bins.

$$k = 1 + 3.3 \, (\log n), \tag{12.1}$$

where:

   k is the number of bins

   n is the number of observations

However, this rule is an approximate guideline.

Step 4: Determine the limits of the bins and tally the observation in each bin. The tally gives the frequency distribution of the data. Calculate the percentage of data in each bin, which will give the percentage frequency distribution.

Step 5: Draw a graph with bin limits on the x-axis and the percentage frequency on the y-axis using convenient scales.

Example 12.1

The data below represent the length of bolts in millimeters. Draw a histogram of the data (Table 12.2).

SOLUTION

   Step 1: The number of observations, $n = 30$

   Step 2: Largest value = 98, smallest value = 50

$$\text{Range, } R = 98 - 50 = 48$$

   Step 3: Number of bins: $k = 1 + 3.3 \, (\log n) = 1 + 3.3 \, (\log 30) = 5.9 \cong 6$

   Step 4: Bin width = 48/6 = 8

   Now, select a convenient value slightly less than the lowest score as the starting point and add the bin width to the starting point to get the bin limits. Represent each value by a tally mark in the appropriate bin. Total the tally marks to find the total frequency for each bin (Table 12.3).

   Step 5: Draw a graph with bin limits on the x-axis and the percentage frequency on the y-axis using convenient scales.

Table 12.2 Example Data

85	52	70	69	65	79	71	78	93	95
96	72	81	61	76	86	79	68	50	92
83	84	77	65	71	87	72	93	57	98

Table 12.3 Histogram Table

Number of bins	Bin limits	Tally	Frequency distribution	% frequency distribution	% cumulative frequency distribution
1	45–53	| |	2		
2	54–61	|	1		
3	62–69	|	1		
4	70–77	| |	2		
5	78–85	| | | | | | |	7		
6	86–94	| | | |	4		
7	95–102	| | | |	4		
		Total			

Many computer programs can prepare a histogram for a given set of data. Here, we solve the previously mentioned example using Minitab.

After you have entered your data into the Minitab spreadsheet and named the column Bolt Length, select Graph → Histogram. The default selection Simple should be highlighted. Click OK. You will obtain the screen shown in Figure 12.1.

Click C1 → Select → OK, which produces the histogram shown in Figure 12.2.

Figure 12.1 Minitab histogram entry.

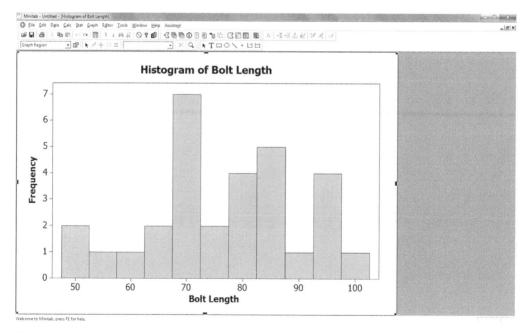

Figure 12.2 Minitab histogram.

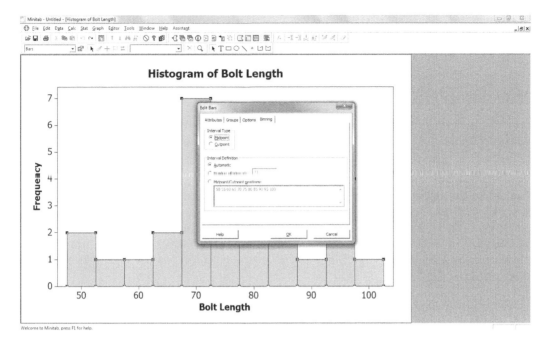

Figure 12.3 Minitab histogram edit.

Minitab chooses bin interval widths based on the data. If you want to specify the interval endpoints, click on any of the bars in the histogram, then right-click and select Edit Bars (Ctrl+T) from the pop-up menu. Now, click on Binning tab and you will obtain the screen shown in Figure 12.3.

Select the cutpoint for interval type, then select Midpoint/Cutpoint and enter the endpoints of your intervals in the box. Separate the endpoints by using a space. When you click OK, Minitab will automatically update the graph with custom bins.

SHAPES OF THE HISTOGRAM

When histograms are developed several shapes are encountered. The most commonly encountered shape is the bell-shaped symmetric histogram which is identical on both sides of its central point. This bell-shaped histogram is called the normal distribution or the Gaussian distribution. Other shapes include skewed (skewed to the right or skewed to the left) histograms, which have longer tails and rectangular or uniform histograms that has the same frequency for each bin.

From the frequency distribution, the measures of interest are the mean and standard deviations. The mean (also called as average) represent the location of the distribution in the x-coordinate around which the data are distributed. The standard deviation represents the spread or dispersion or variability in the data around the mean.

$$\text{Average, } \bar{X} = \frac{\sum_i^n X_i}{n}. \tag{12.2}$$

$$\text{Standard deviation, } s = \sqrt{\frac{n \sum_i^n \left(X_i - \bar{X}\right)^2}{n-1}}. \tag{12.3}$$

When the sample size is large, such as $n \geq 50$, the sample mean and sample standard deviations can be considered as population mean and standard deviation, respectively.

12.3.3 Other graphical methods

Besides the histogram, two other graphical methods are useful in describing populations with variability.

12.3.3.1 Scatter diagram

To control the dependent variable, the correlation is identified using these diagrams. A cause-and-effect relationship and also relationships between causes are examined to create one. They reveal the degree of dependency, skewness, arrive at root causes to problems, and interpolate it to a line if possible. They test for autocorrelation before drawing control charts. Scatter Diagram Procedure is as follows (excerpted from Tague 1995, pp. 471–474):

- Collect pairs of data where a relationship is suspected.
- Draw a graph with the independent variable on the horizontal axis and the dependent variable on the vertical axis. For each pair of data, put a dot or a symbol where the x-axis value intersects the y-axis value. (If two dots fall together, put them side by side, touching, so that you can see both.)
- Look at the pattern of points to see if a relationship is obvious. If the data clearly form a line or a curve, you may stop. The variables are correlated. You may wish to use regression or correlation analysis now. Otherwise, complete steps 4 through 7.
- Divide points on the graph into four quadrants. If there are X points on the graph, count X/2 points from top to bottom and draw a horizontal line.
- Count X/2 points from left to right and draw a vertical line.
- If number of points is odd, draw the line through the middle point.
- Count the points in each quadrant. Do not count points on a line.
- Add the diagonally opposite quadrants. Find the smaller sum and the total of points in all quadrants.
  - A = points in upper left plus points in lower right
  - B = points in upper right plus points in lower left
  - Q = the smaller of A and B
  - N = A plus B
- Look up the limit for N on the trend test table.
- If Q is less than the limit, the two variables are related.
- If Q is greater than or equal to the limit, the pattern could have occurred from random chance.

A regression line is to be calculated to find the relation between the independent and dependent plots.

$$\text{Correlation coefficient can be given by } r = \frac{\sum_{i=1}^{n}(X - \overline{X})(Y - \overline{Y})}{\left[\sum_{i=1}^{n}\left(X - \overline{X}\right)^2\right]\left[\sum_{i=1}^{n}\left(Y - \overline{Y}\right)^2\right]}. \quad (12.4)$$

12.3.3.2 *Stem and leaf diagram*

The stem and leaf diagram is a graphical tool which provides additional information about the data. In constructing this stem and leaf diagram, each numerical value is divided into two portions—a stem and a leaf, in which the leaf is the last digit to the right of the numerical value and the stem is the remaining digits in the numerical value when the leaf is dropped.

The advantage with this method is that it is easy to apply with minimal computations. The limitation with the stem and leaf diagram is that it cannot be used for attribute data and becomes too crowded for large data sets.

Example 12.2

For the bolt length example, plot the stem and leaf diagram.

Stem	Leaf
5	2 0 7
6	9 5 1 8 5
7	0 9 1 8 2 6 9 7 1 2
8	5 1 6 3 4 7
9	3 5 6 2 3 8

Note: Observe that the values for stem are arranged in an ascending order.

USING MINITAB

Enter the given data into the Minitab spreadsheet and rename the column as Bolt Length. Select Graph → Stem and Leaf. Click on the variable, C1, and Select. If you wish potential outliers to be displayed, uncheck the Trim Outliers box. Click OK and the Stem and Leaf plot appears in the session window of Minitab (Figure 12.4).

Figure 12.4 Stem and leaf plot.

Minitab automatically identifies the leaf units and employs splitting stems, that is, each set of 10 numbers is broken into two sets of five numbers.

With the stem and leaf diagram, it is easy to measure the position of a single value in relation to other values within the data set. There are many measures of position; however, only quartiles, percentiles, and percentile rank are discussed in this text.

QUARTILES AND INTERQUARTILE RANGE

Quartiles are the summary measures that divide a ranked data set into approximately four equal parts. The data are ranked in an ascending order before the quartiles are determined. The second quartile (Q_2) divides the ranked data set into two equal parts. Hence, the second quartile and the median[*] are the same. The first quartile (Q_1) is the middle value among the observed values that are less than the median and the third quartile (Q_3) is the middle value among the observed values that are greater than the median.

The interquartile range (IQR) is the difference between the third and first quartiles.

$$\text{Interquartile range, IQR} = Q_3 - Q_1. \tag{12.5}$$

Example 12.3

Calculate the three quartiles and the IQR for the bolt length example

SOLUTION

Step 1: Rank the data set in an ascending order.

50 52 57 61 65 65 68 69 70 71 71 72 72 76 77 78 79 79 81 83 84 85 86 87 92 93 93 95 96 98

Step 2: The median is calculated as the average of the fifteenth and sixteenth terms. Consequently, Q2 = 77.5
Step 3: The value of Q1 is given by the average of the seventh and eight terms. So, Q1 = 68.5
Step 4: The value of Q3 is given by the average of the 23rd and 24th terms. So, Q3 = 86.5
Step 5: The IQR is the difference between the third quartile and first quartile.

$$\text{Therefore, IQR} = Q_3 - Q_1 = 86.5 - 68.5 = 18.$$

MINITAB EXAMPLE

After entering the data into the Minitab spreadsheet, select Stat → Basic Statistics → Display Descriptive Statistics. Then, select the variable C1 Bolt Length and click on Statistics button. A new window pops-up where you can select the quartiles and IQR. Click OK. The output will be displayed in the Session Window (Figure 12.5).

PERCENTILES AND PERCENTILE RANK

Percentiles are the summary measures that divide a ranked data set into 100 equal parts. The data should be ranked in an ascending order to compute the percentile. The *k*th

---

[*] Median is also called as the truncated mean.

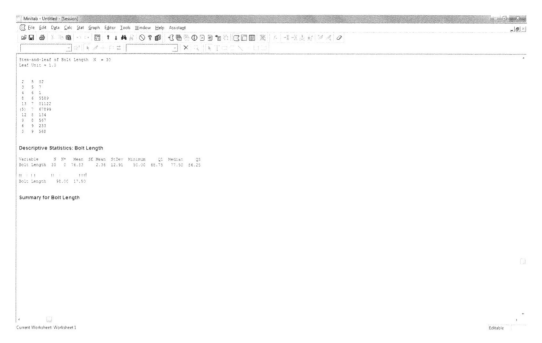

Figure 12.5 Minitab descriptive.

percentile is denoted by X_k, where k is an integer from 1 to 99. The approximate value of the kth percentile in an n sample-sized, ranked data is computed as

$$X_k = \text{value of the } \frac{kn}{100} \text{th term in the ranked data set.} \tag{12.6}$$

We can also calculate the percentile rank for a particular value X_i of a data set by using the following formula:

$$\text{Percentile rank of } X_i = \frac{\text{number of values less than} X_i}{\text{sample size}} \times 100. \tag{12.7}$$

The percentile rank of X_i gives the percentage of numerical values in the data set that are less than X_i.

12.3.3.3   Box and whisker plot

The box and whisker plot is another graphical method used to represent the observations using five measures: the smallest value, the first quartile, the median, the third quartile, and the largest value within the upper and lower inner fences. This plot helps to visualize the location, spread, and skewness of the data set, and detect any outliers within the observation. The box and whisker plot is useful in comparing different distributions.

The steps to construct a box and whisker plot are illustrated as follows:

Step 1: Rank the data in an ascending order and calculate the first quartile, the median, the third quartile, and the IQR.
Step 2: Find the points that are 1.5 × IQR below the first quartile and above the third quartiles. These two points are called the lower and upper inner fences, respectively.

Step 3: Determine the smallest and the largest values in the given data set within the two inner fences.

Step 4: Draw a line and mark the values on it from the ranked data set. Then, draw a box with its left side at the first quartile and the right side at the third quartile. Draw another line at the median inside the box.

Step 5: By drawing two lines, join the points of the smallest and the largest values within the two inner fences to the box. These lines are called whisker. Any value that falls outside the two inner fences is an outlier and is marked as an asterisk.

12.4   Module 2—Statistical models for describing populations

Often we make inferences about population variability using generalized mathematical models. These models are based on the probability of occurrence and are termed as "probability distribution" models. The most commonly employed models in quality include binomial, normal, Poisson, and Weibull. In this textbook, we provide an overview of the above distributions and discuss the normal distribution only. Readers are strongly recommended to read any fundamental textbook on statistic for a complete understanding of the theory of probability and the other probability models.

12.4.1   Binomial distribution

The binomial distribution is a widely used discrete probability distribution in which the outcome of a random experiment will be dichotomous that may be labeled success or failure. The probability of the outcome occurrences are constant and do not affect the outcome of another experiment. For example, if a lot of 1000 bolts with 20% defect rate are inspected for defects using random sampling of 100 bolts, then the probability of success can be modeled as a binomial distribution.

$$f(x;n,p) = {}_nC_x\, p^x\, (1-p)^{n-x}, \quad x = 0,\, 1,\, 2,\ldots,\, n. \tag{12.8}$$

The probability distribution of exactly x successes in n independent experiment is given by the distribution will be symmetric if the probability of success (p) is 0.5 and will be skewed otherwise (Figures 12.6 and 12.7).

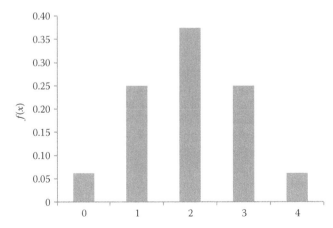

Figure 12.6 Symmetric binomial distribution.

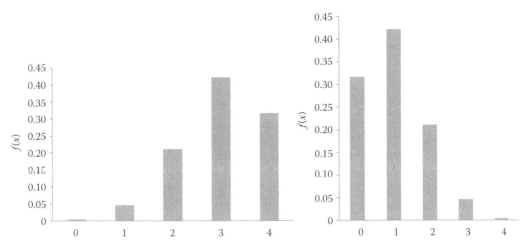

Figure 12.7 Skewed binomial distribution.

12.4.2   Poisson distribution

The Poisson distribution is an example of a probability model. It is usually defined by the mean number of occurrences in a time interval and this is denoted by λ.

The probability that there are x occurrences in a given interval is given by

$$P(X = x) = \frac{e^{-\lambda}\lambda^x}{x!}. \tag{12.9}$$

The Poisson probability is mostly used in attribute chart development. The mean and variance of a Poisson random variable is λ. An attribute is something that either exists or does not exist; it is a peculiar and essential characteristic. There are five basic types of attributes:

1. Structural attribute
2. Sensory attributes
3. Performance attributes
4. Temporal attributes
5. Ethical attributes

Structural attributes have to do with physical characteristics of a particular product or service. Consider an example, hotel with or without balconies for all the rooms. Sensory attributes relate to the basic senses such as touch, smell, taste, and sound. In the case of products, these attributes relate to form design or packaging design for the creation of products that might please the customer. Similarly, ambience in hotels, restaurants, etc. is important for customer experience. Performance attributes check whether a particular product performs the service it is supposed to efficiently or not. Does the lawn mower engine start and so on? Temporal attributes are those, which are in relation to time. Was the parcel delivered on time? Ethical attributes are those which deal with professional ethics. Is everything transparent? Is the teacher kind? Are some of the attributes that one would expect?

- Conditions for Poisson distribution
    - X is a discrete random variable
    - The occurrences are random
    - The occurrences are independent

- Property of Poisson distribution
  - The number of occurrences in one interval is independent of the number of occurrences in any other disjoint interval. This property is called memoryless property of Poisson distribution.

12.1.3 Normal distribution

De Moivre developed the fundamental mathematical equation of normal curve in 1733. Carl Friedrich Gauss (1777–1855) individually developed the normal distribution based on the errors in repeated measurements on the same quantity. This normal distribution is a bell-shaped distribution used extensively in quality control and management. This distribution is symmetric about the mean and is the only distribution which has all the measures of central tendency situated around the mean. For this reason, the normal distribution is extensively used in quality control.

The probability function is a univariate distribution function that is given by

$$f(x; \mu, \sigma^2) = \frac{1}{\sqrt{2\pi\sigma^2}} e^{-\frac{(x-\mu)^2}{2\sigma^2}}. \tag{12.10}$$

where:
  μ is the process mean
  σ^2 is the process variation

The area under the curve is 1. For example, let a particular type of car's miles per gallon (MPG) be normally distributed with a mean of 24.3 and a standard deviation of 0.6. Let X denote the MPG for one tank of gas. The probability of a randomly selected car from the particular population having less than 23 MPG is given by

$$P(X < 23) = \int_{-\infty}^{23} \frac{1}{0.6\sqrt{2\pi}} e^{-\frac{(x-24.3)^2}{2(0.6)^2}} \, dx = 0.0151. \tag{12.11}$$

The challenge in calculating using a complex formula can be easily overcome by transforming the data to a standard normal distribution. The distribution of a normal random variable with $\mu = 0$ and $\sigma^2 = 1$ is called a standard normal distribution. The transformation of any normal PDF to the standard normal PDF can be achieved using the formula

$$Z = \frac{X - \mu}{\sigma}. \tag{12.12}$$

For the car MPG example, the calculation using the standard normal transformation is shown below.

$$P(X < 23) = P\left(\frac{X - \mu}{\sigma} < \frac{23 - 24.3}{0.6}\right) = P[Z < (-2.1667)] \approx P[Z < (-2.17)] = 0.015. \tag{12.13}$$

The normal distribution in also associated with sampling distribution which describes the probabilities associated with a statistic when a random sample is drawn from a population.

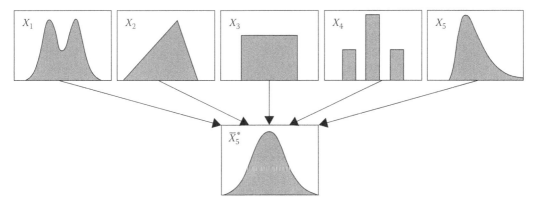

Figure 12.8 Central limit theorem.

Sampling distributions of sample statistics are the results of repeated experiments over time. The sample statistics will result in a different value for each occurrence of the experiment (Figure 12.8).

12.4.4   Application of central limit theorem

12.4.4.1   Central limit theorem

If $\bar{X}$ is the mean of a random sample of size n taken from a population with mean μ and finite variance σ^2, then the limiting form of the PDF for

$$Z = \frac{\bar{X} - \mu}{\sigma/\sqrt{n}}.$$ (12.14)

No matter what the PDF for the X_i ($i = 1, \ldots, n$), $\bar{X}$ has approximately a standard normal PDF provide the sample size, n, which is sufficiently large. As a general rule of thumb, for sample size greater than 30, the central limit theorem (CLT) is applicable for most of the PDFs. The CLT has greater application toward hypothesis testing.

12.4.5   Hypothesis testing

Hypothesis testing is a type of inferential statistics in which the population parameters are estimated from the sample statistics. This procedure is performed with a confidence level to reject the proposed hypothesis based on sample data. This testing is relevant to quality as we perform sampling to infer about the population characteristics. It is relevant to the topics discussed in the following pages. But before we start a discussion on the application of this technique, we must understand terminology in terms of statistics. Hypothesis testing is analogous to a legal trial. In the US legal system, a person begins the trial presumed innocent (really, presumed "not guilty"). That is the null hypothesis (H_0)! The prosecutor accuses the defendant of being "guilty," the alternative hypothesis (H_a). Similarly, we always judge with respect to the null of "not guilty": reject the null (found "guilty") or cannot reject the null (found "not guilty").

There are risks of two incorrect verdicts present in every trial. First, there is the risk of convicting an innocent man (risk of *rejecting* a null hypothesis that describes the *truth*). The second is the risk of letting a guilty man free (risk of *accepting* the null hypothesis that is *false*). In the US court example, the risk or probability of convicting an innocent person (α, or alpha) is deemed of critical concern. This system is based on the belief it is much more evil to convict an innocent man than it is to let a guilty man go free. To minimize the risk of convicting the innocent, very strong evidence is required to decide "guilty." To put it in hypothesis testing terms, we want to be as close as possible to 99.99% certain that we will not convict the innocent.

12.4.5.1   US legal system hypothesis tests
H_0: Defendant is Not Guilty (presumed "innocent"; No different than any other)
H_a: Defendant is Guilty (evidence beyond reasonable doubt to reject innocence)

The two risks on incorrect verdict are β which is setting a guilty person free (failing to reject H_0 when H_a is true) and α is convicting an innocent person (i.e., *rejecting* H_0 *when it is true*).

In technical practice, there has been similar aversion to falsely declaring "difference found" (thus small a), but it has often been at expense of (large, unknown) B!

12.4.5.2   Why and when are hypothesis tests used?
Hypothesis tests are decision aids. They do not make the decisions, but they aid in making good decisions in uncertain situations. For instance, you produce a product on a number of lines, and the overall performance has not been good. You calculated the mean and standard deviation of the output, its conformance (or nonconformance) to the customer specifications. A process improvement project is begun on one of the lines, and management wants to know whether the new process is indeed better than the old one (upgrading the other lines would be a substantial capital cost).

Hypothesis testing would be used to decide whether the performance of the "improved" line is statistically different (i.e., better) than the old way of doing things, given the risk (uncertainty) the business is willing to take.

Hypothesis tests always compare two mutually exclusive (contradictory) hypotheses.

12.4.5.3   Null hypothesis
The null hypothesis (H_0) says, "There is no difference between the two sets. 'If you think there is, you've got to prove it.'" It is the formal statement against which experimental evidence is tallied. Its source is the overall problem that it addresses. A null hypothesis, H_0, can be rejected or failed to accept. The statement that a hypothesis is accepted is never used as it is ambiguous and against the belief that there is not enough evidence to support the statement due to unavoidable circumstances. With enough data if it is decided that we have failed to reject a null hypothesis statement. The null hypothesis (denoted H_0) is referred to as "H-naught." Null hypotheses are almost always set up to be "straw men." We generally hope or expect to disprove or "knock over" the null hypothesis.

12.4.5.4   Alternate hypothesis
The alternative hypothesis (H_a) says, "there is a difference between the two sets of data." The alternative hypothesis (denoted as H_a, but occasionally as H_1) is sometimes referred to as "the research hypothesis." The alternative hypothesis describes the situation that we hope or expect to "prove" or demonstrate as "the truth."

12.4.5.5   Translation

"If the *p*-value is less than our chosen (a) value for the significance level {typically, a = 0.05} then one *can reject the null* hypothesis. Therefore, there is a "statistically significant" difference between samples."

"If the *p*-value is greater than our chosen value for the significance level {typically, a = 0.05} then one *cannot reject the null* hypothesis."

Other forms of the alternative hypothesis are called as one-sided tests. We use them when we want to test whether one sample mean is greater than (or less than) another, for example:

- Are Region One's sales less than Region Two's?
- Is the chemical process yield improved with the new catalyst?

In these situations, the alternative hypotheses are no longer "not equal to ($\neq$)," but "less than ($<$)" or "greater than ($>$)."

- H_a: Region One < Region Two
- H_a: Yield New > Yield Old

The alternative hypothesis says, "there is a difference," and it could be one of the following:

- Samples are *not equal* (we used this H_a in example, above)
- Sample one is *greater than* sample two
- Sample one is *less than* sample two

In hypothesis testing, we "stack the deck" in favor of the null hypothesis. Evidence is required to reject the null hypothesis, that is, to conclude that there really is a difference.

It brings us to a discussion on one-sided and two-sided tests.

- If the value of risk is placed at one side of the test then it is termed as a "one-sided" test. It is also termed as "one-tailed" test.
- If the population has shifted on either direction then a two-sided test. It is also termed as "two-tailed" test.

The "equal/not-equal" test is called a "two-sided" test, because it does not matter whether the difference is greater than zero or less. The "greater than" or "less than" tests are called "one-sided" tests. It should be noted that the null hypothesis almost always remains the same. It states, "There is no difference" between the groups being compared for the statistic of interest. We could get "tricky" with the null hypothesis and use a non-zero Test Mean to test whether the difference in means was some specified (nonzero) value. For example,

- H_0: (Avg_1–Avg_2) = 3
- H_a: (Avg_1–Avg_2) $\neq$ 3

The major difference in the report of one-sided test results is in the statement of the 95% confidence interval for the means' difference. In the one-sided test, the 95% CI is a single lower bound on the difference. Let us consider an example, the lower CI bound is +0.10. We can be 95% confident that the difference between means is no less than +0.10. The best estimate for the means' difference is +3.19, and the minimum estimate for the difference is also positive (+0.10); because the confidence interval of means' difference does not include the value of 0.00, we conclude that there is a difference in the direction indicated by the alternative hypothesis (i.e., ">").

In other words, any test has risk associated with it. This risk (α) determines the level of confidence which is (1–α). Thus, it determines the critical value of test statistic. The two-sided test uses the alternative hypothesis that states that the two sets of data are *not equal*. In the one-sided test, the alternative hypothesis is either *greater than or less than*.

The examples above required comparing two-samples against one another (i.e., "two-sample t-tests"). However, some situations require comparing a single sample against a "known" (or reference) value. Thus, in *one-sample* t-tests we are not comparing two samples but we compare a single sample against the reference. One-sample tests can be one-sided tests (equal vs. greater-than, or equal vs. less than) as well as two-sided tests (equal vs. not equal).

Example 12.4

- Comparing a sample against a historical mean
- Comparing a sample against an industry standard
- Comparing a sample against a claim or goal

The one-sample t-test compares sample results against a known historical mean or a standard of some kind. If we have a specific goal or claim to compare a sample against, we also use the one-sided test.

12.4.6   Regression

Regression analysis is probably the most widely used technique in modeling and data analysis. In the term "simple regression," the word "simple" refers to the fact that there is a single independent (x) variable used to predict the value of the output variable, y. This differentiates simple regression from "multiple regression," in which multiple input variables [conceivably many x(s)] are used to predict the value of y. The term "linear" simply means a transfer function that is composed of input variables that are only raised to the first power. (Recall that $x^1 = x$) In the case of simple (a single x) linear regression, the function is a straight, one-dimensional line. The familiar convention for the equation of a straight line is

$$y = mx + b, \tag{12.15}$$

where:
   m is the slope
   y is the intercept

Simple regression is the technique to model and analyze relationship between a continuous response variable, y, and one continuous predictor variable, x. Simple linear regression produces a model of an output's behavior. Statistical significance of the modeled relationship between y and x is described quantitatively by the p-value. The value of y is expressed, as a linear function of the value of x, by the transfer function which is defined below:

$$y = \beta_0 + \beta_1 x_1 + \varepsilon, \tag{12.16}$$

where:
  y is the predicted response, or dependent variable
  x_1 is the predictor, or independent variable
  β_0 and β_1 are the regression coefficients for the transfer function
  $\beta_0 = y$-intercept of regression line
  $\beta_1 =$ slope of regression line
  ε is an error term

Assume an analysis of a constant variable. Gather data that summarize it according to the constant variable. An equation can be developed according to the function of the variable using simple regression. Note that both the input and output variables are continuous, indicating that we are able to apply regression for the analysis.

12.4.6.1   Simple linear regression model

Simple linear regression model has a straight line that represents set of n data points in a manner such that sum of the squared residuals of the model as small as possible. The slope of the fitted line between will be equal to correlation between y and x modified by the ratio of standard deviation of the given variable. The mathematical equation for the simple regression model is given below (Quality Council of Indiana Inc. 2007).

$$y = \beta_0 + \beta_1 x,$$

where:
  β_0 is the y-intercept when $x = 0$
  β_1 represents the slope of the line

X-axis does not tend to zero such that y-intercept looks to be very high. The actual value of random error will be the difference between the observed value of y and the mean value of y with the specified value of x. Main assumption for the observation value of y changes in a random manner and has a normal probability distribution for the given x value.
  Probabilistic model equation for the observed value is given as follows:

$$y = \left(\text{mean value of } y \text{ with respect to the } x \text{ value}\right) + \text{random error}, \qquad (12.17)$$

$$y = \beta_0 + \beta_1 x + \varepsilon, \qquad (12.18)$$

where:
  y is the predicted response, or dependent variable
  x_1 is the predictor, or independent variable
  β_0 and β_1 are the regression coefficients for the transfer function
  $\beta_0 = y$-intercept of regression line
  $\beta_1 =$ slope of regression line

12.4.6.2   The method of least squares

Method of linear equation is used to approximate sets which have more equations than unknown variables. The statistical method of determining the best fit straight line is in many aspects the method of one fits a line by eye. The goal is to reduce the deviations of the points from the lines. If it denotes the predicted value of y determined from the fitted line as $\hat{y}$.

$$\hat{y} = \beta_0 + \beta_1 x, \tag{12.19}$$

where:

β_0 and β_1 represent estimates of β_0 and β_1, respectively

One should determine the meaning of best if we want to reduce the deviation of points in choosing the best fitting line.

The principle of least square is nothing but the best fit criterion of goodness. To reduce the sum of squares of the deviation of the observed values of y from those predicted. The mathematical expression is shown as follows; it reduces the sum of the squared errors.

$$SSE = \sum_{i=1}^{n}(y - \hat{y})^2. \tag{12.20}$$

Substituting the value of $\hat{y}$ in the above equation we obtain an altered equation,

$$SSE = \sum_{i=1}^{n}\left[y - (\beta_0 + \beta_1 x)\right]^2. \tag{12.21}$$

The least-squared estimator of β_0 and β_1 are estimated as follows:

$$Sx^2 = \sum_{i=1}^{n}X_I^2 - \left\{\left[\sum_{i=1}^{n}X_I^2\right]/n\right\}, \tag{12.22}$$

$$S_{xy} = \sum_{i=1}^{n}X_I Y_i - \left\{\left[\left(\sum_{i=1}^{n}X_i\right)\left(\sum_{i=1}^{n}Y_i\right)\right]/n\right\}, \tag{12.23}$$

$$B_1 = S_{xy}/Sx^2, \tag{12.24}$$

$$B_0 = y - \beta_1\bar{X}. \tag{12.25}$$

Once β_0 and β_1 are computed, substitute the values into the equation of a line to obtain the least squares on regression line (Quality Council of Indiana Inc. 2007).

Things to be noted are as follows:

- Carefully round up errors. The calculation should have a minimum of six significant digits in estimating the sum of squares of deviations.
- Plot the data against the model to ensure accurate modeling and do not project outside of regression line. If it does not provide a reasonable fit for the given data set, then there is a calculation error.
- Do not project outside the regression line.

12.4.6.3 Residuals

In fitting a line to the observed data, the "residual" is the model error. Its magnitude is the vertical distance between the observed value and the predicted value (the fitted line). Residuals can be analyzed to provide information about the model, including:

nonnormality of errors, nonrandom variation of errors, nonconstant variability of errors, and nonlinear relationships.

Normal plot of residuals: this type of residual normally forms a straight line.

Histogram of residuals: it should appear roughly normal symmetric with one peak. Bars located away from the main group (outliers) may indicate unusual observations.

I chart of residuals: any out-of-control points should be investigated as indications of a nonrandom pattern in the data that may be the result of special cause variation.

Residuals versus fits: if points appear to form a curve, there may be evidence of nonlinear relationships. If the spread (variability) of the data points increases or decreases with the fits, the data may have nonconstant variance.

12.4.6.4   *Multiple regression*

Multiple regression is a technique for modeling and analyzing the relationship among a continuous response variable, y, and more than one continuous input variable, $x(s)$. y is expressed, as a function of the $x(s)$, by the transfer function developed from the observed data:

$$y = \beta_0 + \beta_1 x_1 + \beta_2 x_2 + \ldots + \beta_k x_k + \varepsilon, \tag{12.26}$$

where:
   y is the predicted response, or output, variable
   $x_1, x_2, \ldots, x_k$ are the predictor, or input, variables
   $\beta_0, \beta_1, \beta_2, \ldots, \beta_k$ are the regression coefficients (constants)
   ε is the error term

The statistical significance of the regression equation is again determined by the p-value. The topic of multiple regressions covers a range of techniques that can be used to investigate possible relationships between a response variable and two or more potential $x(s)$. A p-value is computed for each input variable, x_i, to determine the statistical strength it contributes to predicting the value of the response, y. Only the significant factors are kept in the prediction equation—the trivial $x(s)$ are dropped from the model.

Best subsets regression is an iterative process that examines potential models with one predictor, then two, then three, and so on, through a model that includes all possible predictors. The best subsets regression identifies the best models (prediction equations) for various numbers of predictors. Only models with the highest R-squared values are retained in the analysis. Model selection is made based upon the Minitab output. A good model should have

- High R-squared
- High adjusted R-squared
- A C–p metric value that is approximately equal to the number of predictors in the specified model
- A low s (s is the standard deviation of the error term in the model)

Many potential $x(s)$ can be screened, simultaneously. There is a danger when $x(s)$ are included that are correlated with each other (called multi colinearity). This risk is minimized with techniques such as best subsets regression and stepwise regression. Discrete variables, for example, operator and method, can sometimes be coded (1-2-3, 0-1, etc.) for use as logical, quasi-continuous variables. As always, correlation does not prove causation.

12.4.7 Analysis of variance

Analysis of variance (ANOVA) is a statistical procedure for testing whether there is a significant difference among *means* of multiple subsets of data. The one-way ANOVA is an extension of t-test:

- t-test compares two means
- One-way ANOVA compares multiple means

12.4.7.1 P-value

The *p*-value tells you whether the various level means are significantly different from each other: If *P* is less than or equal to the significance (a) level chosen, one or more means are significantly different.

As with other inferential statistics tests, we also compare the confidence intervals for the various "levels" or subgroup samples. The two-line ANOVA test clearly shows the nonoverlapping confidence intervals around the sample means: further confirmation of the difference between means. The significant *p*-value in the one-way ANOVA does not indicate that all the lines have different means. The correct interpretation is that there is a statistically significant difference between one (or more) of the lines and the others.

12.4.7.2 Tukey's method

Tukey's method compares the means for each pair of factor levels using a family error rate (often called family-wise error rate) to control the rate of type I error. The family error rate is the probability of making one or more type I errors for the entire set of comparisons. Tukey's method adjusts the error rate for individual comparisons (individual error rate), based on the family error rate you choose.

12.4.7.3 Fisher's least significant difference

Fisher's least significant difference (LSD) method compares the means for each pair of factor levels using the individual error rate you select. Note that the family error rate, which is the probability of making one or more type I errors for the entire set of comparisons, will be higher than the error rate for each individual comparison.

12.5 Module 3—Quality tools

12.5.1 Flowcharting

Flowcharts are effective means of communication medium which can be easily understood. Team members can clearly identify what could be done to the process at each level. Nonvalue-added activities present in the process can be easily determined and can be separated from the process (Figure 12.9).

Steps involved in constructing process maps are given as follows:

- Determine the process limits: In order to chart the process, it is essential to determine the start of the process and end of the process.
- Define the process steps: Use brainstorming to determine the steps for new process.
- Sort the steps in the order of their occurrence in the process.
- Place those steps in the order of the symbols and create a chart.
- Evaluate those steps for efficiency and problems.

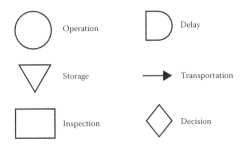

Figure 12.9 Symbols used for flow charting.

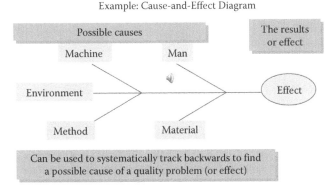

Figure 12.10 Example of cause and effect.

12.5.2 Cause-and-effect diagram

The cause-and-effect diagram is an important quality tool that is used to study the potential causes for an effect. This tool helps in identifying all the causes for problem that may be obvious or unobvious. This tool is performed after brainstorming session by the collective team members seeking to improve the process. The steps involved in developing this tool are given as follows (Figure 12.10):

- Define problem to analyze
- Form a team to perform the analysis
- Brainstorm
- Draw effect box and center line
- Specify major causes and join them as line to the center line
- Identify possible causes and categorize them
- Rank order the causes to identify the most important ones
- Take corrective action

12.5.3 Check sheets

An important activity carried out in the *measure* phase is data collection. Voluminous amount of data are required for this purpose. The check sheet is a useful tool for this purpose. Check sheets consist of events occurring in accordance with a particular time frame. Check sheets are commonly used for analyzing data that are not available for use in taking key decisions. For example, to reduce waste the manufacturer may know the data for the

Table 12.4 Check Sheet Example

Defect	Week				
	1	2	3	4	5
Incorrect address		1			1
Incorrect SSN	1		1	1	3
Incorrect work history	1			1	2
Incorrect salary history	11	1	11		5

quantity of the waste but it is essential to know the various sources of waste within the operation. A check sheet implemented at different work levels may enable documenting the capture of waste from the operations carried out (George 2002) (Table 12.4).

12.5.4   Pareto diagrams

The Pareto diagrams are used to prioritize problems so that the major problems can be identified. The diagrams are used to

- Analyze a problem from a new perspective
- Focus attention on problems in priority order
- Compare data changes during different time periods
- Provide a basis for the construction of a cumulative line
- Prioritize problems
- Segregate the significant problems from the insignificant problems

Pareto analysis is based on 80/20 rule. Dr. Juran states that 80% of the problems came from 20% of causes. Pareto analysis is used to segregate vital few problems with many trivial problems. For the Six Sigma professionals, it is important to identify the most important opportunity for improvement in order to get large returns from the project.

  Pareto chart is similar to histogram. Histogram is the graphical interpretation of the distributed data in a quantitative manner into a class. Pareto chart is the frequency bar chart which represents quantified data by quantitative characteristics. The steps for constructing Pareto chart are given as follows:

- Choose the subject for the chart. The subject can be anything which can be causing problems to any particular line or any product.
- Identify the data that needs to be collected. Identify if percentage or cost or quantity is going to be tracked.
- Collect data related to the quality issues. The quality issues are collected over the time period.
- Use check sheet as the input to collect data. Gather the total number of defects and nonconformities in each category.
- Based on the information from the check sheet calculate the percentage of total nonconformities.
- Specify the cost incurred for the defects or nonconformities.
- Plot the data setting up scale for chart.
- X-axis represents the subject of study usually nonconformities or defects or item of interest. Y-axis represents the number of defects, number of occurrence, percentage, or cost incurred per category.

- Draw the chart by organizing the data in descending order, that is data from the largest category to the smallest category.
- The last step is to analyze the chart. Check for the largest bar, it represents the vital few problems.

12.5.5 Control charts

Control charts are used to manage variation. In this chapter, we will discuss the various statistical charts that can be used to monitor variation. A basic control chart has an upper limit, a lower limit, and a central line. In this chapter, we introduce the various control charts step by step. However, there is a generalized method for each process control chart. This is very important to understand as the process for establishing all the charts is the same. The only difference will be the formulae used for determination of upper limit, lower limit, and the central line.

12.5.5.1 Variable and attribute process control charts

To select the right process chart we should be able to differentiate between variables and attribute. A variable is a continuous measurement such as weight and height whereas a variable is a result of some binomial process which results in an either/or solution. Some of the examples from attributes may be as follows: the inspector is either bald or not bald. In this section, we will discuss the various types of process control charts for variables and attributes.

The charts for variables are X-, $\bar{x}$, R-, MR-, and s charts. The charts for attributes are p, np, and c and u charts. In the following pages, we introduce different kinds of charts and the method of selecting which chart is to be used. There are four basic requirements before we select a process chart:

1. Understanding the process for implementing the chart
2. Interpretation of the charts is to be known
3. Realizing the usage of different charts
4. Knowing how to compute the limits for the different types of process charts

12.5.5.2 Understanding the process of implementing the chart

The process of developing a chart is almost the same for all process charts; the only difference lies in the actual statistical computations. The following are the steps used in developing process control charts (Figure 12.11).

The upper control limit, the central limit, and the lower control limit are computed statistically. Each point represents data from a sample that is plotted sequentially.

- Critical operations in the process in which an inspection might be needed are to be identified. These are the operations that may result in negative effects on the product if not performed properly.
- The characteristics of the product should to be known on a precise level, to ensure the functioning of the product is good or bad.
- Determine if the characteristics of the critical product are variable or attribute.
- Selection of the appropriate process control chart among the many types of charts is done. The decision process and the various charts are discussed later in the chapter.
- Once this is done the control limits for the chart is decided for continual monitoring and improvement.
- As and when improvement or changes occur the limits are to be updated.

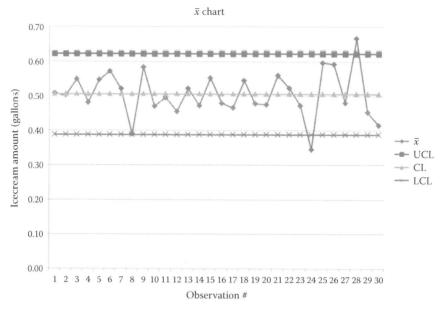

Figure 12.11 Example of a control chart.

12.5.5.3 x̄ and R charts

Now that we have understood the procedure to develop a chart let us discuss the various types of charts. We will first discuss $\bar{x}$ and R charts as they go hand in hand. We developed an $\bar{x}$ chart previously. When we have to monitor the average and the range of some data we use the $\bar{x}$ and R chart. The $\bar{x}$ chart is used in monitoring the average of the process whereas the R chart is used for determining the range of the process or the dispersion of the process. The $\bar{x}$ chart is used to monitor the average of the characteristic that is being measured. To prepare this chart, samples from the process are collected for the characteristic that is being measured and then rational groups are formed from the sample. The average value of each sample is determined by dividing the sums of the measurements by the sample size, and the value on the chart is plotted thereafter.

The R chart is used to monitor the range as the name itself suggests. It is used in conjunction with the x chart when the process characteristic is a variable. To prepare an R chart, samples are collected and segregated into subgroups usually of up to six items. The difference is then computed by finding the difference of the high value of the subgroup from the low value of the subgroup and then the R chart is plotted. Now that we have collected the data to prepare the charts, the next step would be to determine the upper control line, the central line, and the lower control line. Employing certain formulae accomplishes this. The upper control line and the lower control line are usually three standard deviations away from the central line. It may also be noted that there are no formulae for the lower limit of the R chart. This is because the lower limit is usually zero for samples having six points (Figure 12.12).

12.5.5.4 Median charts

Sometimes $\bar{x}$ charts are very time consuming, though they are preferred for variable data. They are also inconvenient sometimes for computing subgroup averages. Also, there may be concerns about the accuracy of computed means. In these cases, the median chart, that

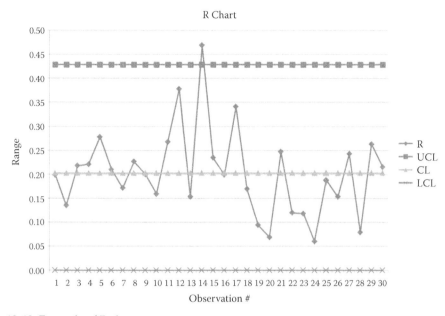

Figure 12.12 Example of R chart.

is, a $\tilde{x}$ chart can be used. The main limitation with this is that you will have to use an odd sample size to prevent calculation of the median. Like the other charts, small sample sizes are used in this case too, although the larger the sample size, the better is the sensitivity of the chart to detect a nonrandom event.

The preparation of a median chart involves determination of the subgroup size and finding out how often the sampling takes place. The rule of thumb to create a median control chart is to use 20 to 25 subgroups and a total of at least 100 measurements. The equations for computing the limits are as follows:

$$\text{Mean of medians} = \text{sum of the medians/number of medians} \qquad (12.27)$$

12.5.5.5 $\bar{x}$ *and s charts*

When dispersion of the data is the primary concern, a range chart might not suffice. In this case, the $\bar{x}$ chart is employed with something called the s charts or standard deviation chart. The standard deviation chart is used when the variation is small. For example, s charts are used in monitoring the silicon chip production for computers. Unfortunately, when using the s chart, because the range is not computed, new formulae are employed to compute the upper control limit and the lower control limit. The formulae are introduced because of their importance in high-technology production.

$$\text{UCL}_s = \mu + 3\sigma/\sqrt{n}. \qquad (12.28)$$

$$\text{LCL}_s = \mu - 3\sigma/\sqrt{n}. \qquad (12.29)$$

After the limits are computed, the sample means are plotted to check if the process is in control. If the chart is not in control, the cause for the out-of-control points is determined and eliminated, and the control limits are again recomputed by eliminating the out-of-control points. If the cause cannot be identified the out-of-control points should not be eliminated.

12.5.5.6 *Attribute process control charts*

An attribute is something that either exists or does not exist; it is a peculiar and essential characteristic. There are five basic types of attributes:

1. Structural attribute
2. Sensory attributes
3. Performance attributes
4. Temporal attributes
5. Ethical attributes

Structural attributes are concerned with the physical characteristics of a particular product or service. Consider an example, a hotel with or without balconies for all the rooms. Sensory attributes relate to the senses such as the basic senses of touch, smell, taste, sight, and sound. These attributes relate to the form of design or packaging design for the creation of products that might please the customer. For example, the ambience inside hotels and restaurants, etc. is important for customer experience. Performance attributes check whether a particular product performs the service it is supposed to efficiently or not. Does the lawn mower engine start and so on? Temporal attributes are those, which are in relation to time. Was the parcel delivered on time? Ethical attributes are those, which deal with professional ethics. Is everything transparent? Is the teacher kind? Are some of the attributes such that one would expect?

The process for developing an attribute chart is given as follows:

- Identifying the vital operations that are required for the process and where inspection might be required. These are operations that can have a negative impact on the product.
- Determining the critical product characteristics. These will result in good or bad form of the products.
- Determining if the characteristic is a variable or an attribute
- Selecting the appropriate process cart
- Establishing the control limits
- Altering the limits as changes occur

12.5.5.6.1 *P-charts for proportion defective.*

The p-chart is an attribute process control chart that is used to proportionally graph the items of a defective sample. This is very useful when there is change in proportion to the defect of a particular sample for a product or a service. A few of the common defects for which a p-chart can be used for monitoring are late deliveries, incomplete orders, calls not getting dialer tones, transaction errors, clerical errors, or parts that do not fit together as desired.

Like every other process chart, a p-chart also has to be divided into subgroups typically between 50 and 100 units. The subgroups may be of different sizes. A constant subgroup is however best to be held. Twenty-five subgroups are established at least for a p-chart. The formulae to determine the limits for a p-chart are as follows

$$P = \bar{p} \pm 3\sqrt{\left(\frac{(\bar{p})(1-\bar{p})}{n}\right)},$$

(12.30)

where:
P is the proportion of defective items
$\bar{p}$ is the estimate of long-term process mean
n is the sample size

12.5.5.6.2   np charts.   The np chart is a graph that gives us the number of defective products in a subgroup. The requirement of an np chart is that the sample size of each sub-group is required to be the same each time a sample is drawn. When subgroup sizes are equal, either the p or the np chart can be used. They are actually the same chart but some people find the np chart relatively easier because it reflects the integer numbers rather than proportions. The p and np chart have essentially the same use.

Subgroup sizes for the np chart are similar to that of the p-chart, that is, they are between 50 and 100. Usually at least 20 subgroups are used in the development of the np chart. Again, the only requirement in this case is that the subgroup sizes should be equal. The computation of the control limit is done by the following formula:

$$CL_{np} = n(\bar{p}) \pm 3S_{np},\tag{12.31}$$

where:
n is the sample size
S_{np} is the standard error

12.5.5.6.3   C and u charts.   C and u charts are a graph, which gives us the number of defects per unit. The units must be of the same sample size, which includes all physical factors such as height, weight, length, and volume. This means that the chance of finding the defects in an area must be the same for each unit. Several units can be formed to make one large group. When multiple units are used it should be noted that all the subgroups should be of the same size. The control limits are computed based on Poisson distribution.

Like other process these charts are also used to determine nonrandom events in the life of a production process. A few of the common applications of the c charts include flaws in auto finish and errors in typing and incorrect responses for a standardized test. The c chart is used when the same size sample space is considered. When this is varied a u chart is used. The u chart is a graph of the average number of defects per unit. This is in contrast with the c chart, which shows the actual number of defects per standardized unit. The u chart has flexibility for different numbers in each sample space. The functions of the u and c charts are the same. The formulae for control limits of u and c charts are the following:

$$CL_c = \bar{c} \pm 3\sqrt{\bar{c}},\tag{12.32}$$

$$CI_{\cdot u} = \bar{u} \pm 3\sqrt{\frac{u}{n}},\tag{12.33}$$

where:
n is the average sample size
$\bar{c}$ is the average number of nonconformities
$\bar{u}$ is the average number of nonconformities per unit

References

George, M. L. (2002). *Lean Six Sigma: Combining Six Sigma Quality with Lean Speed*. New York: McGraw-Hill.

Krishnamoorthi, K. S. (2006). *A First Course in Quality Engineering*. Upper Saddle River, NJ: Pearson Prentice Hall. 2006.

Quality Council of Indiana Inc. (2007). *Certified Six Sigma Black Belt Primer*. West Terre Haute, IN: Quality Council of Indiana.

Tague, N. R. (1995). *The Quality Toolbox*. 2nd edn. Milwaukee, WI: ASQ Quality Press.

Further Readings

Mann, P. S. (2005). Introductory Statistics Using Technology. 5th edn. Hoboken, NJ: John Wiley & Sons.

Rath and Strong Management Consultants (eds.). (2000). *Rath & Strong's Six Sigma Pocket Guide*. Lexington, MA: Rath & Strong Management Consultants.

chapter thirteen

Define

Identifying and organizing
Six Sigma projects for success

Even a journey of one thousand miles begins with a step.

—Lao Tzu

Overview	
Steps	Substeps
Define	• Quality projects (sponsorship) • Define specific project (black belt)
Measure	• Identify critical characteristics • Clarify specifics and targets • Measurement system validation
Analyze	• Determine baseline • Identify improvement objectives • Evaluate process inputs
Improve	• Formulate implementation plan • Perform work measurement • Lean execute the proposed plan
Control	• Reconfirm plan • Set up statistical process controls • Operational process transition

Source: PWD Group LLC.

		Define	Measure	Analyze	Improve	Control
Stage 1	Yellow belt	Interview process		X-Y Map/ QFD/house of quality	Evolutionary operations (EVOP)	Control plans
		Language processing		Fishbone diagram/RCA		Visual systems
		System map (Flowchart)				5-S
						SPC/APC

Recommended Tools by Belt

(Continued)

		Define	Measure	Analyze	Improve	Control
		\multicolumn{5}{c}{Recommended Tools by Belt (Continued)}			
Stage 2	Green belt	Prioritization matrix	Hypothesis testing	FMEA	Design of experiments	TPM
		Thought process map	Flowdown	Multi-vari chart	Fractional factorials	Mistake proofing
		Stakeholder analysis	Measurement system analysis	Chi-square	Data mining	
		Thought process map	Graphical methods	Regression	Blocking	
			Analysis of variance	Pareto analysis		
Stage 3	Black belt	Value stream map	Process behavior charts	Buffered tolerance limits	Theory of constraints	
		Gantt chart development	Quality function deployment	Process capability analysis	Multiple response optimization	
		Project management			Response surface methodology	

Source: PWD Group LLC.

Notes: Black belts are expected to know the continued knowledge obtained throughout green belt and yellow belt certifications. The same goes for green belt and yellow belt. Empty cells represent that skill is not required for that belt stage; however, a general knowledge of the step is expected.

5-S, sort, set in order, shine, standardize, and sustain; APC, advanced process control; EVOP, evolutionary operations; EVOP, evolutionary operations; FMEA, failure modes and effect analysis; QFD, quality function deployment; RCA, root cause analysis; SPC, statistical process control; TPM, total productive maintenance.

13.1   Introduction

In a Six Sigma program project selection is very important. To identify and work with the process that yields more revenues is one of the key aspects to be considered while doing a Six Sigma project. As discussed earlier, not all companies profit from Six Sigma implementation. Hence, the *define* phase provides the necessary tools to identify and locate the variables to work on in Six Sigma in order to increase the revenue of the organization. The important base for any business enterprise is the customers and understanding them completely is crucial. Peter Drucker has stated that (Figure 13.1)

> What people in business think they know about customers and markets is more likely to be wrong than right. There is only one person who knows: the customer. Only by asking the customers, watching them, by trying to understand their behaviors, can one find out who they are, what they do, how they buy, how they use what they buy, and what they expect.... The customer rarely buys what the business thinks it sells.... (Drucker 1986)

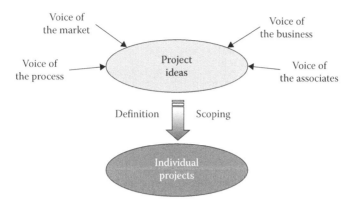

Figure 13.1 Project selection.

The entire process now starts from the customer's perspective. The organization must fit into the shoes of the customers and realize these following aspects as a customer for their own enterprise. Think about some of your recent purchases. Think about one of the following: your last purchase of either a vehicle or major appliance. What were you really buying? The features, characteristics, benefits, etc. What did the seller think you were buying? Pick a product or service from your operation. What do you think the customer is buying? These questions from the customer's perspective will give tangible ideas on what is wrong with the product. To be more precise and helpful the identification of "critical to satisfaction" might give more useful insight on who really are the customers. Who are your customers? (Include: current, lost, potential and competitors' customers). What do you provide your customers? What is critical for the satisfaction for your customers? What is the relative importance of the CTS's? Understanding and listening of voice of the customers is discussed in detail in the later part of this chapter.

The importance and purpose of project selection is very clear. The purpose is to understand the issues related to the customer and the business. The ultimate objective of *define* phase is to

- Gather and interpret voice of customer
- Identify issues important to business
- Map high-level business system
- Complete the project selection process

The source for project idea is to be very clear. The ideas generated from an unreliable source might lead to catastrophe. Hence the following can be considered while listening to the source (Figure 13.1):

- Voice of the market
  - Voice of the customer (Are they happy with everything we offer?)
  - Comparison with competitors (Where are we behind competition?)
- Voice of the business
  - Annual operating plan (Where are the gaps?)
  - Financial analysis (Where are the opportunities?)

- Voice of the process
  - Waste, defective
  - High costs and consumption
  - Problems with new products
- Voice of associates
  - Knowledge of issues and opportunities

13.2   *Project definition worksheet*

The project definition worksheet (PDW) is a guide for further clarifying the project definition. A completed PDW also provides the key elements needed for the Project Charter. The purpose of the PDW is to ask the appropriate questions to clarify the scope and support for the project. Traditionally, not all the answers to the questions can be answered at this point of time. However, the goal is to direct whoever is filling out the form to narrow the scope of the issue, and view the problem from many directions (Figure 13.2; Tables 13.1 through 13.5).

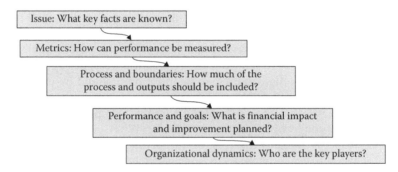

Figure 13.2 Process definition worksheet.

Table 13.1 Project Definition Worksheet Key Facts

Action	Element	Definition	Actual
1.1	What is the specific problem you want to solve or defect you want to eliminate?	The specific condition that is affecting the success of the business.	Crown area separation in BMT tires
1.2	Which customer is most affected by this problem?	The entities most affected by this problem. Can be internal or external customer.	Dealers, fleet, end users
1.3	What is the output associated with this problem?	For example, a service, a transaction, a document, or a physical product.	Bias truck tires
1.4	Where is the problem occurring?	For example, a geographical location, or a part of the organization.	India, Malaysia, Thailand, Taiwan
1.5	When did the problem occur?		10+ years

Table 13.2 Project Definition Worksheet Metrics

Action	Element	Definition	Actual
2.1	Identify the business metrics	The business metric, CTS or Big Y, is the specific outcome of the process targeted by the project.	No. of crown area separation adjustments
2.2	Estimate current performance	The current performance is the quantitative level of business metric for a time prior to the start of the project. If possible the reference period should be past 12 year of performance.	53% of 24,000 tires adjustment through 3Q02 (9 months)
2.3	What is potential consequence metric?	Any other business measurement that may be impacted negatively as a result of improving the primary metric	Product cost, plant, output/capacity
2.4	How do you know about the extent and duration of the business problem?	The primary data source you used to identify business problem. If possible, include title, author, and date of the report and any other sources that can be used to support this claim.	Asia adjustment data 3Q02 2002 GI s350 ABFP 62056

Table 13.3 Project Definition Worksheet Boundaries

Action	Element	Definition	Actual
3.1	What are the key activities that have an impact on the output of the interest?	List the activities that have influence on output of the process	Identifying the system
3.2	Develop a high-level business process map.	High-level map with only three to seven major steps	Process map
3.3	List the includes and excludes.	What: product line, Where: region, storage, condition, When: steps on the process, Who: channel, customer. How: shipment method.	Includes/excludes
3.4	Determine the intended scope and scale of the project.	Project scope defines the part of the process that is to be included in the project. The project scope statement also should consider any time, resources, equipment, and facilities that may be significant.	Project scope and scale
3.5	What are the current and past projects relevant to this project?	What other work you aware of that might be closely related to the project?	Various changes over past 5 years. Data/information is available.
3.6	Write the problem statement.	This item is really a summary of information entered earlier in the worksheet. It should include answers to these questions. Problems? Where does it occur? How often does it occur? Magnitude of the problem? Product or process affected by this problem?	53% of BMT adjustments are categorized as CAS. It is most prevalent in India and Malaysia. This impacts the good year in the region and therefore price.

Table 13.4 Project Definition Worksheet Performance and Goals

Action	Element	Definition	Actual
4.1	Identify the cost categories that are impacted by this problem.	They may be financial or process impacts to a cost center that can be expressed in terms of items such as labor, material, or inventory.	Country por L. region P or L. Inventory working capital. Net selling price
4.2	Identify the total financial impact of the problem.	This refers to tangible financial impact. See guidelines for reporting the Six Sigma project report for more information.	$1.68 million
4.3	What is the amount of improvement planned for this project?	This is a general statement of the project objective. This statement should be structured as follows: improve some metric from some current metric level to some goal by some timeframe.	Goal statement
4.4	Estimate the potential financial benefits of the project.	A forecast of the potential value/benefit of this project. This is an estimate based on current performance and the established goal.	Expected benefits
4.5	Estimate how long will it take to complete the project.	Be aggressive but realistic if a completion time of 6 months or less is not realistic, then the project should be re-scoped.	Timeline: 6 months

Table 13.5 Project Definition Worksheet Organization

Action	Element	Definition	Actual
5.1	Identify the process owners.	Person who has the primary responsibility of ongoing operation of the process.	Plant business centers, Plant and region Qtech: designers and compounders; sales.
5.2	Identify the stakeholders.	Identify the key individuals with the stake in this problem/process/project.	All above, plus: customers, sales force, country, and SBU management.
5.3	Identify the project sponsor.	Who is the individual most suited in terms of authority. Influence the stake in the outcome to sponsor the process.	Technology directors.
5.4	Which black belt should be selected to lead this project?	Consider the project scope. Impact and complexity and expertise likely to be required.	Black belt: To be determined.
5.5	Identify the core project team members or function.	The core team is typically 3–6 people who have complementary skills and expertise likely to be required for the project success.	Team members/function: Design, compound plant BC, Plant Qtech regional customer service and engineering.
5.6	Identify possible resources or constraints.	Identify resources that might be needed. For example, access to database, travel to customer plant sales sites. List potential constraints, including limitation on use of resources.	Possible resource constraint.

	Includes	Excludes
What		
Where		
When		
Who		
How		

Figure 13.3 Includes/excludes.

One way to clarify the scope and scale of the project is to explicitly state what the project includes, and what it excludes. The "four W's and one H" are a useful guide in considering includes/excludes (Figure 13.3):

- What—product line, production equipment
- Where—geographic region, storage condition, site
- When—stage of the BDP, steps of the production process
- Who—channel, customer segment
- How—shipment method, order entry type

Placing an element in the "excludes" category does not mean that it cannot be used for comparison or to benchmark an "included" category. Excluded elements may always be addressed in a follow-up "translation" or transfer project (Figure 13.4).

	Includes	Excludes
What	Crown area separation. Bias medium truck tires	All other conditions and product types
Where	India (first), Malaysia (second), Indonesia, Thailand, Philippines, Taiwan	Other locations
When	Design, manufacturing, selling	Molds, logistics
Who	Fleet and replacement market	Other
How		

Figure 13.4 Includes/excludes example.

13.3   Voice of the customer

Voice of the customers can be classified as the following topics:

- Customer identification
- Customer feedback
- Customer requirements

13.3.1   Customer identification

The concept of Six Sigma was evolved for the customers. In an organization, everything starts and ends with customers. The customers' expectations and needs are set as quality. The customers expect performance, reliability, value for their money, and on time delivery (Harry and Schroeder 2000). The customers are not necessarily the one who ultimately buy the product but Lean Six Sigma (LSS) concepts evolved in such a way that the operator in the next station is the customer. The primary customer of the process will or should have the highest impact on the process (Pande et al. 2000). The *define* phase concentrates on defining the primary customers who make most of the revenues for the organization. Every business has many potential customers and the customers have their own business criteria. The organization considers cost, quality, features and availability factors for weighing the potential customers. The CQFA value grid helps the organization succeed in one way or the other. To define the customers and to analyze customer data is very much required and obtaining wrong customer data can cause a flaw (Pande et al. 2000). Hence customer data is vital. To have proper customer data, customer surveys are required without pitfalls, so the organization or the company can produce and design product for the right market, that is, the right customers. Hence in any market, the customers can constitute

- Current, happy customers
- Current, unhappy customers
- Lost customers
- Competitor's customers
- Prospective customers

Any organization must want to work as any of the above-mentioned customers to improve their market, which is nothing but customer retention and customer loyalty that is discussed in detail later in this chapter. In order to obtain the right customer data, the following methods have to be followed (Eckes 2001):

- Surveys
- Focus groups
- Interviews
- Complaint systems
- Market research
- Shopper programs

The traditional methods of obtaining customer data are

- Targeted and multilevel surveys
- Targeted and multilevel interviews

- Customer scorecards
- Data warehousing
- Customer audits
- Supplier audits
- Quality function deployment

As mentioned earlier, various belts are certified to follow and implement Six Sigma methodologies. The challenge lies in getting as many valuable data as possible. Hence in belt certifications, the apt tools to use for defining customers are taught and discussed later in this chapter. Also, the certified belt users help in defining and distinguishing this internal and external customers who are important to that particular project in an organization. This helps the sponsors or the executive management to understand the basic customers better and work toward effective deliverables for the customers.

13.3.2   Internal customers

Internal customers are those in the company who are affected by the product or service as it is being generated. The internal customers are often forgotten in the process of concentrating on the process to satisfy the external customers. The concept of LSS is to involve every employee and make him/her responsible for the process. This affects employee satisfaction and involvement on a positive scale which on the other hand affects customer satisfaction. For employee satisfaction, sound communication is important which can be improved by (Lowenstein 1995)

- Company newsletters
- Story boards
- Team meetings
- Staff meetings

To involve employees in a better and effective way, training and education by black belts are very important. The black belts and master black belts are vested with a crucial responsibility of selecting right people for the Six Sigma methodology. The process is called stakeholder analysis. When Six Sigma projects are initiated there might be resistance from the people involved for the proposed change. It is obvious for the resistance to arise. No matter how brilliant the idea and how obvious the benefits, any effort to change something will trigger resistance. People have many different reasons to resist change, often quite legitimate. Resistance may take many different forms, depending on the perspective, position and personality of the person. Resistance to change is one of the most frequent reasons why projects ultimately fail. Hence sponsors and black belts are left with no option but to accept the resistance. The success lies in using this resistance as an opportunity to improve the proposed methodology more effectively. The stakeholder analysis is performed by the following:

- Identify key stakeholders of the project
- Identify current level of support/resistance
- Define needed level of support
- In case of gaps, develop strategy to move each stakeholder to the needed level of support.

The first and foremost step is to identify the key stakeholders of the projects. The question "who are the key stakeholders?" is very crucial and can possibly be answered by the following:

- Owners of the process
- Anyone contributing to the process
- Anyone affected by the process
- Those who benefit from process output
- All who see themselves as stakeholders

The next step is assessing the commitment of the stakeholders for the proposed project. If the stakeholders have a hidden agenda, it is important to resolve it before the commencement of the project (Figures 13.5 through 13.7).

The stakeholder analysis can be refined by performing two-dimensional stakeholder analyses. The refinement is done by adding the power/influence dimension. This gives the black belts a map showing where people stand regarding the project, and helps in focusing

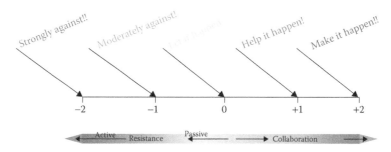

Figure 13.5 Stakeholder's commitment.

Key stakeholder	What is valued by the key stakeholder	Commitment					Recommended actions
		−2	1	0	+1	+2	
Executive						X O	
TTL			X			O	
BTM			X		O		
Production director				X	O		
Purchasing manager				O	X		
Finance manager			X	O			

X Current stakeholder position (present level).

⊙ Where the stakeholder needs to be (required level).

Figure 13.6 Stakeholder commitment data collection example.

TBM-uptime—Stakeholder analysis							
Key stakeholder	What is valued by the key stakeholder	Commitment					Recommended actions
		(X) present level/(○) required level					
		−2	−1	0	1	2	
Plant manager	He gives strong support, he is also the sponsor of the project.					X ○	
Production manager	He will give support, because he will get benefit, but he may be influenced by the BT managers.			X ——▶ ○				He will be involved during the next meetings.
QTEC manager	He will give support, because he will get benefit in view of waste, but he one of his technology managers is involved in the project as Black/Green belt.				X ○		
Production control manager	His support is required if trials are required (later in the project). If we can improve the uptime he will get benefit from the project.		X ——————▶ ○					"Some evening discussions"
BTM truck	If we improve the situation he will get benefit. But he may lose the possibility to hide some of his own problems (DANGEROUS).		X ——————————▶ ○					Involve him before "starting" of the project. Small meeting before project launch. Always try to get informations about his feeling (during walkover to parking, etc.).

Figure 13.7 Stakeholder analysis—example.

on the efforts for project. Some people may be opposed to the change, but having little influence on the success/failure of your project. The goal is to move everyone with success/failure influence toward positive or least neutral. The success/failure influence of those who remain stubbornly opposed must be minimized. The two-dimensional stakeholder analyses are illustrated in Figure 13.8.

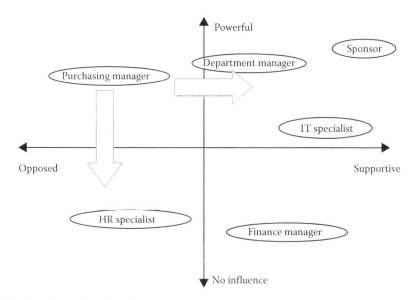

Figure 13.8 Two-dimensional analyses.

13.3.3 External customers

External customers are not part of the organization but are impacted by it. External customers play a crucial role as they make most of the money. The external customers can be end users, intermediate customers, and impacted parties. End users are the customers who buy the product for own utility. The intermediate customers are those who buy the product and then resell. Retailers, distributors, and wholesalers are all some of the entities of intermediate customers. Impacted parties are those who did not actually buy the product but are impacted by it. The *define* phase concentrates mostly on identifying the external customers and it is more complex. In the following sections, the identification customers and data analysis are discussed based on the belt certifications. The external customers in general can be identified by

- Are the customers interested in lowest possible price?
- Are the customers interested in highest quality imaginable?
- Are the customers interested in sparing no expense?

13.3.4 Customer service

Customer service is an important entity nowadays. The best way to identify customers and their retention is by providing best customer service. The following can be considered when the organization is trying to improve its customer service:

- Listen to customer
- Define service strategy
- Set performance standards
- Select and train employees
- Recognize and reward accomplishments

The above-stated points are vital because about 70% of the customers who are not willing to buy the product not because of the quality of the product but due to service quality. In addition, the information from noncustomers is also important because there will be a better feedback about the product in the negative scale which can be used as an opportunity to improve the product.

13.3.5 Customer retention and loyalty

Most of the organizations concentrate on acquiring new customer base, but the cost of retaining a current customer is only one-fourth of the cost of acquiring a new customer. It has been found that a current customer is worth five times than a new customer. The life cycle of customers can be defined in five stages:

1. Acquisition—high cost
2. Retention—cost one-fourth of acquisition
3. Attrition—enthusiasm fades as dissatisfaction increases
4. Defection—loss of customer
5. Reacquisition—highest cost

The company must make efficient steps in retaining the customers. The customers buy happiness from not just the products and services. Happy customers are known as "apostles," who spread goodwill about the product. The effects of unhappy customers are more adverse and they are given the term "terrorist," who spread a negative message about the product. Customer loyalty is not measured in bulk purchase but repeated purchase. Customer retention grows loyal customers. The customers must now be seen as loyal partners.

13.3.6 Customer metrics selection

Metrics are usually developed in the *measure* phase, but metrics in *define* phase are the established measure of customers. The primary metrics are

- Suppliers
- Internal process
- Customers

The basic metrics are quality, cycle time, cost, value, and labor. The nine dimensions of quality are (Garvin 1988) (Figure 13.9)

1. Performance: Primary features of the product
2. Features: Secondary features added to the product
3. Conformance: Obtaining a product that meets fit, form, and function
4. Reliability: The dynamic quality of a product over time
5. Durability: Useful life
6. Service: Ease of repair
7. Response: Human interface
8. Aesthetics: Product appearance
9. Reputation: Based on past performance

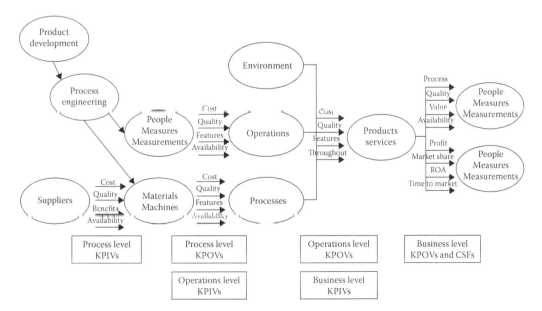

Figure 13.9 Process performance metrics.

Table 13.6 Customer Data Collection

	Business level	Operational levels	Process level
Customers	• Shareholders • Top management	• Whoever purchased the product (external) • Whoever manages the product (internal)	• Employees "next" in the process • Any employee affected
Interests	Financial data • Stock price • Market share • Earnings • ROI	• WIP • Sigma level • Throughput • Yield • Operational effectiveness (external)	• KPV's • Job satisfaction • Advancement fairness • Pay • Training
Time frame	• Quarterly • Annually	• Daily or weekly (internal)	• Hours (production rates)

13.3.7   Customer data collection

As mentioned earlier, feedback data from the customer is important for analysis. When collecting data, it is important to know the various levels in the organization which get affected. The levels can broadly be classified as follows:

- Business level
- Process level
- Operations level

The customers and their impacts are provided in Table 13.6.

13.3.8   Customer surveys

Better understanding of customer satisfaction is done through customer surveys. The customer survey sample sizes and frequency will have significant cost implications. The customer survey should be within the available resources and the need for change is the environment. The most-often-used tool for customer surveys is L-type matrix. The matrix uses numbers 1 to 10 corresponding from very dissatisfied to very satisfied. The surveys can be developed into questionnaire form with 25 to 30 questions (Table 13.7).

Table 13.7 L-type Matrix

	Customer satisfaction									
	Very dissatisfied						Very satisfied			
Task	1	2	3	4	5	6	7	8	9	10
On schedule										
Good product										
Friendly										
Prompt										

13.3.9 Customer data analysis

The data collected through surveys and feedback is used for analyzing the changing attitude of the customers over a period of time. This helps the organization to constantly meet customer requirements without losing their goodwill. The tools utilized are described below.

13.3.9.1 Line graph

The line graph shows the discrete or continuous features of the product over a period of time. The charts help in finding the changes and visually determine whether the products are the same, worse, or better. A discrete chart is illustrated in Figure 13.10.

13.3.9.2 Control charts

Control charts are used in order to determine variation in the product's feature. The calculation of control limits provides an advantage over line charts. An attribute chart is illustrated in Figure 13.11.

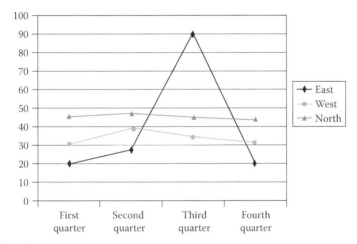

Figure 13.10 Line chart.

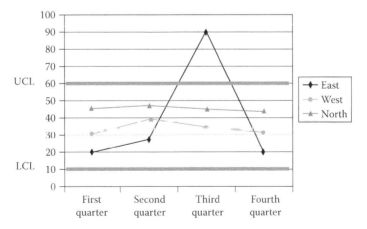

Figure 13.11 Control chart.

Customer	Defect type								Total
	A	B	C	D	E	F	G	H	
1	1	12	1	12	12	1	3		42
2	5			9	12				26
3	3	5		6	13			2	29
4	1	2	3	5	8			3	22
5	1			7	9	1	1	5	24
6	2	5	6	9	1			1	24
Total	13	24	10	48	55	2	4	11	167

Figure 13.12 Matrix diagram.

13.3.9.3 *Matrix diagrams*

A variety of matrix diagrams can be used for examination of customer defects. The data from matrix diagram can be used for project selection. The matrix diagram is illustrated in Figure 13.12.

13.3.10 *Voice of customers*

Good understanding of the needs of the customer is critical for the survival of most companies. VOC actually enables the organization to

- Make a decision on products and services
- Identify product features and specifications
- Focus on improvement plans
- Develop baseline metrics on customer satisfaction
- Identify customer satisfaction drivers

Understanding and listening to the voice of the customers is very important for any company. There are various tools to perform this and these are discussed in later sections according to belt certifications.

13.3.10.1 *Methods of hearing the VOC*
Passive techniques:

- Existing data (customer complaints, phone logs, sales reps reports)
- Advantages (low cost, historical comparison, data available)
- Disadvantages (depend on customer to communicate)

Active techniques:

- Requires going to the customer (interviews, direct observations, surveys)
- Advantages (deeper understanding, target audiences)
- Disadvantages (data not available, resources required)

13.3.10.2 *Process to understand VOC*
- Plan
  - Understand the purpose of the VOC process and its alignment with business objectives
  - Identify customers to be studied and prepare questions

- Gather data
  - To gather voice of customer data
  - Gather voice of customer data
- Understand VOC
  - Translate customer language into business requirements and metrics
- Deploy
  - Take action based upon the voice of the customer

13.3.10.3 *Critical to quality*

This tool actually focuses on the key metrics of customer satisfaction. A CTQ tree will translate the initial customer requirements to numerical or quantified requirements for the product or service. The creation of the CTQ tree involves these steps:

- Identify the customer
- Identify the customer's need
- Identify the set of basic requirements of the customer
- Progress further with more levels as needed
- Validate the requirements with the customer

The tools for analyzing the VOC differs based on belt certifications and is been discussed accordingly in the later sections of this chapter.

13.4 *Project charter and teams*

The *define* step involves a team and its sponsors reaching an agreement on what the project is and what is the goal to be achieved from them. Presuming that the project charter is already in place, the most important function of *define* is for the team to decide what the team has to accomplish and confirm their understanding with the sponsorship. The sponsor and team should agree on the following:

- Decide and determine what the problem is and the agreement on the problem: which customers will be affected, what their "voices" are saying, how the current process or outcomes fails to meet their needs, and so on.
- The link of the project between corporate strategy and its expected contribution to return of investment.
- Decide on the boundaries of the project.
- Know what indicators or metrics are used for the evaluation of the success of the project.

In service environments, the last two points that are mentioned above often prove particularly more important. When the process that has to be done is studied and mapped out then deciding the start and end points of the project becomes a simple matter. Most services have not been mapped prior to improvement; there is often an argument between the team and the sponsors in the early stages of the project improvement as the team creates a SIPOC or value stream map and then the direction to identify exactly what they should include as part of their project and what they should not.

13.4.1 Setting project boundaries

Projects that are too big will end up with floundering teams who have trouble finishing in a reasonable frame of time. Choosing a small or insignificant team and you will never convince anyone that Six Sigma is worth the investment. Projects that do not significantly contribute to financial payback and everyone from line managers to the senior executives will quickly lose interest. Another concern to be considered when we think about project boundaries is the level of the certification the employees possess. During their training period black belts work on a project of limited scope only. The metrics for success in the *define* phase might be as follows:

- Customer satisfaction, to make sure all customers segments are represented
- Speed/lead time
- The team to determine the defects and opportunities for improvement. To have a sigma level improvement
- How will the above processes help financially?

There are two key issues in *define*:

- Making sure the right people are on the bus. The decision should be determined by not only the kind of people that are representative of the work area(s) affected by the project and those who possess the knowledge, experience, and training that can help the team reach the project goals but also by an evaluation of the dynamics of the team.
- It is important to make sure that everyone in the team is starting from the same point have the same goals from the project. This includes all the members of team, belt-certified professionals, champions, black belts, and the other staff who are working for the process but might not be part of the team.

Of course, granting authority to employees does not guarantee that people will work together or necessarily achieve all the lofty goals that are espoused in this approach. Many issues surround empowerment and teamwork that must be addressed. These issues range from operations and behavior to organizational design. For example, if the existing culture does not reward this type of activity, it is doubtful that participatory approaches will work until the cultural issues are resolved. However, using teams can lead to cultural changes that facilitate improvement. This chapter focuses on the issues related to managing projects and teams to help make the transition succeed.

From a behavioral perspective, empowerment is a tool to enhance organizational learning. Organizational learning implies change in organizational behavior in a way that improves performance. This type of learning takes place through a network of interrelated components. These components include teamwork, strategies, structures, cultures, systems, and their interactions. Cooperate learning relies on an open culture where no one feels threatened to expose opinions or beliefs—a culture where individuals can engage in learning, questioning, and not remain constrained by "taboos" or existing norms. This strategy includes continuous improvement projects as a governing principle for all team members.

13.4.2   Project charter

The project charter is represented in the following four areas:

1. Content of the charter
2. Negotiation of the charter
3. Project management
4. Project measures

13.4.2.1   Content of the charter

The development of a charter is a vital element for establishing an improvement team. The charter is a document that defines the teams' mission, the boundaries of the project, the consequences, and the time frame. It is usually the top management that creates the charter and presents it to the teams or the teams can also create a charter and present it to the top management and obtain their approval. Either way, the top managements are responsible to give the team the support and direction of working.

It begins with creating a purpose statement. This may be a single- or double-line statement explaining the purpose of formation of the team. The purpose statement should be in correlation with the organization's vision and mission statements. The objective to be achieved also should be defined in this purpose statement.

The objective should be defined in such a way that it can be measured. The scope or the boundaries of the project should also be given in the project charter. This is to determine the organizational limit within which a team is permitted to operate. Time delaying and energy draining can be prevented if the boundaries of the project are defined.

Teams are supposed to know what is expected of them. The team has the permission, authority, and blessing from the various levels of management to operate, conduct research, consider, and implement changes that may be of need for the process. A charter provides the following advantages:

- Confusion elimination
- Determines the subject boundaries
- Areas that are not to be addressed
- Determines the deliverable
- Provides a basis for the team to set a goal
- Authorizes the team to collect relevant data

A team project charter should contain the following:

- The financial impact
- Problem definition
- Scope or boundary of the project
- Goal statement
- Role of team members
- Milestones or deliverables
- Resources required

13.4.2.2 Financial impact

This is a short summary of the reason for carrying out the project. It would normally involve quality, cost of delivery of a product with a financial justification. There are four basic activities:

1. Design of a new product
2. Redesign of an existing product
3. Design of a new process
4. Redesign of an existing process

A common problem for most companies is the lack of the measurement of the impact. A project improvement team should carry out whatever project in accordance with financial department justification guidelines. According to Eckes (2001), for example, if the existing quality defective rate is at 5,000 defectives per million opportunities, the possible justification is a reduction to 250 defectives per million opportunities with a cost saving of $1,000,000. The advantages and shortcomings of the project should be looked into. There should be an involvement of the entire organization, if necessary, to determine the key costs and their resources for a successful project. Projects that do not provide financial augmentation should be eliminated right away. A common problem for many projects is the lack of a company impact measurement.

13.4.2.2.1 Problem statement.   A problem statement will give a detailed statement on the issue that has to be improved. The problem statement should be crafted so that it describes as much as it can. Such as, for how long has the problem been there, how it has affected the business, what is the gap in performance and the measurable item that might be affected? The problem statement should not be such that it makes one jump to conclusions. A sample problem statement would be "The ABC Company, in 2007, has experienced a 25% drop in sales, with a 40% drop in net profit." The problem statement should contain a reference to a baseline measure for guidance. The collection of good data and process performance measurements will provide a picture of the areas in the company that need improvement the most. In addition, the foundation of the work will provide a measure for other teams working on other projects as well.

13.4.3   Goal statement

The goal statement is created and agreed to by the team and the champion. The goal is hoped to be attained in a 120- or 160-day period. According to the Six Sigma metric, it is required to have 50% reduction in some initial metric. For example, by reducing the collectibles from 120 days to 60 days, reduces the scrap from 25% to 2.5%. One of the most efficient ways of formulating the goal statement is the theory of constraints.

13.4.3.1 Step 1: Identify

Concentrating on a nonconstraint resource would not increase the throughput (the rate at which money comes into the system through sales) because there would not be an increase in the number of orders fulfilled. There might be local gains, but if the material ends up waiting longer somewhere else, there will be no global benefit.

- In order to manage a constraint (bottleneck), it is first necessary to identify it.
- Constraint (bottleneck): resource whose capacity is less than the demand.
- This knowledge helps determine where an increase in "productivity" would lead to increased profits.
- To increase throughput, flow through the constraint must be increased.

13.4.3.2   Step 2: Exploit

- Once the constraint is identified, the next step is to focus on how to get more output within the existing capacity limitations.
- Because the constraint is what limits the system's throughput, we have to make it work to the maximum.

13.4.3.3   Step 3: Subordinate

Subordination usually involves significant changes to current (and generally long-established) ways of doing things at the nonconstraint resources.

- Subordinate the nonbottlenecks to the system constraint.
- All the other components of the system must work so as to guarantee full-speed functioning of the constraint.

13.4.3.4   Step 4: Elevate

After the constraint is identified, the available capacity is exploited, and the nonconstraint resources have been subordinated, the next step is to determine if the output of the constraint is sufficient to supply market demand. If so, there is no need at this time to "elevate" because this process is no longer the constraint of the system. In that case, the market would be the constraint, and the TOC thinking process should be used to develop a marketing solution.

- If, after fully exploiting this process it still cannot produce enough output to meet demand, it is necessary to find more capacity by "elevating" the constraint.

13.4.3.5   Step 5: Go back to Step 1

- Once the output of the constraint is no longer the factor that limits the rate of fulfilling orders, it is no longer the constraint.
- Step 5 is to go back to Step 1 and identify the new constraint—because there always is one. The five-step process is then repeated.

13.4.4   Milestones or deliverables

A well-organized project is bound to have set of short-term goals or deliverables that are used to keep the project on track and help bring it to completion. It has been pointed out that the initial team projects should be at the 120-day length. Only half of the project time is supposed to be allocated to *define* and *measure* phase. Assigning teams and the right kind of people for the project is very important. The success rate of the project decreases as the length of time assigned to complete the task increases. A typical milestone chart might be

- Day 0: start team activities
- Day 1: start the *define* portion of the project

- Day 3: begin the *measure* portion of the project
- Day 80: start the *analysis* of the project
- Day 120: start the improvement phase of the project
- Day 160: bulk of project control elements in progress

Resources required for a project is very important to be noted and detailed down. Typical resources might be as follows:

- Qualified people
- Machine time
- Machinery
- Lab or office space
- Phones and faxes
- Computer equipment
- Utilities

The Six Sigma *define* phase should provide the top management the following information:

1. Importance of carrying out the project
2. Goal of the project
3. Skills of the champion and other leaders
4. Boundaries of the project
5. The key process
6. Metrics
7. Customer requirements

13.4.5   Charter negotiation

The team to the top management can present the project charter. However, the project team might be closer to the actual facts through another approach toward the problem. So, there are bound to be charter negotiations. They might be as follows:

- Objectives—change in the design or final product due to customer feedback
- Scope—boundaries the organization provide might require further expansion. Requirement for more people to carry out the project.
- Resources—requirement of more resources. A complete accurate requirement of resources can never be provided. Management may be required to prioritize certain resources beyond the team's control.
- Project transition—the transition of a project to normal company controls might require a time extension.
- Project closure—project closure date might be required to be moved up because of diverse events or changes in customer preference.

13.4.6   Project management and its benefits

Project managers are hired by most organizations to ensure that complexities of a project move in an organized fashion and make the proper transitions in a timely and economical manner. This type of "matrix management" has proved very effective in providing deliverables on time. The project management roles and responsibilities include the following:

- Leading the cross-functional team
- Possessing excellent communication skills and ability to convey the message clearly
- Schedule meeting to check progress at regular intervals
- Sustaining the team and its motivation
- Development of a detailed project plan
- Letting the team know the benefits of the project to all the share holders
- Tracking of the progress of the tasks and deliverables
- Maintaining flow of information between financial and information management.

Projects need charters, plans, and boundaries. A project may be selected from a broad range of areas including

- Customer feedback
- Improvement in process capabilities
- Cost reduction chances
- Defects reduction
- Employing Lean principles
- Growth in market share
- Reduction in cycle times
- Improvement in services

The project should be consistent with the strategies of the company for survival and/or growth. The project should rather be specific.

13.4.7 Project measures

It should be noted that the vital project measurements are hard to decide until the project charter and its processes are not complete. The accurate selection of project measure ensures the overall success of Six Sigma implementation. Since most projects deal with time and money issues, most project measures will also be related to time and money. The project measures provide information, which is required to analyze and improve the business process, as well as manage and evaluate the impact of the Six Sigma project. After a list of activities of the project is prepared, the budget of the project is determined. During the project, the actual costs are collected as inputs and used as inputs to determine the estimated costs. The project manager or the team leader compares the revised estimated costs with the actual costs and determines the progress of the project. A project should be reasonable, attainable, and based on estimates of the tasks to be accomplished.

The various revenue factors included in the analysis are

- Income from additional sales generated due to the revised cost of the product, and the changes in the quality, features, and availability to the customer
- Reduced amount of defects, scrap, returns, warranty claims, cost of poor quality, and poor market phase

The cost factors included in the budget are

- Labor cost
- Administrative expenses
- Equipment costs

- Subcontracted work
- Overheads
- Contingency funds

The timing of the revenues and costs also play a very important role. Sometimes the revenues projected to be obtained might not be obtained because the funds were not available in the right time frame. The precision and detail of the project-planning phase plays a very important role in the success of the project. The costs associated with each project are obtained based on historical data, quotes, standard rates, or similar activities performed previously. Estimates of project revenues are described based on four types of measurements:

1. Budget—the plan of the total costs and cash inflows expressed in dollar amounts for the project. The plan also includes timing and of the revenue and costs and a cost–benefit analysis.
2. Forecast—the predicted total revenues and costs, adjusting to include the actual information at the point of completion of project
3. Actual—revenues and costs that have actually been occurred for the project.
4. Variance—the difference between the budgeted and actual revenues and costs. A positive variance shows the project will be favorable while a negative variance shows that a loss will be incurred.

13.4.8   Teams

Flattening hierarchies for improved effectiveness. Along with emphasis on teamwork and empowerment, there has been a move toward flattening hierarchies in organizations. Led by consultants such as Tom Peters and others, top managers have eliminated layers of bureaucratic managers in order to improve communication and simplify work. Having many layers of management can have the effect of increasing the time required to perform work. For example, it has been reported that in the 1980s, one of the largest automobile manufacturers in the United States required 6 months to determine its standard colors for office phones. Probably this decision required many, many meetings and proper authorization. However, such decisions were needed to be made. The time required to make this decision was excessive.

Too many layers of management also can impede creativity, stifle initiative, and make empowerment impossible. With fewer layers of management, companies tend to rely more on teams. When Lee Iacocca took the reins at Chrysler Corporation, one of his first acts was to eliminate several levels of management. Iacocca credits this move with making other needed changes easier within the organization.

13.4.8.1   Team leader roles and responsibilities

Quality professionals are unanimous—to be successful in achieving teamwork and participation; strong leadership both at the company level and within the team is essential. However, what is not always clear is what it means to be an effective team leader. We know that leaders are responsible for setting the team direction and seeking future opportunities for the team. Leaders reinforce values and provide a system for achieving desired goals. Leaders establish expectations for high levels of performance, customer focus, and continuous learning. Leaders are responsible for communicating effectively for evaluating organizational performance, and for providing feedback concerning such performance.

An important aspect of leadership is the organization's preparedness to follow the leadership. The best general is probably not going to be successful if the troops are not well trained or prepared. Hersey and Blanchard proposed a theory called a situational leadership model that clarifies the interrelation between employee preparedness and effectiveness of leadership. According to Hersey and Blanchard, situational leadership is based on interplay among the following:

- The amount of guidance and direction a leader gives (task behavior).
- The amount of socioeconomic support a leader provides (relationship behavior).
- The readiness level that followers exhibit in performing a specific task, function, or objective.

Therefore, if team members are trained and prepared and prepared so that they are "task ready," leadership will be more effective. Readiness, in this context, is the "extent to which a follower has the ability and willingness to accomplish a specific task." Readiness is a function of two variables. These are ability and technical skills and self-confidence in one's abilities. Therefore, effective leadership helps employees become competent and instills confidence in employees that they can do the job.

As it relates to quality management, leadership is especially difficult. Leaders are told that they should empower employees. Having too many leaders implies laissez-faire or hands-off approach to management. In other words, many leaders feel that they have to provide resources but that they should not be involved in overly controlling employee behavior. Although the literature contains examples of companies that have been successful in delegating authority to this extent, quality management is not a vehicle by which leaders abdicate their responsibility.

In most organizations employees want leaders who provide clear direction, necessary information, and feedback on performance, insight, and ideas. Skilled team leaders need to demonstrate this ability to lead. The single most important attribute of companies with failed quality management programs is lack of leadership. A close second is poor communication, which is related to leadership. Effective leaders are people who are able to provide vision, idea, and motivation to others to achieve the greater good.

Besides team leaders, there are a variety of roles which individuals occupy in teams. Also, team roles can be defined functionally. Often teams require different functional talents such as management, human resources, engineering, operations, accounting, marketing, management information systems, and others. In these cases, the managers overseeing the project help to identify the talents needed and then search for the team members to provide these talents (Figure 13.13).

13.4.8.2   *Team formation and evolution*

The way a team is formed depends—to an extent—on the objectives or goals of the team. Regardless of the type of team your firm employs, teams experience different stages of development. These stages include the following: *Forming*, where the team is composed, and the objective for the team is set; *storming*, where the team members begin to get to know each other, and agreements have not yet been made that facilitate smooth interaction between team members; *norming*, where the team becomes a cohesive unit; and *performing*, where a mutually supportive, steady state is achieved. And in successful projects, the final stage is *mourning*, where the team members regret the ending of the project and breaking up of the team.

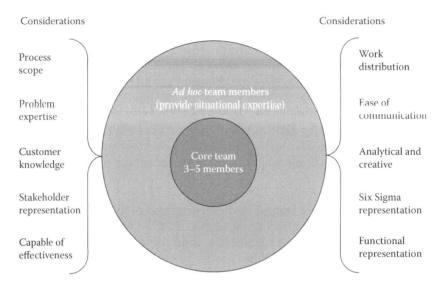

Figure 13.13 The core and *ad hoc* team.

13.4.8.2.1 Team rules.   During the norming stage, teams develop ground rules. Such ground rules can forestall conflict. It is often useful to establish ground rules for a team to be functional. If a team is functional, individual participation enhances the group's effectiveness. If the team is dysfunctional, such participation reduces the effectiveness of the group. Acts of commission include talking behind the backs of other team members or otherwise acting out one's feelings. There are also acts of omission in such passive aggressive behavior as forgetting to attend meetings or withholding information. Counteractive behavior improves the group's effectiveness by negating dysfunctional behavior. Counteractive behavior can be enacted either by the team, the facilitator, the team manager, or even the offending individual.

13.4.8.3   Types of teams

At this point, we will pause to define the various types of teams that are used in improving quality. Continuous process improvement often requires small teams that are segmented by work areas. Projects with multiple departments in a company require cross-functional teams. Large projects require teams with large budgets and multiple members. Smaller projects, such as "formulating a preventive maintenance plan for oiling the metal lathes," probably will require a much smaller team. In the following sections, we list and define a number of teams.

1. *Process improvement teams*: Process improvement teams are teams that work to improve processes and customer service. These teams may work under the direction of management or may be self-directed. In either case, the process improvement teams are involved in some or all of the following activities: identifying opportunities for improvement, prioritizing opportunities, selecting projects, gathering data, analyzing data, making recommendations, implementing change, and conducting post implementation reviews. Many process improvement teams are on outgrowth of quality-related training. These teams use the basic tools and plan–do–check–act cycle to effect change relating to processes.
2. *Cross-functional teams*: Cross-functional teams enlist people from a variety of functional groups within the firm. In the real world, problems often cut across functional borders. As a result, problem-solving teams are needed that include people from a

variety of functions. These cross-functional teams often work on higher-level strategic issues that involve multiple functions. Such teams often work on macro level, quality-related problems such as communication or redesigning companywide processes.

3. *Tiger teams*: A tiger team is a high-powered team assigned to work on a specific problem for a limited amount of time. These teams are often used in reengineering efforts or in projects where a specific problem needs to be solved in a very short period of time. The work is very intense and has only a limited duration.

4. *Natural work groups*: Natural work groups are teams organized around a common product, customer, or service. Many times these teams are cross-functional and include marketers, researchers, engineers, and producers. The objective of natural work groups include tasks such as increasing responsiveness to customers and market demand. Improved work design and implementation of natural work improves work life for employees. The key, elemental impact of natural work groups is to improve service by focusing work units in an organization on the customer. A by-product is improved communication with customers. Often a natural work group will be established for a specific customer.

5. *Self-directed work teams*: A self-directed work team is a team chartered to work on projects identified by team members themselves. There is little managerial oversight except to establish the teams and fund their activities. Self-directed teams are identified as either little s or big S teams. Little s self-directed work teams are made up of employees empowered to identify opportunities for improvement, select improvement projects, and complete implementation. Big S self-directed teams are involved in managing the different functions of the company without a traditional management structure. These types of teams contain totally self-directed employees who make decisions concerning benefits, finances, pay, processes, customers, and all the other aspects of running the business. Often big S self-directed work teams hold partial ownership of the companies they work for so that they participate in the benefits of their teamwork.

6. *Technology and teams*: New tools for teamwork are constantly emerging. Also, team effectiveness is a precursor to project task performance. Integrated IS tools involve integrated information systems such as CAD/CAM and CIM. This aids in achieving improvement in efficiency and effectiveness. Process technology is used in helping to improve task performance. Process standardization methods such as the tools of quality and customer input methods complete the model. This model amplifies that more and more, team effectiveness is assisted by integrated tolls and technologies, and the impact of technology should increase. As software becomes cheaper and easier to use, more tools will be used by everyone involved with the project.

7. *Virtual teams*: The term *virtual* team is emerging as more companies become "virtual organizations," loosely knit consortia that produce products and services. Virtual teams are teams that rarely or never physically meet, except in electronic meetings using group decision software. Among virtual organizations, projects often cross organizational boundaries. Today, internet and intranet-based applications called teamware are emerging that allow us to access the world wide web and build a team, share ideas, hold virtual meeting, brainstorm, keep schedules, and archive past results with people in far-flung locations around the world. Hectic schedules and the difficulty in finding convenient times to meet to solve problems will make teams of this type more important in the future.

13.4.8.4 *Implementing teams*

The teams in our examples have something in common. The performance of the team is essential to their individual success, and in some cases, even lives hang in the balance. If the NASCAR team performs ineffectively, the driver loses. If the Massachusetts General Hospital team is ineffective, people die. If the SEALs do not function properly, lives are lost and the mission fails. How do we engender this sense of urgency in quality improvement teams? How do we create a momentum or team ethic that will help us beat the odds and be successful? Accomplishing this often requires facilitation and team building. Facilitation is helping or aiding teams by maintaining a process orientation and focusing the group. Team building is accomplished by following a process that identifies roles for team members and then helps them to become competent in achieving those roles.

The role of the facilitator is very important in managing teams, particularly when team members have little experience with teamwork. The role of the facilitator is to make it easy for the group to know where it is going, know why it wants to get there, know how to get there, and know what it is going to do next. A facilitator focuses the group on the process it must follow. Successful facilitation does not mean that the group always achieves its desired results. The facilitator is responsible for ensuring that the team follows a meaningful and effective process to achieve its objectives.

How is this accomplished? The facilitator should plan how the group will work through a task, help the group stay on track and be productive, draw out quiet members, discourage monopolizes, help develop clear and shared understanding, watch body language and verbal cues, and help the group achieve closure. Again, facilitators must remain neutral on content. Facilitators cannot take sides or positions on important areas of disagreement. However, facilitators should help key members reach points of agreement. Meeting effective management is an important skill for a facilitator of quality improvement teams. Often quality improvement involves a series of meetings of team members who meet to brainstorm, perform root-cause analysis, and carry out other activities. Tools for a successful meeting management include an agenda, predetermined objectives for the meeting, a process for running the meeting, processes for voting, and development of an action plan using these tools requires outstanding communication skills as well as human relations skills. The steps required for planning a meeting are (Figure 13.14):

- Defining an agenda
- Developing meeting objectives
- Designing the agenda activity outline
- Using process techniques

Date: Time:

In attendance:

    Project sponsor(s), black belt, core team members, master black belt agenda:

    – Review business case for the project

    – Review and discuss Project Charter

    – Develop team-behavior contract

    – Develop project plan, including team information and project schedule

Assignments:

What   Who   By when

Next meeting:

Figure 13.14 Meeting notes.

Structured processes, a set of rules for managing meetings, work well in conducting meetings. It is paradoxical that structured processes are inhibiting, time consuming, and unnatural-which is why they work. Why do we use processes in meetings? The answers are clear. We wish meetings to stay focused, to involve deeper exploration, to separate creative from evaluative activities, to provide objective ground rules defensiveness, and to separate the person from the idea.

Tools such as flip charts, sticky dots, whiteboards, and sticky notes are used commonly in structured process activities. The focus of team meetings moves from clarifying to generating ideas, to evaluating ideas, and to action planning. Some of the techniques, such as silent voting and idea writing, help team members reach consensus rapidly.

Another useful meeting management tool that was pioneered by Hewlett-Packard is the "parking lot." The parking lot is a flip chart or whiteboard where topics that are off the subject are parked with the agreement that these topics will be candidates for next meeting's agenda. At the end of the meeting, the group agrees on the agenda for the following meeting, and the parking lot is erased.

Conflict resolution in teams. As people work closely together in teams, conflicts arise. Conflict resolution is a hugely important topic for team leaders and members. Conflicts are endemic to all kinds of team projects. Using team processes, assumptions are questions, change is brainstormed, and cultures are challenged. This type of creative activity results in possible conflict. It is claimed that team leaders and project managers spend more than 20% of their time resolving conflict. If this is true, then conflict resolution resounds as one of the very important under-discussed topics in team building.

There are many sources of conflict. Some conflicts are internal, such as personality conflicts or rivalries, or external, such as disagreements over reward systems, scarce resources, lines of authority, or functional differentiation. Teams bring together individuals from a variety of cultures, backgrounds, and functional areas of expertise. Being on a team can create confusion for individuals and insecurity as members are taken out of their comfort zones. It is interesting to note that these are also some of the reasons teams are successful. Some organizational causes of differences are more insidious: faulty attribution, faulty communication, or grudges and prejudice. Four recognizable stages occur in the conflict resolution process:

- Frustration. People are at odds, and competition or aggression ensues.
- Conceptualization and orientation. Opponents identify the issues that need to be resolved.
- Interaction. Team members discuss and air the problems.
- Outcome. The problem is resolved.

One of the things a leader must be able to do is manage conflict in the organization. In order to foster a wee-run workplace, leasers must be able to resolve conflict effectively in the organizations. Leaders resolve conflict in a variety of ways:

- Passive conflict resolution: Some managers and leaders ignore conflict. This is probably the most common approach to working out conflict. There may be positive reasons for this approach. The leader may prefer that subordinates work things out themselves. Or the conflict may be minor and will take care of itself over time. Leader feels that some issues are small enough to not merit micromanagement.
- Win–win: Leaders might seek solutions to problems that satisfy both sides of a conflict by providing win–win scenarios. One form of this is called balancing demands for

the participants. This happens when the manager determines what each person in the conflict wants as an outcome and looks for solutions that can satisfy the needs of both parties.

- Structured problem solving: Conflicts can be resolved in a fact-based manner by gathering data regarding the problem and having the data analyzed by a disinterested observer to add weight to the claims of one of the conflicting parties.
- Confronting conflict: At times it is best to confront the conflict and use active listening techniques to help subordinates resolve conflicts. This provides a means for coming to a solution of the conflict.
- Choosing a winner: In some cases, where the differences between the parties in the conflict are great, the leader may choose a winner of the conflict and develop a plan of action for conflict resolution between parties.
- Selecting a better alternative: Sometimes there is an alternative neither of the parties to the conflict has considered. The leader then asks the conflicting parties to pursue an alternative plan of action.
- Preventing conflict: Skilled leaders use different techniques to create an environment that is relatively free of conflict. These approaches are more strategic in nature and involve organizational design fundamentals. By carefully defining goals, rewards, communication systems, coordination, and the nature of competition in a firm, conflict can be reduced or eliminated. Conflicts often are the results of the reward systems in the firm. A systems approach will focus attention on organizational design rather than individual interactions.

13.5   Project tracking

Project management is an essential process in establishing a scheduled plan and allocating available resources judiciously. The strings of activities that constitute a project are to be monitored for straying off the timeline for its proper implementation. Thus the objectives of a project are adhered to: Utilizing the allocated resources within time and cost constraints at the desired level of performance to reach specified goal. The phases of project management include the following:

- Planning
- Scheduling
- Controlling

Tracking is the controlling phase of project management to ensure the requirements of the undergoing project are met. It involves an exercise of making rightful decisions (Figure 13.15):

- Identifying sources of bottlenecks
- Delegating duties
- Choosing the right tradeoffs
- Measuring results against expected outcomes
- Applying timely corrective plans
- Establishing tolerance level in timeline
- Future developments to current work
- Efficient communication and continuous tracking

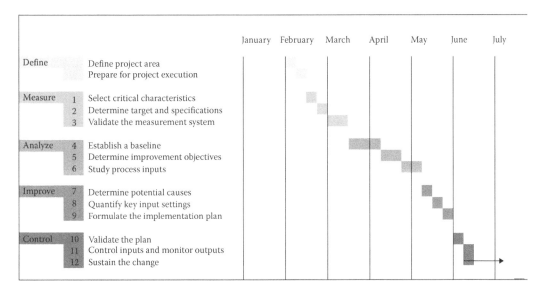

		January	February	March	April	May	June	July
Define		Define project area						
		Prepare for project execution						
Measure	1	Select critical characteristics						
	2	Determine target and specifications						
	3	Validate the measurement system						
Analyze	4	Establish a baseline						
	5	Determine improvement objectives						
	6	Study process inputs						
Improve	7	Determine potential causes						
	8	Quantify key input settings						
	9	Formulate the implementation plan						
Control	10	Validate the plan						
	11	Control inputs and monitor outputs						
	12	Sustain the change						

Figure 13.15 Project timeline.

13.5.1 Work breakdown structure (WBS)

It is a descriptive document on the constitution of the project, the assigned responsibilities for each work group. Each responsibility is converted to a string of activities and further broken down to subtasks or elements that are set into motion by a team. From subtask to activity each is having a constricting time frame which should be flexible enough to accommodate variations in implementation. Interrelationships between predecessor events and following events are to be streamlined and scheduled. Offsetting is required to avoid losses by arranging for a safety stock of resources and to accommodate any increase in cost incurred; however, the deadlines are fixed in nature. For adherence to such fixed time requirements, resources have to be fixed and aplenty. There should be arrangements for coordinating between different elements in the organization that source their equipment, material, etc. from the same pool. Seamless scheduling becomes essential for parallel tasks with slack time. Planning for such activities requires tools that can recognize the amounts of resources, time, and costs. These include the following:

1. PERT
2. CPM
3. Gantt chart

The WBS is useful for pulling out relevant activities and the investment required. As is evident, these tools are helpful in the *analyze* phase in the Six Sigma process which we shall review in detail later.

The network planning rules are based on these tools and these are

- All activities must begin on completion of preceding events
- Logicality of event precedence is implied using arrows
- The direct event connection can be established through only one activity
- Network must begin and end at single events
- Event numbers must be unique

The PERT and CPM tools are similar except that CPM is activity oriented and PERT is event oriented. Gantt Charts are representative of activities or events against time. The purpose, frequency, method (written reports, summary, and forms), prioritizing among random events, feedback loops, and contingency plans must be kept in mind. The feedback loop has methods for contingency measures. It is an autogenerated control plan. Success or failure is to be measured against

- Fulfillment of goals
- Adherence to time limits
- Boundary variance
- Utilization index of resources

A properly executed plan must have a high rate of success for all these measures. Contingency plans must be at no extra costs and wastage of resources. Unanticipated events, complexity of a project that poses challenges in terms of better technologies are not acceptable excuses for occurrence of failure. Performance standards for results are set high and issues must be resolved justifiably by the team leader to maintain the schedule for the given budget levels.

The unit of measurements for timeline is specified as days, weeks, months, etc. according to convenience. The methods for control range from planning and monitoring using manual methods such as paper, graph, and markers to software.

13.5.1.1   Advantages of manual methods
- Easy to use
- Apt for scheduling
- Reduced cost
- Customizable
- Flexible to needs
- Hands-on feel of status

13.5.1.2   Advantages of computer software
The advantages of computer software are that it

- Monitors random events
- Impacts alternatives
- Gives detailed project status reports
- Automatically calculates time frames
- Easily reports
- Quickly generates summaries
- Plans in real time
- Automates data collection activities

13.5.2   Milestones reporting

Significant events in the project timeline are termed as milestones. They help keep pace with the project schedule and in reporting the progress and status level of the project. They act as a decision to point to proceed with the project state. Presentations are made at these pit stops to check for further expected bottlenecks. Plans for dealing with the conditions that arise due to these potential setbacks are discussed. The time is already set during the project planning phase as to which activity can be termed as an important milestone. For any well-managed project, a set of stages or milestones are used to keep the project on track and to help bring a project to completion. Only half of the project would be allocated to define and measure stages. Assigning teams a first project with lengths of more than 160 days will lower their success rate.

13.5.3 Project report

Project report is a progress summary that states the performance vis-à-vis the anticipated benefits and costs incurred with the budget planned initially. Also the tasks completed against the level of milestone are analyzed for status. The postmortem analysis is the next step that tells what went wrong and is used as a base for building and improving on further plans. The benefit from these effective processes are given as follows.

13.5.3.1 Document archiving

Document archiving mean documenting the data, source of the materials, process parameters and reports generated during the project. The organization of these files must be coherent and have special storage conditions.

These are

- Security with special access
- Retrievable
- Protection from damage
- Traceability of these documents using indicators
- Duplication of files for safety
- Using reliable mediums

13.6 Yellow belt

Tools used in yellow belt are SMART objective, SIPOC, and translate VOC to CTQ.

13.6.1 SMART objective

The objectives framed by the organization should follow SMART way so that the company can thrive to achieve the goal. SMART is defined in Table 13.8.

Table 13.8 SMART Objective

Specific	The objective need to be more specific and clearly identify the issue.
Measurable	It is possible to measure whether the objective has been achieved.
Achievable	The objective should be achievable.
Relevant	The objective should be relevant to company's strategy.
Timed	There should be a target date for completion.

Figure 13.16 SIPOC process.

13.6.2 SIPOC

SIPOC is abbreviated as Supplier Inputs Process Outputs Customers. It is a high level process map with four to seven steps displayed. The following steps should be followed to develop the SIPOC:

- Have the team create the process map
- The process may have four to five key steps
- List the output of the process
- List the customers of the output of the process
- List the inputs of the process
- List the suppliers of the process
- Identify some preliminary requirements of the customers

Involve the team leader, champion, and other stakeholders for verification of the project (Figure 13.16).

- Suppliers: Suppliers are the provider of inputs to your process.
- Inputs: Inputs are the materials, resources, or data required to execute your process.
- Process: Process is defined as structured set of activities that transform a set of inputs into specified outputs.
- Outputs: Outputs are products or services that result from the process.
- Customers: Customers are the recipients of the process output.

13.6.3 *Voice of the customer in yellow belt*

The primary driver for any Six Sigma initiative is voice of the customer. Market trends trigger the performance of any system. These two factors are the strategy for any system that initiates Six Sigma. For any firm to be successful, they need to understand what the customer requires from the company's product or services. Voice of customer can be defined by the following three terms:

- Customer's needs
- Customers perception
- Customer's attributes

The attributes which drive toward the achievement of specific goals is defined as the customer's needs. The customer's opinion to evaluate the product or services in a favorable manner or unfavorable manner is defined as the customer's attributes. The manner which the customer presumes and interprets a company is known as customer's perception.

Companies work on these attributes to attract purchasers. They look into the primary function as well as the needs, wants, and expectations of the product to achieve high serviceable level. They develop products or services based on the following functions:

- Service levels that favorably influence customer attitudes.
- Advertising objectives that positively impact customer perceptions.
- Product specifications that meet customer needs.

Primary targets for any firm increase in market share and revenue growth. For this purpose, the firm should identify a strategy for retaining the current customer base and acquisition of a new customer base. The firm should not only listen to its own customer's requirement but should also get information from the customers who currently do business with its competitors. This enables increase market share and revenue growth. If the company considers only the voice of the customer it loses the chance of attracting its competitor's customer.

The strategic use of Six Sigma is to process the value gap. Value gaps can either be closed or can be increased to that of the competitors. Value gaps can be customized based on voice of customer and competitor's customer. Inputs from this process can be used for changes in strategy and it should be incorporated into the system. It is not helpful only to determine the changes in value stream map if the inputs obtained from the above process are not incorporated into the information system.

Various sources of voice of customer in yellow belt are listed as follows:

- Internal quality metrics—rework, scrap
- Customer complaints
- Customer inspection meetings
- Telephone calls
- Performance relative to alternatives
- FAL reports/Feedback
- Questionnaires
- Personal visits
- Surveys
- Focus groups
- Interviews

Voice of customer (VOC) can be analyzed by two methods. The first method is to tabulate all the expectations from the product or services by the firm. This analysis is done from the customer's perspective. All the defects in the product or services are tabulated in the flip chart. The second method is to analyze the product or services individually. The facilitator is responsible for translating the results in the flip chart. Then the details about the results are discussed by having a meeting among the team.

Voice of customer can be assessed based on two techniques—qualitative research and quantitative research. Qualitative research is done to obtain results of customer's perception. Qualitative research is based on structured questions. It is also called survey research. This method is used for collecting data from the customer. This quantifies the customer's expectations, perceptions, requirements, etc.

Primary target or metrics for any firm is to achieve customer satisfaction. Customer requires highest possible quality for a product than emotional commitment toward the firm. It is not just concerned about the tradeoff between price and quality. Customer satisfaction parameter can be measured by analyzing all the attributes till the product is purchased by the customer.

Value is also considered to be the parameter for customer satisfaction. Suppose that customer is willing to pay high for getting a quality product then the price is sublimed in this process and quality takes first place in this condition. Hence, the value of the product should be considered an important parameter for customer satisfaction.

13.6.3.1   *Critical to quality*

Any measurable product or service characteristic that is important to the customer from the customer's point of view is critical to quality. This tool actually focuses on the key metrics of customer satisfaction. A CTQ tree will translate the initial customer requirements to numerical or quantified requirements for the product or service.

13.6.3.1.1 Translation of VOC to CTQ.   VOC drives the firm to achieve customer perception and needs. Critical to quality quantifies the voice of customer. Voice of the customer should be clearly studied and the following steps need to be analyzed:

- Customer needs
- Define the metric
- Target
- Specification/Tolerance limit

Voice of the customer is analyzed thoroughly and detailed sketch for the customer needs, metrics, and targets are charted out (Table 13.9).

- Customer's needs: Customer's needs is defined as the attribute which drives the firm to the specific goal. It also specifies about consistent delivery against agreement.
- Define the metric: Metric is defined as the difference in the number of days between the customer's requested date and actual date received.
- Target: Target should be striving to achieve zero metric.
- Specification/Tolerance limit: Tolerance limit specifies one or two days ahead are behind the metric.

Basic insights from this process are specified as follows:

- It is impossible to measure whether some customer needs are being met. In those cases, find an alternative to measure such that it will specify whether customer satisfaction is achieved.
- Target value is the point at which the customer is most satisfied.
- Customer satisfaction level becomes unstable if the value goes above or below the target value.

Table 13.9 Translation of VOC to CTQ

VOC	Customer need	Define the metric	CTQ target	CTQ tolerance

- Tolerance is the point between the customer's satisfaction or not.
- Critical to quality has no impacts on the capacity of the process.

13.7   Green belt

The tools used in green belt are Pareto diagrams, benchmarking, brainstorming process, thought process mapping, process mapping, and value stream mapping.

13.7.1   Pareto diagrams

Pareto diagrams are used to prioritize problems so that the major problems can be identified. The diagrams are used to

- Analyze a problem from a new perspective
- Focus attention on problems in priority order
- Compare data changes during different time periods
- Provide a basis for the construction of a cumulative line
- Prioritize problems
- Segregate the significant problems from the insignificant problems

Pareto analysis is based on 80/20 rule. Dr. Juran states that 80% of the problems come from 20% of causes. Pareto analysis is used to segregate vital few problems with trivial many problems. For the Six Sigma professionals, it is important to identify the most important opportunity for improvement in order to get large returns from the project.

Pareto chart is similar to histogram. Histogram is the graphical interpretation of the distributed data in a quantitative manner into a class. Pareto chart is the frequency bar chart which represents quantified data by quantitative characteristics. Steps for constructing Pareto chart are

- Choose the subject for the chart. The subject can be anything which can be causing problems to any particular line or any product.
- Identify the data that needs to be collected. Identify if percentage or cost or quantity are going to be tracked.
- Collect data related to the quality issues. The quality issues are collected over the time period.
- Use check sheet as the input to collect data. Gather total number of defects and non-conformities in each category.
- Based on the information from the check sheet calculate the percentage of total nonconformities.
- Specify the cost incurred for the defects or nonconformities.
- Plot the data setting up scale for chart.
- X-axis represents the subject of study usually nonconformities or defects or item of interest. Y-axis represents the number of defects, number of occurrence, percentage or cost incurred per category.
- Draw the chart by organizing the data in descending order that is data from largest category to smallest category.
- The last step is to analyze the chart. Check for the largest bar, it represents the vital few problems.

13.7.2   *Benchmarking*

Any enterprise should know their strongest competitors on a process-by-process basis. Sometimes it is necessary to look outside the industry to find good benchmarking data. Six Sigma firms often compare their performance with other companies in order to improve their performance in the marketplace. The company compares its standards with the best practices of other companies. Benchmarking is done by visual judgment by the reviewers, documentation, and interviews with people who are directly involved.

With the above inputs about the comparisons, company identifies the steps for the improvement for its own performance. The areas where benchmarking can be done for an organization are listed as follows:

- Procedures
- Processes
- Quality improvement efforts
- Marketing
- Operational strategies

Another type of benchmarking is by investigating the effectiveness of the system implemented by the company with its designed performance level. This is known as effectiveness benchmarking.

The third type of benchmarking is known as continuous benchmarking. This enables continuous improvement as a vital feature for an organization. This also puts day-to-day improvement as a means for continuous improvement. The areas where continuous benchmarking can be done are listed as follows:

- Operations
- Procedures
- Processes
- Performance
- Project
- Strategies

Benchmarking is done by the following sequence:

1. Determine current practices
   - Select problem area
   - Identify key performance factors
   - Understand own processes and the processes of others
   - Select performance criteria based on needs and priorities
2. Identify best practices
   - Measure the performance within the organization
   - Determine the leaders in the criteria areas
   - Find an internal or external organization to benchmark
3. Analyze best practices
   - Visit the organization as a benchmark partner
   - Collect information and data of the benchmark leader
   - Evaluate and compare current practices with the benchmark

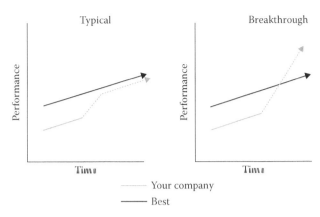

Figure 13.17 Benchmarking.

4. Model best practices
   - Drive improvement changes to advance performance
   - Extend performance breakthroughs within the organization
   - Incorporate new information in the business decision making
   - Share results with benchmark partner
   - Seek other benchmark leaders for further improvement
5. Repeat
   Examples of benchmarking
   - Customer satisfaction
   - Actual customer desires
   - Current competition
   - Best in related industries
   - Best in the world

Benchmarking process provides path for continuous improvement process. The information provided can be used to investigate root causes and reduce variation in the process followed (Figure 13.17).

13.7.3 Thought process map

The purpose of thought process map is to document questions and decisions made during Six Sigma project. Thought process map helps in assuring that the questions raised during the project are not ignored. This tool is used for evaluating the document. It is also used for updating the document during the course of the project.

Thought process map is used to raise questions as the black belt works through the project and it also concentrates on assumptions underlying the questions. It also provides answers to those questions and methods to answer the questions.

13.7.3.1 Value stream mapping

Value stream mapping (VSM) is a Lean tool that is used to determine the nonvalue-added activities and also to eliminate those activities to reduce the cost associated with those processes. The activities are analyzed and investigation of their purpose is done. Various questions associated with VSM are given below (Figure 13.18).

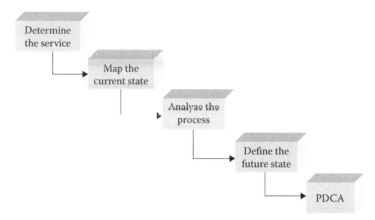

Figure 13.18 Value stream mapping.

- Whether the activities are required
- Whether the activities are performed efficiently
- Whether the activities give useful result

Steps involved in VSM are given as follows:

1. Determine the service process
   In determining the types of service activities, it is best to start with the end process. End work product will be consumed by the customer. Six Sigma professionals will do mapping by this process.
2. Map the current state
   Starting with the end process in a system, Six Sigma professionals will follow the guidelines listed as follows:
   - Begin with physical activation
   - Follow the sequential work order
   - Map the process in backward flow
   - Document the data obtained. Document the physical flow of information. Determine the actual start and end times based on these values which calculate the cycle time
   - Map the entire flow of work sequence
3. Analyze the process
   Value stream process analysis consists of cycle time measurements, slack time at each level of the value stream, quantification of resources required to produce the product, and actual time taken. Additional information like rework, process with high bottlenecks, and excessive accuracy are checked for the process. If wasteful activities are determined, the following four techniques are used to reduce the costs. In activity elimination if the activity is not used to achieve the required customer service level then stop performing that process. Activity sharing is concerned about exploiting the advantage of economies of scale. Combine process in efficient manner for effective operation of the services. Activity reduction reduces the time and other resources involved in that particular activity by scaling back the activity. If there is voluminous amount of data involved in the value stream, there is a possibility of automation.
4. Define the future state
   The future state value stream map is the altered workflow by making the activities and process Lean. This improves the efficiency of the system. Each member in the

team individually recommends process modification, activity changes, structural changes, and technological improvements. These inputs will serve as the communication tool for the improved solution with its benefits.

13.7.3.2   Process mapping

Process maps are a vital tool in green belt which provide a clear insight into how processes and activities are carried out in an organization. Process maps are the graphical tools that represent each process in pictorial representation. Creating the process maps helps the Six Sigma team to understand current processes and also aids in improving the current process. When the modification is made to the processes, process maps are useful tools in communicating the proposed changes to the process in the system.

Process maps use flowcharts and flow diagrams. Flowcharts are effective means of communication which can be easily understood. The team members can clearly identify what could be done to the process at each level. Nonvalue-added activities present in the process can be easily determined and can be separated from the process.

Steps involved in constructing process maps are given as follows (Figures 13.19 and 13.20):

- Determine the process limits: In order to chart the process, it is essential to determine the start of the process and end of the process.
- Define the process steps: Use of brainstorming to determine the steps for new process.

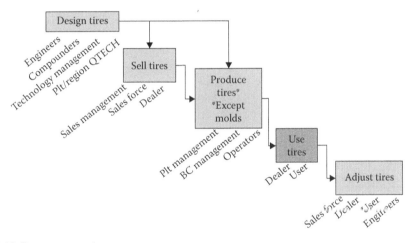

Figure 13.19 Process mapping.

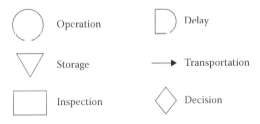

Figure 13.20 Symbols used for process mapping.

- Sort the steps in order by their occurrence in the process.
- Place those steps in order according to the symbols and create chart.
- Evaluate those steps for efficiency and problems.

13.7.4   Brainstorming

To determine the root cause of the problems, Six Sigma improvement team looks into brainstorming as the input to resolve. The need for brainstorming is to develop the use of opportunities, problems, and ideas from the team. Each member present in the team needs to participate in it. Brainstorming enables the team members to generate different ideas that will suit certain processes. Effective brainstorming involves the recording of all ideas developed during the brainstorming session. The leader should facilitate everyone in the team to participate during the course of brainstorming. Brainstorming involves no negativism, no criticism, and no evaluation of ideas.

13.8   Black Belt

13.8.1   Kano model

Kano model can also be referred to as Kano analysis. It used for the analysis of the requirements of the customer. Noriki Kano is a Japanese engineer and consultant whose work is been employed by Japanese and American engineers for analysis. It is based on three categories.

13.8.1.1   Dissatisfiers (basic requirements)

The expectations of the customer involve these basic requirements as a part of the total package. If these requirements cannot be completed, the customer will not be satisfied. For example, an American traveling to China will expect to have traveling facilities like in America. He is let down because travel facilities cannot be as good all over the world. It was noted that the Americans found the travel facilities to China were as good as those in America during the Beijing Olympics.

13.8.1.2   Beijing Olympics

The more the requirements of the customer is met the more pleased is the customer. The tourist taking a tour to the Caribbean on a cruise expects a week of entertainment, food, and relaxation at a reasonable price. The more personal attention given to the tourist, the happier the tourist is. This makes the travel experience a great one for the crew and staff as well.

13.8.1.3   Delighters

These are the features or the services that are provided to the customer which are beyond the expectations of the customer. For the tennis, it would be great if a customer could see the tennis players when he purchases a ticket at the U.S. Open, not just get sunburned and watch a thrilling match. Imagine meeting these players for lunch and also having to watch the match. This would be a delighter.

Competition to provide the best to the customer is constantly increasing the basic expectations of the customer. The standards of a happy customer are rising. What was once considered to be a delighter will soon become a basic satisfier to the customer. It is important to know the changing needs of the customer and move to improve their own performance.

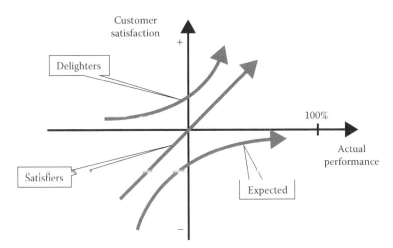

Figure 13.21 Kano model.

Improvement projects, most of the time are selected between the satisfier and the delighter categories. Most companies in a competitive environment would not stay long enough to tackle a basic issue (Pande et al. 2000; Rath & Strong Management Consultants 2000) (Figure 13.21).

13.8.2   Quality function deployment

Quality function deployment is a tool that is sometimes referred to as the voice of the customer or house of quality. It is defined as the tool which determines the customer's wants and needs and converts them into technical characteristics.

The technical characteristics are taken care by the company through the design function, or better still, through a team which is cross functional and includes domains of sales, marketing, design engineering, manufacturing engineering, and operations. The basic goal of this activity is to ensure the product or service is focused on meeting the customer's needs and requirements. It is a tool which is to be used for the entire organization. It is flexible and customizable for the wish of the customer and also functions well for products that are manufactured and in the service industry.

QFD saw its first application in the KOBE shipyards in 1972 by Yoji Akao and his team. It met great success and was introduced to America by Don Clausing in the mid-1980s. Various companies in America have applied the principles of QFD to their product design process. Hauser and Clausing (1988) provide an illustration concerning the position of the emergency-parking brake for a sport car. Engineering initially wanted to place the brake between the seat and the door, but this caused a problem for the women driver wearing a skirt. Could she get out gracefully? Would this eventually cause dissatisfaction?

As an advantage of QFD (Besterfield 1999), we can learn that Toyota and Honda reduced the new product cycle times to about 2.5 to 3 years by means of QFD. The carmakers were on a cycle time of 5 years. QFD could be considered as a concurrent engineering tool.

It provides a graphic method of expressing relationships between customer's wants and design features. It is a matrix that lists the attributes of the customer's requirements and compares it to the features of the design (Hunter 1994).

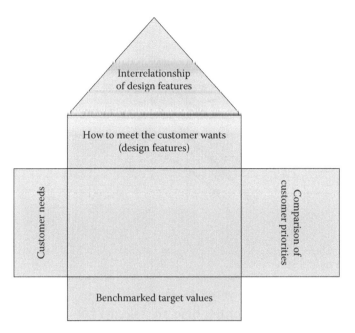

Figure 13.22 House of quality.

The customer's collection of wants and expectations are expressed through the available methods to most any organization: surveys, focus groups, interviews, trade shows, hot lines, etc. The house of quality is a technique used to organize the data. The house of quality is so called because of the image used in the construction. The use of matrices is the key to building the house. The primary matrix shows the relationship between the requirements of the customer and the design features and requirements.

The construction of the house is shown as follows (Hauser and Clausing 1988) (Figure 13.22):

- The left side comprises the needs and wants of the customer.
- The ceiling has the design features and technical requirements.
- The right side contains the priorities of the customer.
- The foundation contains the benchmarking, target values.
- The roof of the house contains the matrix describing the relationship between design features.

The possible benefits of using QFD (Figure 13.23) are that it

- Creates a customer focuses environment.
- Reduces the cycle time of the process.
- Uses engineering methods that are concurrent.
- Reduces design to manufacture costs.
- Increases communication through teams that are cross functional.
- Creates data for proper documentation of engineering knowledge.
- Makes priority requirements and quality improvement.

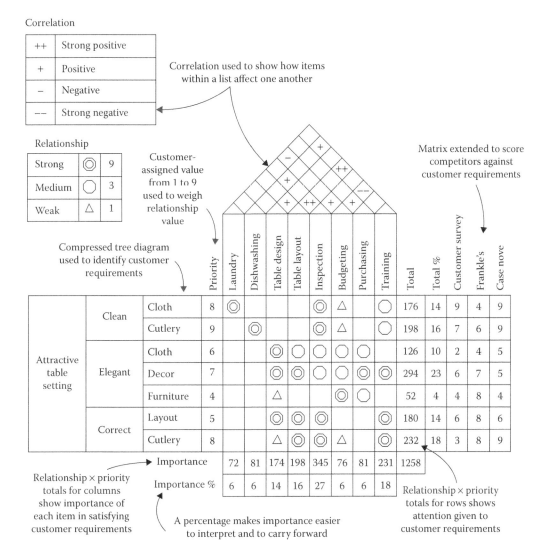

Customer requirements for table-setting versus effort in restaurant procedures

Figure 13.23 Example for house of quality.

Cascaded matrices are used to propagate the customer's voice through more detailed parts of the design and solution stages (Figure 13.24).

The house of quality is flexible and customized to each situation. Each organization has its own guidelines that will modify the above image. However, the basics remain the same, to reach out to the customer's needs and requirements and to be proactive in the design of products to meet the customer's needs.

After the primary design characteristics are set, Hauser and Clausing (1988) suggest that using the "how's" from the house of quality as the "what's" of another house that depicts the detailed product design. This process is repeated with a process planning

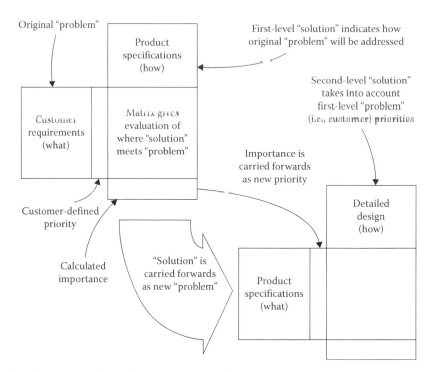

Figure 13.24 Hypothetical black belt house of quality.

house and then production planning house. In this way, the voice of the customer is carried through every stage from design to manufacturing.

While it is easy to get caught up in the process of building the house of quality and feeding in the date, one should not lose sight of the objectives of the house of quality methodology. Hauser and Clausing (1988) states that QFD is a kind of map with concepts that provides the means for inter-functional planning and communications. The principal benefit of QFD is quality in-house. It gets people thinking in the right direction and thinking together.

The voice of the customer, internal and external both, is presented in the form of house of quality. The various organization groups, engineering marketing, manufacturing, and so on are able to see the effect of design and planning changes to maintain and keep the customer happy, costs and engineering characteristics in the development of new or improved products and services.

13.8.3 Affinity diagram

Affinity diagram is one of the tools which is used in black belt. Affinity diagram is used for the purpose of organizing data and ideas. Affinity diagram is used for organizing inputs obtained from the brainstorming into groups based on the relationship and analysis.

Affinity diagram is used for the following purposes:

- Refine the successful thinking to generate more volume of idea.
- Sort the ideas based on different groups according to the relationship in notes or cards.

- Generate idea which is apart from typical meeting.
- Proper brainstorming is required and it is further organized as a large number of ideas.
- Summarize the ideas according to their groups and figure out the depth of the problem.
- Raise issues, questions, and review. Prioritize the ideas and record them.

The following can be used for performing affinity diagrams (Figures 13.25 through 13.28):

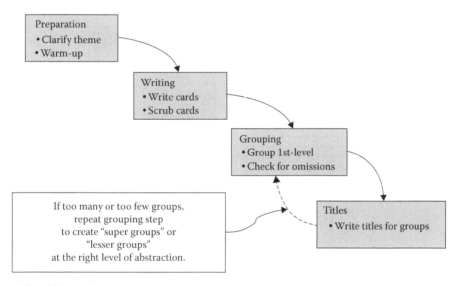

Figure 13.25 Affinity diagram.

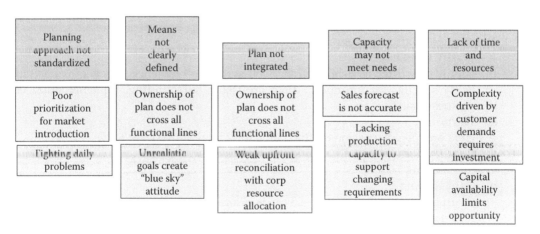

Figure 13.26 Affinity diagram for business plan.

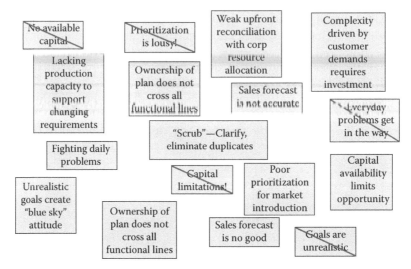

Figure 13.27 Generate ideas for affinity diagram.

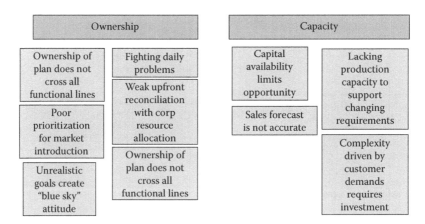

Figure 13.28 Identification of groupings.

1. Preparation of affinity diagram
   a. Clarify the theme
   b. Should be written as a short, clear, "well-scrubbed" sentence, stated as a question
   c. Fairly specific, for example, "What are issues surrounding implementation of business plan?"
   d. Warm-up
   e. Hold an open discussion on the theme for a total of five minutes
   f. Time is allocated equally among the participants
2. Write the Cards
   a. Card Writing
      i. Initially, plan for about 20–25 cards, allocated equally among the participants.
      ii. Each member records idea on self-stick note.
      iii. No talking while writing cards

      b. Hints
        i. Responses should not be solutions.
        ii. Express associated problems or weaknesses
      c. Checklist
        i. One idea per card
        ii. Complete sentence—in the form of a statement, not a question
3. Scrub the Cards:
      a. Team leader reads each note to the group.
      b. If the card is not clear to all, edit as needed.
      c. Repeat until all cards are scrubbed.
      d. Place all of the "scrubbed" notes randomly onto a flip chart.
      e. Checklist
      f. Cards are at a low level of abstraction.
      g. Each card represents only one idea.
      h. The cards answer the question given in the theme.
      i. Everyone understands the specific (concrete) meaning of the card.
      j. Note: Approximately 1/3 to 1/2 the time you will spend in your LP session should be spent scrubbing cards.
4. First-Level Grouping:
      a. With everyone standing at the chart, team silently arranges notes into 5–10 related groupings based on similar meanings.
      b. Keep the clusters to 3–5 notes
      c. Some cards may not fit in any group—leave them by themselves.
      d. If a note keeps getting moved from cluster to cluster, it's OK to create a duplicate.
5. Write titles:
      a. Team writes title header card for each cluster of notes to convey their meaning.
      b. Hints
        i. Purpose of titles is to extract the essential thought of the group of cards.
        ii. Title should not be simply addition of the thoughts of cards in the group.
        iii. Avoid overgeneralization—the title should only be one level of abstraction above the group.
6. Finish the affinity diagram:
      a. Draw the affinity diagram connecting all finalized title header cards with their groupings.

13.8.4 Interrelationship diagram

The interrelationship diagram is a tool used to prioritize those issues with the greatest impact. The interrelationship diagram aids in making distinctions between causes (or drivers) and effects (or outcomes). The affinity diagram was used to identify the key underlying issues of a problem. Main purpose of this tool is for better understanding relationships between complex issues.

Main uses of this tool are (Figure 13.29)

- Identify or select Six Sigma project area.
- Identify key drivers that influence customer needs.
- Identify less significant drivers that do not greatly influence customer needs.

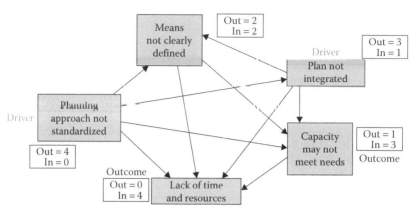

Figure 13.29 Interrelationship diagram.

References

Besterfield, D. H. et al. (1999). *Total Quality Management*. Upper Saddle River, NJ: Prentice Hall.

Drucker, P. F. (ed.) (1986). *Managing for Results: Economic Tasks and Risk-Taking Decisions*. New York: Harper & Row.

Eckes, G. (ed.) (2001). *The Six Sigma Revolution*. New York: John Wiley & Sons.

Garvin, D. (ed.) (1988). *Managing Quality: The Strategic and Competitive Edge*. New York: The Free Press.

Harry, M. and Schroeder, R. (2000). *Six Sigma, The breakthrough Management Strategy*. New York: Currency/Doubleday.

Hauser, J. R. and Clausing, D. (1988). "The House of Quality." *The Harvard Business Review*. May–June, pp. 1–13.

Hunter, M. R. and Vanlandingham, R. D. (1994). "Listening to the Customer Using QFD: Don't let the simplicity of QFD scare you away. Quality Progress 27(4): 55–59.

Lowenstein, M. (1995). *Customer Retention*. Milwaukee, WI: ASQC Quality Press.

Pande, P. S., Neuman, R. P., and Cavanagh, R. R. (eds.). (2000). *The Six Sigma Way*. New York: McGraw-Hill.

Rath & Strong Management Consultants (eds.). (2000). *Rath & Strong's Six Sigma Pocket Guide*. Lexington, MA: Rath & Strong Management Consultants.

Review Questions

Question 1: A work schematic that details the relationship between inputs, processes, outputs, customers, and feedback is called

(a) A subsystem
(b) A SIPOC diagram
(c) A system diagram
(d) An activity network diagram

Question 2: On a process flow diagram, the component that indicates need for correction is called

(a) Inputs
(b) SIPOC
(c) Feedback
(d) Outputs

Question 3: Any business process is composed of three main levels for the purpose of Six Sigma strategy development. Which of the following is NOT included?

(a) Operations
(b) Process
(c) Shareholder
(d) Business

Question 4: The process map known as SIPOC provides team members an understanding of the process. It is a view of the process taken at

(a) The customer's perspective
(b) Floor level
(c) A very detailed level
(d) A very high level

Question 5: The SIPOC process map stands for suppliers, inputs, process, outputs, and customers. It provides a view of the process that contains approximately how many steps?

(a) 21–40
(b) 4–7
(c) 8–15
(d) 16–20

Question 6: A work schematic that details the relationship between inputs, processes, outputs, customers, and feedback is called

(a) A subsystem
(b) A SIPOC diagram
(c) A system diagram
(d) An activity network diagram

Question 7: The team's charter describes the team's

(a) Meeting dates, milestones, and targets
(b) Leader, facilitator, recorder, and timekeeper
(c) Mission, scope, and objectives
(d) Members, sponsors, and facilitators

Question 8: One advantage of project management is that it does NOT require

(a) Planning
(b) People
(c) Objectives
(d) Unlimited resources

Question 9: The project charter will be useful in many ways, including

(a) Ensuring the team members will support the charter
(b) Permitting the team leader to develop milestones from it
(c) Assuring the champion will assign responsible team members
(d) Providing a consistent target for the team

Question 10: A commonly reported problem with Six Sigma projects deals with

(a) A desire to complete projects on time
(b) A failure to complete any project charter documentation
(c) A lack of business impact for the company
(d) A requirement that projects must be at least $100,000 in value

Question 11: The relevant stakeholders in an important project would typically include all of the following EXCEPT:

(a) Owners and stockholders
(b) Potential suppliers
(c) Potential competitors
(d) Hourly employees

Question 12: Upper management typically supports the team process best by

(a) Reinforcing positive team results
(b) Punishing negative team members
(c) Providing direction and support
(d) Allowing teams to establish the company's mission statement

Question 13: Upper management typically supports the team process best by

A. Reinforcing positive team results
B. Punishing negative team members
C. Providing direction and support
D. Allowing teams to establish the company's mission statement
  (a)   A
  (b)   A C
  (c)   A B C
  (d)   A B C D

Question 14: Team success is most dependent on which of the following?

(a) The team leader
(b) Team members having full knowledge of the fishbone technique
(c) Active support by mid-managers
(d) Policy support by top management

Question 15: The term variables can be described in the following way:

(a) A definable attribute or characteristic of a product
(b) A quality which is absent in a product in one or more specifications
(c) A quality which can assume several (more than two) values
(d) A quality which can be absent or present in a product

Question 16: A customer satisfaction program has started on the right foot. It has gone very well for the last year or so. The company should

  A. Look to improve the program, with new customer input.
  B. Do nothing with the program, it's not broken.
  C. Form a manager's group to add new wrinkles to the program.
  - (a) A
  - (b) B C
  - (c) B
  - (d) C

Question 17: Benchmarking should be done in the following sequence:

  A. Measure competitive performance
  B. Implement significant improvement
  C. Understand your own process
  D. Identify improvement criteria
  - (a) A C D B
  - (b) C D A B
  - (c) D C B A
  - (d) A B C D

Question 18: A graphical display of the total percentage of results below a certain measurement value is called a

  - (a) Cumulative distribution function
  - (b) Histogram
  - (c) Probability density function
  - (d) Expected value

Question 19: Which of the following are principal reasons for utilizing process mapping?

  A. To identify where unnecessary complexity exists
  B. To visualize the process quickly
  C. To eliminate the total planning process
  D. To assist in work simplification
  - (a) A B
  - (b) A B C
  - (c) A B D
  - (d) A B C D

Question 20: A SWOT analysis is an easy way for a company to evaluate itself and plan a strategy. A key concern is

  - (a) Lack of resources
  - (b) Situation changes over time

(c) Lack of objectivity in the analysis
(d) Lack of planned change

Question 21: Which of the following process mapping symbols would NOT be associated with a decision point?

(a) Hexagon
(b) Square
(c) Triangle
(d) Diamond

Question 22: Process flow improvement steps normally do NOT include

A. Asking why we do it this way
B. Asking what would make it "perfect"
C. Analyzing each step in detail
D. The use of Pareto diagrams
E. A comparison with processes different than your own
  (a) A E
  (b) D
  (c) C
  (d) B D

Question 23: Which of the following are principal reasons for utilizing process mapping?

A. To identify where unnecessary complexity exists
B. To visualize the process quickly
C. To eliminate the total planning process
D. To assist in work simplification
  (a) A B
  (b) A B C
  (c) A B D
  (d) A B C D

Question 24: How should the value of a customer best be measured?

(a) Retention length
(b) Current volume contribution
(c) Loyalty
(d) Lifetime worth

Question 25: The organization's customer service program can be enhanced in many ways. One of the ways would be

(a) Restrict access to customer data
(b) Utilize employee involvement

(c) Provide better procedures for customer service personnel
(d) Have supervisors available to answer more questions

Question 26: Ideally, customer feedback should satisfy organization needs such as

A. Identifying customer requirements
B. Fulfilling ISO 9000 requirements
C. Spotting upcoming trends
D. Having data for analysis
  (a) A C
  (b) A B C D
  (c) B C
  (d) A

Question 27: Flowcharting of activities and systems is most helpful in detecting

A. Inappropriate use of resources
B. Deficiencies in the organizational structure
C. Holes or gaps in the control system
D. Improper use of statistical methods
  (a) A C
  (b) A B D
  (c) A B
  (d) A

Question 28: Understanding customer needs is a constant requirement for organizations. Customer data and information is often collected in various amounts. The organization should resolve to

A. Use more proactive approaches
B. Focus on improvement plans
C. Identify customer satisfaction drivers
D. Sort out the unneeded customer data
  (a) A B C D
  (b) A
  (c) B C
  (d) A B C

Question 29: Customer expectations follow which hierarchy of needs, from low to high?

A. Expected
B. Basic
C. Unanticipated
D. Desired
  (a) B A C D
  (b) C A B D
  (c) D B A C
  (d) A B C D

Question 30: The benefits of market segmentation include

(a) Finishing a consistent marketing plan
(b) Providing uniform product manufacturing
(c) Spotting market trends
(d) Generating high volume, low cost production

Question 31: A SWOT analysis is an easy way for a company to evaluate itself and plan a strategy. A key concern is

(a) Lack of resources
(b) Situation changes over time
(c) Lack of objectivity in the analysis
(d) Lack of planned change

Question 32: In the preparation of a project, efforts should be made to identify and involve various parties affected by the planned changes. These other parties are known as

(a) Stakeholders
(b) Champions
(c) Team leaders
(d) Process owners

Question 33: Benchmarking should be done in the following sequence:

A. Measure competitive performance
B. Implement significant improvement
C. Understand your own process
D. Identify improvement criteria
   (a) A C D B
   (b) C D A B
   (c) D C B A
   (d) A B C D

Question 34: Which of the following process mapping symbols would NOT be associated with a decision point?

(a) Hexagon
(b) Square
(c) Triangle
(d) Diamond

Question 35: EVOP is

(a) Evolutionary operations
(b) Environmental operating procedure
(c) Experimental variation on operations or processes
(d) Exploratory variable operation process

Question 36: Benchmarking should be done in the following sequence:

  A. Measure competitive performance
  B. Implement significant improvement
  C. Understand your own process
  D. Identify improvement criteria
    (a) A C D B
    (b) C D A B
    (c) D C B A
    (d) A B C D

Question 37: (Analyze/Improve) Benchmarking should be done in the following sequence:

  A. Measure competitive performance
  B. Implement significant improvement
  C. Understand your own process
  D. Identify improvement criteria
    (a) A C D B
    (b) C D A B
    (c) D C B A
    (d) A B C D

chapter fourteen

Measure

Identifying obtainable data for realistic use

When you can measure what you are speaking about and express it in numbers, you know something about but when you cannot express it in numbers, your knowledge is of a meager and unsatisfactory kind; it may be the beginning of knowledge, but have scarcely, in your thoughts, advanced to the stage of Science, whatever the matter may be.

—Lord Kelvin

Overview

Steps	Substeps
Define	• Quality projects (sponsorship) • Define specific project (black belt)
Measure	• Identify critical characteristics • Clarify specifics and targets • Measurement system validation
Analyze	• Determine baseline • Identify improvement objectives • Evaluate process inputs
Improve	• Formulate implementation plan • Perform work measurement • Lean execute the proposed plan
Control	• Reconfirm plan • Set up statistical process controls • Operational process transition

Source: PWD Group LLC.

Recommended Tools by Belt

		Define	Measure	Analyze	Improve	Control
Stage 1	Yellow belt	Interview process Language processing System map (flowchart)		X–Y map/QFD/ house of quality Fishbone Diagram/RCA	EVOP	Control plans Visual systems 5-S SPC/APC

(Continued)

Recommended Tools by Belt (Continued)

		Define	Measure	Analyze	Improve	Control
Stage 2	Green belt	Prioritization matrix	Hypothesis testing	FMEA	Design of experiments	TPM
		Thought process map	Flowdown	Multi-vari chart	Fractional factorials	Mistake proofing
		Stakeholder analysis	Measurement system analysis	Chi-square	Data mining	
		Thought process map	Graphical methods	Regression	Blocking	
			Analysis of variance	Pareto analysis		
Stage 3	Black belt	Value stream map	Process behavior charts	Buffered tolerance limits	Theory of constraints	
		Gantt chart development	QFD	Process capability analysis	Multiple response optimization	
		Project management			Response surface methodology	

Source: PWD Group LLC.

Notes: Black belts are expected to know the continued knowledge obtained throughout green belt and yellow belt certifications. The same goes for green belt and yellow belt. Empty cells represent that skill is not required for that belt stage, however, a general knowledge of the step is expected.

5-S, sort, set in order, shine, standardize, and sustain; APC, advanced process control; EVOP, evolutionary operations; FMEA, failure modes and effect analysis; QFD, quality function deployment; SPC, statistical process control; TPM, total productive maintenance.

14.1   Introduction

The DMAIC process is an operational Six Sigma methodology. The previous chapter dealt with the first and critical phase of any Six Sigma project, that is, "Define." The voice of the customers is now translated in terms of quality and business. The purpose and objective of the project is defined. Now, the current processes involved are to be studied. The studies are based on certain standard metrics which is aimed at describing the performance of the processes involved. The establishment of metrics is vital to understand the need for improvement. Hence, this chapter presents various metrics and measurement systems that the certified Six Sigma practitioners ought to know, which help in measuring the variables chosen for improvement.

The measure phase consist of three basic classifications and these are

1. Selection of critical characteristics
2. Determination of targets and specifications
3. Validation of measurement systems

This chapter is narrated in such a way that only basic and general areas are covered and it also explains certain statistical thinking concepts, assuming that the reader is aware of statistics and probability.

14.2 Critical characteristics

The main aim of measure is to create a clear and measurable project that is identified in define phase. The initial objectives are to determine the critical to quality (CTQ) and create a flow down that clearly relates various processes and its measurements to the chosen project. Let the project chosen be denoted as Y. The flow down helps in recognizing and identifying the various tools required for measuring the process metrics and process performance (Figure 14.1).

The three critical characteristics that sets the stage for setting up quality are as follows (Figure 14.2):

1. Critical to satisfaction (CTS)
2. Critical to quality
3. Critical to process (CTP)

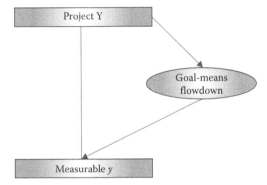

Figure 14.1 Project flow down.

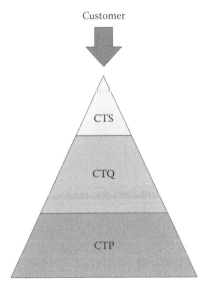

Figure 14.2 Stage for quality.

14.2.1 Critical to satisfaction

The main attributes of CTS is to study the critical variables that satisfy the customers' needs. This may be considered the project big "Y." The needs of customers keep changing. When the basic needs of the customers are fulfilled they look for new advancements. Hence satisfaction is critical and vital for every project that clearly measures the current needs of the customers. The customer's needs can be of the following (Juran and Godfrey 1999):

- Stated needs
- Real needs
- Perceived needs
- Cultural needs
- Unintended needs

14.2.2 Critical to quality

CTQ is used to for identifying the measurable characteristics or features of a product or a service to verify if the customers are satisfied (big Ys), though the precise measurement of metrics is not done in this stage (Eckes 2001). The measurements and its techniques are discussed later in this chapter.

14.2.3 Critical to process

The above two topics have now provided with the information on customers' needs and variables. More precisely, the actions CTP are carried out to measure the output of the processes involved in improving the quality of the product based on customers' needs. The variables in the processes can be termed "little y's" which provide controlled output measures to indicate satisfaction to CTQs. Figures 14.3 and 14.4 represent the importance of each critical thinking function.

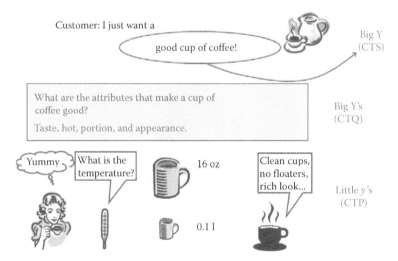

Figure 14.3 CTS, CTQ, and CTP.

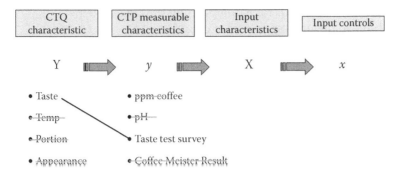

Figure 14.4 Flow down from "Y" to "x."

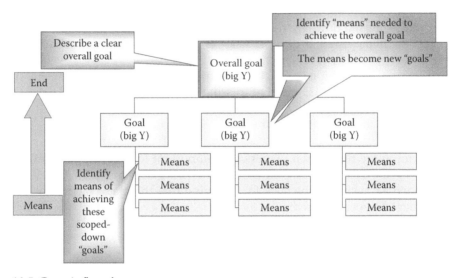

Figure 14.5 Generic flow down process.

14.3   Flow down process

The flow down process is done in order to narrow down the various attributes of the product into specific features that are important in improving quality. Flow down process thus helps in identifying the CTS, otherwise known as output "Y." If the "Y" is too broad in scope, the features are to be funneled down until a manageable "little y" is achieved. The "little y" thus obtained should be measurable. The following is the generic flow down process transforming "Y" to "little y" (Figure 14.5).

14.3.1   Construction of a flow down process

The construction of flow down process is very important. The variables are narrowed down based on the CTS and if any important variables are missed from the process it could lead to catastrophe and make the Six Sigma project void. There are two approaches for constructing the flow down diagram: (1) Top–Down and (2) Bottom–Up.

1. Top–Down
   a. State single high-level (Level 1) objective.
   b. Identify next-level (Level 2) means for accomplishing the objective.

      c.   Identify Level 3 means for accomplishing the Level 2 items.
      d.   Continue to the necessary level of detail.
  2. Bottom–Up
      a.   State single high-level (Level 1) objective.
      b.   Identify the specific actions needed to accomplish that objective. (Working at the most detailed level of the diagram.)
      c.   Group the specific actions into natural clusters (using the Affinity Diagram process). Write the descriptions of these groups.
      d.   If necessary, create one more level of grouping, leading to the Level 1 objective.

Once the flow down diagram is drawn and the variables are zeroed in the diagram, it should be reviewed for validation. The review should follow these steps:

- Read the completed flow down diagram, beginning at Level 1.
- Look at the items listed at Level 2. Ask, "If we accomplish all these items (Means), will we achieve the Level 1 objective (goal)?"
  - If there is a missing item, decide whether it is relevant to the problem area you are addressing. If it is, add that missing item and the missing elements of that branch of the tree.
- Continue this checking down to the lowest level of necessary detail.

While performing the flow down process, it is important to keep the purpose in mind. The objective is not trying to build an exhaustive listing of items, to the smallest detail. The whole purpose is simply to check for verifying that the major factors touching the project area have been considered.

14.3.2 *Characteristics of good "Little y"*

As mentioned earlier at this point in the Six Sigma process, the goal is to only look at the process outputs (little y). Care must be taken such that the flow down process is not flown down too far and get into the "x" (those things that cause the output to change). This task lies with the black belt and the project team members must identify a measurable "y" that the project will address. It must be clear how the project y—a critical characteristic—links to higher-level requirements of the customer and the business. The goal-means flow down tree diagram is a good tool for showing this linkage. The characteristic of good "little y" is as follows:

- Measurable
  - Continuous scale (e.g., length or temp.)
  - Ordinal scale (e.g., "good," "fair," "poor")
  - Count (e.g., number of typos in document)
  - Classify (e.g., "good"/"bad")
- Not money, but something that drives dollars
  - (e.g., a resource, storage, delay, defect)
- Can be collected quickly and efficiently

Hence, it is clear from the above characteristics that ideally there should be only *one* "little y" for the DMAIC project. If there is a feeling among the certified belt holders that there must be two or more for the project, the following should be answered to draw a conclusion:

- Are these actually distinct outcomes, with multiple projects required?
- Are these actually subelements of a slightly larger y, with the larger y as the correct project y?

14.4   Two dimensions

Six Sigma projects tend to fall into one of two broad categories (Figure 14.6), although it is rare to have a problem that is completely into one category, with no effect in the other.

1. "Quality" or Effectiveness—"Doing the right things"
   - The ability of the process to deliver products or services according to specification.
   - Typical measures involve conformance to specification.
2. "Lean" or efficiency—"Doing things right"
   - Typical measures involve cycle time and resources consumed.
   - The steps of the DMAIC process are identical for both categories, although the approach and methods differ at some points.

14.5   Determination of targets and specifications

The first phase in the measure phase of DMAIC process has critically identified the customers' needs and satisfaction. Now, the targets and specifications must be set by measuring the current processes to improve the selected "little y." Targets and specifications are defined by the customer—either explicitly or implicitly. A "target" is the customer-defined ideal value for the CTQ. It is the ideal level, from the customer's perspective, for the process output *y*. The customer's target should not be confused with the business' goal for improvement from the project. "Specifications" are the customer-defined range of

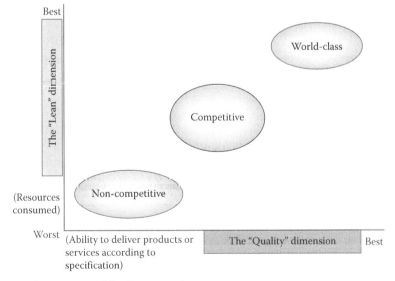

Figure 14.6 Two dimensions of Six Sigma projects—measure.

acceptable (or tolerable) values (minimum, maximum, or both) for the CTQ. Specifications should be so precise that everyone comes to the same conclusion as to their meaning. The precision and clarity can be achieved from an operational definition. The components of operational definition are as follows:

- Test method that provides data on the criteria
- Criterion (or criteria) for determining conformance

Operational definition can well be defined by the following example. The question "How many people are in the conference room?" will simply give us the answer of how many people are actually sitting in the room. But, this intuition does not help in Six Sigma projects. The right thinking would be to consider all the people sitting, standing, entering and exiting; stopping by to hear a presentation; delivering lunch; and so on. Hence, it is clear that the answer depends on the purpose of the question and operational definition helps solving the purpose of the question. How could the "test" (the description part of the operational definition) be described for these different purposes?

- Deciding how many lunches to order
- Deciding how many workbooks to print
- Deciding how many teams to form for the next working session

14.5.1   The need for targets and specifications

The target and specification limits provide a reference point—a "line in the sand"—against which performance and improvement can be judged. In some cases, the target and specifications are clear such that the customers have clearly stated what is required. In situations such as that listed further, the criteria provided for calculating a limit are somewhat arbitrary, but they have proven to be useful in determining a reference point. Some of those criteria are

- Throughput/productivity
- Level of "slow-moving inventory"
- Cycle time of a process step
- Amount of unplanned downtime
- Errors in test requests

Hence at a minimum, an upper or lower limit for continuous data or a target for discrete data is required for the little y. For continuous measure if the standard is not available for the current process, average can be used as the upper or lower limit. To do this, data about the performance of the process is required. If the measure is an attribute, such as the number of "defects" or proportion of "defective" and if the natural target is zero, zero can be set as the process target. It is not necessary to determine an "upper specification limit" for defects. The measurement techniques for various processes are later discussed in this chapter.

When the target and specification are known, the issues are mostly around finding out what they are and making sure they are correct. Even in cases in which the target and specifications are known, it is important to question this information. Ask questions such as "how do we know?" and "is there any uncertainty about the validity of the target or specifications?" Also the following questionnaire can help:

- What is the target? What is the specification?
- How do you know?
- Is there any uncertainty about the meaning of the specification or determining conformance?
- What other specifications or performance requirements are related?
- Are any of the relationships conflicting? (Improving one metric degrades performance of another.)

When processes do not have specifications defined, or at least they are not very clear, the following questionnaire can help:

- Who will identify the target and specification?
- What indication do you have at this point of the target and specification?
- What approaches can be taken to gather information to define the target and specification?
- Are there consequential metrics to be concerned about?

14.6   Sponsorship level

14.6.1   Balanced scorecard

The balanced scorecard (BSC) is a strategic planning tool which was proposed by Robert Kaplan and David Norton (1992).

It could be used to facilitate the learning process to bring about improvements in the organization by measuring the number of financially advantageous outcomes for a given input via a key process. It helps understand this through customer's, processes', personnel's, and shareholder's perspectives.

For every beneficiary, the BSC helps you build means to measure performance toward a goal (Prabhu 2012).

However, it turns out that it is more useful for management purposes and helps streamline the working of the team toward the missionary goal. It helps capture value for future purposes and for evaluating and re-focusing of the team.

This method has been found to have some shortcomings for the lean approach. It is observed that

1. A lot of data is required to arrive at conclusion on performance levels using forecasting.
2. A strategy map is built using cause-and-effect relationships. But it is harder to zero in on a cause if the process varies above a threshold.

A remedy to this has been proposed to bring about a Lean-based BSC (Toppazzini 2012). This can be done by devising a method for collecting data at a faster pace and using stable processes to build on the strategy map consisting of cause-and-effect relationships.

14.7   Yellow belt

First step in the measure phase is to develop the current state process map, which then follows data collection for measuring attributes. Using the collected data, it is possible to figure out the baseline process performance. Tools used in yellow belt are process mapping,

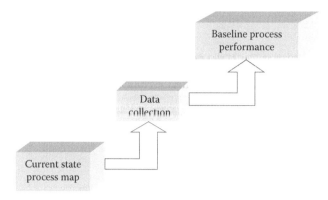

Figure 14.7 Measure process.

swim lane flowchart, data collection, and it also has Lean tools such as 5S's and 7 wastes, value analysis, and 5 why's (Figure 14.7).

14.7.1 Process mapping

Process maps are a vital tool in yellow belt that provides clear insight into how processes and activities are carried in an organization. Process mapping involves all the teams within the activity. Analyze the process within the system. Any insignificant, irrelevant, or trivial information should be analyzed. Process maps are the graphical tool that represents each process in pictorial form. Creating the process maps helps the Six Sigma team to understand current processes which also aids in improving the current process. When the modification is made to the processes, process maps are useful tools in communicating the proposed changes to the process in the system (Taylor 2009).

Process map or flowchart is a useful tool for the people who are familiar with the process. Process maps use flowcharts and flow diagram. Flowcharts are effective means of communication medium which can be easily understood. Flowchart can describe the sequence of the products, paper work, administrative procedures, containers, and operator actions. Flowchart is the initial step for process improvement for the Six Sigma team. Team members can clearly identify what could be done to the process at each level. Nonvalue-added activities present in the process can be easily determined and can be separated from the process. The main advantage of using flowchart is that it can be easily visualized. Most of the flowcharts use standard symbols.

Flowchart can be used to determine the improvement strategy as shown in the following sequence:

- Gather a team to examine the process
- Construct a flowchart to describe each process step
- State and analyze all steps in a brief manner
- Always ask the question "why" to question the method for the current procedure
- Compare the actual process with the improved process
- Analyze for unnecessary complexity in the process
- The process is checked for the way it should be done
- Improvement ideas may generate from different process sources

Symbols used for process mapping are shown in Figure 14.8.

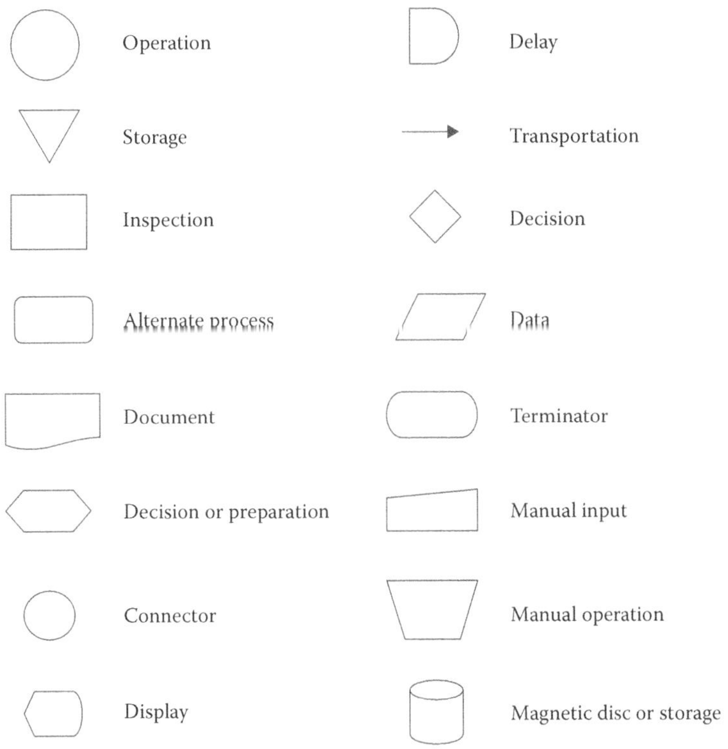

Figure 14.8 Common flowchart symbol.

The examples of process mapping are listed below:

- Customer calls and requests a quote
- Sales person researches options for aircraft configuration
- Salesperson builds configuration
- Send to contract department
- Verify configuration
- Sales person builds configuration
- Verify configuration with customer
- Build quotes
- Send to sales person
- Forward to customer
- Send to contracts department
- Customers give brief description of what they want

There are a number of ways to represent the flowchart styles. Some of the styles are conceptual, person to person, and action to action. The following example is taken from *The Quality System Handbook* by Neville Eden Borough (Figure 14.9).

Process map is shown in Figure 14.10. Steps involved in constructing process maps are given below:

- Determine the process limits: In order to chart the process, it is essential to determine the start of the process and end of the process.
- Define the process steps: Use of brainstorming to determine the steps for new process

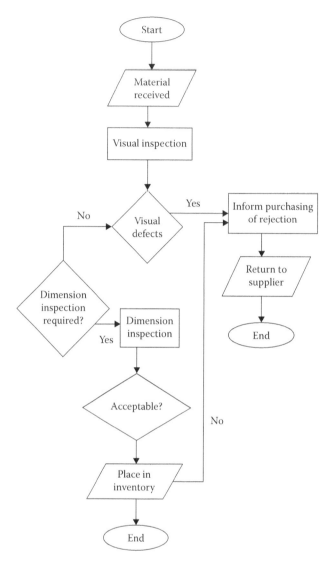

Figure 14.9 Action to action flowchart example.

- Sort the steps in order by their occurrence in the process
- Place those steps in order according to the symbols and create chart
- Evaluate those steps for efficiency and problems

14.7.2   Data collection

The team should collect data in a standard procedure for measuring and logging data. Three tools used for data collection and data accuracy are checksheets, control charts, and gage R&R.

In order to verify the accuracy of the data collection and data logging carried out by the team, it has to be checked for its accuracy. Data collection and accuracy are used for

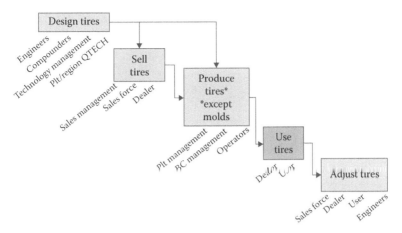

Figure 14.10 Process map.

accounting the variation in the measurement and measuring system. Checksheets are the most effective data-recording technique. As the event occurs, a check mark is placed in the checksheets. As a number of events occur, the observer uses the same check mark to mark the occurrence of the event (George 2002).

14.7.2.1   Control charts

Control charts are used to determine the variation of the product's feature. Control chart is also known as attribute chart. The calculation of control limits provides an advantage over line charts.

14.7.2.2   Check sheets

An important activity carried out in the measure phase is data collection. Voluminous amount of data are required for this purpose. Checksheet is a useful tool for this purpose. Checksheets consists of events occurring in accordance with a particular time frame. They are commonly used for analyzing data that are not available which could be used for taking key decision. For example, in order to reduce waste the manufacturer may know the data for the quantity of waste but it is essential to know various sources of waste within the operation. Checksheets implemented at different work levels may enable capturing waste from the operations carried out (George 2002; Table 14.1).

Table 14.1 Check Sheet

	Week				
Defect	1	2	3	4	5
Incorrect address		1			1
Incorrect SSN	1		1	1	3
Incorrect work history	1			1	2
Incorrect salary history	11	1	11		5

Source: George, M. L. *Lean Six Sigma: Combining Six Sigma Quality with Lean Speed.* McGraw-Hill, New York, 2002.

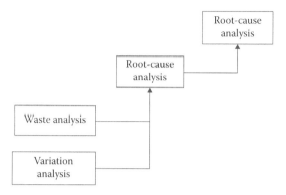

Figure 14.11 Waste and variation analysis.

14.7.3   *Waste and variation analysis*

Waste analysis and variation analysis include root-cause analysis and root-cause verification. Determining the nonvalue-added activity in the process will enable reduction of waste. Waste can be determined using 5S tool.

Total operational process time consists of value added and nonvalue added. Value added usually focuses on adding people, complexity, or equipment, and work longer. Nonvalue-added time needs to be eliminated to improve the value stream. It is observed that for a world class company nonvalue-added time is 80% of total throughput time and typical nonvalue-added activity at aerospace and defense industries is 99% (Figure 14.11).

Value added could be considered vital because it physically alters the product or services. They are carried out in right sequence or place in the process. It provides a genuine, sustainable high quality of treatment. It would be more intrinsic since it delivers the value it seeks from the customer. Nonvalue-added activities can be categorized into subdivisions because they are waste and require nonvalue-added activity. Nonvalue-added activities are required because it supports the company requirement and it also aids in reporting. These activities help in reducing risk, defect, cost, etc. It also allows subsequent work for the customer to be performed more accurately or correctly. This satisfies good business practices, government, legal, or regulatory requirement.

An activity which does not fall in the category of nonvalue-added activities is waste. Process or activities are considered to be waste because they do not change or augment to the product or services to be delivered as the final output. These actions are carried out of sequence and they are done to correct the prior activities. These activities would not be seen as most intrinsic to deliver value to the customer. These would not be willingly paid by the customer (Figure 14.12).

The 5S process is based on concepts that are easy to understand and it requires discipline to follow. The five stages of 5S are sort, set in order, shine, standardize, and sustain. Sort, set in order, and shine are essential to ensure smooth and efficient flow of activities. Standardize is the method for maintaining the first three stages. Sustain means making a habit (culture) of properly maintaining correct procedures. The principles of 5S can be applied in sales, service and transactional processes, and manufacturing. The waste analysis and variation analysis are shown in Figure 14.11.

14.7.3.1   *First S—Sort (organization)*

Sort is used to distinguish between what is needed and what is not needed. Sort is defined as to remove all items from the work place that are not needed for current operations.

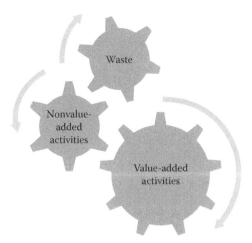

Figure 14.12 Types of activities in a process.

Some examples where sort can be used are unnecessary inventory, unnecessary transportation, and unnecessary equipment.

14.7.3.2 Second S—Stabilize (orderliness)

Stabilize is nothing but a place for everything and everything in its place. Set in order is defined as arranging needed items so that they are easy to use and further if they are labeled they are easy to find and place in required spot. Examples for orderliness are tools hung on shadow boards, lines on floor, and forms color coded.

14.7.3.3 Third S—Shine (cleanliness)

Shine refers to cleaning and looking for ways to keep things clean. Shine ensures that everything in the workplace stays clean. In other words, cleaning is a way of inspection. Examples of cleaning are sweeping of floors, wiping of equipment, and cleaning and filing of papers in an organized manner.

14.7.3.4 Fourth S—Standardize (promote adherence)

Standardization is a method to maintain first three stages that is sort, set in order, and shine. Examples for standardization are standard operating procedures, track lists, and area monitoring.

14.7.3.5 Fifth S—Sustain (self-discipline)

Sustain refers to making a habit of properly maintaining the correct procedures. Without the sustain stage other stages will not last. Examples of sustain stage are communication boards, leadership support, and reward systems.

14.7.3.6 Value analysis

Value is the attribute which is specified by the customer for the amount paid and the cost to produce it. In many cases, the manufacturer will not be able to provide the appropriate level of customer service for the product. They desire customers to purchase products or services at a lower price. They try to increase the profitability of the firm by reducing the cost for the product.

Value is seen as a different entity according to the culture in different countries and also on perspectives. Germans view high on value as product feature and process orientation. Technical people are in control of the businesses. Hence it concentrates on enhancements and improvements in product features. But one disadvantage about the German culture is that customers are not sophisticated enough to understand new features.

In Japan, value is defined as the context where value is created. The greater the features are manufactured in Japan the greater will be the value for that feature. Customers view value as what they require are satisfied with or not (Womack and Jones 1996).

For long-term planning, value must be determined by the customer. The customer wants specific products with specific features at a reasonable price. The initial step in Lean thinking is to specify value. New methods must be taken to communicate with the customer to get closer with them.

The cost of the product may be defined by determining the customer value. The cost of the product is more than the market cost of the product. Market cost is the manufacturing cost with selling expenses and profit. In Lean thinking, value analysis involves the cost of the product to be the combination of current selling price by the competitors, and examination and the elimination of waste through Lean methods and value analysis (Womack and Jones 1996).

14.8   Green belt

14.8.1   Pareto analysis

Pareto diagrams are used to prioritize problems so that the major problems can be identified. The diagrams are used to

- Analyze a problem from a new perspective
- Focus attention on problems in priority order
- Compare data changes during different time periods
- Provide a basis for the construction of a cumulative line
- Prioritize problems
- Segregate the significant problems from the insignificant problems

Pareto analysis is based on 80:20 rule. Dr. Juran states that 80% of the problems come from 20% of causes. Pareto analysis is used to segregate a few vital problems with many trivial problems. For Six Sigma professionals, it is important to identify the most important opportunity for improvement in order to get high returns from the project. Pareto chart is similar to a histogram. A histogram is the graphical interpretation of the distributed data in a quantitative manner into a class. Pareto chart is the frequency bar chart which represents quantified data by quantitative characteristics.

The steps for constructing a Pareto chart are

- Choose the subject for the chart. The subject can be anything which can be causing problems to any particular line or any product.
- Identify the data that needs to be collected. Identify if percentage or cost or quantity are going to be tracked.
- Collect data related to the quality issues. The quality issues are collected over the time period.
- Use check sheet as the input to collect data. Gather total number of defects and non-conformities in each category.

- Based on the information from the check sheet, calculate the percentage of total nonconformities.
- Specify the cost incurred for the defects or nonconformities.
- Plot the data setting up scale for chart.
- The *x*-axis represents the subject of study usually nonconformities or defects or item of interest. The *y*-axis represents the number of defects, number of occurrences, percentage, or cost incurred per category.
- Draw the chart by organizing the data in descending order that is data from the largest category to the smallest category.
- The last step is to analyze the chart. Check for the largest bar, it represents the few vital problems.

14.8.2 Data collection

The most important aspect of measuring the quality of a product or process is to carefully sift through the data relevant to the measure at hand. This is dependent on the method of sourcing this data. As we look at the types of data and the various scales of measurability, it shall be easier to weigh these methods for the condition at hand.

14.8.2.1 Measurement scales

Table 14.2 displays the four measurement levels.

Table 14.2 The Four Measurement Levels

Scale	Description	Example
Nominal	Data have names or categories only.	A bag of candy contained the following colors: Yellow 15 Red 10 Orange 9
Ordinal	Data are arranged in some order but differences between values cannot be determined.	Product defects, where A type defects are more critical than D type defects and are tabulated as follows: A 16 B 32 C 42 D 30
Interval	Data is arranged in order and differences can be found. However, there is no inherent starting and ratios are meaningless	The temperature of three ingots was 200°F, 400°F, and 600°F. Note that three times 200°F is not the same as 600°F as a temperature measurement.
Ratio	An extension of the interval level that includes an inherent zero starting point. Both ratios and differences are meaningful.	Product A costs $300 and product B costs $600. Note that $600 is twice as much as $300.

Source: Quality Council of Indiana Inc. *Certified Six Sigma Black Belt Primer.* Quality Council of Indiana Inc., West Terre Haute, IN, 2007.

14.8.2.2 Types of sampling

14.8.2.2.1 Random sampling. Sampling is done due to economic advantages and time. It requires randomness in sample selection. It is clear that random sampling gives every part an equal chance of being selected for the sample.

The sample is representative of the lot. Hence it requires some thought process and planning to obtain sample. Sampling that does not have randomness does not gives effectiveness for the plan. Sampling plan should be of independent random plan (Quality Council of Indiana Inc. 2007).

14.8.2.2.2 Sequential sampling. Sequential sampling is similar to that of multiple sampling plan but sequential sampling be continuous indefinitely. These plans end after the number exceeds three times the sample size. This type of testing is used for expensive and destructive testing with the number of sample as one (Wald 2004).

14.8.2.2.3 Stratified sampling. One of the assumptions made in sampling is that the sample is selected randomly from a homogeneous lot. Sampling may not be homogeneous, that is the product can be produced from different machines or under different conditions.

The idea behind stratified sampling is to select several samples randomly from each group or process that is different from other similar groups. The resulting mix cannot be biased since it has relative frequency of the group.

14.8.2.3 Ensuring data accuracy

Data that is not accurate not only affects the cost of collecting data but it also corrupts decision making. Data accuracy can be ensured by having the following insights:

- Do not be biased with the final value and tolerance when measuring or counting.
- Rounding minimizes the measurement sensitivity hence avoid rounding. Averages should be calculated to one more decimal position than individual reading.
- Record the data in chronological order.

14.8.3 Data collection methods

Collection of information is more expensive. Thought process is required to analyze the expected result and to ensure the collected data is relevant data. Data collection methods consist of both manual technique as well as automatic method. Some traditional or manual data collection method includes tally sheets, checklists, and check sheets. Automatic data collection method consists of data coding and automatic measurement.

Data collection guidelines are provided as follows:

- Define the problem statement clearly.
- Determine the quantity that needs to be measured.
- Appropriate measurement techniques need to be selected that is vital in this process.
- Assign task to person who will be collecting data.
- Choose appropriate sampling technique.
- Decide who will interpret and report results (Quality Council of Indiana Inc. 2007).

Manual data collection technique can be done using printed form. Manual data collection technique relies upon human accuracy for collection of data.

Automatic measurement system is used for sorting process which is used to sort parts based on various attributes such as dimension and shape. Computers are primarily used with automatic measurement system which always provides accurate result than the manual measurement system. Automated measurement system has the characteristics of speed and volume and hence it can process voluminous data at high speed.

Data coding involves assigning distinctive code by adding or subtracting or multiplying or by dividing the given value with a factor. Data coding is vital for measurement process. If coding is not done, it results in insensitivity in analytical results due to large data sets.

14.8.3.1 Check sheets

Voluminous amount of data are required for the purpose of data collection. Check sheet is a useful tool for collecting and organizing data. Check sheets are used to collect data to determine the location of data from where it is generated. Check sheets consist of events occurring in accordance with a particular time frame. Check sheets are commonly used for analyzing data that is not available which could be used for taking key decision.

Data can be defined in the check sheet by marking the characteristic of check in regions according to the events. Data are read by observing number of marks on the check sheet and location of those checks.

Check sheets are used to determine 5W's. They are

- Who completed the check sheet?
- What data was gathered from the check sheet?
- Where the data was collected (location of data collection).
- When was the data gathered?
- Why was the data collected?

For example, in order to reduce the waste, manufacturer may know the data for quantity of waste but it is essential to know various sources of waste within the operation. Check sheet implemented at different work levels may enable to capture waste from the operations carried out (George 2002). By collecting data, teams or individuals can make better decisions (Table 14.3).

Recording check sheet can be used to get measured data or counted data. Data are gathered by making a tick mark on the check sheet. Check sheet can be fragmented to

Table 14.3 Check Sheet

Samples of 1000 solder joints	Part number X-1011	Part number X-2011	Part number X-3011	Part number X-4011	Part number X-5011
Cold solder	||||			⋊̸	
No solder in hole	⋊̸		||	||	
Grainy solder	⋊̸	|		|||	
Hole not plated through	⋊̸			|||	
Mask not properly installed	⋊̸		||||	⋊̸	
Pad lifted	|				

Source: http://www.qualityamerica.com/images/checksheetdefectsstratified.jpg.

Table 14.4 Meeting Process Chart

Scale (1 = poor, 10 = Excellent)	Member #1	Member #2	Member #3	Member #4	Member #5
On track					
Participation					
Listening					
Leadership					
Decision quality					

Source: Quality Council of Indiana Inc. *Certified Six Sigma Black Belt Primer.* Quality Council of Indiana Inc., West Terre Haute, IN, 2007.

Table 14.5 Housekeeping Checklist Example

Date: Area	Visual inspection	Washed once per week	Supervisor (operator)
Preheat house			
# 2 Tub room			
# 3 Tub room			
A press room			
B press room			
Finishing room			
Roll mill room			

Source: Quality Council of Indiana Inc. *Certified Six Sigma Black Belt Primer.* Quality Council of Indiana Inc., West Terre Haute, IN, 2007.

indicate either day or month or shift. These data can be summarized in the form of tally sheet (Wortman 2010). Table 14.4 shows a process check sheet used to analyze the productivity of the meeting.

14.8.3.2 Checklists

Next type of check sheet is checklist. Checklist is used for informational aid for reducing the probability of failure of completing a task. It helps in completing a task. Checklists are usually used for learning how to operate complex or delicate equipment. Checklists is usually used for inspecting machinery or product. Checklist is presented in form of check boxes, as the event is occurred a tick mark is placed in the checklist. Table 14.5 illustrates an example, which shows the housekeeping checklist. It is used for the purpose of mistake proofing while performing multi-step procedure.

14.8.4 Types of data

Data is the lowest level from which objective information that everyone agrees upon is derived (Quality Council of Indiana Inc. 2007). It can be classified into three types of which two are frequently used as described below:

1. Attribute data: It is discrete and has "raw" values of the type integers for quantities.
2. Variable data: It is continuous and has "raw" values of the type real as they are varying with each measurement and are procured using appropriate devices/instruments that measure time, volume, distance, and so on.

It is desirable to use variable data as it provides more knowledge and, hence, proper conversion techniques have to be applied. Instead of binary attribution to represent data as good or bad, a degree of fitness within tolerance level can be represented. Cost of collection has to be considered as measurement using instruments is expensive due to variability, slow speed, and analyze by secure storage of the overload of values like mean, standard deviation, and other statistical estimates of given population. The *attribute* data type has fewer values to store but manually procuring data requires skill. The type of data and purpose of collection is a significant factor to choose the data collection method. Data could be nominal or ordinal type. Where nominal has binary values like Pass/Fail and Ordinal Data has multiple classes, with a ranking or ordering of the classes.

14.8.5 Measurement methods

Measurement system is used for measuring variables that are of our interest. Measurement system consists of measurement methods, metrology, and enterprise measurement system and measurement system analysis. Measurement methods are listed below:

- Transfer tools
- Attribute gages
- Variable gages
- Reference/measuring surfaces

14.8.5.1 Measurement system analysis
For clear understanding of the quality of data collected, certain parameters are taken into consideration to examine the system (Figure 14.13). They are

- Sensitivity
- Reproducibility: To be able to produce similar data on repetitive measurements with similar environmental conditions in near time.
- Accuracy: The data has to be unbiased and nearest to the actual value. It is reported as a mean difference between the measured data points and true value.
- Precision

"Calibration is necessary to maintain accuracy but not precision" (Quality Council of Indiana Inc. 2007, VI-70). Precision is more of a measure of how the values obtained from continuously repeated measurements within a given time fall close to each other. A cataloged testing method that is statistically stable is necessary to ensure good

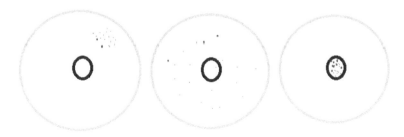

Figure 14.13 Precision and accuracy.

values for these parameters. To quantify the variability in repeatability and reproducibility, we have methods like the range, average and range, and analysis of variance (ANOVA).

It has to be noted that repeatability is the reliability of the collection on re-testing. It describes the "variation in measurements obtained with one measuring instrument when used several times by an appraiser while measuring the identical characteristics on the same part" (Automotive Industry Action Group 1995), whereas reproducibility refers to the variation in the average of the measurements made by different appraisers using the same gage when measuring a characteristic on one part. *Gage* refers to a measurement system collecting continuous, numeric, type data while *attribute* refers to a system in which measurement revolves around classification of characteristics.

Let us look at an example to understand subtle differences: The Shore a Durometer is a device used to measure the hardness of rubber products. In this example, we submit the same sample 10 times to 2 appraisers, Jack and Marie. (This is a high volume process and many samples are tested every shift. The appraisers have been told that the measurement system study is being conducted, but they are not told which sample is the study sample.) On repeated measures, it is not unusual for an appraiser to report somewhat different values. These differences represent the repeatability of the measurement system. The averages of measurements of the two appraisers are different. This difference is an indication of the measurement system reproducibility.

This example shows two operators for the determination of reproducibility. It is also possible that the measurement system uses multiple machines instead of test operators. In that case, the differences in machine averages provide the measure of reproducibility.

We assume that accuracy and sensitivity levels of the devices are less prone to error. Error can be measured as average values or variability and it significantly impacts how we view the data. If measurement error is large, the true process values will be lost, or masked, by the measurement system contribution. On repeated measures, it is not unusual for an appraiser to report somewhat different values. These differences represent the repeatability of the measurement system.

14.8.5.2 *Range method*

It calculates the error in reproducibility and repeatability of a measurement system. For individual components in the error, that is, those contributed by repeatability and reproducibility (between inspectors) separately, we use methods like average and range and ANOVA that allow calculation of total as well as errors on repeatability and reproducibility. These methods also emphasize on part or process variation due to interaction with the quality testers.

This step is about quantifying the capability of the measurement system. This is often a major hurdle in the project, but it is not possible to move ahead without an adequate measurement system. As a general guideline, it must be possible to have four or more "in-control" range values in order to have adequate discrimination. Inadequate measurement discrimination: When the process standard deviation is smaller than the measurement unit, the measurements will begin to be rounded to the same values. The rounding will contaminate the estimate of the process standard deviation by deflating the average range. Moreover, the smaller the process standard deviation is relative to the measurement unit, the greater the contamination will be.

Table 14.6 Guidelines

Subgroup size	Minimum number of possible values for range within limits
2	4
3	5
4	5
5	5

Source: Wheeler, D. C. *Advanced Topics in Statistical Process Control.* SPC Press, Knoxville, TN, 1995.

The range chart of the R&R study—or a range chart of actual process measurements—can be used to assess measurement discrimination. To do this, subdivide the vertical axis of the range chart, beginning with a value of zero, and continuing up to the value which exceeds the upper control limit. Count the number of tick marks which fall within the control limits on the range chart. Use these guidelines to determine if there is adequate measurement discrimination (Table 14.6).

In certain cases, the measurement system analysis may require a more logical approach as opposed to one of statistical vigor. For example, if a project metric (little y) were cycle time as measured by date or time of day, it is not necessary to validate the calendar and wrist-watch of every involved associate. What *is* required is some discussion as to the logic applied to ensure the data collected is uniform and reliable. For example, if time of day were to be recorded, a written plan which defines how the time would be recorded (military time, from a central clock, from a computer clock, etc.) and to what level of detail (nearest hour, nearest minute, nearest second, etc.) may suffice to ensure accuracy and uniformity of data collection.

Thus the Deliverable for this analysis for the purpose of statistically or logically determines the adequacy of the measurement system is

- A quantification or justification of the adequacy of the measurement system
- If necessary, a record of actions taken to improve the measurement system

Relying upon historic data without considering the validity of that data can send your project down a wrong path from the very beginning.

Often in transactional projects, the question will be more around whether or not effective data collection systems exist. Since many of these projects tend to focus on cycle times, number of material hand-offs and the like, the questions will focus on whether effective means exist to capture that information rather than whether we can measure the workday to the nearest fraction of a second.

14.8.5.3 *Analysis of variance*

It is the most accurate of all three methods listed and requires a standard procedure to show the partitioning of errors. A prerequisite of statistical fundamentals is a must before we continue on to describe the procedure (Table 14.7).

1. Select any five parts at random identified by numbers and a quality characteristic to measure.
2. From the workforce a bunch is chosen to measure with the same testing conditions.
3. There must be at the least two repetitive measures per inspector.

Table 14.7 Cell Interaction Squares

Part No	Inspector A	Inspector B	Inspector C	Sum, n	Average	Row Sq	A	B	C
1									
2									
3									
4									
5									
Sum, n									
Average									
Col Sq									

Note: Correction for the mean is given by $(\sum X)^2/N$ and denoted as "CM."

4. Based on the data collected from these measurements a table is formulated that has a frame as shown in Table 14.1.
5. Certain formulae are required for the partitioning of variance.
   a. ColSq = summation of column^2/column number
   b. RowSq = row total^2/row number
   c. Cell interaction Sq = cell total^2/sample size are given for each inspector A, B, C.

Summing up the values in RowSq, Interaction Cells, and ColSq to TotR, TotI, and TotC, we now remove the correction for the mean value from these summations to get TechSS, InterSS, and PartSS, respectively, using the following equations. Tot I gives the grand mean and mean from other elements is here, SS depicts the sum of squares between groups. It is nothing but group variance and is dependent on assumptions like the samples must be normally distributed, independent in nature, and have equal variance (Table 14.8).

$$\text{TechSS} = \sum \text{TotR} - \text{CM}. \tag{14.1}$$

$$\text{PartSS} = \sum \text{TotC} - \text{CM}. \tag{14.2}$$

Using results from Equations 14.1 and 14.2, we have

$$\text{InterSS} = \sum \text{TotI} - \text{CM} - \text{TechSS} - \text{PartSS}. \tag{14.3}$$

Table 14.8 ANOVA Table

Source	SS	DF	MS	F	F(α)	Variance	Adjusted	Percent
Technician								
Part								
Interaction								
Error								

Furthermore, we require values for the degrees of freedom (DF) pertaining to the technicians, parts, and interaction between the inspectors and parts. They can be calculated as

- DF of technician or Tech DF = number of technicians − 1
- DF of parts or Part DF = number of parts − 1
- DF for interaction or Inter DF = Tech DF × Part DF
- Total DF = $N − 1$
- Error in deriving the DF
- Error DF = Total DF − Tech DF − Part DF − Inter DF
- The mean squared error (MSE) is a ratio of the sum of squares and DF.
- Thus, MSE = SS/DF

The *F*-test statistic is also a ratio of between group variance and the within group variance. The DF for the numerator are the DF for the comparison between groups (k–1) and the DF for the denominator are the DF for the comparisons within group (N–k) where k is the number of samples and n is one of many sample sizes, but N is the total sample size. It can also be given by the following equation:

$$F = \frac{\text{MS (for each element)}}{\text{Error MS}}.$$ (14.4)

The variance is thus derived from these values, that is, the difference between Effective MS for each element and Error in MS is divided by the coefficient of variance. All negative results are adjusted to zero. A percentage of contribution of each element is also computed.

Table 14.8 introduces a lot of values that can help arrive at the parameters to measures repeatability. The squared root of Error MS or Sigma of error is a sign for repeatability bias. This *F*-statistic is used for hypothesis testing where the null hypothesis is that there is no inter-operator differences. Process variations are always weighed against the specifications.

14.8.5.4 Measurement correlation

Measuring against an accepted standard by having a tolerance level that encompasses the variation that comes with it is correlation. It could also be the comparison of data derived from various measurement systems or statistical values from similar devices as well as similar measurands. Calibration is done to avoid such differences through a technique called proficiency testing or round robin testing. The components of variability error are collected as contributed by the measurand and the instrument individually by repeated testing. Measuring under different conditions for the various techniques can also be a contributor. These errors can be classified into five categories for a broader perspective. These are bias, repeatability, reproducibility, stability, and linearity (*Measurement System Analysis Reference Manual*) (AIAG 1995). A combination of these gives a resultant correlation variance.

14.8.5.4.1 Bias. It is the difference between the result and the actual value and is weighed against the standard value to get the sign of bias. A negative sign implies underestimation and a positive implies overestimation. The following expression is used:

$$\frac{\sum_{i=1}^{n} X}{n} - \tau.$$ (14.5)

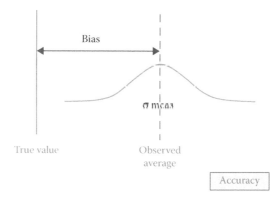

Figure 14.14 Bias.

Here, *n* is the number of tests, *X* is the measurand, and τ is the standard value. It is usually represented as a percentage of process variance or tolerance. The significance of this bias in our analysis is separately deduced using the discriminatory power of *t*-test. The best measurement systems are traceable to some standard. The difference between an observed average value (i.e., the average of the values generated by the measurement system) and the standard is bias.

   Bias can be evaluated using tests to determine if there is a statistical difference between the observed average and zero (0) or against established rejection criteria. If unacceptably large bias levels are observed, the measurement system may require re-calibration, or an offset correction factor (Figure 14.14).

   14.8.5.4.2 Linearity.   Bias derived is plotted against the reference values in the plotting range of an instrument. The default procedure is to measure 10 parts 5 times each (Quality Council of Indiana Inc. 2007). The percentage linearity is the slope of the straight line that curves through all the data points. The equation gives us the linearity as a product of bias and process variation. It can be represented as in the Figure 14.15.

$$L = b(\text{P.V}).\tag{14.6}$$

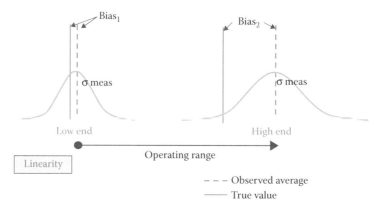

Figure 14.15 Linearity.

14.8.5.4.3 Stability. Stability is also known as drift and refers to a change in bias over time. Stability may be of particular concern when measurement systems include devices that are prone to deterioration over time such as short-lived radioactive isotopes, or devices which contain frequently consumable parts such as some spectrophotometers and sensors. A stable measurement process is in statistical control with respect to location (AIAG 1995).

Percentage Agreement: The correlation coefficient which is used for calibration is the basis to arrive at the impact of change in the independent variable on a dependent variable. The two extremities are given below:

1. If the coefficient is ±1 then 100% in agreement
2. If the coefficient is 0 then 0% in agreement

Attribute measurement systems are those that tend to classify outputs into two or more categories without actually assigning measured numeric values. Notice that a larger sample size is required or recommended when using attributes measures. This is due to the fact that the attribute data itself is not as information rich as continuous data. Instead of knowing any precise information about your process or product, you only know fairly large categories of information.

- Guidelines suggested are: For studies with two levels (e.g., pass/fail).
- If at all possible, try to have a minimum of 20 "good" samples and 20 "bad" samples.
- Try to balance—approximately 50% "good" and 50% "bad"

Also, try to have a variety of degrees of "good" and "bad."

Example: The manufacturers of "Blue Moon" ice cream bars have received complaints from their customers that the ice cream bars are not as blue as their jingle suggests …

If it isn't BLUE, we won't sell it to you …

Management demands that "this problem be investigated and the first step is to evaluate our measurement system for BLUENESS."

Twenty samples of the product were selected by the standards-setting group, such that 10 samples were judged as "acceptable" items and 10 samples were judged as "unacceptable" items. These 20 samples were passed through the normal inspection process, in a random order, so that each of three inspectors evaluated each sample two times. The example may seem trivial but it is not too far from the truth in many situations. There are a variety of attribute measures for which no strictly analytical methods exist. These parameters are no less important to a customer and must be in some way evaluated. The concluding remarks are: *Could improve*: Agreement with standard; training, SOPs, visual standard. Inspector Agreement at 75% agreement, inspectors seem to have differing definitions of "BLUE." In attribute measurement systems, the effectiveness will often boil down to how consistently the investigators are able to distinguish good from bad. This consistency is a function of the training they receive and the tools they have available.

14.8.5.4.4 Precision/tolerance. This ratio of precision quantified as the error in measurement and tolerance of the measured parameter is required to be small in order to minimize the net effects of glitches that contribute to variability.

$$P/T = \frac{6\sigma}{\text{Total tolerance}}. \tag{14.7}$$

Another useful ratio is the P/TV and is given by (Adapted from AIAG 1995)

$$P/TV = \frac{6\sigma}{\text{Total tolerance}} = \frac{\text{Error in measurement}}{\text{Product variation} + \text{measured error}}.$$

14.8.5.5 *Enterprise measurement systems*

Key enterprise measures are service and/or transactional in nature, that is, belonging to the realm of marketing, customer satisfaction for both internal and external customers, research and development, manufacturing, engineering and supply. To measure these, we need to delve into the science behind it which is termed as metrology. For the objective, of quality control there is a rule of thumb to work with metrology.

- The standards used for measurement must be internationally recognized and established.
- Use of equipment to check for conformation to specifications is encouraged.
- Calibrating this equipment often to the accepted standards is required.

There are three major Measurement Systems. They are the English, Metric, and the System International D'unites (SI). Conversions and calculations become cumbersome with the English system but the Metric and SI systems are decimal based and have units that can be easily correlated as factors of ten. Measurements can be of seven types in general:

1. Length
2. Time
3. Mass
4. Electric current
5. Temperature
6. Light
7. Amount of substance

14.8.5.5.1 Instrument selection. Instrument selection is vital in case of measure phase. Black belt person should know how to select appropriate instrument. Black belt person is responsible for selection of appropriate instrument used in measure phase. Some measuring instruments used in measure phase are listed below:

- Gage blocks
- Spring calipers
- Dial calipers
- Vernier calipers
- Digital calipers
- Surface plates
- Micrometers
- Ring gages
- Plug gages
- Dial indicators

- Pneumatic gages
- Interferometry
- Laser designed gaging
- Coordinate measuring machine

The standard tests performed to measure the targets and specifications are listed as follows:

- Nondestructive testing (NDT)
- Nondestructive evaluation (NDE)
- Visual inspection
- Ultrasonic testing
- Impact test
- Steel rule

14.8.5.6   Calibration

When considering the impact of the measurement system, a graphical representation reinforces the need to minimize measurement system error and variability. When viewing measurement system performance in light of spec limits, a 99% confidence interval can be developed (5.15*s_{meas}) and compared to the specification range. Ideally, the measurement system variation will take up very little of the range of specifications. The true process variation cannot be measured directly, but can be estimated by subtracting the observed total variance from the measurement variance:

Total Variability in general is given by the equation

$$\sigma_{total}^2 = \sigma_{process}^2 + \sigma_{measured}^2. \qquad (14.8)$$

Confidence intervals of the mean of measurements that includes these errors are deduced using the Central Limit Theorem for *n* number of readings which gives us an idea of tolerance. In addition to these criteria, the AIAG states that the number of distinct categories (ndc) "ought" to be ≥5. The ndc is the number of distinct 97% confidence intervals that will span the expected product variation. It is also important to remember that this fairly short-term evaluation of the measurement system is only part of the picture. The long-term performance of the measurement system may also require review using graphical analyses over time.

14.8.5.6.1   Comparison to process performance.   Although it should not be necessary in a Six Sigma project (because the specification has been defined in step 3), there may be situations when you want to make some assessment of a measurement system adequacy, but there is no specification for comparison. In such a case, compare the measurement standard deviation to the process total standard deviation. Use the 10%, 30% values as (very general) guidelines.

14.8.5.6.2   Process.   The types of variation that can be found from measurements are given in these categories:

- Inter-operator variability
- Human error of the operator
- Variability in equipment functioning
- Variations in material quality
- Variations in procedural steps or methods

- Variation in software versions or formulae
- Variations in environment

14.8.5.6.3 Calibration interval.   The interval in which the device has to be calibrated is based on three aspects:

- Stability
  - Ability to consistently maintain the metrological characteristics over time proven from recorded data.
  - Purpose.
  - Objective determines the method.
- Degree of usage
  - Environmental conditions when the device is being used.
  - Intervals should be shortened or lengthened in accordance with previously recorded calibration and usage records of the equipment.
  - Verification methodology is the technique that uses shorter frequencies for checking and costs for cyclical calibration and use of equipment can be reduced.

The practices that are to be followed are

- Comparison with all possible standards must be considered for all data points.
- Easily maintained and replaceable instruments are applicable recipients to this system.

14.8.6   *Graphical methods*

These methods are used to represent data and analytically deduce meaningful relationships among the data points.

14.8.6.1   *Boxplots*

A simple technique to summarize data where a median divides these points into upper and lower quartile sections is called a box and whisker plot. The largest and smallest observations are drawn as extensions to the box called end of line or whiskers. They are complex and can be notched to imply variance and the width of these notches could be variable. It is equivalent to logarithm of the sample size. It must be ensured that the means are different at a 5% significance level if medians do not overlap. Outliers can also be 1.5 times the interquartile distance from each quartile. The inter-quartile range (IQR) can be illustrated in the Figure 14.16. The box illustrates spread and skewness in the data.

14.8.6.2   *Scatter diagrams*

To control the dependent variable the correlation is identified using these diagrams. A cause-and-effect relationship and also relationships between causes are examined to create one. They reveal the degree of dependency, skewness, arrive at root causes to problems and interpolate it to a line if possible. They test for auto-correlation before drawing control charts. Scatter diagram procedure is as follows (Tague 2004):

- Collect pairs of data where a relationship is suspected.
- Draw a graph with the independent variable on the horizontal axis and the dependent variable on the vertical axis. For each pair of data, put a dot or a symbol where

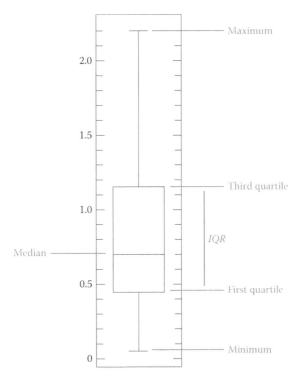

Figure 14.16 A simple box plot.

the *x*-axis value intersects the *y*-axis value. (If two dots fall together, put them side by side, touching, so that you can see both.)

- Look at the pattern of points to see if a relationship is obvious. If the data clearly form a line or a curve, you may stop. The variables are correlated. You may wish to use regression or correlation analysis now. Otherwise, complete steps 4–7.
- Divide points on the graph into four quadrants. If there are *X* points on the graph,
- Count *X*/2 points from top to bottom and draw a horizontal line.
- Count *X*/2 points from left to right and draw a vertical line.
- If number of points is odd, draw the line through the middle point.
- Count the points in each quadrant. Do not count points on a line.
- Add the diagonally opposite quadrants. Find the smaller sum and the total of points in all quadrants.
    - *A* = points in upper left + points in lower right
    - *B* = points in upper right + points in lower left
    - *Q* = the smaller of *A* and *B*
    - *N* = *A* + *B*
- Look up the limit for *N* on the trend test table.
- If *Q* is less than the limit, the two variables are related.
- If *Q* is greater than or equal to the limit, the pattern could have occurred from random chance.

A regression line is to be calculated to find the relation between the independent and dependent plots (Table 14.9).

Table 14.9 Relationship Table

$R = -1$	Strong negative	When X increases, Y decreases
$R = 0.5$	Slight negative	When X increases, Y decreases
$R = 0$	No correlation	Two variables are independent
$R = +0.5$	Slight positive	When X increases, Y increases
$R = +1$	Strong positive	When X increases, Y increases

Correlation coefficient can be given by

$$r = \frac{\sum_{i=1}^{n}(X-\bar{X})(Y-\bar{Y})}{\left[\sum_{i=1}^{n}(X-\bar{X})^2\right]\left[\sum_{i=1}^{n}(Y-\bar{Y})^2\right]}.$$

14.8.6.3 *Stem and leaf plots*

It is a manual method to display variable and categorical data sets (Tukey 1977). The data is grouped by class intervals and are represented as stems. Increments in data are represented as leaves. They are deemed better than histograms at representing data values as individual data points are displayed clearly.

14.8.6.4 *Run (TREND) charts*

To represent data as dynamically varying with time using trend charts is a useful way to analyze if the change in important aspects of a process is desirable or not with respect to a given process. It makes processing large amounts of data easier. The trend could be upward, downward, random, cyclic, and also have an increasing variability.

14.8.6.5 *Histograms*

Histograms are a static representation of data and are grouped according to frequency of occurrence. A given interval is termed as a bar and these bars can be joined through their mean points to form a curve. The curve tracing these points can imply if the process is stable or not. If it is bell-shaped or unimodal it is predictable. Other distribution shapes like gamma, beta, Weibull, Poisson, hypergeometric, exponential, etc. are also termed to be stable. In other words, it shows the probability distribution of an underlying variable that can have various values and estimates its density (Figure 14.17).

It must be noted that small variations around a bell curve is a result of natural variation. All other skewness is assignable. The areas under these bars are corresponding to these frequencies. There are two types of histogram: ordinary and cumulative that counts the cumulative number of observations. It can be represented mathematically as a function m that counts the frequency of data.

$$N = \sum_{i=1}^{k}M,\tag{14.9}$$

where k is the difference between maximum and minimum values of x or the continuous variable; N is the number of observations.

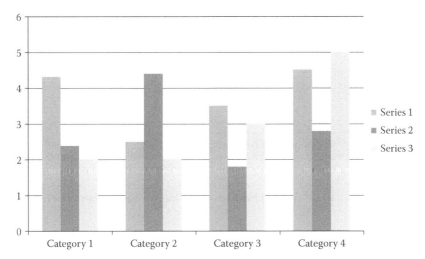

Figure 14.17 Sample histogram.

14.9 Black belt

A few of the tools that can be used in the measure phase of DMAIC which qualify to be used for black belts are as follows:

- Control charts
- Attribute control charts
- Gage R&R

Below we will discuss each of the above tools for the implementation of measure in a Six Sigma project.

14.9.1 Control charts

Control charts are used to manage variation. In this chapter, we will discuss the various statistical charts that can be used to monitor variation. A basic control chart has an upper limit, a lower limit, and a central line. In this chapter, we introduce the various control charts step by step. However, there is a generalized method for each process control chart. This is very important to understand as the process for establishing all the charts is the same. The only difference will be the formulae used for determination of upper limit, lower limit, and the central line.

- Variable and attribute process control charts

To select the right process chart we should be able to differentiate between variables and attribute. A variable is a continuous measurement such as weight and height, whereas a variable is a result of some binomial process which results in an either-or solution. Some of the examples from attributes may be as follow: the inspector is either bald or not bold. In this chapter, we will discuss the various types of process control charts for variables and attributes. Table 14.10 shows the different kinds of attribute and variable charts.

The variable charts are X, $\bar{x}$, R, MR, and s charts. The attribute charts are p, np, and c and u charts. In the following pages, we introduce the different kinds of charts and how

Table 14.10 Variables and Attribute Charts

Variables	Attributes
X	P
$\bar{x}$	N-p
R	C
MR	U
S	

X, process population average; $\bar{x}$, mean or average; R, range; MR, moving range; S, standard deviation; P, proportion defective; N-p, number defective; C, number conforming; U, number nonconforming.

to select which chart is to be used. There are four basic requirements before we select a process chart.

1. Understanding the process for implementing the chart
2. Interpretation of the charts is to be known
3. Realizing the usage of different charts
4. Knowing how to compute the limits for the different type of process charts.

14.9.1.1 *Understanding the process for implementing the chart*

The process of developing a chart is almost the same for all process charts, the only difference lies in the actual statistical computations (Figure 14.18). Following are the steps used in developing process control charts:

The upper control limit, central limit, and lower control limit are computed statistically. Each point represents data from a sample that is plotted sequentially.

- Critical operations in the process where inspection might be needed is to be identified. These are the operations in which, when the operations are not performed properly might result in the product being negatively affected.
- The characteristic of the product, which may determine that the products functioning are good or bad, is to be known.
- Determination if the critical product characteristics are a variable or attribute.
- Selection of the appropriate process control chart among the many types of charts is done. The decision process and the various charts are discussed below in the chapter.

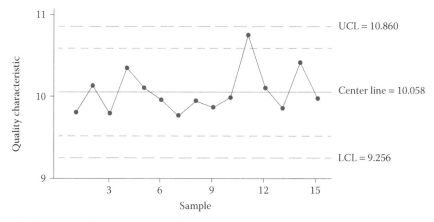

Figure 14.18 Example of a control chart. (Data from http://en.wikipedia.org/wiki/Control_chart.)

- Once this is done, the control limits for the chart are decided for continual monitoring and improvement.
- As and when improvement or changes occur the limits are to be updated.

14.9.1.2 $\bar{x}$ and R charts

Now that we have understood the procedure to develop a chart let us discuss the various types of charts. We will first discuss $\bar{x}$ and R charts as they go hand in hand. We developed an $\bar{x}$ chart previously. When we are interested in monitoring the average and the range of some data we use the $\bar{x}$ and R chart. The $\bar{x}$ chart is used in monitoring the average of the process whereas the R chart is used for determining the range of the process or the dispersion of the process. The $\bar{x}$ chart is used to monitor the average of the characteristic that is being measured. To prepare this chart, samples from the process are collected for the characteristic that is being measured and then rational groups are formed from the sample. The average value of each sample is determined by dividing the sums o measurements by the sample size and the value on the chart is plotted.

The R chart is used to monitor the range as the name itself suggests. It is used in conjunction with the $\bar{x}$ chart when the process characteristic is a variable. To prepare an R chart, samples are collected and segregated into subgroups usually of up to six items. The difference is then computed by finding the difference of the high value of the subgroup from the low value of the subgroup and then the R chart is plotted. Now that we have collected the data to prepare the charts, the next step would be to determine the upper control line, central line, and the lower control line. Employing certain formulae does this. The upper control line and lower control line are usually three standard deviations away from the central line. It may also come to notice that there are no formulae for the lower limit of the R chart. This is because the lower limit is usually zero for samples having six points (Figures 14.19 and 14.20).

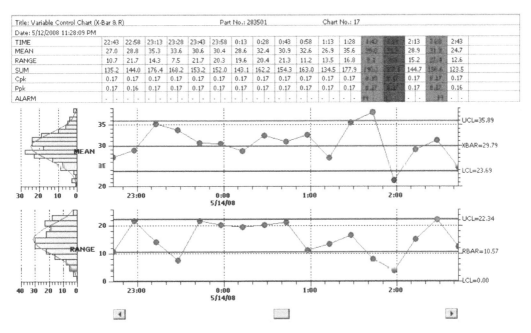

Figure 14.19 Example of an $\bar{x}$ and R chart. (Data from http://www.quinn-rtis.com/QCSPCChart CFProdPage.htmmKjzo2RM&hl=en&sa=X&ei=DGumUJLPK4jQ2AXGoCQ&ved=0CFQQ9QEwB Q&dur=345.)

Calculations for average and range control charts

1. Calculate basic averages from measured values

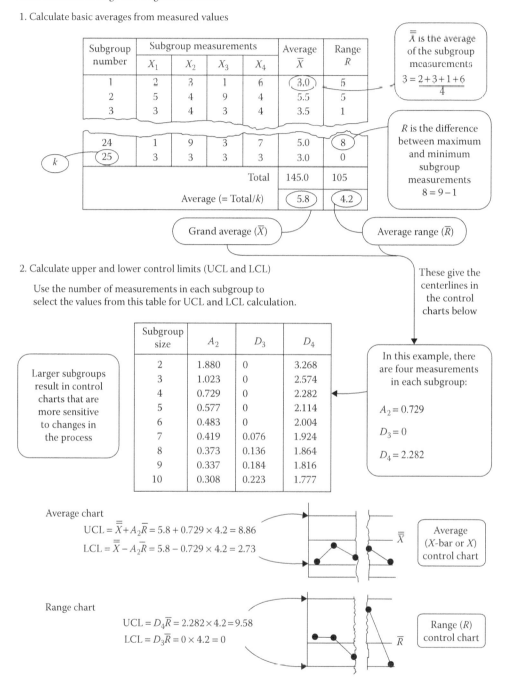

Figure 14.20 Calculation worksheet for $\bar{x}$ and R chart, $\bar{x}$-MR chart. (Data from http://www.syque.com/quality_tools/tools/Tools71.htm.)

14.9.1.3 Median charts

Sometimes x charts are too consuming though they are preferred for variable data. They are also inconvenient sometimes to compute subgroup averages. In addition, there may be concerns about the accuracy of computed means. In these cases, the median chart, that is, a $\tilde{x}$ chart can be used. The main limitation with this is that you will have to use an odd sample size to prevent the calculation of the median. Like the other charts here also small sample sizes are used, although the larger the sample size, the better is the sensitivity of the chart to detect the nonrandom event.

The preparation of a median chart, involves the determination of the subgroup size and how often the sampling takes place. The rule of thumb to create a median control chart is to use 20–25 subgroups and a total of at least 100 measurements. Equations for computing the limits are as follows:

$$\text{Mean of medians} = \frac{\text{Sum of the medians}}{\text{Number of medians}}.$$

14.9.1.4 $\bar{x}$ and s charts

When the dispersion of the data is the primary concern, range chart might not be enough. In this case, the $\bar{x}$ chart is employed with something called the s charts or standard deviation chart. The standard deviation chart is used when the variation is small. For example, s charts are used in monitoring the silicon chips production for computers. Unfortunately when using the s chart, since the range is not computed new formulae are employed to compute the upper control limit and the lower control limit. The formulae are introduced because of their importance in high technology production.

$$UCL_s = \frac{\mu + 3\sigma}{\sqrt{n}}. \tag{14.10}$$

$$LCL_s = \frac{\mu - 3\sigma}{\sqrt{n}}. \tag{14.11}$$

After the limits are computed, the sample means are plotted to see if the process is in control. If the chart is not in control, the out of control points cause is determined and the cause is eliminated and then the control limits are again recomputed by throwing the out of control points. If the cause cannot be identified, the out of control points should not be eliminated.

14.9.2 Attribute process control charts

An attribute is something that either exists or does not exist; it is a peculiar and essential characteristic. There are five basic types of attributes:

1. Structural attributes
2. Sensory attributes
3. Performance attributes
4. Temporal attributes
5. Ethical attributes

Structural attributes have to do with physical characteristics of a particular product or service. Consider an example, hotel with or without balconies for all the rooms. Sensory attributes relate to basic senses such as touch, smell, tastes, and sounds. For products these attributes relate to form design or packaging design for the creation of products, which might please the customer. Ambience in hotels, restaurants, etc. is important for customer experience. Performance attributes check whether a particular product performs the service it is supposed to efficiently or not. Does the lawn mower engine start and so on? Temporal attributes are those which are in relation with time. Was the parcel delivered on time? Ethical attributes are those which deal with professional ethics. Is everything transparent? Is the teacher kind? Are some of the attributes that one would expect?

Process for developing an attribute chart:

- Identifying the vital operations that are required for the process and where inspection might be required. These are the operations that can have a negative impact on the product.
- Determining the critical product characteristics. These will result in the products good form or bad form.
- Determining if the characteristic is a variable or an attribute
- Selecting the appropriate process cart
- Establishing the control limits
- Altering the limits according to the changes occurring

14.9.2.1   P-charts for proportion defective

The p-chart is an attribute process control chart that is used to graph the items of a sample that are defective proportionally. They are very useful when there is a change in the proportion defective of a particular sample for a product or a service. Few of the common defects that a p-chart can be used for monitoring are late deliveries, incomplete orders, calls not getting dial tones, transaction errors, clerical errors, or parts that do not fit together as desired.

Like every other process chart, a p-chart also has to be divided into subgroups typically between 50 and 100 units. The subgroups may be of different sizes. A constant subgroup is however best to be held. Twenty-five subgroups are established at least for a p-chart. The formulae to determine the limits for a p-chart are as follows:

$$P = \bar{p} \pm 3\sqrt{\frac{(\bar{p})(1-\bar{p})}{n}}, \tag{14.12}$$

where:
$\bar{p}$ is the proportion defective
n is the sample size

14.9.2.2   np charts

np chart is the graph which gives us the number of defectives in a subgroup. The requirement of an np chart is the sample size requires of each subgroup to be the same each time a sample is drawn. When the subgroup sizes are equal, either the p or np chart can be used. They are actually the same chart but some people fin the np chart relatively easier because it reflects the integer numbers rather than proportions. The p and np chart have essentially the same uses.

Subgroup sizes for the np chart are similar to that of the p chart that is between 50 and 100. Usually at least 20 subgroups are used in the development of the np chart. Again, the only requirement here is that the subgroup sized should be equal. The computation of the control limit is done by the following formula:

$$CL_{np} = n(\bar{p}) \pm 3s_{np},\qquad(14.13)$$

where:
   n is the sample size
   s_{np} = standard error

14.9.2.3   C and u charts
C and u charts are a graph, which gives us the number of defects per unit. The units must be of the same sample size, which include all physical factors such as height, weight, length, and volume. This means that chance of finding the defects in an area must be the same for each unit. Several units can be formed to make one large group. When multiple units are used it should be noted that all the subgroups should be of the same size. The control limits are computed based on Poisson distribution.

Like other process these charts are also used to determine non-random events in the life of a production process. Few of the common applications of the c charts include flaws in auto finish and errors in typing and incorrect responses for a standardized test. The c chart is used when the same size sample space are in consideration. When this is varied a u chart is used. The u chart is a graph of average number of defects per unit. This is in contrast with the c chart, which shows the actual number of defects per standardized unit. The u chart has the flexibility for different numbers in each sample space. The functions of the u and c charts are the same. The formulae for control limits of u and c charts are given as follows:

$$CL_c = \bar{c} \pm 3\sqrt{\bar{c}},\qquad(14.14)$$

$$CL_u = \bar{u} \pm 3\sqrt{\frac{\bar{u}}{n}},\qquad(14.15)$$

where:
   n is the average sample size
   $\bar{c}$ is the average number of nonconformities
   $\bar{u}$ is the average number of nonconformities per unit

References

Automotive Industry Action Group (AIAG) et al. (1995). *Statistical Process Control (SPC): Reference Manual*. Southfield, MI: Automotive Industry Action Group.

Eckes, G. (ed.) (2001). *The Six Sigma Revolution*. New York: John Wiley & Sons.

George, M. L. (2002). *Lean Six Sigma: Combining Six Sigma Quality with Lean Speed*. New York: McGraw-Hill.

Juran, J. M. and Blanton Godfrey, A. (eds.) (1999). *Juran's Quality Handbook*. 5th edn. New York: McGraw-Hill.

Kaplan, R. and Norton, D. (1992). "The balanced scorecard—measures that drive performance." *Harvard Business Review*, January–February, 71–79.

Prabhu, C. L. N. (2012). "American Society for Quality." http://asq.org/learn-about-quality/balanced-scorecard/overview/overview.html

Quality Council of Indiana Inc. (2007). *Certified Six Sigma Black Belt Primer*. West Terre Haute, IN: Quality Council of Indiana Inc.

Tague, N. R. (2004). *The Quality Toolbox*, 2nd edn. Milwaukee, WI: ASQ Quality Press, pp. 471–474.

Taylor, M. G. (ed.) (2009). *Lean Six Sigma Service Excellence*. Fort Lauderdale, FL: J. Ross Publishing.

Toppazzini, K. (2012). "Kaplan & Norton Got It Wrong-Lean Six Sigma Balanced Scorecard." Toppazzini and Lee Consulting Inc., December 16, 2012. http://tleecorp.com/blog/bid/199989/Kaplan-Norton-Got-it-Wrong-Lean-Six-Sigma-Balanced-Scorecard

Tukey, J. W. (ed.) (1977). *Exploratory Data Analysis*. Reading, MA: Addison-Wesley.

Wald, A. (2004). *Sequential Analysis*. Mineola, NY: Courier Dover Publications.

Wheeler, D. C. (1995). *Advanced Topics in Statistical Process Control*. Knoxville, TN: SPC Press.

Womack, J. and Jones, D. (eds.) (1996). *Lean Thinking: Banish Waste and Create Wealth in Your Corporation*. New York: Simon & Schuster.

Wortman, B. (2010). *The Quality Engineer Primer*. Terre Haute, IN: Quality Council of Indiana.

Further Reading

Quality America Inc. (2012). "Check sheet defects stratified." http://www.qualityamerica.com/images/checksheetdefectsstratified.jpg

Review Questions

Question 1: Kaplan and Norton have outlined a business planning process that gives consideration to factors other than financial ones. It provides more perspectives for stakeholder interest. This approach is referred to as

(a) Balanced scorecard
(b) Strategic planning
(c) Five forces of competitive strategy
(d) Quality function deployment

Question 2: Quality cost analysis has shown that appraisal costs are apparently too high in relation to sales. Which of the following actions would NOT be considered in pursuing this problem?

(a) Work sampling in inspection and test areas
(b) Adding inspectors to reduce scrap cost
(c) Pareto analysis of quality costs
(d) Considering elimination of some test operations

Question 3: Which three of the following four techniques could easily be used to display the same data?

A. Stem and leaf plots
B. Boxplots
C. Scatter diagrams
D. Histograms
  (a)   A C D
  (b)   A B D
  (c)   B C D
  (d)   A B C

Question 4: Which of the following measures of variability is NOT dependent on the exact value of every measurement?

  (a) Variance
  (b) Range
  (c) Standard deviation
  (d) Mean deviation

Question 5: Basic assumptions underlying the analysis of variance include:

  A. Observations are from normally distributed populations
  B. Observations are from populations with equal variances
  C. Observations are from populations with equal means
     (a) A B C
     (b) A C
     (c) B C
     (d) A B

Question 6: If one chose to look at any business enterprise on a three main level basis: process, operations and business, which of these categories would have both KPIV's and KPOV's?

  A. Process
  B. Operations
  C. Business
     (a) A
     (b) A B
     (c) B C
     (d) A B C

chapter fifteen

Analyze

Evaluating data to determine root causes

The more you know the luckier you get

—J. R. Ewing of *Dallas*

Overview	
Steps	Substeps
Define	• Quality projects (sponsorship) • Define specific project (black belt)
Measure	• Identify critical characteristics • Clarify specifics and targets • Measurement system validation
Analyze	• Determine baseline • Identify improvement objectives • Evaluate process inputs
Improve	• Formulate implementation plan • Perform work measurement • Lean execute the proposed plan
Control	• Reconfirm plan • Set up statistical process controls • Operational process transition

Source: PWD Group LLC.

Recommended Tools by Belt						
		Define	Measure	Analyze	Improve	Control
Stage 1	Yellow belt	Interview process Language processing System map (flowchart)		X–Y map/QFD/ house of quality Fishbone diagram/RCA	EVOP	Control plans Visual systems 5-S SPC/APC
Stage 2	Green belt	Prioritization matrix Thought process map	Hypothesis testing flowdown	FMEA Multi-vari chart	Design of experiments Fractional factorials	TPM Mistake proofing

(Continued)

Recommended Tools by Belt (Continued)

		Define	Measure	Analyze	Improve	Control
		Stakeholder analysis	Measurement system analysis	Chi-square	Data mining	
		Thought process map	Graphical methods	Regression	blocking	
			Analysis of variance	Pareto analysis		
Stage 3	Black belt	Value stream map	Process behavior charts	Buffered tolerance limits	Theory of constraints	
		Gantt chart development	QFD	Process capability analysis	Multiple response optimization	
		Project management			Response surface methodology	

Source: PWD Group LLC.

Notes: Black belts are expected to know the continued knowledge obtained throughout green belt and yellow belt certifications. The same goes for green belt and yellow belt. Empty cells represent that skill is not required for that belt stage; however, a general knowledge of the step is expected.

5-S, sort, set in order, shine, standardize, and sustain; APC, advanced process control; EVOP, evolutionary operations; FMEA, failure modes and effect analysis; QFD, quality function deployment; RCA, root cause analysis; SPC, statistical process control; TPM, total productive maintenance.

15.1   Introduction

Analyze is an important phase in DMAIC. Having the metrics established and measured, and also having identified the key process input variables that affect the key process output variables, analyze phase utilizes different statistical and hypothesis testing tools to quantify the data obtained in the measure phase to be used in later phases, that is, improve and control. The analyze phase is classified as the following:

- Measuring and modeling relationships between variables
- Hypothesis testing
- Additional analysis methods

The above-mentioned methods utilize tools which are discussed later in the chapter based on belt certifications.

15.2   Establishing baseline

The purpose of establishing baseline is to quantify the performance level of "little y" that was zeroed down in measure. The main objectives are to determine and document the baseline overall process performance and the baseline potential process capability. The baseline results should consist of Z-scores (both long-term and short-term) for the baseline process and also the confidence interval estimates. It is nothing but the targets and specifications identified in the previous measure phase. Hence, this step of the project produces estimates of baseline (i.e., current) process performance for the process output(s). Confidence intervals for those estimates are determined to bind their imprecision. Additional process mapping also documents the configuration of the process that

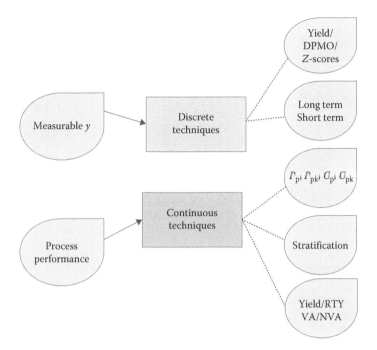

Figure 15.1 The plan—analyze.

generates the output of interest. We require greater detailed understanding of the process operations and flow than was needed at the outset of the project.

Generally, the analyze phase can be represented as shown in Figure 15.1.

Improvement of any process requires the precise understanding of the current process. Hence, baseline is established to measure and report the ability of the current process to perform to the required levels of target and specification. Baseline is the current state of the process (its basis or starting point) before beginning the Six Sigma process improvement project. Baseline refers to both

- Current overall process performance
- Current potential process capability

Before initiating the process improvements, it is vital for any belt-certified member to understand how the process is performing. Establishing the current process performance includes estimation of current defect rates, defective rates, overall and stepwise yields, mean outputs, and output variances. Generally, the baseline process metrics will be the following for any processes (Figure 15.2):

- Defects, defectives, and related metrics
- Six Sigma process metrics and Z-scores
  - Overall process performance
  - Potential process capability
- Related quality engineering process performance metrics
- Process output yield and related metrics

Hence, the Six Sigma metrics can be defined and examined in terms of Z-scores by combining process statistics and customer targets and specifications. In the process, the

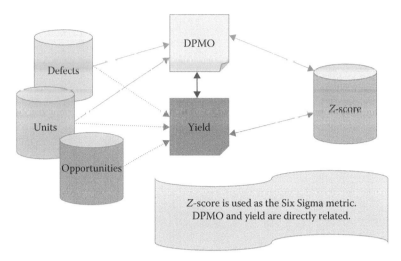

Figure 15.2 Six Sigma process metrics.

distinction between short-term and long-term variation will also be considered to arrive at the metrics of overall process performance and potential process capability.

15.2.1   Baseline performance metrics

15.2.1.1   Counting "defects" or "defectives"
Its examples are

- Specific sizes of new product not available at market introduction date (number not available/number planned).
- Errors in test requests submitted for road test evaluation (number of requests with errors/total number planned)—or—(total number of errors on all requests/total number of "opportunities for error" on all requests).

15.2.1.2   Measuring characteristic or variable
Its examples are

- The cycle time, in seconds, to evaluate the uniformity characteristics of a tire.
- The viscosity of a polymer.
- The length of a conveyor belt.

15.2.1.3   Baseline process metrics
Baseline process metrics quantify the ability of the current process to meet specifications. There are two primary baseline process metrics:

- Overall process performance—a measure of overall, or "long-term," process results
  - Estimated from summary statistics using all the data
  - Resulting Z-score reported as $Z_{\text{long-term}}$, or Z_{LT}
- Potential process capability—a measure of within-group, or "short-term," process potential
  - Estimated from data subdivided into rational sets or subgroups
  - Resulting Z-score reported as $Z_{\text{short-term}}$, or Z_{ST}

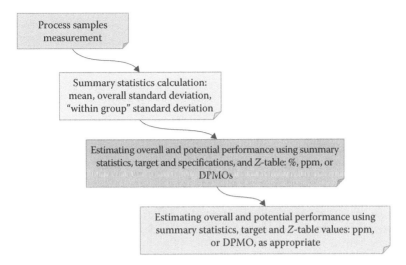

Figure 15.3 Determination of baseline process metrics.

The estimation of baseline process metrics are given below and in Figure 15.3.

- Gather data about the process output, y
  - Retain process-associated information "tags"
  - No special data handling or follow-up
- Analyze the data
  - Calculate summary statistics
  - Look at process behavior
- Calculate the $6s$ performance metrics
  - Overall process performance, Z_{LT}
  - Potential process capability, Z_{ST}

The baseline process metrics can be classified as discrete and continuous outputs. The background on this topic was discussed in the previous chapter and the reader can also read the appendix. This section provides with the different discrete and continuous data and its mode of measurements.

15.2.2   Discrete outputs

Discrete data gives only certain specific values possible. Following are some examples:

- Pass?   (yes, no)
- Color?   (black, gray, white)
- Grade?   (A, B, C, D, F)
- Opinion rating?   (good, OK, poor)

Each question above has only one particular distinct value possible. The two common discrete output variables are

- Defective   (yes or no)
- Defect   (type-1, type-2, type-C, etc.)

The definitions of "defective and defect" are given below:

- Defective—Any unit with one or more defects.
- Defect—An imperfection, blemish, nonconformance. Any instance where the product or service fails to meet a specific requirement. One unit of output may contain more than one defect.

The most commonly encountered process output "defects" are defect of output quality, defect of process cycle time, and defect of process cost component. Hence, it is not possible to assess process performance without the assessment of when the process meets, and does not meet, the customer requirements (i.e., knowing what constitutes defects and defectives). Therefore, the project teams, the sponsor, the process owner, etc. must know and agree upon critical, measurable process output characteristic(s), as well as the relevant targets and specifications. To count and measure the defects and defectives, the following terminologies are important to be known:

Unit is a specified quantity of process output; it is the basis for process performance analysis. Units are operationally defined to count defects and defectives and the rational definition varies by situation. Some examples of unit are lot; container (truck trailer, pallet, box); piece (tire, bale, belt, document); and time (shift, day, week, month).

Opportunity is any property of the process output (product or service) or the process that is or can be chosen for improvement based on the inputs from define and measure. Some examples of opportunities are product opportunities such as parts, features, materials, and process opportunities such as machines, procedures, and tools. The terminologies being made clear about discrete data, the important part is to know the procedure for determining performance statistics for discrete data (Figure 15.4).

- Count the number of defects or defectives.
- Determine the total number of "opportunities for defect" or "opportunities for defective."
- Calculate the ratio of defects (or defectives) to opportunities. The ratio may be expressed as a percentage, parts-per-million, and defects per million opportunities (DPMO).
- Determine the Z-score that is equivalent to the ratio just calculated.

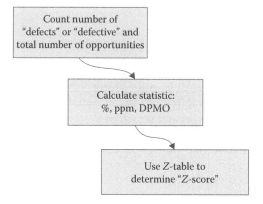

Figure 15.4 Baseline statistics for discrete data.

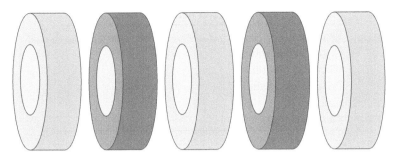

Figure 15.5 Tires for inspection.

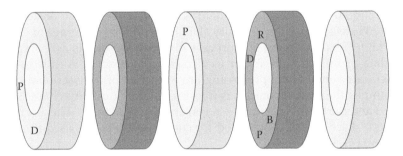

Figure 15.6 Results of tires for inspection.

The following example will demonstrate the above steps in detail for better understanding. There are number of tires provided for inspection for testing. The question is to find how many are blemished tires? (Figure 15.5).

Notes: A blemish is the presence of any of six different conditions (coded as D, B, P, G, R, and F). It is possible for the sidewall to have more than one of these conditions.

The specification for above figure is "No blemishes on the tire sidewall." Since the process output has a stated specification of a quality characteristic, "no blemishes," it can be evaluated for defects (and defectives). The tire is the unit of interest in this example; each tire is one unit of output. The "nonserial side" of each tire is carefully examined for the presence of six different blemishes, coded as D, B, P, G, R, and F. Once again consider the same example for blemishes on tires and blemishes on the tires but with defects (Figure 15.6).

To perform the first step in the baseline statistics, the defective and defect level are to be counted by performing the following questionnaire:

- Defective level
  - What is the unit? How many units are there?
  - What is a defective? How many defectives? What is the DPMO?
- Defect level
  - What is the defect? How many defects? How many defects per unit (dpu)?
  - What is the opportunity? How many opportunities per unit?
  - How many DPMO?

Table 15.1 Test Request Form

Requestor
Development code
Blocking condition
Performance standard
Gradient
Functional tests

Counting the units with defects (3) shows three *defectives* in the five units, or 60% defective. This can also be expressed as 600,000 DPMO. Counting the *defects* (7) gives seven defects for five units. But there is more than one "opportunity for defect" for each unit. In this case, there are six opportunities per unit. There are five units (NSS sides of five tires), so there are 30 opportunities for defects. Seven defects were observed, so there are 7/30, or 0.233 defects per opportunity (DPO). Expressed in terms of "DPMO" this is 233,000 DPMO. The description of a process in which the inputs and outputs are predominantly data or information, rather than a physical product is called transactional process. Order entry and claims processing are examples of transactional processes. In the above example, the process output is a test request. Six quality characteristics have been identified and the target is defined—no errors in critical fields. The test request is the "unit of interest." Each request is checked for conformance to the six different information requirements (Table 15.1).

An error is defined as missing or incorrect information corresponding to any of the six different information fields entered in test requests (coded as D, B, P, G, R, and F). It is possible for one request to have more than one of these "defects." Five test requests are reviewed (in the actual project, the number was over 200, but let us keep it simple for now).

We now have process output samples that have been graded for "defective" or "not-defective." But the output attribute (little y) is unknown and also the defect is unknown. All that is known is that the discrete attribute grading was done with a validated measurement system. The "defective" units were "taped" at the time that *lot* samples were graded and the *unit* is either the production lot or a pallet of tires representing that lot. Thus, there are 20 units in the process output sample, and three of them are clearly indicated as defective. Seventeen of the units are not defective, so there is an estimate of 0.85 proportion-conforming, or 85% yield.

Process output pass/fail rates can be expressed in many ways. The basic measurement is the observed raw count. Here are some other pass/fail expressions and their units, along with corresponding values for our example (Table 15.2).

Process "yield" can be a useful metric, depending on its basis and definition. It is an aggregate measure of pass rate. The complement to process yield is process defective or defect rate. You may often see these rates expressed as ppm or PPM. PPM is identical to DPMO when there is one opportunity per unit.

The spreadsheet given as follows shows the cumulative probability value as a function of the Z-value or Z-score. This spreadsheet table provides a quick way to look up the approximate value of Z-score for a given process output. The required information needed, to look up a Z-score, is an observed or estimated process yield (proportion conforming). Conversely, one can also use the complement of process yield, that is, the proportion of nonconforming process output (Figures 15.7 and 15.8).

Table 15.2 Discrete Process Output Metrics

	Pass	Fail
Raw counts	17	3
Percentages	85	15
Proportions	0.85	0.15
Fractions	17/20	3/20
Ratios	17:3	5.7:1
Parts per million	850,000	150,000

Proportion conforming	Rt. tail	Percentage out	PPM	Z-score
0.50	0.50	50%	500,000	0.000
0.55	0.45	45%	450,000	0.126
0.60	0.40	40%	400,000	0.253
0.65	0.35	35%	350,000	0.385
0.70	0.30	30%	300,000	0.524
0.75	0.25	25%	250,000	0.674
0.80	0.20	20%	200,000	0.842
0.85	0.15	15%	150,000	1.036
0.90	0.10	10%	100,000	1.282
0.91	0.09	9%	90,000	1.341
0.92	0.08	8%	80,000	1.405
0.93	0.07	7%	70,000	1.476
0.94	0.06	6%	60,000	1.555
0.95	0.05	5%	50,000	1.645
0.96	0.04	4%	40,000	1.751
0.97	0.03	3%	30,000	1.881
0.98	0.03	3%	25,000	1.960
0.98	0.02	2%	20,000	2.054
0.99	0.01	1%	10,000	2.326

Figure 15.7 Converting yield, % defective, DPMO, etc., Z-score.

As the Six Sigma culture grows within the business organization, it will become more common to hear process performance referred to in terms of Z-score. The spreadsheet table can be used to quickly convert to the process yield, PPM, etc. given a stated Z-score. For example, the process yield corresponding to a Z-score of 3.0 is about 99.83 (%) by visual interpolation. The process defective rate corresponding to a Z-score of 3.0 is approximately 1700 (ppm), again by visual interpolation (Table 15.3).

The process performance can be evaluated using several approaches. In the first approach, we assess performance using lot as the unit of process output. Further, we will measure the process output using a discrete quality variable—defective status—this output variable has two possible values, "defective" or "not-defective." Above are the tabulated observations of discrete process output values for the 20 lots. The table displays output values of "yes" and "no." For our analysis, we will usually code this type of discrete values (i.e., textual values) in a Minitab worksheet with numeric values, such as "1" and "0" in this case.

Proportion conforming	Rt. tail	Percentage out	PPM	Z-score
0.991	0.009	0.90%	9,000	2.366
0.992	0.008	0.80%	8,000	2.409
0.993	0.007	0.70%	7,000	2.457
0.994	0.006	0.60%	6,000	2.512
0.995	0.005	0.50%	5,000	2.576
0.996	0.004	0.40%	4,000	2.652
0.997	0.003	0.30%	3,000	2.748
0.998	0.002	0.20%	2,000	2.878
0.999	0.001	0.10%	1,000	3.090
0.9991	0.0009	0.09%	900	3.121
0.9992	0.0008	0.08%	800	3.156
0.9993	0.0007	0.07%	700	3.195
0.9994	0.0006	0.06%	600	3.239
0.9995	0.0005	0.05%	500	3.290
0.9996	0.0004	0.04%	400	3.353
0.9997	0.0003	0.03%	300	3.432
0.9998	0.0002	0.02%	200	3.540
0.9999	1E-04	0.01%	100	3.719
0.9999966	3.4E-06	0.0003%	3	4.508

Figure 15.8 Converting Z-score to yield.

Table 15.3 Overall Performance by Defective Lots

Lot	Defective
1	Yes
2	No
3	No
4	No
5	No
6	Yes
7	No
8	Yes
9	No
10	Yes
11	No
12	No
13	No
14	Yes
15	No
16	No
17	No
18	No
19	No
20	Yes

15.2.2.1   *Process capability for binomial process outputs*

Binomial data are usually associated with recording the *number of defective* items out of the *total number* of items sampled. For example, if you are a manufacturer, you might have a go/no-go gauge that determines whether an item is defective or not. You could then record the number of items that were failed by the gauge and the total number of items inspected. Or, you could record the number of people who call in sick on a particular day, and the number of people scheduled to work that day. These examples could be modeled by a binomial distribution if the following conditions are met:

- Each item is the result of identical conditions.
- Each item can result in one of two possible outcomes ("success/failure," "go/no-go," etc.).
- The probability of a success (or failure) is constant for each item.
- The outcomes of the items are independent of each other.

15.2.2.2   *Process capability for Poisson process output*

The *Poisson* distribution characterizes data for which you can only count the nonconformities that exist—it is impossible to count nonconformities that do not exist. The Poisson distribution characterizes defects data, which are nonconformities that affect part of a product or service but that do not render the product or service unusable.

- The data are counts of discrete events (defects) that occur within a finite area of opportunity.
- The defects occur independently of each other.
- There is an equal opportunity for the occurrence of defects.
- The defects occur rarely (when compared to what could be).

Now for one set of data three different values of overall performance (Z-score) are available and the question which metric is correct.

- Defectives, with lot as the unit: $Z_{LT} = 0.5$
- Defectives, with tire as the unit: $Z_{LT} = 2.3$
- Defects, with six opportunities per tire: $Z_{LT} = 2.8$

The correct metric depends on the nature of the project and whichever approach chosen, stick with it throughout the project.

- Defectives/Lot:
  - (+) Historic data generally more available
  - (−) Cannot improve process via defectives if multiple defects
- Defects/Tire:
  - (−) Historic data generally less available
  - (+) Can better identify source(s) of defect, to target solution(s)

Hence, with the previous illustrated example the overall process performance for discrete outputs can be conducted by the following steps:

- Choose appropriate test measurement and level of detail, that is, defectives, total defects, or specific defect(s)
- Choose unit to measure, for example, a tire, a pallet of tires, an order of tires, a service order, a service hour, a service form
- Decide which segments, groups, types, regions, etc., to include in the study

- Conduct the study; collect the data
  - Important: If you will collect data specifically for the 6s project, gather data for as many process-related elements as practically/economically reasonable
- Analyze process data
  - Use either Binomial (pass/fail counts) or Poisson (rates) capability analysis
- Report the DPMO and the Z-score, with their confidence limits

15.2.3   *Continuous outputs*

Continuous outputs are within a given interval, any real value possible (at least in theory) and some of the examples are temperature, time, mass, pressure, viscosity, etc. Continuous data sets provide descriptive statistics, sample mean and sample standard deviation (Figure 15.9).

The typical plot for continuous data is given in Figure 15.10.

The metrics for continuous data is also established by means of Z-scores. Z-scores are the process performance metrics to report for both overall process performance (Z_{LT}) and potential process capability (Z_{ST}). The subscript LT is for long-term project and ST is for short-term project. The computation of the overall process performance metric, Z_{LT}, is simple algebra. It is a mere substitution of the proper specification limit, and the long-term process mean and standard deviation.

Calculation of one-sided Z-scores: Figure 15.11 shows the one-sided Z-score calculation.

The practice of adding 1.5 to the overall process performance (Z_{LT}) is completely arbitrary. The assumption is that process performance will have smaller variance over "short-term" than overall (long-term). With a smaller magnitude for S_{ST} than S_{LT}, more standard deviations can now fit between the process mean and specification limit. In fact, the popular "add 1.5 units to Z_{LT} to estimate Z_{ST}" rule of thumb is a special case! Here's one scenario when this would be a good estimation method:

- Measured $Z_{LT} = 4.5$
- Measured $Z_{ST} = 6.0$
  - Or (independent of mean and specs)
- $S_{ST} = 0.75\, S_{LT}$ (approximately) and measured $S_{LT}^2 \sim 2S_{ST}^2$

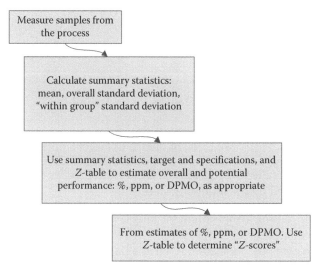

Figure 15.9 Baseline statistics for continuous data.

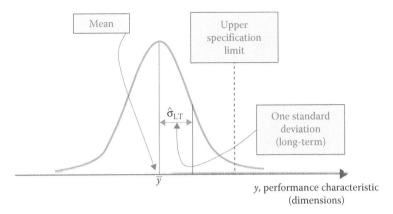

Figure 15.10 Continuous data.

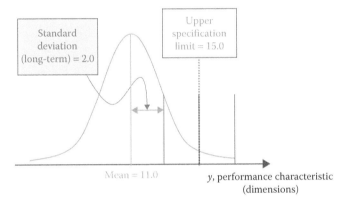

Figure 15.11 One-sided Z-score.

Calculation of two-sided Z-score: Figure 15.12 shows the calculation for two-sided Z-score.

Here is the extension to the initial, one-sided specification process output example, as promised earlier. Conceptually, we add the individual probabilities of nonconforming process output (from the two tails) into a total nonconforming proportion.

The mechanics used to compute the overall process performance metric (Z_{LT} or Z_{bench}), for processes with two-sided specifications, are

- Compute the predicted probability of the nonconforming output in the lower tail
- Compute the predicted probability of the nonconforming output in the upper tail
- Sum the individual probabilities for a total predicted probability of nonconformance
- Convert the total probability to a Z-score

15.2.3.1 *Process capability analysis for normal process outputs*

It is known that process output will not be normally distributed. Regarding the nonnormality of the data, there are methods that can be used to normalize certain outputs, for example, Box-Cox transformation. However, no ordinary methods are available to normalize bimodal data. The bimodal process output suggests that different sources of variation exist within the process. If the previous methods to assess process stability such as process

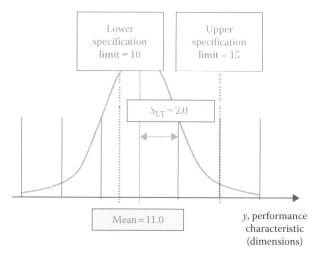

Figure 15.12 Calculation of two-sided Z-score.

behavior charts are used, the output will be unstable and nonnormal. At this point, the initial process capability analysis can be considered as a gross estimate of overall performance, only. The baseline report is sometimes called the process scorecard. The actual overall process performance is reported, with the Z-score (Z_{LT} or Z_{bench}) and the DPMO both shown. The above section provides the computations of only the process capability. The next section focuses on "short-term" performance, that is, potential process capability.

15.2.4   Potential process capability

Overall process performance metric, or "long-term" performance, includes variance across all observed output. Overall data can be stratified, or grouped, into shorter time periods or other rational subgroups. Calculation based on the performance within subgroups will give us an estimate of the potential capability of the process. "Within-group" variation also referred to as "short-term" variation. Estimates of short-term variation are critical in determining baseline potential process capability, rational sub grouping (Figures 15.13 through 15.16).

In the estimation of potential process capability, there is a need to hypothesize a "rational subgroup" by which to stratify, or group, the overall process output results. There are three general classes of rational subgrouping, or variation, though any rational difference/factor can be used:

- Positional (location): Variation within a single unit, or across a unit with many parts, for example, variation from district-to-district, machine-to-machine, operator-to-operator, etc.
- Sequential: Variation among consecutive pieces, batches, groups, etc. For example, variation from step-to-step.
- Temporal (time): Variation within short time spans, for example, shift-to-shift, week-to-week, etc.

Each subgroup contains multiple elements, such that an estimation of within-(sub)group variation is possible.

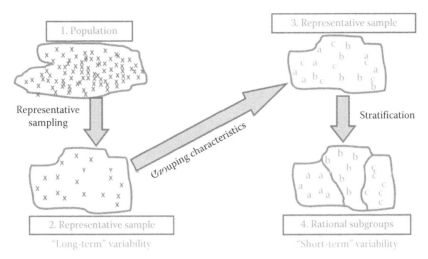

Figure 15.13 Rational subgrouping.

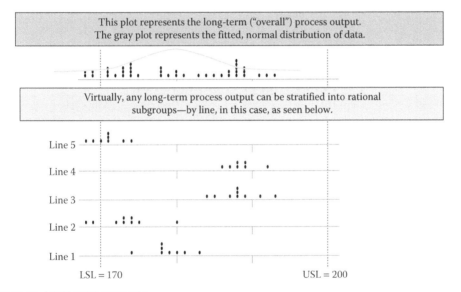

Figure 15.14 Long-term variation.

15.2.4.1 Short-term variation

- The purpose of obtaining an estimate of "short-term" variation is to get an estimate of the capability of the process.
- The total—or overall—variation includes all sources for variability.
- Based on your knowledge of the process, identify various ways to stratify or "slice" the overall data.
- As the description "short-term" implies, one way is to look at shorter slices of time.
- But time is not the only way to group the data, as shown in the following example.

The blue distribution represents the result of merely shifting the mean outputs from all subgroups (the five individual lines) to the overall mean. This outcome does not

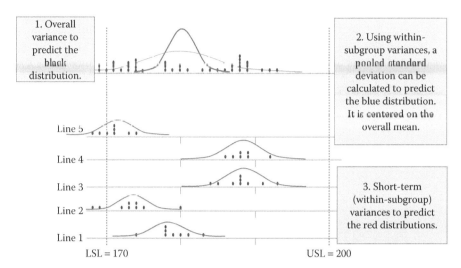

Figure 15.15 Determination of short-term variation.

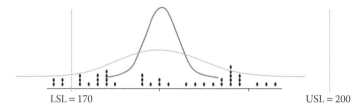

Figure 15.16 Short-term (within subgroup) variation.

require reducing the lines' individual variances. But there is even greater opportunity for improvement than shown in this diagram, if the means can be shifted to the process target. The difference between the Z_{LT} (overall process performance) and Z_{ST} (potential process capability) is called Z_{shift}.

$$Z_{shift} = Z_{ST} - Z_{LT} \tag{15.1}$$

"Long-term" data are segmented by some rational factor to compute the "short-term" variability. An estimate of "short-term" standard deviation for a three subgroup variation is given by

$$\sigma_{ST} = \sqrt{\frac{\sigma_{ST1}^2 + \sigma_{ST2}^2 + \sigma_{ST1}^2}{3}} \tag{15.2}$$

The short-term (or within-group) standard deviation is the basis for the estimate of potential process capability. This simple formula for the short-term standard deviation holds true when subgroup sizes are equal. When not equal, a more general formula for "pooled standard deviation" is used. The pooled standard deviation is a weighted-average of the subgroups' standard deviations and is given by

$$S_p^2 = \frac{(n_1 - 1)s_1^2 + (n_2 - 1)s_2^2 + (n_3 - 1)s_3^2}{(n_1 - 1) + (n_2 - 1) + (n_3 - 1)} \tag{15.3}$$

Based on the above insight tools are available to analyze the data. Those tools are categorized and explained based on different belts as follows.

15.3   Yellow belt

15.3.1   Cause-and-effects diagram (fishbone diagram)

Cause-and-effect or fishbone diagrams are an effective team-based tool which can be used to determine the potential root cause of a problem.

A cause-and-effect diagram helps in the following ways:

- Breaking the problem down into many bits
- Display of all possible causes in a graphical manner
- May be also called fishbone, 4M, or Ishikawa diagram
- Shows the interaction of various causes
- Follows the idea of brainstorming when ideas are to be generated.

A fishbone session may be divided into three sessions:

1. Prioritizing
2. Brainstorming
3. Action plans development

The problem statement is identified and brainstorming for the categories is done. The next step would be the prioritization of the problem causes, polling is often used. The three most probably causes are encircled from the action plan development. The 4M (manpower, machine, method, and machine) version of the fishbone diagram is usually sufficient. The expanded version must be occasionally used. In a lab environment, measurement is the key. During the discussion of brown grass in the lawn environment is important. A 5M and E schematic is shown in Figure 15.17.

Given below is an example of a cause-and-effect diagram. It shows the 5M and E concept of determining and solving the causes for a quality problem (Figure 15.18).

For additional examples of cause-and-effect or Ishikawa diagram refer to *Guide to Quality Control* by K. Ishikawa (1982).

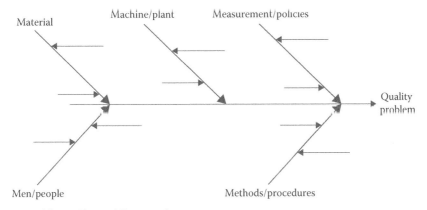

Figure 15.17 Fishbone 5M and E example.

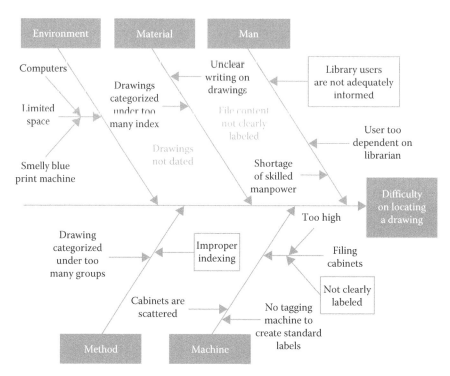

Figure 15.18 Actual fishbone example.

15.3.2   Fault tree analysis

Fault tree analysis (FTA) is a method which systematically defines an event that is unique, undesirable, and determines all possible reasons or failure which could cause the event to occur. This event might constitute the top event in a fault tree diagram and generally is used to represent the complete failure of the product. In comparison to failure mode and effects analysis (FMEA), FTA is a faster and easier method of analysis because its focus is on a selected group of all possible system failures, especially those that could cause a catastrophic event. FMEA works in a sequential manner through all events that could cause possible system failures, regardless of how severe they could be.

On proper application, FTA becomes a very useful tool during initial product design phase and also an evaluation tool for performing preliminary design modifications. Several potential uses of FTA can be as follows (Nicholls 2005):

- Performing the functional analysis of highly complex systems
- Performing the evaluation of subsystem events on top event
- Performing the evaluation of safety requirements and specifications
- System reliability evaluation
- Potential design and safety hazards identification
- Simplification of troubleshooting and maintenance
- Logically removing the causes of an observed failure

FTA over FMEA when

- Safety of public, operation and maintenance is the first priority.
- Identification of a small number of differentiated events can be done.
- Critical importance to the completion of functional profile is given.
- Errors by humans and software provide a high potential of failure.
- Quantified risk evaluation is the primary concern.
- Highly complex or highly interconnected product functionality
- Once initiated product is not repairable.

FMEA over FTA when

- The top events are limited to a small number.
- There is a feasibility for multiple potentially successful functional profiles.
- Determining the possible failure modes is important.
- Software intervention or scant human involvement in product functionality

15.3.2.1   Fault tree symbols

FTA employs the logic gates concept for the determination of reliability of a system. Assessing potential system failure modes can also be done using FTA. There are many FTA symbols which are broken down into two main categories, event and gate symbols. A few examples of the various FTA symbols are shown as follows:

- Basic event: The lowest level of fault that one would wish to study. Used as an input to logic gate (Figure 15.19).
- Fault event: It contains a description of the lower level fault (Figure 15.20).
- Initiator event: An external event that is used to initiate the process (Figure 15.21).
- "OR" gate—The output occurs only when one input event occurs (Figure 15.22).
- "AND" gate—The output event occurs only both input events occur (Figure 15.23).

Figure 15.19 Basic event.

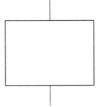

Figure 15.20 Fault event.

Figure 15.21 Initiator event.

Figure 15.22 "OR" gate.

Figure 15.23 "AND" gate.

15.3.3 Root cause analysis

The responsibility of root cause determination is usually given to a team or an individual to fix the deficit or to correct it. Some problems might be very complex and difficult to solve. In other cases, the solution might be known, but time might be required to solve it. The table that follows shows the proposed actions that can be taken to solve the problem (Table 15.4).

Table 15.4 Corrective Actions—Short and Long Term

Situation	Immediate action	Intermediate action	Root cause action
The dam leaks	Plugging the dam	Patching the dam	Determining the cause of leak so it does not repeat again.
Oversized parts	Critical inspection	Installing an oversized kick out device in line	Analysis of the process and taking the actions to eliminate the production of oversized parts.

15.3.3.1   Subjective tools

- Ask why two times
- Brainstorming
- Analysis if the process flow
- Problem solving systematically
- PDCA
- Nominal group technique
- Ishikawa diagrams
- Six thinking hats
- FMEA or FTA
- Employing of teams

15.3.3.2   Analytical tools

- Data collection
- Pareto charts
- Regression analysis
- Check sheets
- Process capability analysis
- Portioning of variation
- Subgrouping of data
- Simple trials
- Analytical tests or hypothesis tests
- Control charts

When a permanent corrective action is determined, the management must determine if the root cause analysis has been performed to its full extent, the corrective action is satisfactory to remove all the failures and if the corrective action is reliable and maintainable.

15.3.3.3   5 Whys

The 5 Whys approach to determining the root cause analysis may also be described as asking the question "why?" five times. It is generally attributed to a Japanese method to root cause analysis. The following is an example of 5 Whys.

Symptoms: the customer shipment was not delivered on time.

1. Why?—Die stamping press broke down resulting in running out of parts
2. Why?—Lack of scheduled maintenance of the press for a period of three months
3. Why?—Reduction of maintenance department staff from 6 to 8.
4. Why?—Maintenance department budget shot up due to overtime costs and the GM required a reduction in costs of overtime for all overhead support departments.
5. Why?—Removal of unnecessary spending by the CEO because company was not reaching profit goals. So the root cause was the CEO being worried about getting fired for poor profit performance.

There is nothing magical about 5 whys. In fact the root cause may be determined during the 3rd or 4th why itself. In other cases one may need to go beyond 5 whys to determine the root cause.

15.3.3.4   5 Ws and H

The 5 Ws and H approach to root cause analysis is described by asking the questions who? what? when? why? where? and how? The 5 Ws and H is an old method used by reporters of newspapers in asking questions to get the full story. In the quality context, responses to these questions can be organized into a fishbone diagram or a cause-and-effect diagram. In some cases, the same basic method is simply referred to as 5 Ws. Note that the order of Ws varies, depending upon the problem. The technique looks at the problem from more than one viewpoint in order to include information as much as possible that might be needed to determine the problem.

15.4   Green belt

15.4.1   Entitlement

It helps in finding out how a process could be improved further from results through previously discussed methods. Finding Entitlement by three approaches for current process performance is the key. Entitlement is defined as *the best that the existing process can perform, often intermittently, and for only short periods of time.*

This definition implies that some empirical evidence of process output exists for the current process. The intermittent nature of entitlement usually indicates that certain process output subgroups are produced by certain process conditions or states that do not exist continuously, throughout the entire process system. When we have empirical measurements for rational subgroups, we can test for differences in performance. These could be differences in output mean, as well as output variance. The tests are called hypothesis tests.

We can also look externally or internally for other empirical evidence of entitlement for the existing process. Internal benchmarking looks for data from within the organization, for instance, an operation with the same process. External benchmarking would look at other companies using the same process. Benchmarking can also identify "best-in-class," the best process performance to produce the same process output. Finding out the design intent to determine theoretical potential is the third approach. We look into understanding hypothesis testing by looking at the types of errors.

15.4.1.1   Types of errors

Errors are to be considered before making conclusions from results of tests. There are two types of possible errors:

- Type I: This type of error occurs when null hypothesis is rejected though it is true. Producer's risk is defined this way and the probability of this risk is termed by a variable alpha (α) and $0 < \alpha < 1$.
- Type II: This type of error occurs when null hypothesis is not rejected when it is false. Consumer's error is defined this way and the probability of this risk is termed by a variable (β) and $0 < \beta < 1$.

The level of risk that the recipient is willing to undertake is decided upon and is used to arrive at results. Of course, smaller values of α and β, that is, smaller risk is desirable. However, it is observed that if you reduce α, β increases. We must find a trade–off region to work with. Also, if we deal with large data points the risk reduces considerably.

To understand risks better, let us consider this simple situation. We make (and accept) decisions based upon incomplete knowledge all the time. Examples of such decisions with incomplete knowledge:

- Judge or jury verdict in a legal proceeding
- Referee's ruling in a professional sports game
- Hypothesis testing of sample means or other statistics

All decisions based upon sample observations risk reaching an incorrect conclusion even if we had perfect measurements and used perfect judgment due to finite sample size (a form of incomplete knowledge). The only decisions made without risk are those in which complete knowledge is available, assuming perfect judgment is exercised. We accept that no person has complete knowledge from a finite sample from a large population.

Hypothesis testing is our interest in decision making. Decisions are almost always made with respect to a null hypothesis (H_0) of "no difference." The alternative hypothesis (H_a) can only be supported with sufficient evidence to overturn (i.e., reject) the assumption of the null. In baseball, the umpire's assumption is that the base runners are "no different than 'safe.'" This is the null hypothesis of baseball. Only when there is clear evidence to reject the null hypothesis, can the umpire rule that the runner is "not safe." Unfortunately, the umpire does not have complete knowledge. For instance, he can only view the play from one point or perspective, and only at "full speed." As instant video replays often show (from different angles, or in slow motion), the umpire's decisions are not always perfect.

Alpha (α) risk is the probability that a sample mean is found by chance to be in the tail of the distribution, while the sample distribution is, in fact, the same as the true population distribution. This could cause a type I error, that is, rejecting the null hypothesis when it is actually true. Again, alpha risk is the risk of thinking that the sample is from a population that is units away from the reference distribution when, in fact, it is actually from the reference distribution. Delta is the minimum difference that the experimenter wants to detect. The possibility of this being the true difference between the means and still going undetected is the beta risk. One way to increase our chance of detecting the difference is to accept a greater risk.

Critical difference (δ): Minimum practical difference you need to detect based on financial or technical consequence, to make a good decision. Type II error (β risk) is the probability of missing a difference of size δ.

$$\delta\,(\text{delta}) \quad \text{The difference between } m_0 \text{ and } m_1 \text{ that we desire to detect.}$$

The experimenter needs to decide how small a difference (δ) he needs to detect, before the testing or experiment since after the test, you already know whether or not you found a significant effect! With enough samples, even very small differences can be detected. The experimenter should have an idea of the minimum size of the difference he would like to detect in the experiment. The probability of there being a difference of this size and of not noticing the difference in the hypothesis test is beta. The "power" of a statistical test is the test's probability of detecting a critical difference of size δ (Figure 15.24).

This graphical presentation of the data suggests that there is a difference between the two lines' output performance. Six Sigma principles require, however, that quantitative evidence be evaluated by quantitative methods (statistics!) in order to make that decision. The difference in the lines' outputs for this example might be intuitively obvious to you, but what if the difference is not so (visibly) apparent? (Figure 15.25).

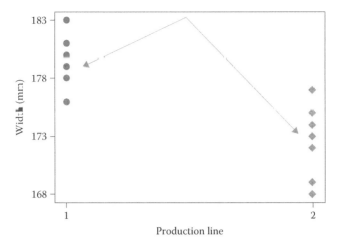

Figure 15.24 Example 1.

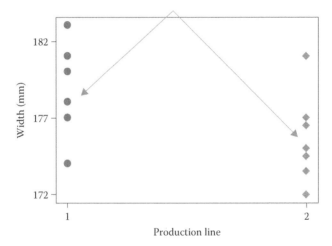

Figure 15.25 Example 2.

This illustrates the difficulty in using visual, intuitive weighting to make decisions. What if there were more observations, what if there were fewer, how large does the difference really need to be, etc. There are two mistakes that can be made in answering the question as to whether the lines being compared are really different:

- You decide that there is a difference between the two lines, but there really is no difference.
- You decide that there is no difference between the two lines, but there really is a difference.

In general, the same two mistakes can be made in any comparative decision of this sort (i.e., a binomial equal/not-equal, greater/not-greater, yes/no, etc. decision). We will revisit this idea in the next section of this module.

15.4.2   *Hypothesis testing*

Hypothesis testing is the analysis based on the observed data points to reach a conjecture in the decision-making process. Here an assumption is made on a statistical parameter and this is asserted to be true by comparing the results with the data. A statistical procedure to *decide* whether there are *differences* among data sets being *compared*, with a predetermined level of confidence. Rejection of a hypothesis means that there is a small probability of obtaining the sample, that is, refuting the theory based on information observed, when in fact, the hypothesis is true (Myers). It is relevant to the topics discussed in the following pages. But before we start a discussion on the application of this technique, we must understand terminology in terms of statistics.

Hypothesis testing is analogous to a legal trial. In the US legal system, a person begins the trial presumed innocent (really, presumed "not guilty"). That is the null hypothesis (H_0)! The prosecutor accuses the defendant of being "guilty," the alternative hypothesis (H_a). Similarly, we always judge with respect to the null of "not guilty": reject the null (found "guilty") or cannot reject the null (found "not guilty").

There are risks of two incorrect verdicts present in every trial. First, there is the risk of convicting an innocent man (risk of *rejecting* a null hypothesis that describes the *truth*). The second is the risk of letting a guilty man go free (risk of *accepting* the null hypothesis that is *false*). In the US court example, the risk or probability of convicting an innocent person (a, or alpha) is deemed of critical concern. This system is based on the belief it is much more evil to convict an innocent man than it is to let a guilty man go free. To minimize the risk of convicting the innocent, very strong evidence is required to decide "guilty." To put it in hypothesis testing terms, we want to be as close as possible to 99.99% certain that we will not convict the innocent.

15.4.2.1   *US legal system hypothesis tests*

H_0: Defendant is not guilty (presumed "innocent"; No different than any other)
H_a: Defendant is guilty (evidence beyond reasonable doubt to reject innocence)

The two risks on incorrect verdict are β which is setting a guilty person free (failing to reject H_0 when H_a is true) and α is convicting an innocent person (i.e., rejecting H_0 when it is true).

In technical practice, there has been similar aversion to falsely declaring "difference found" (thus small a), but it has often been at expense of (large, unknown) b!

15.4.2.2   *Why and when are hypothesis tests used?*

Hypothesis tests are decision aids. They do not make the decisions, but they aid in making good decisions in uncertain situations. Say that you produce a product on a number of lines, and the overall performance has not been good. You calculated the mean and standard deviation of the output its conformance (or nonconformance) to the customer specifications. A process improvement project is begun on one of the lines, and management wants to know whether the new process is indeed better than the old (upgrading the other lines would be a substantial capital cost).

Hypothesis testing would be used to decide whether the performance of the "improved" line is statistically different (i.e., better) than the old way of doing things, given the risk (uncertainty) the business is willing to take.

Remember: Six Sigma uses data-based decision making. The hypothesis test is the statistical test to determine whether there are real differences between data sets.

Hypothesis tests always compare two mutually exclusive (contradictory) hypotheses:

15.4.2.3 Null hypothesis

The null hypothesis (II_0) says, "There is no difference between the two sets. If you think there is, you've got to prove it." It is the formal statement against which experimental evidence is tallied. Its source is the overall problem that it addresses. A null hypothesis, II_0, can be rejected or failed to accept. The statement that a hypothesis is accepted is never used as it is ambiguous and against the belief that there is not enough evidence to support the statement due to unavoidable circumstances. With enough data if it is decided that we have failed to reject a null hypothesis statement. The null hypothesis (denoted H_0) is referred to as "H-naught." Null hypotheses are almost always set up to be "straw men." We generally hope or expect to disprove or "knock over" the null hypothesis.

15.4.2.4 Alternate hypothesis

The alternative hypothesis (H_a) says, "There is a difference between the two sets of data." The alternative hypothesis (denoted as H_a, but occasionally as H_1) is sometimes referred to as "the research hypothesis." The alternative hypothesis describes the situation that we hope or expect to "prove" or demonstrate as "the truth" (Figure 15.26).

For the *mean width—production line* example, given earlier:

- Null (H_0): There is no difference between average widths (i.e., the averages are equal: $avg_1 = Avg_2$).
- Alternative (H_a): There is a difference in average widths (i.e., the averages are not equal: $avg_1 \neq Avg_2$).

> "If p is low, the null must go."

This little ditty is useful to some people in remembering which hypothesis to accept or reject.

15.4.2.5 Translation

"If the p-value is less than our chosen (a) value for the significance level (typically, a = 0.05) then one *can reject the null* hypothesis. Therefore, there is a 'statistically significant' difference between samples."

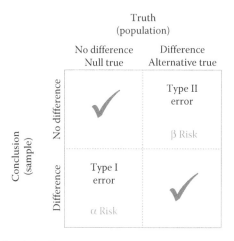

Figure 15.26 General hypothesis testing.

"If the *p*-value is greater than our chosen value for the significance level (typically, a = 0.05) then one *cannot reject the null* hypothesis."

Other forms of the alternative hypothesis are called as one-sided tests. We use them when we want to test whether one sample mean is greater than (or less than) another, for example:

- Are region one's sales less than region two's?
- Is the chemical process yield improved with the new catalyst?

In these situations, the alternative hypotheses are no longer "not equal to, ≠," but "less than, <" or "greater than, >"

- H_a: region one < region two
- H_a: yield new > yield old

The alternative hypothesis says, "There is a difference," and it could be one of the following:

- Samples are *not equal* (we used this H_a in example, above)
- Sample one is *greater than* sample two
- Sample one is *less than* sample two

In hypothesis testing, we "stack the deck" in favor of the null hypothesis. Evidence is required to reject the null hypothesis, that is, to conclude that there really is a difference.

It brings us to a discussion on one-sided and two-sided tests.

- If the value of risk is placed at one side of the test then it is termed as a "one-sided" test. It is also termed as one-tailed test.
- If the population has shifted on either direction then a two sided test. It is also termed as "two-tailed" test.

The "equal/not equal" test is called a "two-sided" test, because it does not matter whether the difference is greater than zero or less. The "greater than" or "less than" tests are called "one-sided" tests. It should be noted that the null hypothesis almost always remains the same. It states, "There is no difference" between the groups being compared for the statistic of interest. We could get "tricky" with the null hypothesis and use a nonzero test mean to test whether the difference in means was some specified (nonzero) value. For example,

- H_0: $(\text{avg}_1 - \text{avg}_2) = 3$
- H_a: $(\text{avg}_1 - \text{avg}_2) \neq 3$

The major difference in the report of one-sided test results is in the statement of the 95% confidence interval for the means' difference. In the one-sided test, the 95% CI is a single lower bound on the difference. Let us consider an example, the lower CI bound is +0.10. We can be 95% confident that the difference between means is no less than +0.10. The best estimate for the means' difference is +3.19, and the minimum estimate for the difference is also positive (+0.10); since the confidence interval of means' difference does not include the value of 0.00, we conclude that there is a difference in the direction indicated by the alternative hypothesis (i.e., ">").

In other words, any test has risk associated with it. This risk (α) determines the level of confidence which is $(1-\alpha)$. Thus, it determines the critical value of test statistic. The two-sided test uses the alternative hypothesis that states that the two sets of data are *not equal*. In the one-sided test, the alternative hypothesis is either *greater than* or *less than*.

The examples above required comparing two samples against one another (i.e., "2-sample t-tests"). However, some situations require comparing a single sample against a *"known"* (or reference) value. Thus, in *1-sample* t-tests we are not comparing two samples but we compare a single sample against the reference. One-sample tests can be one-sided tests (equal vs. greater-than, or equal vs. less-than) as well as two-sided tests (equal vs. not equal).

Examples are as follows:

- Comparing a sample against a historical mean
- Comparing a sample against an industry standard
- Comparing a sample against a claim or goal

The one-sample t-test compares sample results against a known historical mean or a standard of some kind. If we have a specific goal or claim to compare a sample against, we also use the one-sided test.

Now we try an exercise: Accounts receivable (A/R) Six Sigma project.

Exercise 1: Process output (A/R as percent of sales) for the 12-month baseline period averaged 24%. Benchmarking resources indicate that "best-in-class" performance (A/R as percent of sales) in similar businesses is no greater than 18%. The region manager agrees that receivables are not below 18%, but says that, considering month-to-month fluctuations, the results are not "significantly above 18%" and "the trend shows that we should be fewer than 18% soon."

- What one-sample test could you perform to test the manager's hypothesis?
- How could you test the manager's claim about the "trend"?
- What are your conclusions?

Solution: Always consider this question: What are you trying to prove with your hypothesis test? In STEP 5, we must determine the entitlement for the existing process. In this particular example, we are trying to develop statistical evidence as to whether the baseline A/R process output (accounts receivable, as percent of sales) is above the industry standard. If the "industry standard" process and the baseline process are comparable, and the claim of "industry standard" is accurate, we may have achieved entitlement performance already. Use the one-sided, one-sample t-test.

Exercise 2: (Continuation) Process results (A/R as percent of sales) during the six months after implementation of a process change (trial) showed reduction of A/R from 24% to 20.5%. If the difference is real, the payback would be good; however, the proposed process changes would also be costly to implement. Review the data. What are your conclusions and recommendation(s)?

Solution: Again, think about what it is you are trying to prove. In this example, we are trying to develop statistical proof that the new method causes the accounts receivables to be lower than the current method. Use the two-sample t-test. The alternative hypothesis should be greater than current method.

The value of β is large if $\mu = \mu_o$ and small if μ is largely skewed from μ_o. To construct a power curve $1-\beta$ is plotted against alternatives of μ [Certified Six Sigma Black Belt (CSBB)].

There is gain in power if α is low and also when sample size increases. $1-\beta$ is the probability of rejecting the null hypothesis given the null hypothesis is false. It is evident that sample size is dependent on risks, variance, and minimum value to be measured ($\mu-\mu_o$).

15.4.3 Chi-squared test

The goodness-of-fit test using the chi-square distribution is used as a technique for modeling and analyzing the $y = f(x)$ relationship when both the y and the $x(s)$ are discrete. The test determines whether two discrete variables are associated, that is, whether the distribution of observations for one variable differs, depending on the category of the second variable. The data are usually displayed in tabular form and the results are often called cross-tabulations. The statistical conclusions drawn from the p-values are the same as in the other analytical methods. We always attempt is characterize process output characteristics with continuous data, because of their greater power.

There are often situations in which all data process data are discrete, and this test is appropriate in those cases. As part of the retail sales process, the sales representatives are instructed to obtain certain buyer information. The information can be used to provide follow-up information and reminders to the buyer. The BB project is focused on improving the compliance to this instruction. The BB pulls records from the five sales regions. While the processes are the same, sample size from the regions varies. It is desirable to predict whether the number of faults (an order missing the needed information), occur at a similar rate across all five regions. Since both input (region) and output (fault/no fault) are discrete, the chi-square test is appropriate. Since we know the total number of opportunities and the number of faults, we can easily calculate the expected number of non-faults.

The expected number of non-faults is simply

Number of opportunities minus expected number of faults.

The expected number of nonfaults can also be calculated based upon the expected proportion of nonfaults in the same manner by which we previously calculated the expected number of faults.

$$\text{Six Sigma} = \left(f_o - f_e^2\right)/f_e. \tag{15.4}$$

From the Six Sigma information, chi-squared is easily calculated:

$$\chi^2 = SS_{\text{faults}} + SS_{\text{non-faults}}. \tag{15.5}$$

15.4.4 Analysis of variance

Analysis of variance (ANOVA) is a statistical procedure for testing whether there is a significant difference among *means* of multiple subsets of data. The one-way ANOVA is an extension of t-test.

- t-Test compares two means.
- One-way ANOVA compares multiple means.

15.4.4.1 P-value

The p-value (P) tells you whether the various level means are significantly different from each other: If P is less than or equal to the significance (a) level chosen, one or more means are significantly different.

As with other inferential statistics tests, we also compare the confidence intervals for the various "levels" or subgroup samples. The two-line ANOVA test clearly shows the nonoverlapping confidence intervals around the sample means: further confirmation of the difference between means. The significant *p*-value in the one-way ANOVA does not indicate that all the lines have different means. The correct interpretation is that there is a statistically significant difference between one (or more) of the lines and the others.

15.4.4.2   *Tukey's method*

Tukey's method compares the means for each pair of factor levels using a family error rate (often called familywise error rate) to control the rate of type I error. The family error rate is the probability of making one or more type I errors for the entire set of comparisons. Tukey's method adjusts the error rate for individual comparisons (individual error rate), based on the family error rate you choose.

15.4.4.3   *Fisher's least significant difference*

Fisher's least significant difference (LSD) method compares the means for each pair of factor levels using the individual error rate you select. Note that the family error rate, which is the probability of making one or more type I errors for the entire set of comparisons, will be higher than the error rate for each individual comparison.

Overall performance is calculated by using the standard deviation of the entire sample (no subgrouping). The potential capability is calculated by using the pooled standard deviation of the rational subgroups. To study more about ANOVA, refer to the previous chapter.

15.4.5   X–Y *process map*

Deliverables of Lean project is to identify areas of opportunity. The areas of opportunities are listed as follows:

- Rework
- Time traps
- Redundant work
- Unnecessary work

Identify real value added, business value added, and nonvalue-added activities. Process are identified and taken on the quick success. Deliverables of Six Sigma project is shown in Figure 15.27 as follows:

The *x–y* process map shows the inputs and output(s) for the major process steps. The *x–y* process map is the foundation for the further work of deliverables for Lean project. X–Y process map is shown in Figure 15.28.

X–Y process map is the input for FMEA. With the failure mode analysis it is possible to determine variation throughout the time series. The next step will be to analyze the data with respect to data type that is continuous or discrete. Using the information data *x–y* relationship is generated. Potential *x*(s) is determined with the result obtained from the generated *x–y* relationship.

Let us consider an example for *x–y* process map for the tire used in the truck. The inside surface of the finished tire and the liner should have a uniform appearance, with the impression of the curing bladder over the entire surface. Irregular surface appears

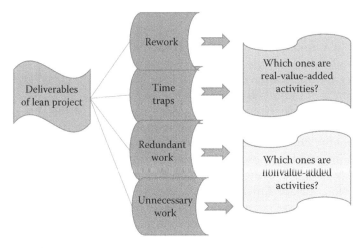

Figure 15.27 Deliverables of Six Sigma project.

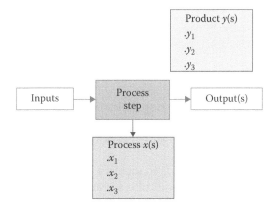

Figure 15.28 X–Y process map.

when the bladder impression is missing. Although it is not an indication of a performance problem with the tire, customers will not accept any tire with this condition.

Major steps for process inputs and outputs for each step are determined. It is essential to analyze the process. Use the observations and the knowledge of process experts, the front line people doing the work. To get an accurate picture of the process as it actually operates today the above process is essential. The x–y map can also be documented in spreadsheet form. In this format, it is easy to transfer the results of the x–y map to FMEA.

15.4.6   *Failure mode and effect analysis*

FMEA is an indispensable tool in identification and prioritization of the x(s). The FMEA helps to identify potential failure modes at each step of the process that may affect the critical to quality. FMEA provides vital information to reliability engineer, design engineer and other tools to analyze the subsystem and system and other things that causes potential failure mode. For each failure mode, the downstream effects and upstream causes are identified. Each failure mode may have multiple potential causes and each failure mode may have multiple effects. This is an important qualification. Then probability of failure mode that will occur and the effect it will create on rest of the system is analyzed. The new technique is known as failure modes, effects and criticality analysis (FMECA). Criticality can be analyzed by rating it for each process at each step. FMEA is generally used to identify all possible failure modes, effects, and causes associated with a product or process. In a Six Sigma project, the application is narrowed to only those elements related to the project y(s) (Quality Council of Indiana Inc. 2007).

Map the process to identify the major steps (about 3–6) of the process. This step is already completed in the x–y map. Next step is to identify the quality characteristics of each step. The quality characteristics measure how well the function is being performed. A function is an intended purpose of the product or process. These are the y(s) of the x–y process map. In a manufacturing process, quality characteristics could include dimensions, weight, color, surface finish, etc. In a service or transactional process, examples could include response time, completed fields of an order, etc.

Potential failure modes, or categories of failure, can then be identified by describing the way in which process steps fail to meet the function required. Failure modes fall into one of five possible failure categories:

- Complete failure
- Partial failure
- Intermittent failure
- Failure over time
- Overperformance of function

Failure mode is a nonconformance at each specific step. Use the process steps from the x–y map as the process steps of the FMEA. Use the y(s) from the x–y map to define the potential failure modes in the FMEA. There can—and generally will—be multiple failure modes for each element or activity (Table 15.5).

The next step in the FMEA process is to identify potential downstream consequences when the failure mode occurs. This should be a team brainstorming activity. After consequences have been identified, they must be fit into the FMEA model as effects. It is assumed that failure mode effects always occur when the failure mode occurs. There may be multiple failure effects for each element or activity. An assessment of the severity of

Table 15.5 Potential Failure Mode

Process step	Potential failure modes	Effect of failure	SEV	Cause of failure	OCC	Current controls	DET	RPN
Build bands	Lamination ridges Liner damage Inner contamination							
Build first stage carcass	Roller Imprint Tom liner							
Build second stage carcass	Cold carcass Carcass sags							
Prepare for cure	Bladder not in full contact with liner Holding pressure too low							

Table 15.6 Effects of Failure

Process step	Potential failure modes	Effect of failure	SEV	Cause of failure	OCC	Current controls	DET	RPN
Build bands	Lamination ridges	Forms void to trap air						
	Liner damage	Forms void to trap air						
	Inner contamination	Low component tack						
Build first stage carcass	Roller imprint	Small voids trap air						
	Torn liner	Slit traps air						
Build second stage carcass	Cold carcass	Resists bladder forces						
	Carcass sags	Resists bladder forces						
Prepare for cure	Bladder not in full contact with liner	Air between bladder and liner						
	Holding pressure too low	Bladder pulls away from liner						

the effect of the potential failure mode is made. Severity applies to the specific effect only. For a given effect and given customer the severity rating does not change (Table 15.6).

15.4.6.1 Failure mode and effect analysis process steps
Failure mode and effect analysis process steps are shown in Figure 15.29.

- FMEA number: Unique numbering is provided for log controlling purpose. It is assigned by the reliability team for tracking purposes.
- Part number, part name, and other appropriate description are provided.

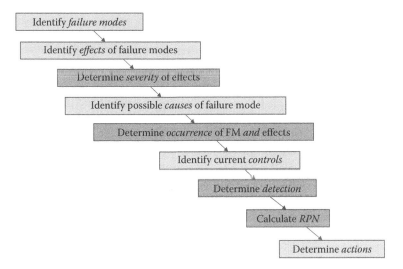

Figure 15.29 FMEA process steps.

- Design responsibility is assigned to each department.
- Job will be assigned to a person for the preparation of failure mode and effect analysis.
- Provide time information to the FMEA and prepare any revision if it is required at any level.
- The part number or subsystem number that is being analyzed needs to be specified.
- Component function is studied thoroughly.
- Potential failure mode is analyzed.
- Potential effects of failure mode are also analyzed.
- Causes for the failure effects are studied.
- Appropriate control measures are sorted to avoid the failure from occurring (Quality Council of Indiana Inc. 2007).

15.4.6.2   *Risk assessment and risk priority number*

Major step involved risk assessment and risk priority number (RPN) are listed below (Table 15.7).

- RPN can be determined by the following equation. This number is the product of indices of probability of failure that occurs, severity of the effect of the failure, and measure of effectiveness of the current control.

$$\text{RPN} = P \times S \times D. \tag{15.6}$$

- Probability of failure mode that will occur is denoted by the letter P. Values are assigned from 1 to 10. If the value is 1 there is no chance of failure to occur and if the value is 10 there is high possibility of occurrence of failure.
- Severity of the failure affects the subsystem and failure effects are denoted by the letter S.
- D denotes the effectiveness of the current control to determine the weakness. This index is numbered from 1 to 10. If the value is 1 then it denotes the highest possibility of getting affected by the failure. If the value is 10 then it denotes the product will make it to final production without detection.
- All decisions are based on the RPN.
- Separate column is provided to note the steps to be taken to reduce the risk (Quality Council of Indiana Inc. 2007).

15.4.6.3   *Types of FMEA*

There are four types of FMEA. These are design FMEA, process FMEA, system FMEA, and functional FMEA (Pelaez and Bowles 1995).

1. Design FMEA: FMEA is performed at the design level of the product/services. Aim of this process is to determine how failure mode affects the system and what are the steps to reduce the failure effects. This process is done before the product is released before the manufacturing phase.
2. Process FMEA: This type of FMEA is done to the manufacturing and production process. This process is done during quality phase during the production. All possible failures during the manufacturing process is accounted and described in this type.
3. System FMEA: All the part level FMEA will be merged to form the whole system. As the system goes to the downstream more failure modes are taken into account.
4. Functional FMEA: This focuses on the functional attribute of the system. This concentrates more on performance of each part rather than the specific part attributes of the part (Quality Council of Indiana Inc. 2007).

Table 15.7 Calculation Risk Priority Number

Process step	Potential failure modes	Effect of failure	SEV	Cause of failure	OCC	Current controls	DET	RPN
Prepare for cure	Bladder not in full contact with liner	Air between bladder and liner	8	Toe vents plugged	2	Operator inspection	4	64
		Air between bladder and liner	8	Bladder position off-center	5	Operator judgment	5	200
		Air between bladder and liner	8	Low shaping pressure	5	Pressure gauge	4	160
		Air between bladder and liner	8	Incomplete air purge	8	Operator judgment	8	512
		Air between bladder and liner	8	Holding air pressure too low	7	Pressure gauge	4	224
		Air between bladder and liner	8	Insufficient spray on bladder	4	Operator judgment	4	128
		Air between bladder and liner	8	Cold bladder	4	Operator judgment	5	160
	Holding pressure too low	Bladder pulls away from liner	7	Air line removed too soon	5	Operator judgment	4	140
		Bladder pulls away from liner	7	Line pressure drops	2	Pressure gauge	5	70
		Bladder pulls away from liner	7	Air leaks from bladder assembly	4	Pressure gauge	5	140

15.5   Black belt

The tools used in black belt are multi-vari studies, process capability, and regression.

15.5.1   Multi-vari studies

Multi vari chart is a visual way of representing variation through a series of charts. Multi vari literally means many variables. It is graphical tool for investigating the effect of up to four variables on the output of interest. Multi-vari analysis can be used to study uncontrolled noise variation with the intent to reduce or eliminate the variation due to noise. The content of the chart is evolved over time. Multi-vari study is statistical process control; it is used to track variables like temperature or pressure. It uses discrete input data. It also concentrates on historical data.

Key assumption of testing with numerical data is that the data is the same across all points but often will vary across the sample. When any measurement is taken, for example let us assume that we are measuring temperature of a cross section of the furnace and temperature of the cross section of the furnace will vary depending on where the measurement is taken. It also varies for different thickness of the part. In that case multi-vari chart is very useful tool for analyzing the variable. Mini tab is the tool which is used for representing the variation in data set. The graphical representation of the data generated in Minitab can quickly identify major sources of variation or noise in the data for further investigation. These charts are used to determine the consistency or stability of a process. It consists of series of vertical lines along a time frame. Length of each line determines the range of values in sample size. Multi-vari sampling plan procedures are listed as follows:

- Choose the process and characteristics that need to be analyzed.
- Choose appropriate sample size and time frequency.
- Build tabulation sheet to record values and time for each sample set.
- Plot multi-vari chart with measured value on the vertical scale and time along horizontal axis.
- Join the observed values with appropriate lines.
- Analyze the chart for variation with sample to sample, data set over the time.
- It may be necessary to do more studies on areas where there is maximum variation.
- It may be mandatory to repeat multi-vari study to verify results (Quality Council of Indiana Inc. 2007).

Data analyzed to determine potential relationships between key process input variables, the x(s), and key process output variables, y(s). It should help identify main sources of variation. It should also be helpful to think sources of variation as positional, sequential, or temporal variables or factors.

Three categories of variation are positional, sequential, and temporal. Understanding these categories of variation is very helpful when submitting the data for analysis and designing data collection.

1. Positional (location): In case of positional variation within a single unit, or across a unit with many parts. In other words, variation can be represented within the piece.
2. Sequential (order): Variation is observed in sequential among consecutive pieces, batches, group, etc. Variation is from step to step. Variation is observed from piece to piece.

3. Temporal (time): Variation in temporal is over short time span against long time span. It is used to track time related changes. Example of temporal: shift-to-shift, week-to-week, etc.

15.5.1.1 Advantages of multi-vari chart

- In multi-vari chart variation can be represented within the piece.
- Variation is also observed from piece to piece.
- It is used to track time-related changes.
- It helps in reducing variation by determining the areas of excessive variation. It also helps to identify areas which do not have excessive variation (Quality Council of Indiana Inc. 2007).

15.5.1.2 Characteristics of multi-vari chart

The quality characteristics measured at two extremes of the measurements are drawn as lowest value and highest value over a time frame. Quality characteristics are represented by three attributes of interest that are given in three horizontal panels which are

- Piece to piece variability
- Time to time variability
- Variability on a single piece

Examples for multi-vari chart are while analyzing furnace temperature, flow of river, and rolled steel thickness. Temperature rises when the measurement is close to the heat source in case of furnace temperature. Flow is slower when it is measured closer to the land. The actual thickness may vary across the sample. Variation is achieved within the piece where the part is tapered and thin, for example, the middle; variation can occur over time and it is repeated over time. Variation in the sample is shown by the line length. Variation from sample to sample is represented by vertical lines (Figure 15.30).

To represent a multi-vari chart, a sample set is taken and plotted from the highest value to the lowest value. Variation will be established by a vertical line. Thickness is measured at four points across the width as shown in Figure 15.31. The illustrations of excessive variability within the plate and less variability are also shown in Figures 15.32 and 15.33.

15.5.2 Regression

After accounting for variation, it requires sorting through the many inputs x(s) that could be driving the output response (s) of interest, y(s). We need to accurately and efficiently identify the inputs that have the greatest effect on our outputs so that we can focus the business resources from which we get the greatest project benefit. The main aim of the analysis is to identify importance of x–y relationship. The two-by-two matrix presented here is a useful reference when determining the proper statistical tool to employ based upon the data types of a given project. Knowing the data types for the input and output under study, use the table to help select the applicable statistical tool. The goal in developing transfer functions is not simply to describe the data collected; rather it is to develop a model that can be used to accurately predict future outcomes. Amidst all the statistical tools it is easy to forget the bigger goal—we want to understand our processes so completely that we are able to control not only during the project but well into the future (Figure 15.34).

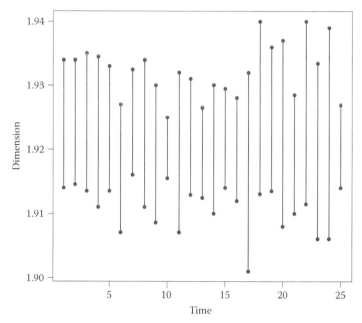

Figure 15.30 Multi-vari chart example. (http://r-resources.massey.ac.nz/161325examples/examples 66x.png)

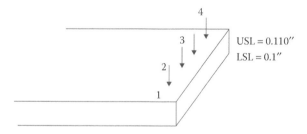

Figure 15.31 Multi-vari measurement of a plastic plate. (CSSBB PRIMER 2007.)

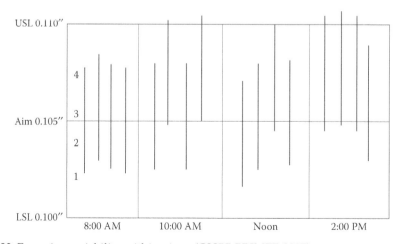

Figure 15.32 Excessive variability within piece. (CSSBB PRIMER 2007.)

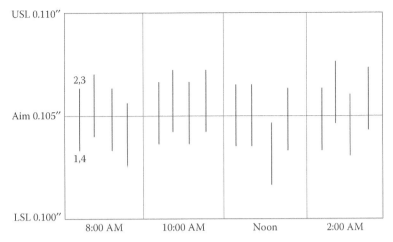

Figure 15.33 Less variability within piece. (CSSBB PRIMER 2007.)

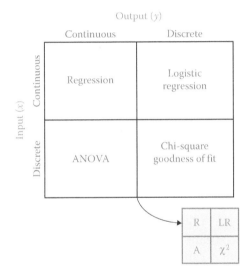

Figure 15.34 Statistical tools for analysis.

Regression analysis is probably the most widely used technique in modeling and data analysis. In the term "simple regression," the word "simple" refers to the fact that there is a single independent (x) variable used to predict the value of the output variable, y. This differentiates simple regression from "multiple regression," in which multiple input variables [conceivably many x(s)] are used to predict the value of y. The term "linear" simply means a transfer function that is composed of input variables that are only raised to the first power. (Recall that $x^1 = x$.) In the case of simple (a single x) linear regression, the function is a straight, one-dimensional line. The familiar convention for the equation of a straight line is:

$$y = mx + b, \qquad (15.7)$$

where:
   m is the slope
   y is the intercept

Table 15.8 Study Time and Performance

Student	Study time (hours)	Test results (%)
1	60	67
2	40	61
3	50	73
4	60	80
5	35	60
6	40	55
7	50	62
8	30	50
9	45	61
10	55	70

Simple regression is the technique to model and analyze relationship between a continuous response variable, *y*, and one continuous predictor variable, *x*. Simple linear regression produces a model of an output's behavior. Statistical significance of the modeled relationship between *y* and *x* is described quantitatively by the *p*-value. The value of *y* is expressed, as a linear function of the value of *x*, by the transfer function which is defined below.

$$y = \beta_0 + \beta_1 x_1 + \varepsilon, \tag{15.8}$$

where:

y is the predicted response, or dependent variable
x_1 is the predictor, or independent variable
β_0 and β_1 are the regression coefficients for the transfer function
β_0 = *y*-intercept of regression line
β_1 = slope of regression line
ε is an error term

Assume an analysis of a constant variable. Gather data that summarize it according to the constant variable. Equation can be developed according to the function of the variable using simple regression. Note that both the input and output variables are continuous, indicating that we are able to apply regression for the analysis.

Let us consider analysis of student individual study time and individual test performance. Table shows the data points for the study time and test performance. Data sets in the table are plotted as graph which is known as scatter plot. In this case, plot consists of study time in horizontal scale or *x*-axis and test results in the vertical scale or *y*-axis. Observe that *y* increases as *x* increases. One technique to predict equation relating *x* and *y* can be determined by placing a ruler and move it through data points and if it passes through the majority of points then it is considered as best fit line. Judging from the "look" of the data, we suspect that a relationship exists. We could estimate the form of the relationship by drawing a visually "best-fit" straight line through the data points (Table 15.8; Figure 15.35).

15.5.2.1 Fitted line plot

The fitted line plot figure will show the scatter plot, the best-fit line, and other useful information about the relationship between the two variables. In case of fitted line plot, we will examine options that provide a visualization of the predictive power of a regression.

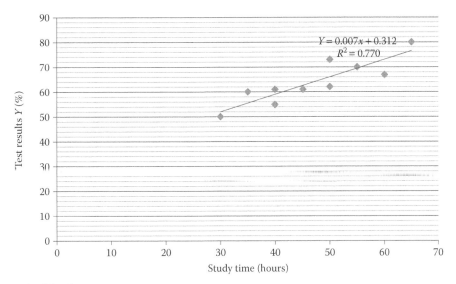

Figure 15.35 Plot for scatter.

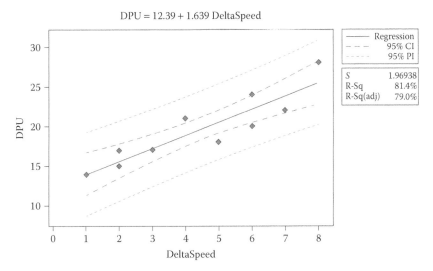

Figure 15.36 Fitted line plot.

Using Minitab, a fitted line plot can be generated, based upon the data that minimizes the distances between the points and the line. Notice that the output also provided the regression equation, as well as values for R-squared and R-squared (adjusted), but no *p*-value. In generating the fitted line plot or the graph of the best regression equation, it is the sum of the squared distances from each actual observation (the plotted points) to the fitted line that are minimized (Figure 15.36).

The *p*-value gives an indication of the significance of the coefficients. The R-square value is a measure of how well the model fits the data; it gives the "percent of total variance" that is "explained" by the model. The R-square (adjusted) "discounts" R-square when there is too much complexity in the model. It provides a better indication as to how well the regression will fit future data. Red line in the plot represents the confidence interval of

```
Regression analysis: DPU vs. DeltaSpeed

The regression equation is
DPU = 12.4 + 1.64 DeltaSpeed
```

Predictor	Coef	SE coef	T	P
Constant	12.389	1.370	9.04	0.000
DeltaSpeed	1.6389	0.2774	5.91	0.000

$S = 1.969$ R-Sq = 81.4% R-Sq(adj) = 79.0%

Analysis of variance

Source	DF	SS	MS	F	P
Regression	1	135.37	135.37	34.90	0.000
Residual error	8	31.03	3.88		
Total	9	166.40			

Figure 15.37 Regression output.

the regression line. The red confidence interval lines represent the area where we are 95% confident that the actual regression line falls. Green line indicates the confidence interval for prediction of individual part. The green predictive interval represents the area where we are 95% confident that future values predicted by the regression will fall (Figure 15.37).

15.5.2.2 Residuals

In fitting a line to observed data, the "residual" is the model error. Its magnitude is the vertical distance between the observed value and the predicted value (the fitted line). Residuals can be analyzed to provide information about the model, including nonnormality of errors, nonrandom variation of errors, nonconstant variability of errors, and nonlinear relationships (Figure 15.38).

Normal plot of residuals: This type of residual normally forms a straight line.
Histogram of residuals: It should appear roughly normal symmetric with one peak. Bars located away from the main group (outliers) may indicate unusual observations.
I chart of residuals: Any out-of-control points should be investigated as indicative of a nonrandom pattern in the data that may be a result of special cause variation.
Residuals versus fits: If points appear to form a curve, there may be evidence of nonlinear relationships. If the spread (variability) of the data points increases or decreases with the fits, the data may have nonconstant variance.

15.5.2.3 Multiple regression

Multiple regression is a technique for modeling and analyzing the relationship among a continuous response variable, *y*, and more than one continuous input variable, *x*(s). *y* is expressed, as a function of the *x*(s), by the transfer function developed from the observed data:

$$y = \beta_0 + \beta_1 x_1 + \beta_2 x_2 + \ldots + \beta_k x_k + \varepsilon, \tag{15.9}$$

where:
is the predicted response, or output, variable
$x_1, x_2, \ldots, x_k$ are the predictor, or input, variables
$\beta_0, \beta_1, \beta_2, \ldots, \beta_k$ are the regression coefficients (constants)
ε is the error term

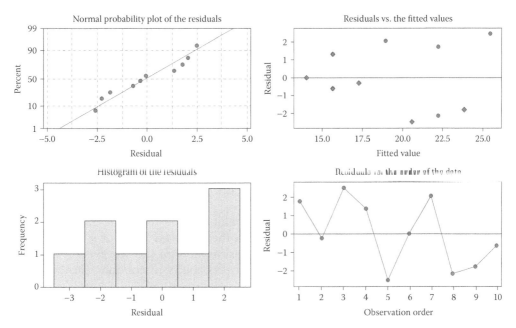

Figure 15.38 Interpreting residuals.

The statistical significance of the regression equation is again determined by the *p*-value. The topic of multiple regressions covers a range of techniques that can be used to investigate possible relationships between a response variable and two or more potential *x*(s). A *p*-value is computed for each input variable, x_i, to determine the statistical strength it contributes to predicting the value of the response, *y*. Only the significant factors are kept in the prediction equation—the trivial *x*(s) are dropped from the model.

Best subsets regression is an iterative process that examines potential models with one predictor, then two, then three, and so on, through a model that includes all possible predictors. The best subsets regression identifies the best models (prediction equations) for various numbers of predictors. Only models with the highest R-squared values are retained in the analysis. Model selection is made based upon the Minitab output. A good model should have

- High R-squared
- High adjusted R-squared
- A *C*–*p* metric value that is approximately equal to the number of predictors in the specified model
- A low *s* (*s* is the standard deviation of the error term in the model)

Many potential *x*(s) can be screened, simultaneously. There is a danger when *x*(s) are included that are correlated with each other (called multicolinearity). This risk is minimized with techniques such as best subsets regression and stepwise regression. Discrete variables, for example, operator and method, can sometimes be coded (1-2-3, 0-1, etc.) for use as logical, quasi-continuous variables. As always, correlation does not prove causation.

15.5.2.4 Simple linear regression model

Simple linear regression model has a straight line that represents a set of n data points in a manner such that sum of the squared residuals of the model as small as possible. Slope of the fitted line between will be equal to correlation between i and x modified by the ratio of standard deviation of the given variable. The mathematical equation for the simple regression model is given below:

$$y = \beta_0 + \beta_1 x, \tag{15.10}$$

where:

β_0 is the y-intercept when $x = 0$

β_1 represents the slope of the line (Quality Council of Indiana Inc. 2007)

X-axis does not tend to zero such that y-intercept looks to be very high. The actual value of random error will be the difference between the observed value of y and the mean value of y with the specified value of x. Main assumption for the observation value of y changes in a random manner and has a normal probability distribution for the given x value (Figure 15.39).

The probabilistic model equation for the observed value is given below:

$$y = \left(\text{mean value of } y \text{ with respect to the } x \text{ value}\right) + \text{random error}, \tag{15.11}$$

$$y = \beta_0 + \beta_1 x + \varepsilon, \tag{15.12}$$

where:

y is the predicted response, or dependent variable

x_1 is the predictor, or independent variable

β_0 and β_1 are the regression coefficients for the transfer function

$\beta_0 = y$-intercept of regression line

$\beta_1 = $ slope of regression line

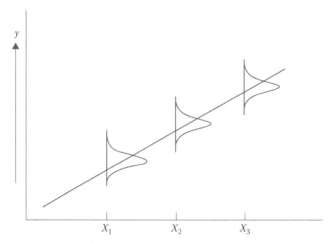

Figure 15.39 Simple linear regression model illustrating variation in y with respect to x.

15.5.2.5   *The method of least squares*

The method of linear equation is used to approximate sets which have more equations than unknown variables. The statistical method of determining the best fit straight line is in many aspects has the method of one fits a line by eye. The goal is to reduce the deviations of the points from the lines. If it denotes the predicted value of y determined from the fitted line as $\hat{y}$ (Figure 15.40).

$$\hat{y} = \beta_0 + \beta_1 x, \tag{15.13}$$

where:

β_0 and β_1 represent estimate of β_0 and β_1, respectively

One should determine the meaning of best if we want to reduce the deviation of points in choosing the best-fitting line.

The principle of least square is just the best fit criterion of goodness. To reduce the sum of squares of the deviation of the observed values of y from those predicted. Mathematical expression is shown below; it reduces the sum of the squared errors.

$$SSE = \sum_{i=1}^{n}(y - \hat{y})^2. \tag{15.14}$$

Substituting the value of $\hat{y}$ in Equation 15.8, we obtain an altered equation,

$$SSE = \sum_{i=1}^{n}[y - (\beta_0 + \beta_1 x)]^2. \tag{15.15}$$

The least squared estimator of β_0 and β_1 are estimated shown as follows:

$$Sx^2 = \sum_{i=1}^{n} X_i^2 - \left(\left[\sum_{i=1}^{n} X_i^2 \right] / n \right), \tag{15.16}$$

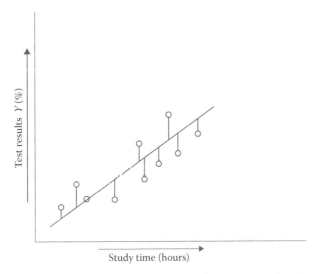

Figure 15.40 Method of least squares illustrating test results versus study time.

$$Sx^2 = \sum_{i=1}^{n} X_i Y_i - \left\{ \left(\left[\sum_{i=1}^{n} X_i \right] \left[\sum_{i=1}^{n} Y_i \right] \right) / n \right\}, \tag{15.17}$$

$$\beta_1 = S_{xy} / Sx^2, \tag{15.18}$$

$$\beta_0 = y - \beta_1 \bar{X}. \tag{15.19}$$

Once β_0 and β_1 are computed, substitute the values into the equation of a line to obtain the least squares on regression line (Quality Council of Indiana Inc. 2007).
    Note:

- It is noteworthy to be careful of rounding off errors. The calculation should have a minimum of six significant digits in estimating the sum of squares of deviations.
- Plot the data against the model to ensure accurate modeling and do not project outside of regression line. If it does not provide a reasonable fit for the given data set, then there is a calculation error.
- Do not project outside the regression line.

15.5.2.6   Logistics regression

Logistic regression is a technique for modeling and analyzing the $y = f(x)$ relationship when the y is attribute (usually binary, but multiple ordinal responses are also acceptable) and the x(s) are continuous. Logistic regression relates single dependent variable to more independent variable. It is similar to that of regular linear regression since it has regression coefficients, residuals, and predicted value. In the case of linear regression, it is assumed that response variable is continuous whereas in case of logistic regression response variable is continuous. Logistic regression is not used as often as ANOVA or regression, as the emphasis in Six Sigma is to find y(s) that are continuous variable. Since the response is generally binary, a logistic transformation is performed to prevent the possibility of negative probabilities as outcomes. The statistical conclusions are drawn from logistic regression. *P*-values are no different than when using the other analytical methods. Ordinary least squares approach is used to determine the regression coefficient for linear regression, and maximum likelihood estimation is used to determine the regression coefficient for logistic regression.
    Logistic regression is used to provide information about analysis of the two value of interest that is good or bad, vote or not vote, pass or fail, enlist or not enlist, yes or no. Since logistic regression model takes only the value of 0 and 1, it is also known as binary regression model. General equation for logistic regression is given below:

$$y = b_0 + b_1 x_1 + e, \tag{15.20}$$

where:
    $y_i = 0,1$

Transfer function before logistic transformation:

$$p = \alpha + \beta_1 x_1 + \ldots + \beta_k x_k, \tag{15.21}$$

where:
    p = the predicted probability of an outcome

The probability of the results can be computed in this certain category by the following mathematical formula:

$$p = 1/\left(1 + e^{-y}\right) = e^{y}/\left(1 + e^{y}\right) = 1/\left[1 + e^{-(b_0 + b_1 x_1)}\right].$$ (15.22)

The above equation is similar to that of linear regression equation model, but the regression model uses equation to calculate b coefficient values. Since the right side of the transfer function could be less than 0 (or greater than 1) for certain values of $x_1, \ldots, x_k$, predicted probabilities of less than 0 or greater than 1 could be obtained, which is impossible. This is why the logistic transformation of p is used as the dependent variable:

$$\log(p) = \ln \frac{p}{1-p} = \alpha + \beta_1 x_1 + \cdots + \beta_k x_k.$$ (15.23)

It can also be expressed as

$$p = \left(e^{\alpha + \beta_1 x_1 + \cdots + \beta_k x_k}\right)/\left(1 + e^{\alpha + \beta_1 x_1 + \cdots + \beta_k x_k}\right).$$ (15.24)

Let us consider the example of piping problem. In this case the effect of temperature on the leaks is analyzed. The predictor (x) or input is joint temperature and it is continuous variable. Output (y) is joint leaks in this case. There are only two possible outcomes: these are "leaks" and "no leaks," which accounts for 100% of all possible results. Data were collected from 24 launches prior to the challenger. The data below illustrates the analysis of whether the temperature is the predictor of leaks (Figure 15.41; Table 15.9).

Binary logistic regression: Status code vs. joint temp
Link function: Logit
Response information

Variable	Value	Count
Status code	1	7 (Event)
	0	17
	Total	24

Logistic regression table

Predictor	Coef	SE coef	z	P	Odds ratio	95% CI Lower	Upper
Constant	15.048	7.278	2.07	0.039			
Joint temp	−0.2325	0.1066	−2.18	0.029	0.79	0.64	0.98

Log-likelihood = −10.144
Test that all slopes are zero: $G = 8.687$, DF = 1, p-value = 0.003

Goodness-of-fit tests

Method	Chi-square	DF	P
Pearson	11.238	15	0.736
Deviance	11.969	15	0.681
Hosmer–Lemeshow	6.898	8	0.548

Figure 15.41 Results of logistic regression.

Table 15.9 Illustration of Logistic Regression Example

Joint temp	Incidents	Joint status	Status code
52	3	Leak	1
56	1	Leak	1
58	1	Leak	1
63	1	Leak	1
70	1	Leak	1
70	1	Leak	1
75	2	Leak	1
66	0	OK	0
67	0	OK	0
67	0	OK	0
67	0	OK	0
68	0	OK	0
69	0	OK	0
70	0	OK	0
70	0	OK	0
72	0	OK	0
73	0	OK	0
75	0	OK	0
76	0	OK	0
76	0	OK	0
78	0	OK	0
79	0	OK	0
80	0	OK	0
81	0	OK	0

15.5.3   Process capability analysis

Process capability analysis provides with the information of the natural process variation with respect to the customers' specifications. Previously the setting of targets, specifications, and its measurements were studied. This section tells us where the current processes or its variation stand with respect to customers' requirements. A capable process is the one in which all natural variation fits within the customers' target range (Eckes 2002). The process variation is compared with the engineering or specification tolerances to check the suitability of the process. It is also important for the process to be stable before performing the analysis and hence it is always performed after the control charts. The process capability analysis has three basic steps (Quality Council of Indiana Inc. 2007):

1 Planning for collecting data
2 Collecting data
3 Plotting and analyzing results

The process capability is carried out in order to reduce the variability of the process so that conformity occurs with the customers' targets. When the natural process limits are compared with engineering or specifications range, the following actions are performed:

- Do nothing
- Change the specifications
- Center the processes
- Reduce variability
- Accept the losses

The identification of characteristics and its specifications based on customers' perspective has already been discussed and the data collection for the same was discussed in green belt section. The important task is to develop sampling plans. The knowledge of sampling and its probability distributions are widely required here. The process capability indices are usually calculated based on probability distributions and generally classified as normal and nonnormal distributions.

Process capability is carried out for three categories: (1) The process that is currently running and is in control. The data from control chart can be utilized for analysis. (2) Analysis of new processes. Pilot run be can used to evaluate the capability analysis. (3) Process capabilities carried out to improve the process. The analysis can be performed by using design of experiments (DOEs). The DOE gives optimum values which provide minimum variations.

15.5.3.1 *Process capability for normal distribution*

The process is said to be stable over time with only common causes of variation than it follows normal distribution (Automotive Industry Action Group 1995). This text provides with information for process capability analysis with normal distribution data. When process capability with only common causes of variation is plotted a bell-shaped curve is obtained corresponding to normal distribution. The area outside the specification is given by the Z value.

$$Z_{lower} = \frac{\bar{X} - LSL}{S},$$

(15.25)

$$Z_{upper} = \frac{USL - \bar{X}}{S}.$$

(15.26)

The Z transformation formula,

$$Z = \frac{X - \mu}{\sigma},$$

(15.27)

where:
X: data value
μ: mean
σ: standard deviation

The Z value hence gives the standard deviations away from the mean and standard normal table can be used to find the area under the curve which gives the probability of occurrences. This book as mentioned earlier assumes that the reader is well aware of the concepts in probability.

15.5.3.2 Process capability indices

The process capability is determined by means of sigma and indices. The capability index can be defined as C_P,

$$C_P = \frac{USL - LSL}{6\sigma_R},$$ (15.28)

$C_P > 1.33$; the process is capable
$C_P = 1.00-1.33$; the process is capable with tight control
$C_P <$ the process is incapable

The capability ratio is defined as C_R,

$$C_R = \frac{6\sigma_R}{(USL - LSL)},$$ (15.29)

$C_R > 0.75$; the process is capable
$C_R = 0.75-1.00$; the process is capable with tight control
$C_R >$ the process is incapable

The ratio between the smallest answer is C_{pk},

$$C_{pk} = \frac{USL - \bar{X}}{3\sigma_R} \text{(or)} \frac{\bar{X} - LSL}{3\sigma_R}.$$ (15.30)

C_{pm} index is based on Taguchi index, which concentrates on centering the process toward its target (Breyfogle 2003).

$$C_{pm} = \frac{USL - LSL}{6\sqrt{(\mu - T)^2 + \sigma^2}},$$ (15.31)

where:
USL: upper specification limit
LSL: lower specification limit
T: target value
μ: process mean
σ: process standard deviation

15.5.3.3 Process performance

The process performance is estimated with the help of sigma estimate. The sigma estimate is given by the following equation:

$$\sigma_i = \sqrt{\frac{\sum(X - \bar{X})^2}{(n-1)}},$$ (15.32)

where:
σ_i is a measure of total data sigma

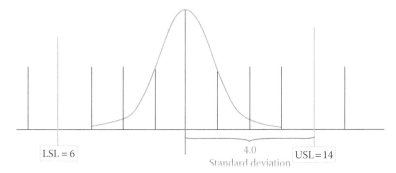

Figure 15.42 Performance index.

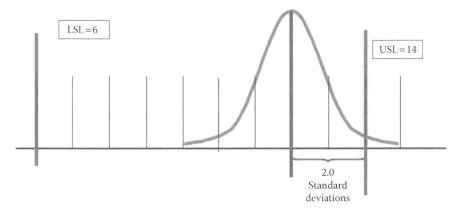

Figure 15.43 Process performance index.

The performance index is defined as P_p (Figure 15.42),

$$P_p = \frac{USL - LSL}{6\sigma_i}. \tag{15.33}$$

The performance ratio is given by P_R:

$$P_R = \frac{6\sigma_i}{(USL - LSL)}. \tag{15.34}$$

The ratio giving the smallest answer between is P_{pk} (Figure 15.43):

$$P_{pk} = \frac{USL - \bar{X}}{3\sigma_i} \ (or) \ \frac{\bar{X} - LSL}{3\sigma_i}. \tag{15.35}$$

15.5.3.4   *Process performance metrics*

The process performance indices are calculated based on the measurements of the systems. The common process performance metrics are discussed here.

The Six Sigma metrics are given by

$$\text{Total opportunities: TO} = \text{TOP} = U \times O$$

$$\text{Defects per unit: DPU} = D/U \text{ also} = -\ln(Y)$$

$$\text{Defects per normalized unit:} = -\ln(Y_{norm})$$

$$\text{Defects per opportunity: DPO} = \text{DPU}/O = D/UO$$

$$\text{Defects per million opp.: DPMO} = \text{DPO} \times 10^6$$

where:
  Defects = D
  Units = U
  Opportunities for a defect = O
  Yield = Y

Yield relationships are given by

- Probability of a unit containing X flaws: $P(X) = \dfrac{e^{-DPU}DPU^X}{X!}$
- First pass yield: $Y = \text{FPY} = e^{-DPU}$
- DPU: $DPU = -\ln(Y)$
- Rolled throughput yield (RTY): $Y_{rt} = \text{RTY} = \prod\limits_{i=1}^{n} Y_i$
- Normalized yield number of steps: $Y_{norm} = \sqrt[n]{\text{RTY}}$
- Total defects per unit: $\text{TDPU} = -\ln(Y_{rt})$

15.5.3.5   Rolled throughput yield

RTY is defined as the cumulative calculation of yield or defects through multiple process steps. The following steps are followed for determining the RTY:

- Calculate the yield for each step and the resulting RTY
- The RTY for a process will be the baseline metric
- Revisit the project scope
- Significant differences in individual yields can suggest improvement opportunities

$$\text{RTY} = \text{yield of step 1 * yield of step 2 * yield of step}$$

Example: RTY = 0.90 * 0.86 * 0.92 * 0.65

15.5.3.6   Chi-squared analysis of contingency tables

Coefficient of contingency is an objective methodology used for analysis of chi-squared test. It measures dependency level between any two attributes. Higher values of C imply higher degree of relationship. The optimal values of C are determined by the dimensions of the matrix of attributes derived from arranging data into a tabular form (Merkin 2001). An attribute representing a row/column is supposed to have the number of traits equivalent to degrees of freedom of the matrix of data.

The coefficient of contingency is estimated using the following equation:

$$C = \sqrt{\chi^2 / \left(N + \chi^2\right)}.$$

References

Automotive Industry Action Group (AIAG) et al. (1995). *Statistical Process Control (SPC): Reference Manual*. Southfield, MI: Automotive Industry Action Group.

Dreyfogle III, F. W. (2003). *Implementing Six Sigma: Smarter Solutions Using Statistical Methods*. New York: John Wiley & Sons.

Eckes, G. (2002). *The Six Sigma Revolution: How General Electric and Others Turned Process into Profits*. New York: John Wiley & Sons.

Ishikawa, K. (1982). *Guide to Quality Control*. Tokyo: Asian Productivity Organization.

Merkin, B. (2001). "Eleven ways to look at chi-squared coefficient for contingency tables." *The American Statistician*, 55(2): 111–120.

Nicholls, D. (2005). *System Reliability Toolkit*. Utica, NY: RIAC.

Pelaez, C. E. and Bowles, J. B. (1995). "Applying fuzzy cognitive-maps knowledge-representation to failure modes effects analysis." *Proceedings of the Annual Reliability and Maintainability Symposium*, January 16–19, Washington, DC.

Quality Council of Indiana Inc. (2007). *Certified Six Sigma Black Belt Primer*. West Terre Haute, IN: Quality Council of Indiana Inc.

Review Questions

Question 1: When testing equal variances should I use Bartlett's test (or the F test) or Levene's test?

Question 2: A Six Sigma improvement team may be required to analyze customer data in order to define a project or the results of an improvement. Which of the following tools could be employed?

  A. Statistical tests
  B. Line graphs
  C. Matrix diagrams
  D. Pareto analysis
    (a) A B D
    (b) A C D
    (c) B C D
    (d) A B C D

Question 3: A null hypothesis requires several assumptions, a basic one of which is

  (a) That variables are independent
  (b) That the confidence interval is 2 standard deviations
  (c) That the sample size is adequate
  (d) That the variables are dependent

Question 4: A graphical display of the total percentage of results below a certain measurement value is called a

(a) Cumulative distribution function
(b) Histogram
(c) Probability density function
(d) Expected value

Question 5: If the probability of a car starting on a cold morning is .6, and we have two such cars, what is the probability of at least one of the cars starting on a cold morning?

(a) .36
(b) .60
(c) .84
(d) .81

Question 6: Which table should be used to determine a confidence interval on the mean when the standard deviation is not known and the sample size is 10?

(a) F
(b) X squared
(c) t
(d) z

Question 7: One would normally describe recorded values reflecting length, volume, and time as

A. Measurable
B. Discrete
C. Continuous
D. Variable
   (a) A B C D
   (b) B D
   (c) A C D
   (d) A C

chapter sixteen

Improve

Utilizing data to predict implementation success

To improve is to change, to be perfect is to change often.

—Winston Churchill

Overview	
Steps	Substeps
Define	• Quality projects (sponsorship) • Define specific project (black belt)
Measure	• Identify critical characteristics • Clarify specifics and targets • Measurement system validation
Analyze	• Determine baseline • Identify improvement objectives • Evaluate process inputs
Improve	• Formulate implementation plan • Perform work measurement • Lean execute the proposed plan
Control	• Reconfirm plan • Set up statistical process controls • Operational process transition

Source: PWD Group LLC.

Recommended Tools by Belt					
		Define	Measure	Analyze	Improve	Control
Stage 1	Yellow belt	Interview process Language processing System map (flowchart)		X–Y map/QFD/ house of quality Fishbone diagram/RCA	EVOP	Control plans Visual systems 5-S SPC/APC
Stage 2	Green belt	Prioritization matrix Thought process map Stakeholder analysis	Hypothesis testing Flowdown Measurement system analysis	FMEA Multi-vari chart Chi-square	Design of experiments Fractional factorials Data mining	TPM Mistake proofing

(Continued)

Recommended Tools by Belt (Continued)

		Define	Measure	Analyze	Improve	Control
		Thought process map	Graphical methods	Regression	Blocking	
			Analysis of variance	Pareto analysis		
Stage 3	Black belt	Value stream map	Process behavior charts	Buffered tolerance limits	Theory of constraints	
		Gantt chart development	QFD	Process capability analysis	Multiple response optimization	
		Project management			Response surface methodology	

Source: PWD Group LLC.

Notes: Black belts are expected to know the continued knowledge obtained throughout green belt and yellow belt certifications. The same goes for green belt and yellow belt. Empty cells represent that skill is not required for that belt stage; however, a general knowledge of the step is expected.

5-S, sort, set in order, shine, standardize, and sustain; APC, advanced process control; EVOP, evolutionary operations; FMEA, failure modes and effect analysis; QFD, quality function deployment; RCA, root cause analysis; SPC, statistical process control; TPM, total productive maintenance.

16.1   Introduction

Improve is an important phase in the entire DMAIC process. Until the analyze phase the key process input and output process variables, the closed confidence intervals, specifications, and major deviations from the targets have been established. The analyze phase played its role effectively in identifying the root causes for major deviation that affect the quality of the existing system. The improve phase helps in realizing the goal of the Six Sigma project; by improving and correcting the variables that causes the deviations. This chapter has its structure as that of other chapters but there is no role of yellow belt in this phase because it involves advanced statistics and tools. The improve phase can be presented in the following topics in belts categorized as

- Design of experiments (DOEs)
- Theory of constraints
- Waste elimination
- Cycle time reduction

16.2   New process architecture

Architecture is the tangible and intangible structures of the system that cause it to behave in a unique manner. Architecture involves the process owners in generating ideas for changes to architecture. The underlying answers for the purposes and assumptions of the current process is studied in order to make the changes in process architecture like, is this purpose really necessary? If yes, how the purpose of the process be accomplished without performing this task? And the assumptions are challenged. If the assumptions are valid, does it offer any cues for alternative process designs. Brainstorming is one of the key tools on deciding the key process architecture. The process redesign can be pursued by the following:

- Analyze "As Is" map and prioritize opportunities
- Identify value-adding, nonvalue-adding

- Look for disconnects
- Create new process architecture
- Create "should be" map

The first step "analyze" is performed in the Chapter 15. The identification of nonvalue-adding is carried out by identifying tasks described by the following or similar words:

- Schedule (especially "re-schedule")
- Find errors and defects in the process
- Correct errors and defects
- Recycle
- Move around
- Store
- Accumulate batches
- Set aside
- Scrap

The next step is to look for disconnects, that is, something out of sequence, inadequately performed, or missing. With above information, the new process architecture is created by identifying creative alternatives for process execution. As mentioned earlier, process benchmarking is used to search for better practices and should be map is created. These stages must consist of people who are vital in implementing the changes. The effects of changes in new architecture with respect to the key stakeholders must be predicted. Finally, develop and discuss multiple alternatives to process design. Jumping immediately to the single "right answer" must be avoided. "Don't rush to the high tech solution" (Meyer 1993). The basic workflow should be first redesigned and then the help of new technologies must be determined. Technology plays a key role, but fast cycle time is more dependent on "social innovation" than it is on technology innovation. Fast cycle time does play a central role in defining research priorities and allocating resources, but Fast does not speed up basic research or the invention process.

Organizations that lack strategic alignment effectively disable others from acting because there is no contextual basis for people to know if their actions help or hinder the achievement of organizational goals. How can one ask people to focus only on value-added tasks if one has not defined what is value added and what isn't? (Meyer 1993). Hence process improvement is incremental, not revolutionary. Increasing the organization's speed results from an iterative process of identifying obstacles, designing a new process to eliminate them, and then most importantly, ensuring that the new way is accepted and implemented by those who do the work. Removing the first layer of obstacles is required to see the second layer.

The above comments explain the need and basic procedure for improve phase. Hence, the following is the plan proposed to carry out the improve phase. The simple comparative experiments describe a few brief example studies to demonstrate an introduction to experiments. The "traditional experimental design" demonstrates the problems inherent in common methods that are taught and used informally in practice. The topic of factorial design is presented more formally, as we focus on this fundamental method of experimentation. Powerful and efficient fractional factorials and multiple level designs are introduced following that. All these powerful tools are prone to the black belt officers. The green belt discusses primary tools and the yellow belt is required only to execute the plan and does not involve in planning (Figure 16.1).

Before continuing the tools for improve phase, there is slight refresher on previous steps. Qualitative and quantitative methods were used to identify and prioritize the top

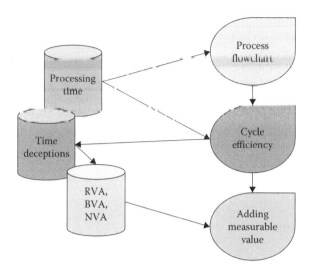

Figure 16.1 Plan—improve phase.

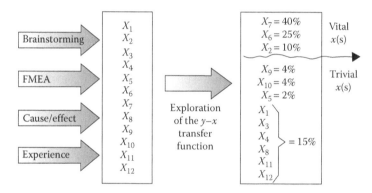

Figure 16.2 Distilling the key process inputs.

potential input variables that affect process output. A list of factors to begin controlled exploration is required. Next, purposeful experimentation is begun with the process inputs to identify the vital x(s), also known as key process input variables (KPIVs). The aim is to separate the key x(s) from the trivial x(s) (Figure 16.2).

Figure 16.3 shows how models and transfer equations help in experimentation.

The next step is to focus on the underlying relationship and that can be established by (Figure 16.4).

- All outputs (y) are a function of many x(s).
- A key is to describe the relationship $y = f(x)$.
- The y is a dependent variable; the x_i(s) are independent factors.
- The y(s) are *not* controlled; only the x_i(s) are.

The above steps have clearly stated the establishment of relationship. Now in order to demonstrate definite improvement to the process, a tool is needed that

- Is structured
- Is prescriptive regarding data collection

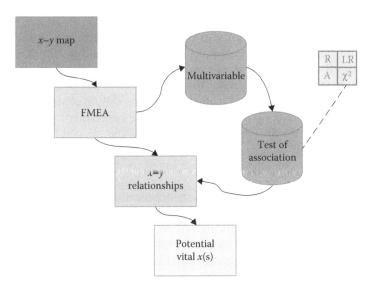

Figure 16.3 Models and transfer equations.

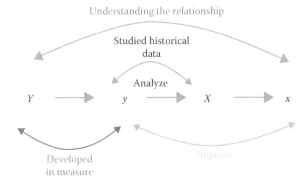

Figure 16.4 Establishing underlying relationship.

- Is statistically sound
- Quantifies the $y = f(x_i)$ relationship
- Segregates the vital few $x_i(s)$

16.3 Green belt

16.3.1 Benchmarking

It is necessary in many firms to determine where they stand in comparison to other firms when it comes to performance in the market. This is called benchmarking. With the information of the data of the comparison company it is easy for the company to determine how and where they can improve the performance of the organization. Benchmarks serve as points of references. The benchmarking process helps to make conclusions about the company's current performance and the necessary improvements that it might need. The typical domains a company might use to benchmark are procedures, operations, processes, quality improvement efforts, and marketing and operational strategies.

16.3.1.1   Purpose of benchmarking

It is a key tool that is used to compare the measures of performance with those of others in order to figure out where improvement opportunities may exist. It is important to realize the factors that motivate the planning of a benchmarking assessment. "Why is the company planning to do, and what do the leaders hope to learn?" Benchmarking measures the organizations position in the market in comparison with the best in field organizations. Company beginning a benchmarking assessment should have the permission to obtain data with the comparison company and should have a plan to use the information generated by the comparison. A major drawback of benchmarking is that the end result of the process cannot be employed as an employment strategy for the organization.

There are several variable reasons for benchmarking. A company might perform benchmarking in order to determine if they can meet the performance standards that the customer may require. Benchmarking will help to determine the areas that require improvements before the company seeks certification. Benchmarking with standards might be done to meet the certification standards and qualifications that customers set. On a larger scale, benchmarking is done to check if quality standards are met according to the ISO 9000 or quality standard awards. Benchmarking answers such questions as:

- Are the company's processes properly constructed and documented?
- Are systems in place to allocate resources and funding appropriately?
- Which areas require the maximum need of improvement?
- What are the needs of the customer, external and internal?

16.3.1.2   Types of benchmarking

- *Perception benchmarking:* In this process of benchmarking, a company hopes to learn the current performance. It can focus on internal issues, seeking to learn what the people inside the company think about themselves, the management, the company, and the quality improvement process. This may be used as a baseline for comparison with future processes.
- *Compliance benchmarking:* In this step of benchmarking, companies seek to attain the quality standards of ISO 9000 certification. Also, they hope to attain the certification. It is an in-depth process of benchmarking which verifies a company's compliance with the standard and qualifications. The question that this type of benchmarking will answer is how a company is performing in comparison with the published standards.
- *Effective benchmarking:* This type of benchmarking makes sure that the company is meeting the requirements and also effective systems are in place to ensure that the requirements are being fulfilled.
- *Continuous improvement benchmarking:* This type of benchmarking verifies if continuous improvement is a faceted and integral part of the process. It determines if the organization is just providing "Lip service" to process improvement or instead is putting into place systems that provide and support continual improvement on a day-to-day basis.

The above four benchmarking processes can be verified through review of the existing business practices. This review should follow an organized format. Activities can be judged on the basis of visual observations by the reviewers, interviews with those directly involved, personal knowledge, and factual documentation.

16.3.1.3 Benefits of benchmarking

The primary benefit of benchmarking is to determine where the organization stands with respect to performance compared to standards set by customers, by the company itself, or by national certification or award requirements. Strategies can be developed with the above knowledge to develop strategies for its own continuous improvement goals. The benchmarking process will help to identify existing assets and also opportunities for improvement outside the organization. Most quality assurance certifications help discover how a company is currently performing, strengthening the companies weaknesses, and then they verify the companies compliance with the certification standards. It is a valuable tool to use throughout the certification process since the assessment provides an understanding of how the company is performing.

16.3.1.4 Different standards for comparison

- Malcolm Baldrige National Quality Award (MBNQA)
- ISO 9000
- Deming prize
- Supplier certification requirements
- Best-in-field organizations

16.3.1.5 How is benchmarking done?

Several plans and methods that exist are available to aid interested companies in the benchmarking process. Flowcharts to depict the various steps of benchmarking are as follows: determining the focus, understanding your organization, determining what to measure, identifying whom or what to be benchmarked, benchmarking, improving performance (Figure 16.5).

1. Determining the focus: It is important to determine at the beginning of the benchmarking process which aspect of the company has to be studied for the people who are involved in it. The focus may be based on customer requirements, standards, or a general continuous improvement. Information gathered during the process should be correlated with the company's goals and objectives. It is important to be aware that benchmarking and gathering information about the processes are of greater

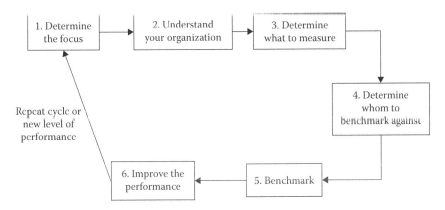

Figure 16.5 Benchmarking process.

value than metrics. A narrow focus on comparison may lead to apples over oranges comparison, whereas a focus on processes encourages improvement and adoption of new methods. To be of greater value, benchmarking should be used as a tool to provide support to a larger strategic objective.

2. Understanding your organization: Understanding all aspects of the situation in which you are working is very important. The team involved in the benchmarking process has to understand the culture of their company. To create a plan and perform the process, it is important for the company to understand the information concerning the internal and external customers and their major inputs and outputs. Often this receives less attention than it should. The company in which the people are working for the benchmarking process is known as entity. Employment of flowcharts is a powerful tool for enhancing everyone's understanding of the system.

3. Determining what to measure: Identifying the measures of performance is the next step once the company's present system is understood. These measures help the individuals involved to make a judgment about the performance of the company. It is at this time when the decision as to what is critical is to be known in order to remain competitive. These critical factors are supported by standards for procedures, processes, and behaviors. Benchmarking will help to determine questions to be answered and issues to be resolved as well as tell us what all procedures and processes are to be improved. A well-developed and prepared list of the items to be benchmarked makes it easier and will result in consistent assessments and comparisons.

4. Identifying whom or what to be benchmarked: Organizations should choose what to benchmark and against whom by considering the various activities and operations under investigation, the size of the company, the number and types of customers, the types of transactions, and even the location of the facilities. Careful attention should be given when selecting appropriate companies. Similarities in size and type of transactions may be more important sometimes than just the competitors. If the company is interested in obtaining ISO 9000 certification, then it should start systematically select areas within its own operations to compare against standards in order to check for compliance.

5. Benchmarking: This is the final step in the process of benchmarking. During this step, all the companies and the areas will be chosen for the assessment. For cooperation, the team should attain authorization from the highest level of the organization. During this step, the investigators perform comparison of data that is collected in step 3 against performance standards. The data that is compared determines the critical success of the company. They verify the company's compliance with the measures and standards and judge the company's ability to perform those measures and standards.

6. Improving performance: This is the final step in the process of benchmarking. Once the reports are summarized and gathered a report is created to determine the strengths and weaknesses of the area that is under study. In this report, documentation of the gap between the existing and desired levels is done. An effective report often concentrates on missing patterns and violation of standards.

In a successful benchmarking process the final report becomes the tool to determine the areas that have to be improved. The information in this is used to solve root causes and prevent the occurrence of nonconformities. It is a power customer feedback tool and should be used accordingly.

Table 16.1 SWOT Analysis Ideas

Internal strengths	Internal weaknesses
Core competencies in critical areas	Too many goals
Solid finances	Lack of strategic focus
Market leader	Outdated technology
Proprietary technology	Inexperienced management
Cost advantages	Manufacturing problems
Good marketing sills	Weak marketing skills
Management excellence	Lack of growth capital
World class manufacturing	Weak cash flow
Good technical and workforce skills	Inadequate R&D
Superior brand names	Cannot implement plans
Web skills	

External opportunities	External threats
Expansion to new markets product lines can be broadened	Global competition
Transfer technical skills to new products	Substitute products are not available
Low industry rivalry	Slow market growth
Minimal regulatory requirements	Legal and regulatory requirements
New emerging technologies	Recessionary cycle
Positive growth cycle	Strong customers or suppliers
Business-to-business on the Internet	New competitors
e-Commerce	Business-to-business on the Internet e-Commerce

16.3.2   SWOT analysis

It is a strategic planning process in which the firm's internal strength, weaknesses, and external opportunities and threats are determined. It is an acronym which means *s*trengths, *w*eaknesses, *o*pportunities, and *t*hreats. SWOT analysis usually requires a comprehensive study of internal and external situations should be done before strategic options can be determined. Strengths of the company and exploiting opportunities determine the quality of the strategy. Differing strengths and weaknesses determine the company's opportunities and threats. Thus each firm will have a different strategy. Table 16.1 is combination of concepts from Thompson (Thompson and Strickland 1995) and Schermerhorn (1993).

16.3.2.1   Strengths and weaknesses

The strength of a firm is that particular domain at which the firm is best. It can be a skill or an expertise or a product of their patent, technology, position in the market, or anything that takes the company ahead. James Brian Quinn (Quinn and Zien 1997) provides a good strength concept:

> If a company is not the best in the world at a critical activity, it is sacrificing competitive advantage by performing that activity with its existing technique.

It is more desirable to have competition on the basis of strengths. If a firm has no strengths, its ability to compete successfully can be difficult. Some of the typical strengths a company can enjoy are

- Engineering expertise
- Technical patents
- Skilled workforce
- Solid financial position
- Reputation for quality

A weakness is something that puts a company or firm in disadvantage or a particular domain which a firm lacks. Either the strength can be overly magnified or the weaknesses minimized, such that the process is negated. The following areas should be covered for the analysis:

- An evaluation report of the business and critical success indicators. Are there winners or losers and is money being made?
- A system of tracking or control systems for the indicators which can include cash flow, growth, staffing, quality, technology, and so on.
- Indication of the firm's level of creativity, risk taking, and competitive approach.
- Determining if the resources are available to carry out as planned. Identifying the current resources and resource plans. Smaller companies work longer to accomplish their objectives.
- Analyzing the current culture of organization and the firm's way of doing the business. Is teamwork being rewarded? Is success being rewarded? And innovation?

Weaknesses include the following:

- Cash flow is poor
- Technology is outdated
- Overhead expenses are high
- Absence of skilled labor
- Customers perception of poor quality

16.3.2.2 *Opportunities and threats*

Along with determining the internal strengths and weaknesses of the firm, it is also critical to understand the situation outside of the company. SWOT analysis gives the ability for management to go beyond the plant boundaries in the strategic planning effort. The external environment can include

- *Economic environment:* Economic environment encompasses economic condition, market trends, Wall Street confidence levels, or global conditions that can affect the firm.
- *Sociopolitical environment:* Sociopolitical environment involves the economic conditions locally, regionally, or nationally, and global units of government, special interest groups, or the legal system on the whole.
- *Social environment:* Social environment implies that the firm's planning can also be affected by the value system, social, and demographic patterns.
- *Technological environment:* Technological environment means identifying the available and anticipated technologies.
- *Competitive environment:* Competitive environment implies the requirement to determine the state of the competition.

A firm's world outside of the boundaries of the plant will provide opportunities and threats. The strategy must match the following:

- Opportunities suited to the firm's capabilities
- Defenses against external threats
- Changes to the external environment

A new economy and the Internet present infinite opportunities to many companies. At the same time, a threat might result if a company lets its competitor seize the opportunity of the Internet.

16.3.3   Theory of constraints

The theory of constraints is the set of framework which is required to improve the organization continuously. The goals of the company are fixed based on the theory of constraint with some critical measures along with its metrics (Goetsch and David 2000). The main theme of the goal is continuous improvement (Goldratt 1986).

   The key points of the theory of constraints are

- Return on investment
- Assumptions
- Cash flow
- Local optimums
- Delivery of results
- Goals
- System thinking
- Common sense
- Jonah
- Socratic way
- Lead times
- Cost accounting
- Fear of change
- Reduction of batch sizes
- Inventory
- Operational expenses
- Bottleneck
- Throughput
- Resistance
- Net profit

According to Goetsch and David there are three constraints that need to be studied in the goal. These constraints directly affect the system performance more than the downtime, balanced plants, machine efficiency, and equipment utilization. They are listed below:

- Inventory
- Operational expenses
- Throughput (Wortman et al. 2007)

Theory of constraints is mostly concerned about bottleneck, inventory, operational expenses, and balanced parts. "Bottleneck" is the term used when the resource capacity is equal to or less than the specified demand. If resource capacity is greater than the specified demand then it is known as nonbottleneck operation. If the equipment or process is said to be the bottleneck then the load has to be reduced from that equipment or the equipment should be made to work only when the part is required immediately. Throughput is defined as the rate at which the system makes money out of its sales. Inventory is the amount invested in procuring items required for the production or services. Operation expenses are the expense required to convert the inventory to throughput. These include all cost parameter including holding cost, scrap, depreciation, etc. The above terms are used to define the cash inflow and cash outflow of the system (Goldratt 1986).

16.3.3.1   Methods for implementing theory of constraints

- Identify the constraint: This constraint causes the organization to reach the optimum goals and performance. These constraints need to be arranged according to the priority for higher impact.
- Exploit: Methods need to be evaluated to exploit the constraints.
- Subordinate: The above-mentioned constraints have some impact; if it goes beyond the limit the methods need to be sorted out to reduce the effects of constraints.
- Elevate: This process thrives in improving the system. Try to reduce or eliminate the problems of the constraint.
- Back to step 1: If the constraints collapse, then go to the step 1 and try to figure out new constraints.

16.3.3.2   Drum buffer rope

The drum buffer rope concepts is apt for the subordinate step as mentioned above. It concentrates on eliminating the effects of the constraint. It also focuses on removing the factor responsible for the bottleneck or by simply increasing the capacity of the bottleneck. If the bottleneck cannot be increased then the capacity of the bottleneck output is increased.

There must be smooth material flow to balance the bottleneck. Various machines have different output rate so it affects the work in process (WIP) rate. It is undesirable to have both high WIP rate and low WIP rate. So it is favorable to figure out optimal WIP rate to keep the production process balanced. To ensure an optimal buffer level, some sort of feedback mechanism is required to control the raw material flow to the downstream machine. For this purpose drum buffer concept is used. Components present in this system are drum, buffer, and rope:

- Drum: Drum represents the pace of the operation or process.
- Buffer: It is required to keep the bottleneck operating in a regular pace. This represents the buffer, inventory, and WIP.
- Rope: This represents the feedback of the system. Feedback acts from the buffer to the inputs. Dispatch point will release an optimal number of materials to maintain smooth flow of inventory (Figure 16.6).

Other methods used in the theory of constraints are

- Effect cause effect
- Evaporating clouds
- Prerequisite clouds

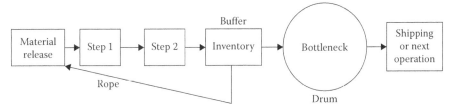

Figure 16.6 Drum buffer rope example.

The first method uses brainstorming technique to find the problem and the cause for it. This technique includes providing assumption for the problem and analyzing the possible route cause for it. Evaporating cloud involves determining the conflicting assumption for the problem. This also includes providing simple solution for the assumption. Solution is obtained by examining the problem.

16.4   Black belts

"A test or series of tests in which purposeful changes are made to the input variables of a process or system so that we may observe and identify the reasons for changes that might be observed in the output response." The plan for an experiment is also called the "design of experiment" or "DOE" (Senecal et al. 2000). The traditional experiments focus on one FAT at two or three levels and hold every other variable constant. Hence, one FAT has complications to be used for experimentation. The shortcomings of traditional experimentation are listed as follows (Wortman et al. 2007):

- Numerous experiments are to be performed to study all the variables.
- The optimum solution for the given variables is never known.
- The interaction between factors cannot be determined.

Hence, DOE proves to be an efficient tool in overcoming these shortcomings. DOE is a methodology of varying the input factors simultaneously, such that their individual and combined effects on the output can be carefully studied. The following are the advantages of using DOE:

- Many factors can be evaluated simultaneously and hence DOE is economical.
- In-depth statistical knowledge is always not necessary.
- The important factors can be distinguished from the less important ones.
- The experimental designs are balanced and hence confidence in the conclusions can be drawn.
- The results can be improved without additional costs.

16.4.1   DOE terminology

Before studying the procedure of performing factorial tests it is important to know the terminology used in DOE. The terminologies and its definitions are taken from CSSBB (2007).

- *Alias:* An alias occurs when two factor effects are confused with each other.
- *Balanced design:* The fractional factorial design is said to be balanced when each factor is given equal number of trials. (The factorial designs are discussed in the later part of the section.)
- *Block:* The experimental plan is always divided into subgroups. A subdivision which has relatively homogenous experimental units is called a block.
- *Blocking:* Block is a dummy factor which does not interact with the real factors. Blocking is done to account for variables the experimenter wishes to avoid.
- *Collinear:* When two variables are totally correlated with each other collinear occurs. In that case one variable must be removed from the experiment.
- *Cofounded:* This condition occurs when the effects of two factors are not separable.
- *Correlation coefficient:* A number between a negative and positive one (−1 and 1) indicates degree of linear relationship between two sets of numbers. Zero (0) indicates no linear relationship.
- *Covariates:* Covariates are things that change during the experiment which were initially considered not to change.
- *Curvature:* Curvature refers to nonstraight line behavior between one or more factors.
- *Degrees of freedom:* DOF is the number of measurements that are independently available for estimating a population parameter.
- *DOEs:* Design of experiments is the order in which the experiments must take place. It is in this stage that the number of levels of one or more factors or its combination is selected to be included in the experiment. The factor levels to be selected are evaluated in a balanced full or fractional factorial design.
- *Efficiency:* Efficiency is stated for estimators. If one estimator has less variance than the other, then the estimator with the low variance is considered efficient.
- *EVOP:* EVOP stands for evolutionary operation. The term describes the ways in which sequential experiments can be planned to adapt to system behavior by learning from the present results and predicting the future treatments for better responses.
- *Experimental error:* Variation in responses of identical or similar test conditions is known as experimental error. This is also known as residual error. These are better explained in the examples in the later sections of the chapter.
- *Fractional:* Fractional represents the fewer experiments required to be performed than that of the full design experiments. An example of three-factor design with two-level, half-fractional designs is shown as follows:

A	B	C
−	−	−
−	+	+
+	−	+
+	+	−

A	B	C
−	−	+
−	+	−
+	−	−
+	+	+

- *Full factorial:* Full factorial represents the experimental design in which all combinations of all levels of factors are considered. No combination factor is omitted. An example of a two-level, three-factor full factorial design is shown as follows:

A	B	C
−	−	−
−	−	+
−	+	−
−	+	+
+	−	−
+	−	+
+	+	−
+	+	+

- *Input factor:* Input factor is an independent variable that may affect the response of the dependent output variable.
- *Inner array:* The factors can be controlled in a process called an inner array. It is followed in Taguchi fractional factorial design.
- *Interaction:* An interaction in the experiment occurs when the effect of one input factor on output is dependent on the level of another input factor. Interactions can easily be examined in full factorial design.
- *Level:* A given factor or a specific setting of an input factor.
- *Nested experiments:* Nested experiments are designs in which the experiments are not completely randomized. There is logical reason for the experiment design.
- *Optimization:* Optimization involves finding the best combination of input factors from the various levels to provide the optimum results for the output.
- *Orthogonal:* A design is orthogonal if the main and interaction effects in a given design can be estimated without confounding the other main effects and interactions. A full factorial is said to be balanced or orthogonal because there is an equal number of data points for each input under each level.
- *Outer array:* The factors that cannot be controlled in a process are called outer array. It is followed in Taguchi style experimental designs.
- *Paired comparison:* It is a technique in which sample-to-sample variations are ignored or considered irrelevant.
- *Parallel comparison:* The experiments on each level are done at the same time. It is the opposite of sequential comparison.
- *Precision:* The closeness of the results with the actual value is termed precision.
- *Qualitative:* This term represents the description of levels considered in the experimental design. It does not give interval or origin. For example, different machines, operators, and materials are all qualitative levels.
- *Quantitative:* These represent order and interval of the levels considered for the experimental design.
- *Randomized trials:* The experiment is completely free from the environment and biases are eliminated completely. The behavior of the system is completely random.
- *Repeated trials:* The trials are repeated in order to judge or estimate trial-to-trial experimental error.
- *Residual error:* Residual errors are the variation or difference between the observed and the predicted value of the result. The predicted values are obtained based on some empirical model.

- *Residuals:* Residuals are the difference between the experimental output and predicted values.
- *Resolution I:* The experiments are conducted by varying one factor at a time in order to get the best result.
- *Resolution II:* The experiment is carried out by cofounding some of the effects.
- *Resolution III:* It is a fractional factorial design in which main effects are not cofounded with each other, but main effects and two-factor interaction effects are cofounded.
- *Resolution IV:* It is a fractional factorial design in which the main effects and the two factor interaction effects are not cofounded but the two-factor effects are cofounded with each other.
- *Resolution V:* It is a fractional factorial design in which only two-factor interactions are cofounded with the three-factor and higher interactions.
- *Resolution VI:* It is a full factorial experimental design with no cofounding.
- *Resolution VII:* The experimental setup with eight blocks of eight runs.
- *Response surface methodology:* It is a graphical representation in which system response is plotted against one or more system factors. The experimental design is used to identify the "shape" of the response surface (the effect of output for the input factors). Later, geometric concepts are used to find the relationships.
- *Response variable:* Response variable is the output of an experimental treatment.
- *Robust design:* A response variable is considered robust to input variables when it is difficult to control it.
- *Screening experiment:* It is a technique in which the most probable factors in an experimental system is identified.
- *Sequential experiments:* The experiments are performed one after another in a sequence. It is the opposite of parallel experimentation.
- *Simplex:* Simplex is a geometric figure which shows that a number of vortexes are equal to one more than the number of dimensions in the factor space.
- *Simplex design:* It is a spatial design used to determine the most important variable combination for the experiment to be carried out.
- *Test coverage:* It is the percentage of all possible combinations of input factors in experimental design.
- *Treatments:* Treatments describe how the experiments are to be carried out for different levels.

16.4.2   Design selection guidelines

The crucial procedure in DOE is selection and scaling of process variables. Process variables include both input and output. The input chosen must have maximum effect on the

Table 16.2 Design Guidelines

Number of factors	Comparative objective	Screening objective	Response surface methodology
1	One-factor completely randomized design	–	–
2–4	Randomized block design	Full or fractional factorial	Central composite or Box–Behnken
Percent or more	Randomized block design	Fractional factorial or Plackett–Burman	Screen first to reduce number of factors

output variable and hence the selection should be done by a team. The design guidelines for the experimental design are given in Table 16.2 (CSSBB 2007).

16.4.3   Various examples for performing DOE

16.4.3.1   One factor, multiple levels

The one factor experiment is explained well with an example. Consider the following example of a golf ball manufacturer. A golf ball manufacturer—a customer of the Chemicals SBU—is evaluating the effects of different golf ball dimple patterns on the distance the balls are driven off the tee. Balls are loaded into, and then struck by, a mechanical device called an "Iron Byron." What graphical methods can be used to explore the data? Looking at the data types, what statistical test(s) can be used to evaluate the effect of the dimple pattern? What are the conclusions? What further questions are needed to be answered about the experiment, data, etc.? The example is best explained with the help of Minitab software. One-way ANOVA is performed. One-way ANOVA is better explained in Chapter 14. Here are the results of one-way ANOVA (Figures 16.7 and 16.8).

The statistical test chosen is one-way ANOVA. The experimental design is one factor with multiple levels (only the factor dimple pattern is studied in this example). From the following screenshot, it can be seen that dimple pattern with a p-value of .000 is significant: dimple pattern has a strong influence on the distance the ball travels. The plot accompanying the ANOVA table indicates that there may well be a significant difference between dimple pattern #1 and patterns #2 and #3. There may also be a significant difference between pattern #4 and patterns #2 and #3. There appears to be no significant difference between patterns

MTB: Stat > ANOVA > One way

Figure 16.7 One-way ANOVA.

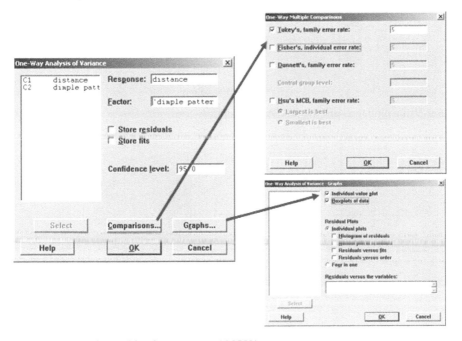

Figure 16.8 Selection of variables for one-way ANOVA.

#1 and #4. From this plot, it is not obvious whether the difference between patterns #2 and #3 is significant. The comparison tests—a portion of the output of the Tukey comparison is shown—can answer that question (Figures 16.9 through 16.12).

A "residual" is the difference between the observed result and the predicted outcome

$$e_{ij} = y_{ij} - y_j \qquad (i = \text{replicate index}; j = \text{subgroup index}).$$

In the case of the one-way ANOVA, the predicted outcome for any member of a subgroup (level) is simply the subgroup average. ANOVA residuals are the differences between the individual observations and their subgroup averages. Nonrandom behavior (patterns) in the residuals plots can indicate problems with data collection, the model (need to transform the data), or other signals. The important step in ANOVA is evaluating the residuals. To obtain the residual plots from Minitab, the following figure can be repeated. By selecting the four-in-one option, the software provides with four different graphical representations of the residuals. These are as follows (Figures 16.13 and 16.14):

- Histogram
- Normal plot
- Versus fits
- Versus order

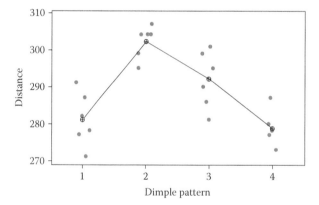

Figure 16.9 Value plot; interpreting ANOVA results.

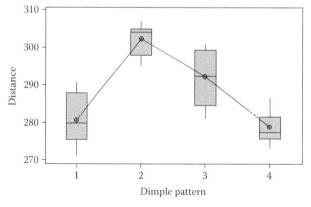

Figure 16.10 Box plot; interpreting ANOVA results.

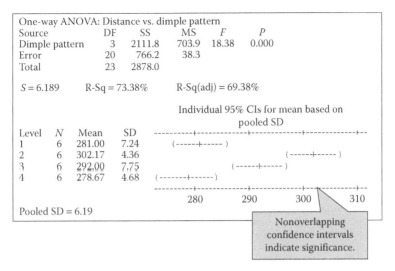

Figure 16.11 One-way ANOVA for dimple pattern.

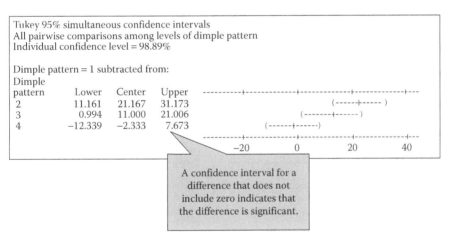

Figure 16.12 One-way ANOVA for dimple pattern.

The figures gave an insight into one factor, single level. When the test is to be carried out with five golfers it makes an example of one factor, multiple levels. Figures 16.15 through 16.17 show the results for one-way ANOVA for five golfers.

The results of one-way ANOVA from the above figure raise the question "is it possible that the differences among the golfers affect the results?" (Figure 16.18).

A plot of residuals versus other variables may reveal a factor or effect that was not considered in the original model. Another factor—golfer—has been added to the system. Although this factor is not the focus of this study, it is possible—and necessary—to eliminate its potential influence. Earlier, the other factor, golfer was not considered. Now the following example considers the other factor. The golfer factor is not something the experimenter can influence—but it does have an effect. In DOE terminology, golfer is a "nuisance factor": known and controllable, at least for the experiment.

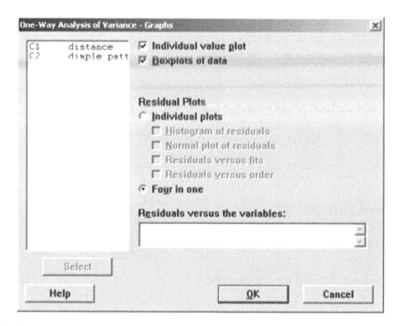

Figure 16.13 Four in one.

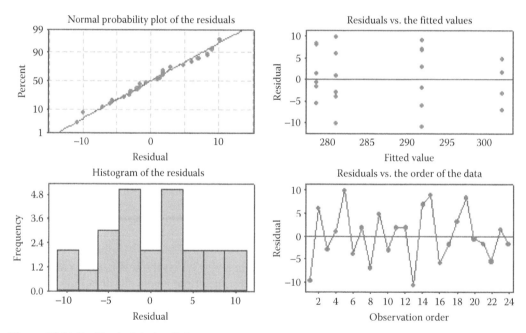

Figure 16.14 Residual plots for distance.

A versatile ANOVA tool that can model multiple factors affecting one or more responses is the general linear model. The general linear model also has the ability to deal with covariance, interactions between factors, and has main effect and interaction plots available. Figure 16.19 shows the results of general linear model.

The main effects plot for general linear plot is shown below. The plot shows a significant effect on pattern and golfer (Figure 16.20; Table 16.3).

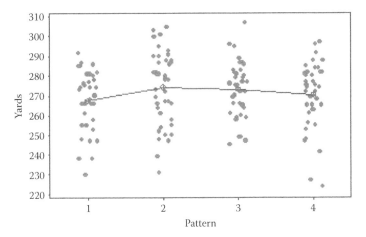

Figure 16.15 Dot plots.

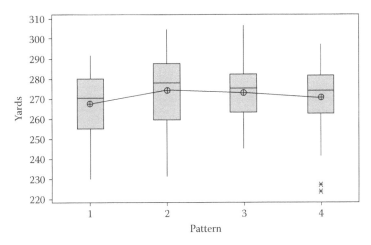

Figure 16.16 Box plots.

From the given figure it is clear that the pattern is significant. The pattern is now separated from golfer. Also the experiment is always carried assuming no interactions.

Nuisance factor: A variable that probably has an effect on the response, but its effect on the output is not of interest. Types of nuisance variables:

- Unknown and uncontrolled factor—handle with randomization.
- Known and controllable factor—handle with blocking.
- Known, but uncontrollable factor   might be able to compensate by using analysis of covariance.

The nuisance factor is known, but uncontrollable. In the golf example, this could be variation within a single golfer, perhaps due to fatigue as the day goes on. The order of dimple pattern by golfer should be randomized to remove potential systematic effects (in the example of the tiring golfers, all Pattern 1 balls need not be hit early and all Pattern 4 balls need not be hit late).

One-way ANOVA: Yards vs. pattern

Source	DF	SS	MS	F	P
Pattern	3	999	333	1.25	0.293
Error	152	40425	266		
Total	155	41424			

$S = 16.31$ R-Sq = 2.41%   R-Sq(adj) = 0.49%

Individual 95% CIs for mean based on pooled SD

Level	N	Mean	SD
1	38	267.47	15.19
2	41	274.00	18.48
3	38	272.97	14.07
4	39	270.31	16.94

```
                                        ------+---------+----------+----------+----
1                                       (------------+------------)
2                                                    (----------+---------)
3                                                 (----------+-----------)
4                                            (----------+---------)
                                        ------+---------+----------+----------+----
                                        265.0   270. 0    275.0     280.0
```

Pooled SD = 16.31

Figure 16.17 One-way ANOVA results for five golfers.

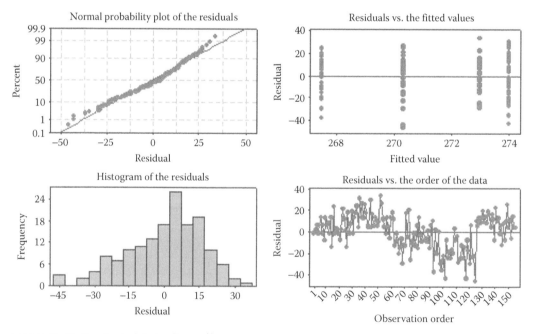

Figure 16.18 Residual plots for five golfers.

Randomized block design can be used for known and controllable nuisance factors. Examples: day, shift, operator, supplier, and machine. It allows the effect of the nuisance variable to be isolated and removed from the effects of experimental factors. Randomized block design is a powerful and necessary approach in industrial experiments. Approaches to experimental design are as follows:

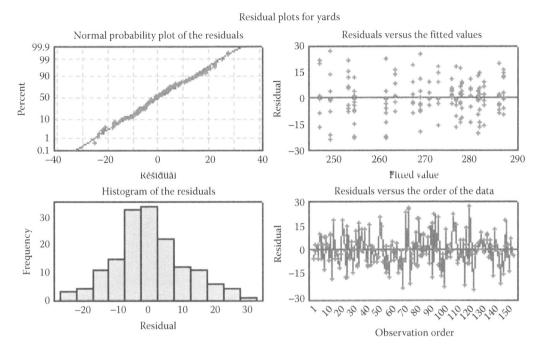

Figure 16.19 Residuals for general linear model.

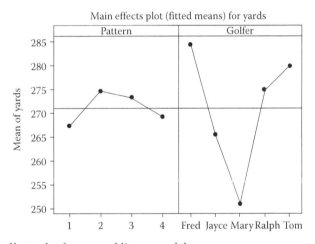

Figure 16.20 Main effects plot for general linear model.

One-factor-at-a-time (OFAT): It focuses sequentially on individual factors. It holds all other factors at nominal constant level until the focused "design" factor produces its optimum response (Figure 16.21).

Stick-with-a-Winner (SWAW): It is another design strategy that sequentially focuses on individual factors. It keeps all other factors at the previous level until the focused "design" factor produces its optimum response (Figure 16.22).

Table 16.3 Results of One-Way ANOVA for General Linear Model

Factor	Type	Level	Values			
		Yards vs. golfer and pattern				
Golfer	Fixed	5	Fred, Joyce, Mary, Ralph, Tom			
Pattern	Fixed	4	1, 2, 3, 4			
Source	DF	Sequential SS	Adjusted SS	Adjusted MS	F	P
		ANOVA for yards using adjusted SS for tests				
Golfer	4	22508.2	22881.0	5720.3	48.26	0.000
Pattern	3	1372.2	1372.2	457.4	3.86	0.011
Error	148	17543.7	17543.7	118.5		
Total	155	41424.2				
S = 10.8876		R-Sq = 57.65%		R-Sq(adj) = 55.65%		

SS, Six Sigma.

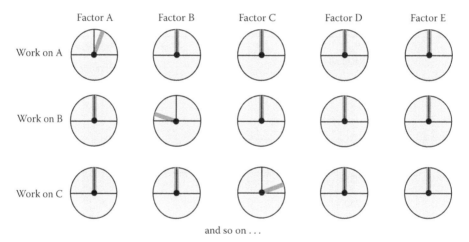

Figure 16.21 Example OFAT.

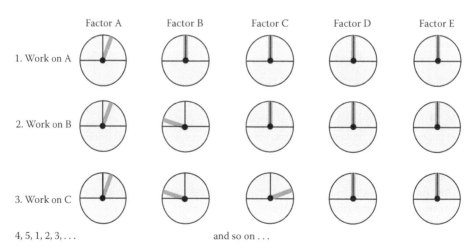

Figure 16.22 Example SWAW.

Table 16.4 Comparison

OFAT	Statistical design
Time consuming and inefficient	Efficient and effective
No guarantee of finding optimum	Optimum is methodically sought, statistically verified, and documented.
Interaction of variables can produce incorrect conclusions	Interactions of variables are incorporated into the design.

Table 16.4 provides comparison between OFAT and statistical design.

16.4.3.2   Planning experiments

Generally, the experiments must be planned in the following sequence. The main aim is to improve and identify the optimized system for the process to enhance quality. Hence, the Six Sigma team can select the procedure from the below available designs.

- Exploratory (Screening)
  - Objective: finding the vital x(s)
  - Types of designs
    - Full factorial
    - 2^k factorial
    - 2^k fractional factorial
- Optimization
  - Objective: finding the set point for x
  - Types of designs
    - Full factorial
    - 3^k factorial
    - Response surface designs
- Confirmation
  - Objective: verifying the set-point for x
  - Implementation

As a thumb rule, a black belt or a champion must follow the following set of instructions time and again to ensure the team is heading in the proper direction. The fatalities of selecting a wrong variable for both input and output could lead to catastrophe.

- Recognition of and statement of the problem
  - Develop objectives of experiment
  - Solicit input from all concerned parties
  - List specific problems or questions to be answered
- Choice of factors, levels, range
  - Potential design factors
    - Design factors, "held-constant factors," "allowed-to-vary" factors
  - Nuisance factors
    - Controllable, uncontrollable, noise
    - Range of factor change—generally broad for screening

- Selection of response variable
  - Multiple responses not unusual
  - Gauge capability (e.g., adequacy/suitability of measurement output)
- Choice of experimental design
  - Sample size, run order, blocking, etc.

The issues to be remembered while performing experimental designs are

- Performing the experiment
  - Do not underestimate the logistical and planning aspects
  - Trial runs advisable
- Analysis of the data
  - Remember simple graphical methods
  - Develop prediction equation
- Conclusions and recommendations
  - Draw practical conclusions about results and decide a course of action
  - Perform follow-up and confirmation testing

While planning an experiment, everything does not go by the book. Experience is an important tool in any Six Sigma project. Hence the role of black belt or champion proves inevitable. The black belt should always consider the practical aspects of the experiment and they are as follows:

- Use your nonstatistical knowledge of the problem.
- Keep the design and analysis as simple as possible.
- Recognize the difference between practical significance and statistical significance.
- Remember the experiments are usually iterative.
  - Best not to conduct a single, large, comprehensive experiment at start of study.
  - Learn from the first experiment about important factors, range of variation, levels, measurement.
  - General rule: Invest no more than 25% of available resources in the first experiment (generally counting the trial run).

The documentation of a planned experiment is an outline of important questions to be answered and decisions to be made in preparing to conduct an experiment. The document can be completed quickly, and the effort can prevent later problems (Table 16.5).
   Difficulties mitigated by statistical methods:

- Experimental error (noise)
  - Variation caused by disturbing factors, both known and unknown
  - May obscure important effects
  - Reduced by planned experimental design and analysis
- Confusion of correlation with causation
  - Design principles, especially randomization, help
- Complexity of effects
  - Can estimate interactive and nonlinear effects

Table 16.5 Documentation of a Planned Experiment

Project:		
Location		
Prepared by:	Date:	
1. Objective:		
2. Background information		
3. Experimental variables:		
A.	Response variables	Measurement techniques
	1.	
	2.	
	3.	
B.	Factors under study	Levels
	1.	
	2.	
	3.	
	4.	
	5.	
	6.	
	7.	
C.	Background variables	Method of control
	1.	
	2.	
	3.	
4. Replication:		
5. Methods of randomization		
6. Design matrix: (Attach copy)		
7. Data collection forms: (Attach copies)		
8. Planned methods of experimental analysis:		
9. Estimated cost, schedule, and other resource considerations:		

References

Goetsch, D. L. and David, S. B. (eds.) (2000). *Quality Management, Introduction to Total Quality Management for Production, Processing and Services.* Englewood Cliffs, NJ: Prentice Hall.

Goldratt, E. M. (ed.) (1986). *The Goal: A Process of Ongoing Improvement.* Revised first edn. Great Barrington, MA: North River Press.

Meyer, C. (ed.) (1993). *Fast Cycle Time: How to Align Purpose, Strategy, and Structure for Speed.* New York: Simon and Schuster.

Quinn, J. B. and Zien, K. A. (eds.) (1997). *Innovation Explosion: Using Intellect and Software to Revolutionize Growth Strategies.* New York: Free Press.

Schermerhorn, J. R. (ed.) (1993). *Management for Productivity.* 4th edn. New York: Wiley.

Senecal, P. K., Montgomery, D. T., and Reitz, R. D. (2000). "A methodology for engine design using multi-dimensional modeling and genetic algorithms with validation through experiments." *International Journal of Engine Research*, 1(3): 229–248.

Thompson, A. A. and Strickland, A. J. (eds.) (1995). *Strategic Management: Concepts and Cases.* Chicago, IL: Irwin.

Wortman, B. et al. (2007). *CSSBB Primer.* Terre Haute, IN: Quality Council of Indiana.

Further Readings

Automotive Industry Action Group (AIAG) et al. (1995). *Statistical Process Control (SPC): Reference Manual*. Southfield, MI. Automotive Industry Action Group.
Merkin, B. (2001). "Eleven ways to look at chi-squared coefficient for contingency tables." *The American Statistician*, 55(2): 111–120.

Review Questions

Question 1: In a hypothetical situation, if all within time, piece to piece, and measurement variation were removed from a process, what would be left?

(a) Time to time variation
(b) Product spread
(c) Nothing, it would all be eliminated
(d) Inherent process variation

Question 2: A process has been experiencing problems lately. The operators charting the process have identified the cause to be due to a change in incoming materials. This problem is

(a) A common cause
(b) A normal event
(c) Attributed to purchasing
(d) A special cause

chapter seventeen

Control

Using data to maintain success

Our task is not to fix the blame for the past, but to fix the course for the future.

—John F. Kennedy

Overview

Steps	Substeps
Define	• Quality projects (sponsorship)
	• Define specific project (black belt)
Measure	• Identify critical characteristics
	• Clarify specifics and targets
	• Measurement system validation
Analyze	• Determine baseline
	• Identify improvement objectives
	• Evaluate process inputs
Improve	• Formulate implementation plan
	• Perform work measurement
	• Lean execute the proposed plan
Control	• Reconfirm plan
	• Set up statistical process controls
	• Operational process transition

Source: PWD Group LLC.

Recommended Tools by Belt

		Define	Measure	Analyze	Improve	Control
Stage 1	Yellow belt	Interview process		X–Y map/QFD/ house of quality	EVOP	Control plans
		Language processing		Fishbone diagram/RCA		Visual systems
		System map (flowchart)				5-S
						SPC/APC

(Continued)

Recommended Tools by Belt (Continued)

		Define	Measure	Analyze	Improve	Control
Stage 2	Green belt	Prioritization matrix	Hypothesis testing	FMEA	Design of experiments	TPM
		Thought process map	Flowdown	Multi-vari chart	Fractional factorials	Mistake proofing
		Stakeholder analysis	Measurement system analysis	Chi-square	Data mining	
		Thought process map	Graphical methods	Regression	Blocking	
			Analysis of variance	Pareto analysis		
Stage 3	Black belt	Value stream map	Process behavior charts	Buffered tolerance limits	Theory of constraints	
		Gantt chart development	QFD	Process capability analysis	Multiple response optimization	
		Project management			Response surface methodology	

Source: PWD Group LLC.

Notes: Black belts are expected to know the continued knowledge obtained throughout green belt and yellow belt certifications. The same goes for green belt and yellow belt. Empty cells represent that skill is not required for that belt stage; however, a general knowledge of the step is expected.

5-S, sort, set in order, shine, standardize, and sustain; APC, advanced process control; EVOP, evolutionary operations; FMEA, failure modes and effect analysis; QFD, quality function deployment; SPC, statistical process control; TPM, total productive maintenance.

17.1   Introduction

The control phase is the final step in the DMAIC (define, measure, analyze, improve, and control) process. The customers' views are translated into quality that is to be improved, the metrics are established for the processes that control the quality and measured, the key process input variables that affect the key process output variables are analyzed, and the variables under study for enhancing the required quality are improved by identifying optimum values. Having performed all of the above-mentioned steps, control is the execution phase. The improvement measures are executed and the process activities are monitored for continuous improvement. This chapter thus provides information about the activities in control phase and the tools used by different belt categories. There is no role for yellow belt holders in control phase except monitoring and documenting the execution. The decisions related to changes are made by black and green belt holders using the tools discussed later in the chapter. The main steps in the control phase can be classified as follows:

- Validating the plan
- Control inputs and monitor outputs

17.2   Validating the plan

17.2.1   Purpose

The purpose of validating is to reevaluate the process performance, to run the process using the new design or settings, and to collect necessary performance data and assure that predicted performance levels are being met.

17.2.2   Objective

The objective is to assess the performance of process being run at predicted optimum configuration/settings.

17.2.3   Deliverable

The main activity in the control phase is to verify the mean, variability, and Z-score levels from the confirmation runs and provide plan for follow-up if the Z-score target has not been met (Figure 17.1).

The steps to be followed in validating the plan are as follows:

- Run the new/revised process.
- Monitor results carefully.
- Calculate performance statistics.
- Compare the actual results with the expected results.

It is important to quantify the noise in the process once the entire x(s) are set. Remember that not all x(s) need to be set, as trivial x(s) that are uncontrollable still vary. If the noise does not lead to an acceptable Z-score level, an enhancement cycle must be performed.

The first step is to run the new or revised process in the "production" environment to ensure compliance, to verify that new practices are being followed, and to determine the reasons for failures to follow. The second step is to carefully monitor the results and to collect data about process performance. The monitoring is done to verify the stability of the process and to deal with the unanticipated cases such as the following:

- Not unusual
- Encouraging those involved to report problems/omissions
- Clarifying or revising process as needed

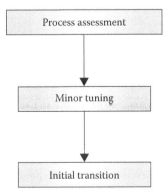

Figure 17.1 The validation plan.

The next step is to calculate and compare results. The process performance statistics, including process average, variability, and Z-scores, is calculated. The obtained results are compared with the results of the Z-score established. (The established Z-score is analyze phase.) If the project goals have not been met, the black belt is responsible for deciding what action is required and an enhancement cycle may be necessary. (Enhancement cycle is performed in improve phase). The final step in validation process is the transition to process owner that includes the following:

- Successful verification of the process improvements marks an important milestone in the transition of control responsibility to the process owner.
- There should be no surprises.
  - The process owner was identified as a key stakeholder at the beginning of the project.
  - The process owner has been involved with the testing and evaluation of the changes.

The process owner can be anyone of the following:

- He does not have a formal title
- He is responsible for and committed to the successful performance of the process.
- He may be the project sponsor.
- His role may be shared among several people.

17.3 Control inputs and monitor outputs

The next part in control activity is to control the key process input variables and to monitor the key process output variables.

17.3.1 Purpose

The purpose is to continue to run process at new settings and collect data on process mean, noise, and Z-score levels. The plan must be defined to ensure that level of x's does not change; but if it does, there should be a corrective control plan available.

17.3.2 Objective

The objective is to identify and design appropriate techniques for maintaining acceptable process performance.

17.3.3 Deliverable

The deliverable for this part would be to provide with a plan for maintaining the chosen level for each vital x.

The previous steps have verified that the right plan has been formulated and that its results meet the target objectives. It now becomes important to not let the input levels change and to continue to monitor that the desired output is maintained. Figure 17.2 shows the hierarchy of control.

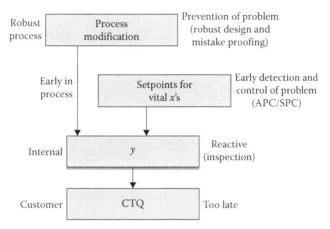

Figure 17.2 The hierarchy of control. APC, advanced process control; CTQs, critical to quality; SPC, statistical process control.

17.4   Controlling x's

Previously, the vital x's have been identified and the best settings for each x have been determined. The key is to not let the settings vary away from the targeted setting. Maintaining the targeted level can be achieved through either detection or prevention. The attribute x's are usually maintained through control plans and the continuous x's are usually maintained through automated process control and statistical process control (SPC). Figure 17.3 shows the natural erosion after process improvement has taken place. So control phase is the one which ensures that the execution of project is in right track. The control phase initiates steps to avoid the natural erosion.

Figure 17.4 shows the use of control in avoiding the erosion process.

Standard is the best currently known way to perform an activity. The standard contains essential information about actions that are vital to perform the process in a way that will deliver in the required levels of process output or to react when there is risk that the performance is not satisfactory. The standard describes performance related to the vital x's. Figure 17.5 shows a schematic representation of standardization.

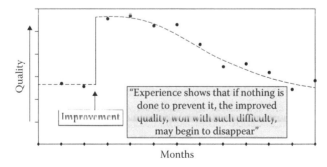

Figure 17.3 Natural erosion after continuous improvement. (Box, G. E. P. and Luceno, A.: *Statistical Control by Monitoring and Feedback Adjustment.* 1997. Copyright Wiley-VCH Verlag GmbH & Co. KGaA. Reproduced with permission.)

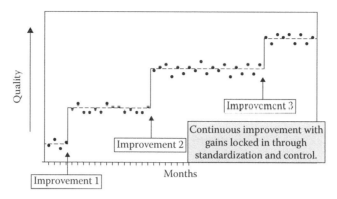

Figure 17.4 Standardization and control.

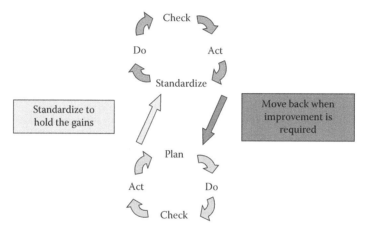

Figure 17.5 Standardization.

17.5   Green Belt

17.5.1   5S (workplace organization)

What is 5S?

- Disciplined approach to housekeeping
- Goal of permanently maintaining an area's orderliness/cleanliness
- Optimized functional aspects of the area

For a company to call itself world class, the fundamental first step it should take is to apply 5S. The implementation of 5S in any organization shows the presence of the senior management in workplace organization, lean manufacturing, and elimination of muda (waste in Japanese). The 5S program makes sure that the resources are provided when needed and available at the right place. The five words of 5S in Japanese are as follows (Figure 17.6):

1. Seiri (proper arrangement)
2. Seiton (orderliness)
3. Seiso (cleanup)

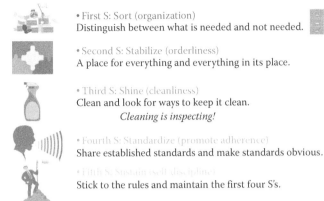

- First S: Sort (organization)
  Distinguish between what is needed and not needed.

- Second S: Stabilize (orderliness)
  A place for everything and everything in its place.

- Third S: Shine (cleanliness)
  Clean and look for ways to keep it clean.
        Cleaning is inspecting!

- Fourth S: Standardize (promote adherence)
  Share established standards and make standards obvious.

- Fifth S: Sustain (self-discipline)
  Stick to the rules and maintain the first four S's.

Figure 17.6 Workplace organization (5S).

4. Seiketsu (standardize)
5. Shitsuke (personal discipline)

For American companies, the translated English equivalents are as follows:

- Sort: separate what is needed from what is not
- Straighten: put everything in place
- Scrub: clean and make the workplace spotless
- Standardize: follow routine for cleaning and checking
- Sustain: follow the previous four steps and improvise

The 5S approach is a method of determination to make the workplace organized, keep it neat and clean, establish standardized questions, and maintain the discipline that is required for the job. Numerous modifications are made for the 5S program. It can become either 4S or 5S + 1S program where the sixth S is safety.

Imai (1997) talks about a visit of a Japanese team to a prospective supplier, and before allowing the supplier to show their grand presentation, the Japanese team insisted on a tour to their shop floor and rejected the supplier in a few minutes before even seeing the presentation. The team knew that the plant was not committed to the highest levels of manufacturing and terminated their visit. It is obvious that on a day-to-day basis there will be some dirt around, but not implementing 5S becomes evident on the very first look.

17.5.1.1 *Details of the 5S program*
Step 1: Seiri (sort)

- Planning a schedule to target each area
- Removing unnecessary items
- Recording everything that is thrown out and having red tags as stuff that is not needed
- Keeping repaired items that will be needed
- Area performs the major housekeeping and cleaning
- Breakages, rust, scratches, problems, and grime inspection in the facility
- Listing items that require repair
- Eliminating the causes of filth and grime or dealing with it

- Grime areas with a red tag show that it requires correction.
- Management review performance on this and other areas

Step 2. Seiton (straighten)

- Everything should have a place and it should be made sure that everything is in its place.
- Analyze the conditions for the existing tooling, equipment, inventory, and supplies.
- Create a name and location for everything and decide where things go.
- Employ labels, tool outlines, and color codes.
- Obey the rules and determine the out-of-stock conditions and everyday controls.
- One person does the reordering and reduction of inventories.
- Determine who has missing items or if they are lost.
- Employ aisle markings, placement for dollies, forklift, and boxes should be given importance.
- Pallet zones should be established for work-in-process (WIP) inventories.

Step 3: Seiso (scrub)

- This does not involve only keeping things clean and it is more of how to keep things clean.
- A commitment is to be established for being responsible in all working conditions.
- Everything in the workplace should be clean, including the equipment.
- Root cause analysis is to be done to remedy machinery and equipment problems.
- Equipment maintenance training should be given high priority
- Division of areas into zones and assigning individual responsibilities should be performed.
- Unpleasant or difficult jobs are to be rotated.
- Three-, 5-, and 10-minute 5S activities should be done.
- Inspection checklists and white glove inspections are to be performed.

Step 4: Seiketsu (standardize)

- Abnormal conditions show up when 5S becomes a routine.
- Important points to be managed are to be determined and where to look should also be figured out.
- To ensure a state of cleanliness, maintenance, and monitoring facilities is important.
- Visual controls to make abnormal conditions are to be made obvious.
- Setting standards, determination of necessary tools, and identification of abnormalities should be done.
- Inspection methods should be determined.
- Short-term counter measures and long-term remedies are to be determined.
- Color coding, markings, and labels should be employed as visual management tools.
- Equipment markings, maps, and charts are to be provided.

Step 5: Shitsuke (sustain)

- Continually improve and commit to the previous four steps.
- The habits of the previous four steps should be used and self-discipline should be acquired.

- Standards for each of the 5S steps should be established.
- Evaluations of each step should be established and performed (Imai 1997; Osada 1991).

The commitment of the management will determine the control and self-discipline areas for the organization. A 5S program can be set up and operational within 5–6 months, but it takes a lot more than just setting it up. The effort to maintain world-class conditions must be continuous. A well-run 5S program will result in a factory that is in control.

17.6   Kaizen

Kaizen is nothing but the Japanese word for continuous improvement. It is taken from the two Japanese words *kai* (change) and *zen* (good). It usually means improvement, which occurs incrementally but on a continuous basis and involves everyone in the firm. Radical innovations have enthralled Western management. Seeing major breakthroughs is being enjoyed and the business is run from home. Kaizen is a term also for the following:

- Productivity
- Total quality control
- Zero defects
- Just-in-time
- Suggestion systems

The strategy of Kaizen is as follows:

- Kaizen management: Management is responsible for maintenance and improving operating standards.
- Process value results: The key to success is the improvement of processes.
- Employing plan–do–check–act (PDCA)/plan–do–study–act (PDSA) cycles: PDCA is an improvement method. The check cycle refers to verification that the process of implementation is being done and is on target to meet goals.
- Quality first: Highest priority is given to quality.
- Speaking with data: Hard data are used to solve problems.
- Customer is the next process: Every step in the process has a customer. It is important to provide the next step with good parts of information.

17.6.1   Kaizen blitz

Most Kaizen activities are considered for long-term processes; a different type of Kaizen strategy can also occur. This has been termed as a Kaizen event, Kaizen workshop, or Kaizen blitz, which involve Kaizen activities in specific areas (planning, training, and implementation) within a short period (Gee and Izadi 1996; Laraia et al. 1999).

The Kaizen blitz employs cross-functional teams for a period of 3–5 days and results in a rapid workplace change on a project basis. This team involves members from various teams such as accounting, marketing, management, maintenance, quality, and production. If the work demands more from a specific department, more members are taken from that department.

Based on the experience levels of groups, a 5-day Kaizen blitz begins involving 2 days of intense sessions on continuous improvement concepts. This is followed by 3 days of complete data collection. Analyzing the workplace and implementation, deep management commitment is required for the last portion of the workshop. The decision-making process as determined by the Kaizen blitz team and facilitator must be trusted by the plant managers.

The implementation stage requires a significant amount of time and money. A final presentation of the project is made by the team and is shown to the plant manager and those who are interested in the presentation. There are possibilities of bringing about immediate changes and benefits by every project.

Tarala et al. (1999) emphasizes that Kaizen blitz events must occur with minimum expense and maximum use of people. The basic changes are concentrated in the process flow and methodology only.

Various metrics that are employed to determine the outcomes of Kaizen blitz are as follows:

- Saved floor space
- Flexibility in the line
- Workflow improvement
- Improvement in ideas
- Increased quality levels
- Safe work environment
- Nonvalue-added time reduction

17.7   Total Productive Maintenance

Total productive maintenance (TPM) is a tool that promotes the coordinated group activities for the equipment to perform better and also requires the operators to share responsibly for routing machine inspection, cleaning, maintenance, and minor repairs. The professional maintenance staff is responsible for the major repairs and serves as coaches for the routine and minor items.

Key features of total productive maintenance (TPM) are as follows:

- Works toward increasing equipment effectiveness
- Maintains a system for productive maintenance for a machine's life span
- Implementation by engineering, operations, and maintenance
- Involves every employee of the firm, from top management to the floor employees
- Operator autonomous maintenance
- Small group activities by the company

The "total" in TPM stands for

- Total effectiveness in pursuing the economic efficiency and profitability.
- Maintenance prevention, maintainability, and preventive maintenance.
- Total participation of all employers from the top management to the shop floor operators.

There are six big losses that can occur due to lack of TPM:

1. Equipment failure: change the mindset to having zero breakdowns. Time losses for lower productivity occur to equipment breakdowns and quality losses due to defective products.
2. Setup and adjustment: time losses that occur due to setup changes. The key is to reduce setup times and have better adjustment periods.

Table 17.1 Six Big Losses

Losses	Goal
Breakdown losses	0
Setup and adjustment losses	<10 minutes
Speed losses	0
Idling and minor stoppage losses	0
Quality defect losses	0
Yield losses	Minimize

3. Idling and minor stoppages: defects in sensors, parts caught on a conveyor, etc., will cause slowdowns and losses. Minor stoppages should be zero for unmanned production.
4. Reduced speed: differences in designed and actual operating speeds result in this.
5. Process defects: malfunctioning equipment might result in scrap and defective material.
6. Reduced yield: product losses occur from machine shutdown and startup.

Table 17.1 shows the goals for each of the big six losses. Elimination of the six big losses, if achieved completely, will lead to improved plant conditions dramatically. Lean manufacturing cannot exist without TPM.

17.7.1   *Designing for maintainability and availability*

There are many situations involving corrective or preventive maintenance, ease of maintenance concerns time, material, and money. Ease of maintainability is a design feature that affects these factors. Areas of poor maintainability can be pointed out using demonstration testing or field use. Schedule and cost increase if modifications are made past initial design.

Moss (1985) provides guidelines for designing for maintainability. The guidelines are as follows:

- Standardization: The number of different parts is to be minimized by looking for compatibility of mating parts. This will result in reduced spare parts inventory.
- Modularization: Standards on sizes, shapes, and modular units are to be maintained This will allow for standardized assembly and disassembly procedures.
- Functional packaging: All needed components of an item are to be placed in a kit or package.
- Interchangeability: Dimension and functional tolerances are to be controlled. This refers to plug-in devices where interchangeable parts can be used to replace failed parts. One part can be used in other units.
- Accessibility: There should be enough space given to workers for them to perform the task properly. A part should be easy to get and replace. To gain access to failed parts, it should be made sure that good parts are not needed to be removed.
- Malfunction annunciation: This helps in notifying the operator when a unit fails. This could include gages, instrument panels, lights, or sound.
- Fault isolation: A malfunction can be traced. This is the most time-consuming task of all maintenance work. Preventive maintenance procedure, built-in-test equipment,

simplicity in design of parts, and trained personnel can be used for minimizing the problem.

• Identification: All components should have a unique identification and a method of recording corrective and preventive maintenance.

17.7.2 TPM metrics

The prime measure used to evaluate TPM is overall equipment effectiveness. There are several formula variations. Plants can adjust the factors according to their needs. They are as follows:

$$\text{Overall equipment effectiveness} = \text{Availability} \times \text{Performance efficiency} \quad (17.1)$$
$$\times \text{Rate of quality products.}$$

$$\text{Availability} = \frac{\text{Operation time}}{\text{Loading time}} = \frac{\text{Loading time} - \text{downtime}}{\text{Loading time}}. \quad (17.2)$$

Loading time is the available time per shift or per unit minus planned downtime. Scheduled maintenance and morning meetings are included in planned downtime. Loading time minus unscheduled downtime is the operation time.

Operating speed rate multiplied by the net operating rate is the performance efficiency. The operating speed rate is given by the ratio of the theoretical cycle to its actual operating cycle time.

$$\text{Operating speed rate} = \frac{\text{Theoretical cycle time}}{\text{Actual cycle time}}. \quad (17.3)$$

The measure of the stability of the equipment, the losses from minor stoppages, small problems, and adjustment losses can be done by the net operating rate.

$$\text{Net operating rate} = \frac{\text{Actual processing time}}{\text{Operating time}}. \quad (17.4)$$

$$\text{Net operating rate} = \text{Processed amount time} \times \frac{\text{Actual cycle time}}{\text{Operating time}}. \quad (17.5)$$

17.7.3 Benefits of TPM

The various factors, which TPM helps improve, are listed as follows:

• Cost reduction
• Inventory reduction
• Accident reduction/elimination
• Control of pollution
• Environment of the workplace

Improved worker and equipment utilization can be brought about by the proper integration of the TPM philosophy such as improving employee attitudes, increasing their skills, and providing a supporting work environment.

17.7.4 Steps to implement TPM

Twelve steps recommended as a path toward achieving TPM results within 3 years are as follows:

- Step 1: Top management commitment to TPM is to be announced.
- Step 2: Company communication programs are to be introduced to TPM.
- Step 3: Every functional level is to be organized to promote TPM.
- Step 4: TPM policies and goals are to be established.
- Step 5: Detailed master plan for TPM is to be prepared.
- Step 7: Improvement of the effectiveness of equipment is to be done to form project teams.
- Step 8: Building skills can be used to develop autonomous maintenance program.
- Step 9: A scheduled maintenance program is to be developed.
- Step 10: Operator and maintenance skills training are to be conducted.
- Step 11: Early equipment management programs are to be developed.
- Step 12: Have TPM implemented and aim for perfection.

17.7.5 Autonomous TPM small group activities

The heart of TPM is nothing but small group activities involving overlapping functions. The department supervisor is usually the team leader. Activities include equipment cleaning, lubrication, bolting, and inspection. The teams will progress through four stages:

1. Self-development: Learning by members
2. Improvement activities: Group improvement activities can be fulfilled
3. Problem solving: The group actively selects problems to solve
4. Autonomous management: High-level goals are selected by the group and manage their work independently.

17.8 Black belts

17.8.1 Statistical process control

SPC is a technique used to measure, monitor, and control the processes selected for Six Sigma projects by applying the statistical analysis. The SPC is performed with the help of control charts. It is a powerful technique to understand the variation of a process. The variation of a process can be classified into two basic types:

1. Chance cause variation
2. Assignable cause variation

SPC helps in monitoring a process and provides information about the stability of the process along with variation and also the cause of the variation. To show the above information, control charts can be used. The following information can be obtained from the use of control charts (Grant and Leavenworth 1988):

- Average level of quality characteristic
- Basic level of the quality characteristic
- Consistency of performance

Hence control charts can be used as a predictive tool for eliminating the cost of inspection by identifying whether the process is in its tolerance specification. Control charts can also be used for monitoring the continuous improvement process changes. If process changes were implemented, the control charts can monitor and evaluate the effectiveness. The SPC procedure can be classified as follows:

- Selection of variables
- Rational subgrouping
- Control chart selection
- Control chart analysis

Each of the above-mentioned topics will be discussed in detail in the following sections.

17.8.1.1   Selection of variables

The selection of variables of control charting is very important. If all the variables of a process is selected, the time and money spent on each variable increases drastically. In course of time, the actual process becomes secondary and charting becomes primary. But that is not the ultimate aim of a Six Sigma project. Hence the variables should be selected such that its monitoring yields profit to the project. The following can be used to select the variables for charting:

- Items that protect human safety
- Items that have a high defective rate
- Key process input variables (KPIVs) that affect the key process output variables (KPOVs)
- Major quality variables from customer complaints
- Items that increase the internal costs
- Variables that help control the process

As discussed in Chapters 11 through 14, the tools such as design of experiments and analysis of variance (ANOVA) can be used to determine the KPIV and KPOV.

17.8.1.2   Rational subgrouping

The control chart provides the information about the variation from sample-to-sample processes and determines if it is consistent with the average variation of the sample. The formation of subgroups can be used to compare the total process variation. The subgroups are nothing but observations of group of samples, which is almost homogeneous as possible. The formation of subgroup should be in such a way that it must provide maximum variation with another subgroup. The size of subgroup is very important. Mostly the size of subgroup is five. When the sample size within the subgroup is too large, there is likely to be variation within the subgroup itself, which is in controversy to the idea "homogeneous."

Before heading into the next topic, the sources of variability are important to be known. The source of variability is vital to eliminate the error and improve the process quality. The common sources of variability are as follows:

- Lot-to-lot variability
- Product spread (long-term variation of a product)

- Stream-to-stream variability
- Within-piece variability
- Piece-to-piece variability
- Inherent error of measurement (human and instrument error)

17.9   Control chart selection

Control charts are the most effective and powerful tool for analyzing the variation in any process. A process which is in statistical control (which exhibits minimum variation) is represented by plots that do not exceed the upper and lower control limits. There are many types of control charts. The two primary types are control charts for variables and attributes.

Figure 17.7 shows the types of charts and their characteristics. The various charts, their characteristics, and their selection have already been discussed in Chapter 14.

17.9.1   Control chart analysis

Control chart analysis is done to analyze the variations of charts established in the previous section. If a process is found to be out of control, then the causes of variation must be identified and eliminated in order to achieve the in-control process. An out-of-control process in a chart is recognized by points on the chart either outside the control limit or by any unusual pattern. Control limits are usually three standard deviations above and below

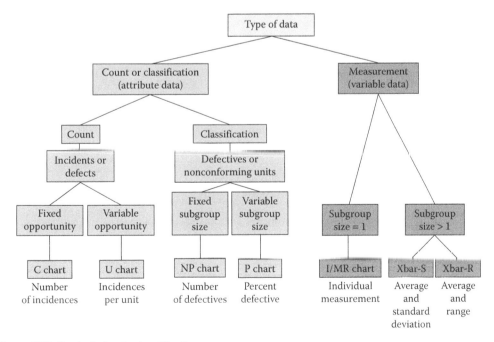

Figure 17.7 Control charts classification.

Table 17.2 Conditions Based on Average and Range

Average	Range	Result
Out-of-control	In-control	Process out-of-control
In-control	Out-of-control	
Out-of-control	Out-of-control	
In-control	In-control	Process in-control

the process average. If the process is in control, 99.73% of the averages will fall inside these limits. There are two components in a control chart: average and range. Based on these two components, there are four possible conditions that can occur in a process. They are tabulated in Table 17.2.

The following provides information for common variations occurring in the process, charts used for analysis, and corrective steps (Quality Council of Indiana 2007).

17.9.2 Trends

- X chart causes
  - Deterioration of machine
  - Tried operator
  - Tool wear
- R chart causes
  - Improvement or deterioration of operator skill
  - Tired operator
  - Change in incoming material quality
- Corrective action
  - Repair or use of alternate machine if available
  - Discuss operation with operator to find cause
  - Rotate operator
  - Change, repair, or sharpen tool
  - Investigate material

17.9.3 Jump-in-process level

- X chart causes
  - Changes in proportions of materials coming from different sources
  - New operator or machine
  - Modification of production methods or processes
  - Change in inspection device or method
- R chart causes
  - Change in material
  - Change in method
  - Change in operator
  - Change in inspection
- Corrective action
  - Keep material supply consistent
  - Investigate source of material

- Check out machine capability
- Examine operator methods and instruction

17.9.4   *Recurring cycles*

- X chart causes
  - Physical environment (temperature and humidity)
  - Tired operator
  - Regular rotation of machine or operator
- R chart causes
  - Scheduled maintenance
  - Tired operator
  - Tool wear
- Corrective action
  - If environment is controllable, adjust it.
  - Service equipment
  - Rotate operators
  - Evaluate machine maintenance

17.9.5   *Points near or outside limits*

- X chart causes
  - Over control
  - Large systematic differences in material quality
  - Large systematic differences in test methods or equipment
- R chart causes
  - Mixture of material of distinctly different quality
- Corrective action
  - Check control limits
  - Investigate material variation
  - Evaluate test procedures
  - Evaluate inspection frequency or methods
  - Eliminate operator over adjustment of the process

17.9.6   *Lack of variability*

- X chart causes
  - Incorrect calculation of control limits
  - Improvement in process since limits were calculated
  - Employee may not be making checks
- R chart causes
  - Collection of a number of measurements from widely differing lots in each sample
  - Improvements in process since limits were calculated
- Corrective action
  - Check control limits
  - Validate rational sample subgroupings
  - Verify checking procedures, gages, etc.
  - Verify proper employee measurement
  - Congratulate someone for improvement

17.10   Poka-yoke (mistake proofing)

"Poka-yoke," which refers to mistake proofing, is the concept developed by Shigeo Shingo. Poka-yoke is another tool that has been borrowed from the Japanese quality programs. It is a way to ensure that mistakes do not occur while recognizing the dignity and intelligence of employees. Translated poka-yoke is to avoid (*yokeru*) inadvertent errors (*poka*). By taking over repetitive tasks or actions, poka-yoke can free a worker's time and mind to pursue more creative and value-added activities. Shingo believed that the human error is not the only reason for the nonconfirming product. Hence he developed procedures or methods that have to be done to reduce mistakes and minimize the chances for translating those mistakes into defective items (Quality Council of Indiana 2007).

The principle of mistake proofing is that repetitive tasks or input actions that depend on constantly being alert are done and worker focuses on process improvement, rather than detection of output defects. "It is good to do it right the first time: it is even better to make it impossible to do it wrong the first time." Various possibilities of mistake proofing are to use colors and color coding, which is nothing but usage of credit card receipts (the customer gets the yellow copy and the merchant gets the white copy), and use shapes to store different types of parts of different shaped bins. Notch a stack of forms so it is easy to tell if the forms are out of order and the last possibility is to use autodetection.

The characteristics of poka-yoke are listed as follows (Figure 17.8):

- They do not require sampling and monitoring.
- They permit 100% inspections.
- They are not expensive to implement (Shingo 1986).

Poka-yoke system can be coupled with other inspection system to obtain zero defects. Possible chances for error to occur are listed as follows:

- Using wrong parts or materials
- Positioning parts in wrong direction during assembly
- Failing to properly tighten the bolt

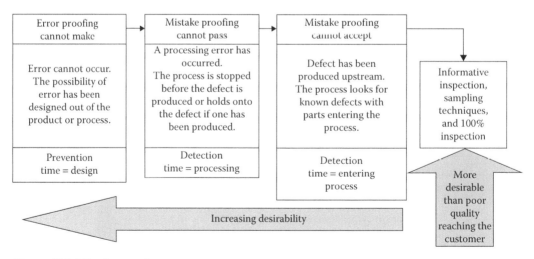

Figure 17.8 Mistake proofing.

- Skipping an operation
- Poor design
- Excessive variation in the process
- Human error
- Excessive variation in raw material (Suzaki 1993)

If we know mistakes are inevitable and inspection is not 100% effective, then the best chance we have to ensure against errors is to not let them happen in the first place. If we cannot prevent the mistakes, at least try for quick detection of errors. Methods are needed which can provide immediate feedback that a deviation has happened or is about to happen. It is impossible to get quality with 100% inspections.

There are various technologies, approaches, and gadgets to halt the process if a worker misses the operation sequence. A dedicated tray can be used during assembly process. Tray consists of all the parts required for the assembly process. The tray is like the visual checklist for the process. Apart from the visual checklist, there is a service-oriented checklist which ensures that all the parts are present during the assembly process.

Apart from the above-mentioned technique, many other mechanical screening techniques are used. Mechanical screening applications are based on length, height, weight, and width. Let us consider an example of supermarket where description about each product is specified. Apart from barcodes used for billing purpose, the physical description of the product reduces the data entry error.

Some sort of warning mechanism can be employed to reduce and indicate human error. Some techniques are given as follows:

- Using tools, jigs, and fixtures that avoid loading ill-positioned parts
- Controlling the process using electric relay
- Designing a part that does not have any defect

There are two terms used for proofing: mistake proofing technique and mistake proofing method. Mistake proofing techniques and mistake proofing methods are different. Mistake proofing technique refers to what is desired to happen in the process. For example, shutdown, warning, and force compliance/control are all mistake proofing techniques. Mistake proofing method is the tool used to achieve the response in the process. Techniques and modes are shown as a 3×2 grid. Two types of modes are prediction and detection. Three types of mistake proofing methods are contact methods, fixed-value methods, and motion-set methods. Contact method is said to have contact with the part highlighting errors. Fixed-value method is a method in which errors are detected through counting. Motion-step method is a method in which errors are detected by motion or lack of it (Table 17.3).

Sometimes the setup operator would set the top and bottom jigs incorrectly resulting in defective parts and possible damage to the die. A guide pin prevents the press from closing unless the proper jig is used and it is correctly set up. Each jig has its own unique guide pin. A forklift truck has a limit switch that will not allow the truck to move forward or in reverse with the mast raised. Another limit switch shuts the forklift down when the load exceeds the maximum weight capacity. A scale, specially designed for the screw and bolt packaging operation, digitally displayed the number of screws in the package based on the weight of the screw. The operator no longer had to convert the weight of the screws and bolts into number of pieces. An enamel spraying operation

Table 17.3 Mistake Proofing Techniques

Technique	Prediction	Detection
Shutdown	Camera will not function without enough light to take a picture.	Laundry dryer has a shutdown device when overheating is detected.
Control	The tank hole for unleaded gas is smaller than that for leaded gasoline.	Governors control speed
Warning	Many cars have warning systems to alert the driver that not all seat belts have been fastened.	Smoke detectors provide warning that smoke has been detected and that there is a possible fire.

Table 17.4 Mistake Proofing Methods

Method	Problem	Solution
Contact	Holes were drilled in furniture kits. When holes are not perpendicular, assembly becomes difficult or impossible.	Guides line up the drill to assure that the drilled hole is 90°.
Fixed-value	Four different hazardous material warning labels had to be applied before shipment. Boxes without the labels would be returned and shipments delayed.	The labels were on four separate rolls. When all four labels were put into one roll, it was easy for the worker to know when a label has been missed.
Motion-step	Seven screws in various sizes were inserted in the final assembly of a CD ROM drive. Often a screw would be forgotten resulting in warranty claims.	The seven screws were put into separate bins with photoelectric switches. When a screw is removed, the beam is broken. The part cannot move until the beam is broken in all seven bits.

required each part to have short sprays of enamel applied before the part was baked. It was determined that five short sprays gave the optimum coverage. A limit switch was put on the pump that would allow only an 0.5-second spray. A counter was also installed on the spray trigger. Only when the spray trigger was squeezed five times, the part would be released to the oven (Table 17.4).

Examples of poka–yoke are as follows:

- Seatbelt buzzer to warn drivers and passengers
- Lawn mower safety shutoff when bar is released
- Product drawings on cash registers at fast food restaurants
- Gas cap attached to a car
- 110 V electrical plugs and polarized sockets
- Gas pump with automatic shutoff nozzles
- Microwave automatically stops when door is opened.
- Barcodes for product identification during distribution

17.11   Kanban pull

Kanban method is introduced by Taiichi Ohno of Toyota motor company. Kanban system is the production scheduling technique that determines what to produce, when to produce, and how much to order. It does not carry any inventory. This system is similar to that of super market in which the product is pulled from the shelf according to customer's order and the product is replenished.

Kanban is a material flow control technique that follows to produce products to the customer with shortest lead time possible. This requires leveling of production. For example, we produce 15 units of product A and 10 units of product B in day 1. We produce 10 units of product A and 20 units of product B. It does not happen in reality; customer can change the size of the order in day 2 so material flow also needs to be maintained or there will be inventory pileup (Quality Council of Indiana 2007).

Kanban system uses cards for the material flow purpose. In order to indicate the supply of product during appropriate time, the following methods are used (Figure 17.9):

- Parts are placed on the assembly line. Withdrawal kanban is used in the designated area.
- For getting additional parts from the previous station, a withdrawal kanban is taken by the worker to previous operation station.
- To produce more parts for the operation, WIP kanban card is used which consists of WIP instruction. This way parts can be pulled from the previous station.
- Kanban card will indicate the next operation to produce more products.
- This process continues for the rest of the process (Quality Council of Indiana 2007).

As mentioned earlier, the worker produces more products only when he/she receives the kanban card from upstream process. Kanban system reduces finished goods inventory, WIP inventory, and paperwork. A sample kanban card is shown in Figure 17.9. Continuous improvement is made by critical thinking and sequencing using kanban system. This process follows certain assumptions such as the machine failure and breakdown, which are not considered at any point of time. Operation halt at any point will jeopardize the entire production system. Care has to be taken that the machine is functional and there is no halt or breakdown at any point.

When the number of production cards is reduced, stock level is minimized. If machine downtime is reduced, the production efficiency is increased and quality level is further increased. This process is apt for the repetitive process and it cannot be used for job shop production system. Normally, cards are employed for the kanban system with flags, space on the floor, and so on. All kanban cards consist of part number, bar codes, delivery frequency, and quantity required (Quality Council of Indiana 2007).

Production instruction KANBAN			
Code			Color
	RZC	5	
Type		Manual	
Quantity		1 Set	
Style		Standard	
Control		4M539ALR	

Figure 17.9 Kanban card. (From Quality Council of Indiana, *Certified Six Sigma Black Belt Primer*, Quality Council of Indiana, West Terre Haute, IN, 2007. With permission.)

References

Gee, G. and Izadi, M. (1994). "Quality problem solving techniques in manufacturing." *Journal of Industrial Technology*, 10(2): 2 5.

Grant, E. and Leavenworth, R. (eds.) (1988). *Statistical Quality Control*. New York: McGraw-Hill.

Imai, M. (ed.) (1997). *Gemba Kaizen: A Commonsense, Low-Cost Approach to Management*. New York: McGraw-Hill.

Laraia, A. C., Moody, P. E., and Hall, R. W. (eds.) (1999). *The Kaizen Blitz: Accelerating Breakthroughs in Productivity and Performance*. 1st edn. New York: Wiley.

Moss, M. A. (1985). *Designing for Minimal Maintenance Expense: The Practical Application of Reliability and Maintainability*. New York: Dekker.

Osada, T. (1991). *The 5S's: Five Keys to a Total Quality Environment*. Tokyo: Asian Productivity Organization.

Quality Council of Indiana (2007). *Certified Six Sigma Black Belt Primer*. West Terre Haute, IN: Quality Council of Indiana.

Shingo, S. (ed.) (1986). *Zero Quality Control: Source Inspection and the Poka-Yoke System*. Stamford, CT: Productivity Press.

Suzaki, K. (ed.) *New Shop Floor Management: Empowering People for Continuous Improvement*. New York: Free Press.

Review Questions

Question 1: (control) A control chart is used to

(a) Detect non-random variation in the process
(b) Measure process capability
(c) Determine causes of process variation
(d) Determine if defective parts are being produced

Question 2: Using TPM metrics, various formulas for machine availability calculations would require information like

(a) Downtime, efficiency, operating time
(b) Operating time, downtime, loading time
(c) Operating time, efficiency, quality rate
(d) Loading time, equipment effectiveness, downtime

Question 3: Consider the following definition: "The best combination of machines and people working together to produce a product or service at a particular point in time." What lean concept is being described?

(a) Future state map
(b) Value stream
(c) Ultimate cycle time
(d) Standard work

Question 4: The essence of Kanban concepts includes all of the following, EXCEPT

(a) Delivery of components and products only when needed
(b) Minimal storage in production areas
(c) Distress through the production system when a machine failure occurs
(d) Wide applicability to repetitive and non-repetitive production plants

chapter eighteen

Lean integration into DMAIC

> Making money is a hobby that will complement all other hobbies
> that you have beautifully.
>
> —Scott Alexander

18.1   Introduction

Lean and Six Sigma are both powerful tools, and since they have surfaced in the quality circles, there has always been ambiguity in selecting between them. To thoroughly enjoy benefits on their implementation, they must be investigated for common grounds of application and exclusivity.

It is known that Six Sigma is useful to detect defects, while Lean is useful for removal of sources of waste. The finer differences can be viewed in Table 18.1.

It must be kept in mind that both the products are aligned toward customer satisfaction and bring about continuous improvement to achieve this. Variation is reduced and process is recovered to be more stable. However, Six Sigma proves to be capable of producing large returns sooner and push the organization to "greatness."

Both these methodologies concentrate on problem solving and identifying to define the problem with Six Sigma having a more orderly hard-set process to determine the root cause. Lean propagates the use of value stream mapping as a visual guide to the flow of process, while Six Sigma favors process mapping that maps the outputs to inputs so the causes can be easily pointed out.

On a decision to use both the methods concurrently, a suggestion is to use Lean first to stabilize the organization and then use both the methods simultaneously for powerful results. This allows the elimination of variation early on the project so that data with higher accuracy can be collated for Six Sigma tool implementation making (Nuteing 2012).

Based on the scenario of the target company, Lean or Six Sigma tools can be applied.

Given below are ways by which Lean addresses problems such as delay, human errors, high inventories, increasing amounts of scrap, and inconsistent workflow:

- It improves the alignment of processes.
- It improves velocity of various motions.
- It reduces the inventory level.
- It eliminates sources of waste.
- It allows mistake proofing.

However, if the organization utilizes Six Sigma, it can

- Reduce aberrations.
- Direct its efforts toward quality problems.

Table 18.1 Comparison of Lean and Six Sigma

Area	Lean	Six Sigma
Effect/Improvement	Reduced waste	Less defects or variation
Means	Speed	3.4 DPMO
Profits	Operating costs	Cost of poor quality
Project selection	Value stream mapping	Many approaches
Length of the project	1 week up to 3 months	2–6 months
Driver	Demand	Data
Complexity	Moderate	High
Learning curve	Short	Long

Source: Quality Council of Indiana, *Certified Six Sigma Black Belt Primer*, 3rd edn., Quality Council of Indiana, West Terre Haute, IN, 2007. With permission.

- Implement technical changes.
- Use an exhaustive approach to solving problems.
- Record through a project charter and streamline work based on it (Crabtree 2006).

18.2   The marriage of Lean and Six Sigma

It must be realized that since an organization can have its own set of custom problems, integrating these ideologies broadens the tool set for the extensive selection criteria. A combinatory approach using principles from both disciplines is in fact a holistic pathway to achieving the organization's goals. This unified methodology has been found to bring about at the least twice the improvement from what is achieved by a singular approach between the two thinking approaches. It must be kept in mind that the scale of improvement is dependent on the measurement system and the application industry type.

Tim Noble, manager of Avery Point Group, has provided an insight on the use of Lean and Six Sigma together. He states that "They are complimentary tool sets and not competing ideologies." Thus multiple combinations can be made out of these tool sets and effectiveness is improved with the use of teams and project management skills to solve problems and apply tools efficiently. The direction of flow for such combinations can be best represented visually by Figure 18.1. It also provides the tool set for each stage in the define, measure, analyze, improve, and control (DMAIC) process.

18.3   Enterprise-wide deployment

A suggested organizational deployment strategy includes a pyramid approach with originally experimental learning, internal capability, organizational capability, and leverage Six Sigma for strategic advantage. All processes must be evaluated. Asking the same questions often leads to the same answers, which leads to the same result. Changing the result requires changing the question (Figure 18.2).

18.3.1   Challenges with deployment

There might be aberrant and unreliable decisions made in organizations that are "flat" due to lower management and nonrobust structure. Similarly, with functional organizations, there is too much specialization to allow for flexibility.

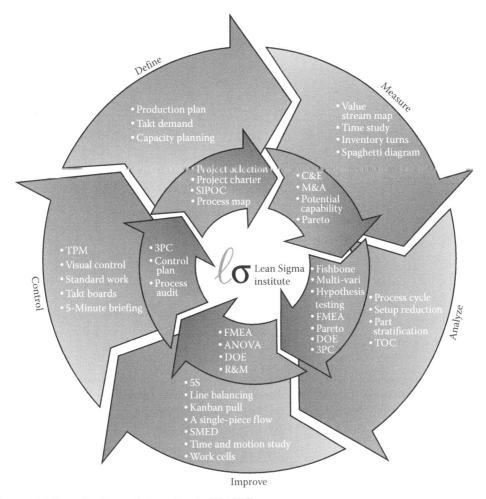

Figure 18.1 Lean Six Sigma integration in DMAIC.

With vertically structured organizations, however, there is an approval process for all decisions by upper management.

In some organizations, the personnel might be too focused on building of one product. Whereas with functional components, spread across boundaries of one country operations can become cumbersome. Also with some matrix-based companies where two or more employees share the same level in the structure.

Cross-functional development is always needed. This is possible with continual improvement efforts and individuals that are capable of bringing about beneficial changes strategically, structurally, or technologically (Quality Council of Indiana 2007).

18.4   Systems and processes

A system is defined as a sequence of elements engaged in component processes working toward a goal. Effectiveness in achieving their goals is termed as performance and is a measure of its reliability. A system consists of many processes.

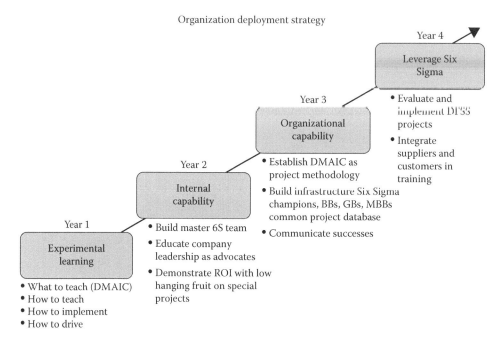

Organization deployment strategy

Year 4

Leverage Six Sigma

- Evaluate and implement DFSS projects
- Integrate suppliers and customers in training

Year 3

Organizational capability

- Establish DMAIC as project methodology
- Build infrastructure Six Sigma champions, BBs, GBs, MBBs common project database
- Communicate successes

Year 2

Internal capability

- Build master 6S team
- Educate company leadership as advocates
- Demonstrate ROI with low hanging fruit on special projects

Year 1

Experimental learning

- What to teach (DMAIC)
- How to teach
- How to implement
- How to drive

Figure 18.2 Deployment flowchart.

Processes are encompassing elements of a system which can be broken down further into sub-tasks. Processes must be continually evaluated for cause-and-effect relation such that change in one does not deter the functioning of the other.

Business processes must make sure that a change does not interfere upstream or downstream.

Places to look for interference are HR, engineering, sales and marketing, finance, product liability, manufacturing, safety and health, legal relations, R&D, purchasing, IT, and more. All these sectors can improve by process optimization methods.

References

Nutting, P. A. et al. (2010). "Journey to the patient-centered medical home: a qualitative analysis of the experiences of practices in the National Demonstration Project." *The Annals of Family Medicine*, 8(1): S45–S56.

Quality Council of Indiana. (2007). *Certified Six Sigma Black Belt Primer*. 3rd edn. West Terre Haute, IN: Quality Council of Indiana.

chapter nineteen

Nontraditional tools for LSS: Work measurement and time studies

If you can't describe what you are doing in a process you don't know what you are doing.

—W. Edwards Deming

19.1 Introduction

Some of the modern and nontraditional tools that are essential for the goal of Lean Six Sigma but are often overlooked are economic evaluation tools: Determining quality costs to stream-line team goals, work measurement, and time study methods. Let us look into them in detail.

19.2 Motion and time study in a Lean framework

The concept of Lean evolved from the scientific management. In a physical world, science is defined in terms of time, motion, and space. Thus, the key to scientific management is the minimization of time and motion that will allow organizational goal realization. The goal of any business is to make money (Goldratt 1984). In order to make more money, organizations employ lean methods such as kanban, just-in-time, and single-minute exchange of die (SMED) to eliminate the seven wastes: transportation, inventory, motion, waiting, overprocessing, overproduction, defects, and mnemonics. However, the objective of these lean methods is to eliminate waste and increase productivity (Figure 19.1).

Productivity can be broadly defined as the ratio between output and some or all of the resources used to produce the output. Generally, productivity is measured in terms of labor productivity, capital productivity, and material productivity. These productivities in addition to other management and supply chain factors contribute to the overall productivity of the organization.

$$\text{Labor productivity} = \frac{\text{Units produced}}{\text{Hours worked}} \tag{19.1}$$

$$\text{Capital productivity} = \frac{\text{Output}}{\text{Capital input}} \tag{19.2}$$

$$\text{Material productivity} = \frac{\text{Output}}{\text{Material input}} \tag{19.3}$$

Over the years, the body of knowledge has evolved to increase the productivity of an organization and the individuals who make up the organization. Time and motion studies focus on the elimination of unnecessary work and design methods that are most effective and suit the person who uses them.

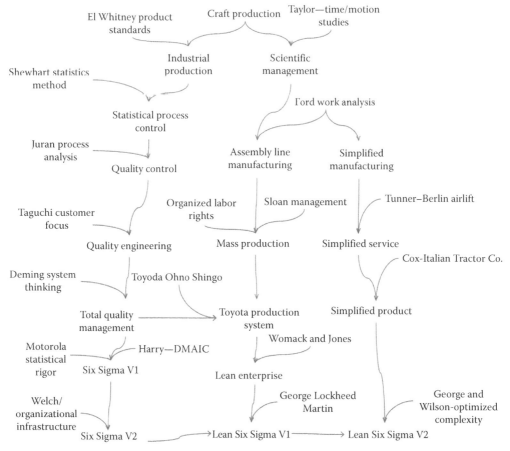

Figure 19.1 Evolution of Lean Six Sigma. (From Upton, M. T. and Cox, C., *Paper Presented at the Pan-Pacific Conference XXI*, May, Anchorage, AK, 2004. With permission.)

19.3   History of time and motion study

Fredrick W. Taylor has been generally considered as the father of scientific management. However, time studies were conducted in Europe many years before Taylor's time. In 1760, Jean Rodolphe Perronet, a French engineer, made extensive time studies in manufacturing, whereas 60 years later, Charles W. Babbage, an English economist, conducted time study in manufacturing. Taylor began his time study work in 1881 at the Midvale Steel Company in Philadelphia. He proposed that the work of each employee be planned out by the management at least 1 day in advance. Workers were to receive complete written instructions describing their tasks in detail and noting the means to accomplish them. Each job was to have a standard time, determined by time studies made by experts. In June 1903, at the Saratoga meeting of the American Society of Mechanical Engineers (ASME), Taylor presented his famous paper "Shop Management," which included the elements of scientific management: time study, standardization of all tools and tasks, use of planning department, use of slide rules and similar time-saving implements, instruction cards for workers, bonuses for successful performance, differential rates, mnemonic systems for classifying products, routing systems, and modern cost systems.

Frank B. Gilbreth and his wife Lillian M. Gilbreth were the founders of the modern motion study technique, which may be defined as the study of the body motions used in performing an operation to improve the operation by eliminating unnecessary motions, simplifying necessary motions, and then establishing the most favorable motion sequence for maximum efficiency. They studied body motions to increase production, reduce fatigue, and instruct operators in the best method of performing an operation. They developed the technique of filming motions to study them, using a technique known as micromotion study. In this micromotion study, the fundamental element or subdivisions of an operation are studied by means of a motion picture camera and a timing device, which accurately indicates the time intervals on the motion picture film. Additionally, they developed the cyclegraphic analysis and chronocyclegraphic analysis techniques for studying the motion paths made by an operator. For the cyclegraphic analysis, the path of motion of an operator is recorded by attaching a small electric light bulb to the finger, hand, or other part of the body and photographed, with a still camera that records the path of light as it moves through space. Such a record is called a cyclegraph. Additionally, if an interrupter is placed in the electric circuit with the bulb, and if the light is flashed on quickly and off slowly, the path of the bulb will appear as a dotted line with pear-shaped dots indicating the direction of the motion. The spots of light will be spaced according to the speed of the movement, being widely spaced when the operator moves fast and close together when the movement is slow. From this graph, it is possible to accurately measure time, speed, acceleration, and retardation, and to show the direction and the path of motion in three dimensions. Such a record is called a chronocyclegraph.

Carl G. Barth developed a production slide rule for determining the most efficient combinations of speeds and feeds for cutting metals of various hardness, considering the depth of cut, size of tool, and life of the tool. He also investigated the number of foot-pounds of work a worker could do in a day.

Henry Laurence Gantt developed simple graphs that would measure performance while visually showing projected schedules. He invented a wage payment system that rewarded workers for the above-standard performance, eliminated any penalty for failure, and offered the boss a bonus for every worker for the above-standard performance. He emphasized human relations and promoted scientific management as more than an in-human "speedup" of labor.

19.4   Continuous improvement

Engineers make things, but industrial engineers make things better. In order to make things better, the industrial engineer applies a logical and systematic approach to solve almost any problem. The steps involved are as follows:

1. Problem definition
2. Analysis of problem (also known as benchmarking)
3. Search for possible alternatives
4. Evaluation of alternatives
5. Recommendation for action
6. Continuously monitor the action

19.4.1   Problem definition

Although we state that the definition or formulation of the problem is the first step in problem-solving procedure, this is often preceded by the problem identification. In most situations, this is accomplished using the Pareto analysis that uses the concept that the major (80%) part of an activity is accomplished by a minority (20%). Thus, the problem presenting the most opportunity receives the greatest attention. Sometimes, it is better to define a subproblem to address a larger problem depending upon the complexity of the problem.

19.4.2   Analysis of the problem

The problem definition usually results in a broad statement or definition. Now it becomes necessary to obtain data to understand the root cause and determine how they apply to the problem. The steps involved are as follows:

1. Specifications or constraints, including any limits on original capital expenditures
2. Description of the current method of operation. This might include process charts, flow diagrams, trip frequency diagrams, man and machine charts, operation charts, and simo charts
3. Determination of activities that man can probably do best and the machine can do best and man–machine relationships
4. Reexamination of problems—determination of subproblems
5. Reexamination of criteria

19.4.3   Search for possible solutions

The basic objective, of course, is to find the preferred solution that will meet the criteria that have been established. This suggests that several alternative solutions be found and then the preferred solution can be found from the alternatives. However, it is important to ask the question: "What is the basic cause that has created the problem?" If the basic cause can be eliminated, then the problem is fixed. Many times, the basic cause cannot be completely eliminated and one should explore a broad and idealistic view in considering possible solution.

19.4.4   Evaluation of alternative

With a set of possible solutions to the problem, there is no one correct answer. Often judgmental factors exist which must be considered over the quantitative evaluation in arriving at the preferred solution. It is desirable to select three solutions: (1) the ideal solution, (2) the preferred solution for immediate use, and (3) the future solution. The evaluation of the alternatives should consider future difficulties such as time and cost to maintain and repair the equipment, the adjustment to widely varying sizes or product mix, and the effects of wear and tear of equipment and the operator.

19.4.5   Recommendation for action

After the preferred solution has been determined, it is communicated to other persons through written and/or oral reports. The written reports become the standard operating procedure. The oral reports are recommendations made as a presentation with simulation models, charts, diagrams, or working models.

19.4.6   Continuous monitoring of the action

Upon implementing the recommended procedure, the system must be continuously monitored using quality charts and periodic reevaluation of the measurable attributes to insure that the improvement has been made and is still being realized.

19.5   Time study

Time study is used to determine the time required by a qualified and well-trained person working at a normal pace to do a specified task. The difference between motion study and time study is that motion study involves largely design and time study involves measurement. The standard time for operations is established using time study. Times studies are mostly utilized to establish wage incentives. The author of this chapter has an extensive experience in conducting time studies for Tyson Foods, Koch Foods, Nebraska Beef, Hutchinson Inc., and other industries. However, it is also used

1. To determine the schedules and planning work.
2. To determine the standard costs and as an aid in preparing budgets.
3. To estimate the product cost before manufacturing it.
4. To determine machine effectiveness and as an aid in balancing assembly lines and work done on a conveyor.
5. To determine the time standards to be used as a basis for labor cost control.

19.5.1   Time study equipment

The equipment needed for time study consists of a timing device and an observation board. The most commonly used equipment are (1) stopwatch or electronic time, (2) video camera, and (3) data collector and computer (Figures 19.2 through 19.4).

Figure 19.2 Stopwatch.

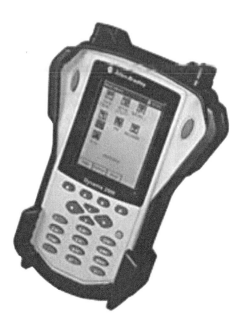

Figure 19.3 Data collector.

Figure 19.4 Video camera.

19.5.2   Important definitions

Normal time: The time required by an average trained operator to perform a task in the true environment and the normal working pace.

Normal pace: The pace of an average trained operator working over an 8-hour shift period

Actual time: The observed time for an operator to perform a task

Allowances: The amount of time added to the normal time to provide for body breaks, personal need, unavoidable delays (not within the control of the operator), and fatigue

The study is completely based on the techniques that help assess the areas of improvement for a worker. It aims at elevating the productivity of a worker and thereby enhances the efficiency of a process on which the employee is working on. It strives to cultivate comfortable work conditions with safety in mind that motivates the workers. It can be seen that the concept of muda or waste of lean thinking is applied here.

Every job is broken down into tasks or the smallest components. The details and timings for each movement are recorded. Thus, a change in a small pocket of work can resonate greatly in terms of overall performance.

19.5.3 Making the time study

The exact procedures differ with the operation under study. However, the generic steps in making a time study are as follows:

1. Secure and record information about the operation and the operator being studied
2. Divide the operation into elements and record a complete description of the method
3. Observe and record the time taken by the operator
4. Determine the number of cycles to be timed
5. Rate the operator's performance
6. Check to make certain that a sufficient number of cycles have been timed
7. Determine the allowances
8. Determine the time standard for the operation

19.5.3.1 Recording information about the operation and the operator

All the information on the top of the observation sheet must be filled prior to the study. Otherwise the study is worthless as a record (Figure 19.5).

19.5.3.2 Dividing the operation into its elements

Timing an entire operation is not practical. Further breaking down the operation into its elements will be beneficial because the elements may be a standard practice for the operation that may be used to establish training programs. Additionally, the time values of the elements will allow the establishment of the standard time. An operator may not work at the same tempo throughout the cycle and need to give a proper rating for each of these elements of an operation. There are certain rules that establish the elements of an operation prior to the study. These rules are as follows:

1. The element should be as short in duration as accurate times.
2. Handling time should be separated from machine time.
3. Constant elements should be separated from variable elements.

Each element should be concisely recorded in the space provided on the sheet.

19.5.3.3 Recording the time

There are three common methods for reading stopwatches: (1) continuous timing, (2) repetitive timing, and (3) accumulative timing.

In the continuous method, the observer starts the watch at the beginning of the first element and permits it to run continuously during the study. The observer notes the readings at the end of each element and records it in the observation sheet.

OBSERVATION SHEET

SHEET 1 OF 1 SHEETS		DATE	
OPERATION		OP.NO.	
PART NAME		PART NO.	
MACHINE NAME		MACH.NO.	
OPERATOR'S NAME & NO.		MALE ☐ FEMALE ☐	
EXPERIENCE ON JOB		MATERIAL	
FOREMAN		DEPT.NO.	

BEGIN	FINISH	ELAPSED	UNITS FINISHED	ACTUAL TIME PER 100	NO. MACHINES OPERATED

ELEMENTS		SPEED	FEED		1	2	3	4	5	6	7	8	9	10	SELECTED TIME
1.				T R											
2.				T R											
3.				T R											
4.				T R											
5.				T R											
6.				T R											
7.				T R											
8.				T R											
9.				T R											
10.	(1)			T R											
11.	(2)			T R											
12.	(3)			T R											
13.	(4)			T R											
14.	(5)			T R											
15.	(6)			T R											
16.	(7)			T R											
17.	(8)			T R											
18.				T R											

SELECTED TIME	RATING	NORMAL TIME	TOTAL ALLOWANCE	STANDARD TIME

SKETCH OF COMPONENTS:	TOOLS.JIGS.GAUGES:
	TIMED BY:

Figure 19.5 Time study observation sheet.

In the repetitive method, the watch is reset to zero at the end of each element. This method gives the direct time without any calculation as required in continuous timing.

In the accumulative method, the direct reading of the time for each element is done with the use of two stopwatches. The stopwatches are set up in such a way that when the first stopwatch is started, the second stopwatch is stopped and vice versa.

In recent days, the operations are videotaped. The videotaped is further analyzed using video editing software such as Multimedia Video Task Analysis (MVTA) and Adobe Premiere. These software packages provide accurate time measurements based on the frame rate and the time stamp on the tape.

19.5.3.4   Number of cycles to be timed

The time required to perform the elements of an operation may be expected to vary slightly from cycle to cycle. As time study is a sampling process, the more number of observations will converge the time results of the representative activity. The number of observations can be computed using the variation formula:

$$\sigma_{\bar{X}} = \frac{\sigma'}{\sqrt{N}}, \tag{19.4}$$

where:

$\sigma_{\bar{X}}$, standard deviation of averages

σ', standard deviation for a given element (generally obtained using quality control methods)

N, actual number of observations of the element

19.5.3.5   Rating an operator

Rating is the process during which the time study analyst compares the speed or tempo as normal performance using his/her judgment. This rating is the most difficult step in time study. It is a common practice to establish the rating using frequency distributions. In many situations, the rating is established using a sample population. However, there are other types of rating which are seldom used:

1. Skill and effort rating
2. Westinghouse system of rating
3. Synthetic rating
4. Objective rating
5. Physiological evaluation of performance level
6. Performance rating (commonly used in the United States)

After the rating has been established, the normal time is calculated as

$$\text{Normal time} = \text{Selected time} \times \frac{\text{Rating (\%)}}{100}. \tag{19.5}$$

19.5.3.6   Determining the allowances and standard time

The normal time for an operation does not contain any allowances. However, the operator can take time-outs for personal needs, rest, and reason beyond his/her control. These allowances are included to establish the standard time.

$$\text{Standard time} = \text{Normal time} + \left[\text{normal time} \times \text{allowances (\%)}\right]. \tag{19.6}$$

19.5.4   Types of time study

The different studies that need to be performed to improve productivity and to eliminate waste include

- Direct time study
- Time study standard data
- Predetermined time systems
- Predetermined time standards system (PTSS)
- Work sampling

The direct time study is a work measurement method in which the actual time is measured by observing the task and the operator using a stopwatch. The measured time is later modified to provide the allowances. However, this study may be complicated with complex tasks. Oftentimes, a task element may be repeated across several operations. A direct time study is not a feasible solution to this method. As a cost-saving activity, organizations study such repetitive task elements to create standard data file. This standard time data may be defined as the normal time values obtained from direct time measurement from a similar operation earlier. The challenge with this method is the cost associated with generating the standard data.

In many situations, a set of jobs or tasks when broken down to finer elements, they may consist of the combination of the same task elements. In such situation, the predetermined times can be calculated by adding the time required to perform individual times.

Changes such as simplification, rearranging the order of task completion and analyzing various profitable combinations can be brought into use. The quality of work life is the priority, and in doing so, motion study is the predecessor to time study. Designs of workstations to reduce unnecessary stress and movement deterrent to the work culture can be incorporated.

Some of the techniques for motion study or study of work methods are as follows:

- Process charts
- Flow diagrams
- Multiactivity charts
- Flow patterns
- Work station design
- Operations analysis chart
- PTSS

19.6   Methods-time measurement

Methods-time measurement (MTM) is a procedure that analyzes any manual operation or method into the basic motions required to perform it and assigns to each motion a predetermined time standard, which is determined by the nature of the motion and the conditions under which it is made.

The PTSSs such as MTM-1, MTM-2, MTM-Universal Analyzing System (UAS), MTM-MEK, MTM-B, Maynard Operation Sequence Technique (MOST), and Modular Arrangements of Predetermined Time Standards (MODAPTS) have been used in establishing labor rates in industry by quantifying the amount of time required to perform specific tasks.

19.7   Time studies and human factors

One of the goals of time and motion studies is to make the work as easy and satisfying for the operator as possible. Human factors is the field of industrial engineering that focuses on people to enhance functional effectiveness and maintain or enhance human health, safety, and satisfaction. In order to enhance human welfare, it is necessary to evaluate the human–machine system from a physiological perspective and the environment in which the person functions. Extensive research (Gnaneswaran et al. 2008, 2011, 2013) is being conducted to address the issues relating to fatigue based on hours of work, rest periods, lighting, heating, ventilation, noise, vibration, and psychology.

These tools bring about the following transformations (Meyers and Stewart 2002):

- Process time: Gilbreths' process charts help answer the Five Why's by charting the flow of processes.
- Improving operations performance through operations chart from the mapping of work patterns in order to analyze various motions and reduce cycle time and setup time
- Worker ergonomics and safety must be considered for consistent performance results.

These tools bring about the following changes (Meyers and Stewart 2002):

- Team and singular performance for every time element is measured to remove elements that do not contribute value.
- Inevitable delay due to fatigue and breaks is allocated for improved pace. The measurement of this amount of time is called work sampling.
- Standardization for future
- Implementing SMED

We can thus use motion and time study to measure performance by line balancing, visual inspection, work sampling, and reporting score through scoreboarding (Meyers and Stewart 2002).

19.8   Challenges with time study

There are no specific limitations with time study. However, there are challenges with unions and the improper training for the time study analyst. Traditionally, time study has been taught in an undergraduate level. However, many graduate students do not get the correct training in performing time studies in their curriculum. Time and motion studies are a staple of any industrial engineering curriculum.

19.9   Economic tools

Tools from the realm of economics are taken for the cost–benefit type of analysis of projects. Timely estimation of costs that might be incurred and forecast performance for allowing decisions by sponsors to undertake such projects are to be done before initiation. However, it is mostly done after completion.

It requires the implementation of the following steps:

- Estimate parameters in the project that contribute to costs called as cost factors
- Express performance improvements in terms of profits

- Evaluate total profit or loss
- Provide suggestions for areas of improvement

We will now identify the tools required for assessment of the net profits. It must be noted that any form of income is termed as positive cash flow and any form of cost incurred or price paid is negative cash flow.

1. Return on assets (ROA): It is the ratio of net positive cash flow and the total value of resources utilized for the project (Quality Council of Indiana 2007).

$$\text{ROA} = \frac{\text{Net positive cash flow}}{\text{Total assets}} \tag{19.7}$$

2. Return on investment (ROI): It is the ratio of net positive cash flow and the investment allocated for the project (Quality Council of Indiana 2007).

$$\text{ROI} = \frac{\text{Net income}}{\text{Investment}} \tag{19.8}$$

3. Net present value (NPV): Time series of cash flow, incoming and outcoming, is defined as the sum of present values.

$$\text{NPV} = \sum_{j=0}^{n} \frac{Xj}{\left(1+k\right)^{j}}. \tag{19.9}$$

The NPV is calculated using Equation 19.9, where Xj represents the net cash flow for the year j, n is the number of years considered for the cash flow, and k is the minimum acceptable rate of return (MARR) value, which is nothing but the MARR, which depends on the type of cash flow. The decision criterion when using NPV as a tool is that if NPV has a value above zero, then the project is approved.

4. Internal rate of return (IRR): It is the discount rate for which the NPV is zero or the net cash flow after applying this MARR interest rate is zero. It is given by the following equation:

$$0 = \sum_{j=0}^{n} \frac{Xj}{\left(1+i\right)^{j}}. \tag{19.10}$$

Here, i represents the IRR. Many iterations are done before arriving at this rate, and for projects, it is normally expected to lie in the range of 5%–25%. The decision criterion for projects is that IRR must be greater than ROI (Quality Council of Indiana 2007).

5. Payback period: It is the number of years required to level out the incoming cash flow with outgoing negative cash flow. It can be calculated using the following equation:

$$0 = \sum_{j=0}^{p} Xj. \tag{19.11}$$

It, however, has disadvantages as it does not consider the time value of money nor the cash flows that may be positive after this period. It is useful while comparing projects for which the module gives faster ROI.

The decision criterion is usually 1–2 years of payback period and is compared with the set limit for all projects undertaken by the organization for approval (Quality Council of Indiana 2007).

19.10   Evaluating quality costs

Quality costs are the costs incurred when quality is not added to the product or service and to reinsert the value to the item. It is not the cost of creation of value that is important for improving quality.

Costs of quality are categorized as cost of poor quality (COPQ) and cost of good quality (COGQ; iSixSigma 2009).

19.10.1   Cost of good quality

It is incurred while preventing the loss of quality due to not confirming to specifications. They are of the following types:

1. Appraisal costs that are the costs associated with evaluating products or services for performance and its conformance to specifications. It includes price for testing and resources for testing: equipment and personnel and calibration.
2. Cost of prevention of processes that lead to poor quality. By following these activities, the cost of prevention increases the following:
   - New product review
   - Quality planning
   - Supplier capability surveys
   - Process capability evaluations
   - Quality improvement team meetings
   - Quality improvement projects
   - Quality education and training (iSixSigma 2009) [American Society of Quality (ASQ) is the governing board for the Six Sigma certifications and when the name ASQ appears it is presenting the idea generated by ASQ.]

19.10.2   Cost of poor quality

It involves the costs incurred to overhaul and refurbish the product to meet requirements. They are of the following types:

- Cost of internal failure
- Cost of external failure

Traditional quality costs are supposed to be tangible and they are incurred on the following activities and are due to internal failures:

- Rework
- Warranty
- Scrap production
- When rejects are made
- Processing again

If quality of the product was intact and as intended, these costs can be avoided.

Additionally, there are costs incurred which are intangible or difficult to measure and these are due to external failures:

- More setups
- Expediting costs
- Lost sales
- Late delivery
- Loss of customer loyalty
- Excessive inventory
- Long cycle times
- Engineering change orders (ASQ)

19.10.3   Total quality costs

The summation of all quality costs represents the loss in sales and income due to failure to meet customer requirements as a result of substandard quality.

19.11   Conclusion

Average COPQ is said to be ~15% of sales. If we do not aim higher than that, we will continue to eat the COPQ, which translates to 15% of sales. As you can see here, this results in a great deal of lost opportunity to the business, most of which has been difficult to measure in the past. Recouping those costs, and reinvesting them in ways that help our business grow and succeed, is critical to future success. Six Sigma is an important means that end this loss and is a major step forward from traditional continuous improvement activities.

With Six Sigma, product and service excellence is identified, measured, and benchmarked. Then, defects are eliminated so that we can deliver customer success and business value, which translates to growth. It involves relentless problem solving on specific projects that bring the most value to the business and customers.

Before approval of a project, such costs must be calculated and preventive measures must be taken.

References

Gnaneswaran, V. and Bishu, R. R. (2011). "Anthropometry and hand performance evaluation of minority population." *International Journal of Industrial Ergonomics*, 41(6): 661–670.

Gnaneswaran, V., Jones, E. C., and Bishu, R. R. (2013). "Endurance limit for periodontal scaling task." *International Journal of Industrial Ergonomics*.

Gnaneswaran, V., Madhunuri, B., and Bishu, R. R. (2008). "A study of latex and vinyl gloves: Performance versus allergy protection properties." *International Journal of Industrial Ergonomics*, 38(2): 171–181.

Goldratt, E. M. and Cox, J. (1984). *The Goal: Excellence in Manufacturing*. Croton-on-Hudson, NY: North River Press.

History of Time and Motion Study, http://ie.emu.edu.tr/development/dosyalar/%7Be_A-az3-Enu%7DCH3.pdf

iSixSigma. (2009). "The 2005 Global 500." December 17, 2012. http://www.isixsigma.com/implementation/financial-analysis/cost-quality-not-only-failure-costs/.

Meyers, F. E. and Stewart, J. R. (2002). *Motion and Time Study for Lean Manufacturing*. Upper Saddle River, NJ: Prentice Hall.

Quality Council of Indiana. (2007). *Certified Six Sigma Black Belt Primer*. West Terre Haute, IN: Quality
    Council of Indiana.
Upton, M. T. and Cox, C. (2004). Lean Six Sigma: A fusion of pan-Pacific process improvement. Paper
    presented at the Pan-Pacific Conference XXI, May, Anchorage, AK.

Further Readings

Barnes, R. M. (1980). *Motion and Times Study Design and Measurement of Work*. 7th edn. New York: John
    Wiley & Sons.
History of Time and Motion Study, http://ie.emu.edu.tr/development/dosyalar/%7Be_A-az3-
    Enu%7DCH3.pdf
Turner, W. C. et al. (1993). *Introduction to Industrial and Systems Engineering*. Englewood Cliffs, NJ:
    Prentice Hall.

chapter twenty

DFSS methods and uses for product- and process-oriented research and development

And I remember misinformation followed us like a plague.

—Paul Simon

20.1   Introduction

Six Sigma methods are recognized by many companies as a means for reducing defects, increasing company productivity, and improving company profitability. Many consider Six Sigma as an extension of total quality management (TQM) initiatives. The advantages that Six Sigma has over other quality initiatives are that it applies statistical techniques not only to product quality but also to many aspects of business operations to improve organizational efficiency.

The label "Six Sigma" originates from statistical terminology. In statistics, the sigma (σ) represents standard deviation. For a normal distribution curve, the probability of falling within a plus or minus Six Sigmas from the mean is 0.9999966. This is more commonly expressed in production processes as a defective rate for a process will be 3.4 defects per million (DPM) units (Yang and El-Haik 2003). Thus, Six Sigma promotes high degree of consistency by designing operations with extremely low variability. Statistically, the goals for Six Sigma methodologies are to reduce operational variation to achieve small process standard deviations.

Although the mathematical history of standard deviation goes back to the eighteenth century, using standard deviation as a measurement standard dates to the 1920s when Walter Shewhart theorized that plus or minus three sigmas from the mean (93.32%) is the point where a process requires correction (Stamatis 2004). This would represent 66,807 DPM units. Six Sigma is a trademark of Motorola who in the 1980s developed the Six Sigma standard and the methodology associated with it. The company has documented over $16 billion in savings using the Six Sigma methodologies.

20.2   Background

The term "Six Sigma" is credited by a Motorola engineer named Bill Smith. Dr. Mikel Harry is the man who originated the phrase "black belt" for Six Sigma purposes among other aspects of Six Sigma (Breyfogle 2003). Harry worked with the Unisys plant in Salt Lake City when he implemented the first versions of Six Sigma. The implementations were on circuit board manufacturing. Clifford Ames, a Unisys facility manager, and Harry documented the methodology that later became Six Sigma (Chadwick 2000). Later, Ames and Harry created the name "black belt" because they were "kicking the heck out of variation" (Maguire 1999). The most important position in any Six Sigma program is the black belt. This is the position that oversees many projects, teaches green belts, coaches other black belts, and serves as the technical expert in Six Sigma tools and statistical methods. Other responsibilities include leading business process improvement projects that have broader impacts on an organization. In addition, they often serve as internal consultants for functional areas.

Other key players in the Six Sigma hierarchy are the yellow belts and the green belts. Each role has a unique job inside of a Six Sigma institution. Yellow belts are typically employees that are management support staff. Their role is to typically report to those that are responsible for a specific line or system. The green belts typically manage an area and are the instigators of Six Sigma projects in that area. In addition, the green belts will manage projects by utilizing yellow belts that are assigned to the Six Sigma team. They can also act as the Six Sigma liaison in an area. By showing their support of the Six Sigma initiatives and showing their team the importance of the project, they play a vital role in the success of a project.

The support for the black belts and the Master black belts is provided for by the executive sponsor. They are trained in enterprise strategy and develop the enterprise Six Sigma deployment. The Master black belt is recruited from the black belts. They are trained to become the enterprise Six Sigma experts. In addition, they become a permanent full-time change agent and are the most highly skilled people within the Six Sigma hierarchy. Often they will be trained in multiple functional aspects of the organization and be able to identify high-leverage opportunities for applying Six Sigma across the enterprise. The Master black belts should report to the executive sponsors. The Master black belts also manage the Six Sigma infrastructure and resources. This centralization of the strategy and resources helps eliminate the traditional roadblocks that can hamper a management system (Breyfogle 2003; Foster 2004; Yang 2003).

According to Breyfogle (2003), black belts have an extensive training in statistics and the Six Sigma methodologies. Black belts then have the responsibility of coaching others in the company about Six Sigma methodologies. This includes participating in Six Sigma training, participating in Six Sigma projects, and acting as a change agent for the project teams.

20.3   *Operational Six Sigma and design for Six Sigma*

Pyzdek's *Six Sigma Handbook* (2003) describes the differences between the two commonly implemented forms of Six Sigma known as DMAIC and DMADV. DMAIC is the acronym for the process improvement methodology that follows the define, measure, analyze, improve, and control steps. DMADV is the acronym for the new product (or process) development methodology that follows the define, measure, analyze, design, and verify steps. The major differences are that DMADV focuses on measuring customer requirements in order to design a product that meets their demand, whereas DMAIC focuses on determining what the root causes of defects are. DMAIC also focuses on controlling the process changes implemented, whereas the final step in the DMADV is the verification that customer requirements are met.

Pyzdek's book also discusses the Deming's learning cycle of plan–do–check–act (PDCA). This cycle can be seen in both DMAIC and DMADV. The first step is to define the issues that are predicted to benefit the company. The next step is to do the necessary experimentation and data collection necessary to determine if the predicted benefit will take place. The last step is where the analyses of the data confirm or reject the predicted benefits. The next step is where the knowledge is used to either design the needed product (or process) and then verify that it is successful or implement the changes and develop a control plan for the changes.

Different methodologies are used with Six Sigma. The primary methodology developed for operational improvement is labeled as the "DMAIC process." Some refer to DMAIC as operational Six Sigma (OSS). Other Six Sigma processes are not utilized for operational improvement, but they are used for product and process developments and labeled as Design for Six Sigma (DFSS).

20.3.1   DMAIC: OSS

OSS is commonly attributed to the steps in the DMAIC process, which includes the following elements:

- Define the project goals and customer (internal and external) deliverables
- Measure the process to determine current performance
- Analyze and determine the root cause(s) of the defects
- Improve the process by eliminating defects
- Control future process performance

20.3.2   Design for Six Sigma

DFSS has been publicized as a new product or process development approach that follows OSS integrations. Some suggest that DFSS cannot be utilized without some maturity of DMAIC processes, tools, and techniques. Others see this methodology as a stand-alone process that does not require prior knowledge of DMAIC. These alternating views have led to multitude of DFSS methodologies.

The main objectives of DFSS are to "design it right the first time," to design or restructure the process in order for the process to intrinsically achieve maximum customer satisfaction and consistently deliver its functions, to avoid painful downstream experiences. Therefore, it is an upstream activity. The term "Six Sigma" in the context of DFSS can be defined as the level at which design vulnerabilities are not effective or minimal.

The best way to differentiate between DMAIC and DFSS is to use the former for improving the existing products and services, and the latter for new products and services. DFSS contrasts with DMAIC in that the phases or steps of DFSS are not universally recognized or defined. It is more an approach than a methodology and typically used to design or redesign a product or service from the beginning.

Many organizations, enthusiastic to build on Six Sigma momentum, generated their own DFSS processes before a standard template emerged. Therefore, organizations have adopted a variety of approaches that resulted in acronyms such as DMADV, IIDOV (invent, innovate, develop, optimize, and verify), CDOV (concept, design, optimize, and verify), IDOV (identify, design, optimize, and validate), DCOV (define, characterize, optimize, and verify), and IDEA (identify, define, evaluate, and activate). Some of the major practices of DFSS are shown in Table 20.1.

Table 20.1 Process Comparison of Six Sigma Methodologies

Six Sigma methodology	DMAIC	DFSS			
		DMADV	IIDOV	CDOV	DFSSR
Scope of application	Process improvement	New process and product development	Technology development	Product design	Environmental testing
Plan	Define, measure	Define, measure	Invention, innovation	Concept	Define, measure
Predict	Analyze	Analyze, design	Develop	Design	Analyze, design, identify
Perform	Improve, control	Verify	Optimize, verify	Optimize, verify	Optimize, verify

DMADV includes the following elements:

- Define the project goals and customer (internal and external) deliverables
- Measure and determine the customer needs and specifications
- Analyze the process options to meet the customer needs
- Design (detailed) the process to meet the customer needs
- Verify the design performance and ability to meet the customer needs

Critical parameter management is a system engineering and integration process that is used within an overarching technology development and product commercialization roadmap. We first look at a roadmap (IIDOV) that defines a generic technology development process and then move on to define a similar generic roadmap (CDOV) that defines a rational set of phases and gates for a product commercialization process. These roadmaps can be stated in terms of manageable phases and gates. These generic phases and gates are useful elements that help communicate how to conduct product development with order, discipline, and structure.

The IIDOV roadmap applies a phase–gate approach to research and technology development. The phase–gate structure of IIDOV for research and technology development follows four discrete project management steps:

1. Invent and innovate: technology concept
2. Develop: technology
3. Optimize: the robustness of the baseline technologies
4. Verify: the platform or sublevel technologies

In the tactical product design engineering process environment, a generic product commercialization process (CDOV) can be segmented into four distinct phases:

1. Concept: develop a system concept based on market segmentation, and product line, and technology strategies
2. Design: design subsystem, subassembly, and part-level elements based on system requirements
3. Optimize: sublevel designs and the integrated system
4. Verify: verification of final product design, production processes, and service capability

For the problems that should be investigated, a rigorous approach is recommended, such as the DCOV model. Ford settled on the four-phase process for all categories of DFSS projects. The DCOV framework aligned with Ford's existing product development system and built on the disciplines of systems, robust and simultaneous engineering. The DCOV model includes the following:

- Define: The purpose of this stage is to identify the critical-to-satisfaction (CTS) drivers, Y, and to establish an operating window for chosen Y's for new and aged conditions.
- Characterize: The goal of this stage is to characterize the robustness of the design. This stage is generally completed through a two-step approach. The first is the system design and the second is the functional mapping.
- Optimize: The goal of this stage is to improve robustness. This stage is also generally completed through a two-step approach. The first is the design for robust performance and the second is the design for productivity.
- Verify: The goal of this stage is to verify that the capability and product integrity over time is as it was designed and as the customer is expecting it to be.

The IDEA process for product and technology portfolio definition and development includes the following:

- Identify: markets, theirs segments, and the opportunities they contain
- Define: portfolio requirements and product portfolio architectural alternatives
- Evaluate: product portfolio alternatives against competitive portfolios
- Activate: ranked and resourced individual product commercialization projects

The IDEA process and its requirements, deliverables, tasks, and enabling tools can add discipline, structure, and measurable results that help ensure that the strategic development and activation of the product and technology portfolios are aligned and capable of supporting the growth goals of the business. Risk can be better managed with this approach as key strategic decisions are made. Problems are prevented on a proactive basis in technology development and product commercialization with this kind of Six Sigma enablement to portfolio development and renewal.

20.3.3 Design for Six Sigma research

Jones and Silveray (2008) introduced the design for Six Sigma research (DFSSR) methodology as a means to support new product development in research laboratories. His research laboratory that focused on radio frequency identification (RFID) technologies developed a DFSS approach to facilitate the common language between industry, academia, and research. The variety of scope projects addressed in a research setting suggests that a consistent experimental approach that includes opportunities for discovery and redesign should be utilized. A Six Sigma approach is a viable option given that it uses statistics to identify the opportunities for improvement. There are two types of Six Sigma approaches: OSS approach, which is referred to by its DMAIC steps, and DFSS. DMAIC is a widely accepted process for operational improvements, but DFSS is used mainly for product and process design. Sometimes DFSS is referred to by one of its original DMADV process steps. Given the characteristics of DFSS, it would be an effective approach due to the fact that is more comprehensive and flexible than the OSS phase, DMAIC. Instead of using a Six Sigma strategy that is based on the PDCA strategies, we are introducing the hybrid version of DFSS. We fuse the traditional research methods with industry's new gold standard Six Sigma into a methodology that is described as DFSSR. Our methodology is based on a strategy to develop operational prototypes and is organized into a plan, predict, and perform (3P) theoretical model shown in Figures 20.1 and 20.2. Figures 20.3 and 20.4 show the substeps and the associated statistical tools.

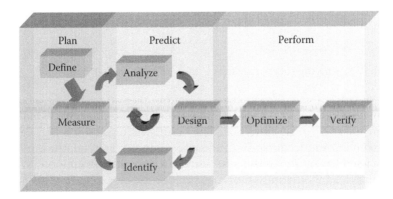

Figure 20.1 DFSSR 3P's methodology framework.

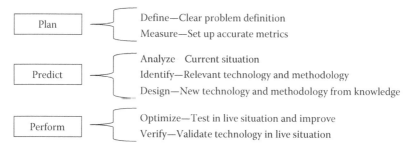

Figure 20.2 DFSSR 3P's phase descriptions.

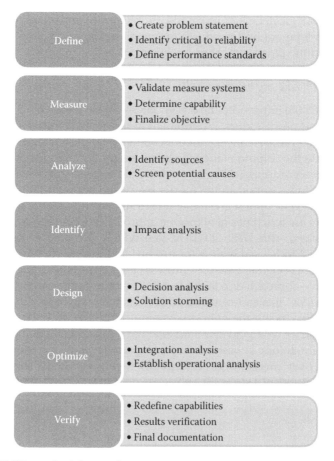

Figure 20.3 DFSSR 3P's methodology substeps.

DFSSR is organized into seven phases—define, measure, analyze, identify, design, optimize, and verify (DMADIOV)—and builds on scientific procedures that lead to optimized solution for real-world scenarios. The results or lessons learned can be used to effectively implement the technology in the environment. Further, the compiled lessons learned can be used to determine the best practices for implementation in the future. The methodology provides the definition of the correct prototype checkpoints and allows for the environmental testing.

Further details are shown in Chapters 10 through 16.

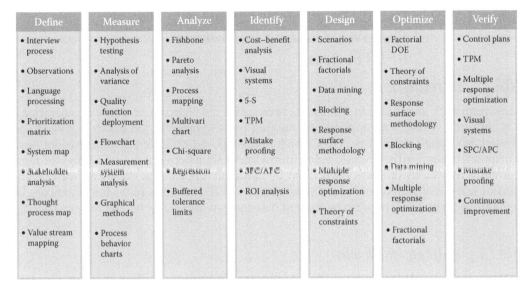

Define	Measure	Analyze	Identify	Design	Optimize	Verify
• Interview process	• Hypothesis testing	• Fishbone	• Cost–benefit analysis	• Scenarios	• Factorial DOE	• Control plans
• Observations	• Analysis of variance	• Pareto analysis	• Visual systems	• Fractional factorials	• Theory of constraints	• TPM
• Language processing	• Quality function deployment	• Process mapping	• Data mining	• Data mining	• Response surface methodology	• Multiple response optimization
• Prioritization matrix	• Flowchart	• Multivari chart	• 5-S	• Blocking	• Blocking	• Visual systems
• System map	• Measurement system analysis	• Chi-square	• TPM	• Response surface methodology	• Data mining	• SPC/APC
• Stakeholder analysis	• Graphical methods	• Regression	• Mistake proofing	• Multiple response optimization	• Multiple response optimization	• Mistake proofing
• Thought process map	• Process behavior charts	• Buffered tolerance limits	• SPC/APC	• Theory of constraints	• Fractional factorials	• Continuous improvement
• Value stream mapping			• ROI analysis			

Figure 20.4 DFSSR 3P's methodology statistical tools.

References

Breyfogle, F. W. III (2003). *Implementing Six Sigma: Smarter Solutions Using Statistical Methods*. 2nd edn. Hoboken, NJ: John Wiley & Sons.

Chadwick, G. C. (2000). "Remembering Bill Smith, father of Six Sigma." Retrieved April 2013. http://www.isixsigma.com/library/content/c030915a.asp

Jones, E. C. and Silveray, J. (2008). "Testing of Gen 2 RFID tags for tracking inventory in the space shuttle." *Journal of Air Transportation*, 12(3): 79–99.

Maguire, M. (1999). "Cowboy quality." *Quality Press Magazine*, October, 32: 27–34.

Pyzdek, T. (2003). *The Six Sigma Handbook: The Complete Guide for Greenbelts, Blackbelts, and Managers at All Levels*. New York: McGraw-Hill Companies.

Stamatis, D. H. (2004). *Six Sigma Fundamentals: A Complete Guide to the System, Methods and Tools*. New York: Productivity Press.

Thomas Foster, S. (2004). *Managing Quality: An Integrative Approach*. 2nd edn. Upper Saddle River, NJ: Prentice Hall PTR.

Yang, K. and El-Haik, B. (2003). *Design for Six Sigma: A Roadmap for Product Development*. New York: McGraw-Hill Companies.

Further Readings

Arveson, P. (1998). "The Deming cycle." Retrieved October 19, 2006. http://www.balancedscorecard.org/bkgd/pdca.html

Baker, J. (2005). "Personal interview." June 10, 2005. Wisconsin: Hudson.

Continuity Central. (2007). "Is Six Sigma a useful business continuity aid?" Retrieved May 2007. http://www.continuitycentral.com/feature0129.htm

Creveling, C. M. (2007). *Six Sigma for Technical Processes: An Overview for R&D Executives, Technical Leaders, and Engineering Managers*. Upper Saddle River, NJ: Prentice Hall PTR.

Creveling, C. M., Slutsky, J. L., and Antis, D. Jr. (2003). *Design for Six Sigma in Technology and Product Development*. Upper Saddle River, NJ: Prentice Hall PTR.

Simon, K. (2000). "DMAIC versus DMADV." Retrieved December 10, 2005. http://www.isixsigma.com/library/content/c001211a.asp

Soderborg, N. R. and Ford Motor Co. (2004). "Design for Six Sigma at Ford." *Six Sigma Forum Magazine*, November.

Appendix 20A: Using Design for Six Sigma—To Develop Real-World Testing Environments for RFID Systems[*]

20A.1   Introduction

Radio frequency identification (RFID) has been called the "barcode of the next generation." It is currently in use, with applications ranging from libraries to toll booth e-passes. The greatest advantage of RFID systems, which are composed of an interrogator (reader) and transponders (tags), is that they do not require a line of sight between the reader and the tags. Currently, thousands of items are tracked by manually updated databases or line-of-sight barcode scans that scan one item at a time. With an RFID system in place, the entire bags of items affixed with RFID tags can be audited in seconds without ever having to open the bag.

20A.1.1   RFID system details

RFID systems consist of an interrogator (also referred to as a reader) and transponders or tags. In other words, the tag can be hidden or embedded within the item and the item will still be identified. In an RFID system, the antenna of the reader emits radio signals. The signals are received by the tag's antenna, which can be powered via a battery or by the radio frequency energy from the reader's pulse. The tags respond with a unique code, which is preprogrammed in the tag's microchip. After the reader antenna receives and decodes the signal, the antenna sends the information to a computer via a standard interface; this information is accessed through a database. Since RFID systems do not require a line of sight between the tag and the reader, a Space Station crewmember could potentially initiate a reader in the general vicinity of a cargo transfer laboratory, which is full of tagged items, and record an accurate count of all items within the bag in seconds. In this chapter, we use a simulated warehouse as an environment for DFSSR techniques.

20A.1.2   Six Sigma methodology

With the advantages illustrated above, a series of experiments and tests will be conducted by Radio Frequency and Supply Chain Logistics (RfSCL) laboratory at the University of Nebraska–Lincoln. A blend of methodologies will help the RfSCL laboratory use information to better develop a solution that best meets the needs of customers. The methodology used in this chapter is the integrated DFSS (Breyfogle 2003). DMADO (De-may-doh) is the acronym for design, measure, analyze, develop, and optimize (Breyfogle 2003). The planning function is to define the objectives and determine the correct measurement parameters. The predictive functions are to analyze viable options, design experiments that lead to the development, identify gaps, reanalyze options based on performance, and create a final design that meets the stated objectives. The performing functions are to optimize the performance of the design and verify that the prototype is operational. This new approach to research will help bridge

[*] By Erick C. Jones and Gao Fei.

the gap between academic organizations and industry. The experiments and analysis will be conducted following the order of the scientific methodology.

20A.2   3P's theoretical model

In this chapter, we introduce a research framework (DFSSR) that is based on a common operational prototype theme that requires the development teams to plan, predict, and perform. This 3P's methodology is utilized to encapsulate our DFSSR framework. For RFID technology, to better serve industrial applications, we need to conduct a series of experiments to validate the principle of facility layout of RFID into real-case scenarios. The Six Sigma methodology helps us build a scientific procedure and makes sure that it is the optimum layout for real warehouse scenarios. We use this framework so operations can identify the status of projects and investigate the detailed processes within the framework. The DFSSR process steps are organized within the 3P framework as shown in Figure 20.1. The results or lessons learned can be used to effectively implement the technology in this environment. Further, the compiled lessons learned can be used to determine the best practices for implementation in the future. The methodology allows for the defining of the correct prototype environment, RFID subsystem testing, and integrated system testing for the prototype environment.

20A.3   Plan

In this phase, we need to identify the critical path for both information and material flow. As a beginning of plan phase, the first thing is to define the problem in real case, for example, what type of product do they use for inventory? What is the frequency of transportation they have everyday? That is all related to our test design for warehouse.

20A.3.1   Define

In this phase, it is necessary to compile the real environmental requirements into the test parameter. This makes it necessary to show the theoretical model in the design and analyze the phases as an explanation and foundation for our future experiments. In this step, we describe the facility layout process, which is based on input data, an understanding of the roles and relationships between activities, a material flow analysis, and an activity relationship analysis. The defining step is shown in the first phase–plan; the clear material flow should be identified in this step as a basis of predict and perform. Figure 20A.1 shows us a clear view of thought process.

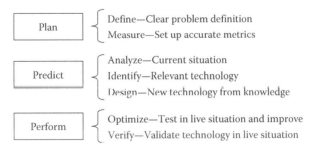

Figure 20A.1 DFSSR 3P's thought process.

20A.3.2 Measure

Multiobjective RF warehouse architecture is the overall RFID warehouse implementing system. It includes three main parts: an RFID edge layer, an RFID physical layer, and an enterprise integration network. The RFID physical layer is the connection of the other two layers. The RFID system is designed to process the streams of tags, or sensor data, coming from one or more readers. The edge layer has the capability to filter and aggregate the data prior to sending them to a requesting application. For example, an action (tag read) is triggered when the object moves or a new object comes into the reader's view. The RFID edge servers filter and collect the tag data at each individual site and send it over the Internet to the third layer—enterprise integration layer. The localized data are identified by moving actions and stationary actions separately. This difference divides the usage of RFID antenna into two types of equipments: one is a fixed reader for warehouse portal door and another is a mobile reader for tracking inventory. The fundamental tenet of the warehouse portal distribution system is that they must be able to accommodate changes that may occur on a network. The portal devices provide real-time positioning access capabilities to user communities, and delivering and searching personal data. They allow external customers and partners access to data secure access.

We can now divide the RFID warehouse system into three parts as discussed earlier: the physical layer, the logic layer, and the system integration layer. Each layer has different components depending on what functions the RFID system needs. By understanding the flow in the warehouse, we can determine the types of tags and antennae needed.

In short, RFID implementation in any process has two or three layers. The physical layer produces log events for RF sensors during process executions. The logic layer records the log event-related data including filter and integrate functions. The analysis of the physical layer activity has been discussed in facility layout research. Previous research with RFID facility layout does not include data flow as a factor that influences the RFID warehouse efficiency and performance.

20A.4 Predict

In this phase, the major issue is to analyze the outcome and process of our RFID operation. As we know, the critical results of experiment or test are very important for the company who wants to implement RFID; the situation may vary from each company. Combining the real environment and site requirements, the test should be conducted in an appropriate and cost-effective way.

20A.4.1 Analyze

Data environment analysis. One of the design components of the RFID warehouse layout is the data flow through the distribution process. This is included in the experimental design phase. The goal of such an activity is to define the input and output data in order to confirm the efficiency of data flow and its physical flow. Data standards can be smoothly exchanged within the supply chain because the data are already formatted and organized. The workflow and the data flow are both generated by the production flow from the physical layer to the logical layer. All the data chosen through the distributing process are generated by RFID equipment including the tags on each pallet or the antenna on the portal. Therefore, the location of the RFID equipment is influential on the accuracy of the distribution process that forms the individual data flow according to the workflow. The location of the RFID antenna, also called a sensor, will be discussed later on.

20A.4.2 Design

According to our analysis of site environment and data types, we choose to use the passive tag as our technology in warehouse, which means low-cost but high-volume information flow.

First, the "sensor" is used to refer to a device that is connected, via network or radio frequency (RF) communication medium, to other sensor devices in the network. The location of the sensors in the warehouse relate to either their environment or their data traffic flow that is detected by a fixed antenna on the portal door. Similarly, the data flow will be employed by sensors specifically in the picking entrance portal and the distributing portal. Therefore, the data traffic through the two portal doors and its layout will be considered in this chapter. The other communication between the nodes in the warehouse will be discussed in future work. In order to measure the accuracy and efficiency of RFID performance in the warehouse, we are using a *ratio* to evaluate the relationship of performance and efficiency of RFID readability. The ratio σ equals the simple relationship between input and output data, which will be related to a regression analysis to show fair performance:

$$\sigma_r = \frac{\alpha_I}{\beta_O},$$

(20A.1)

where:
σ_r = ratio
α_I = input
β_O = output

However, this ratio only gives an average performance for RFID readability. The components of the input require the precise data to evaluate the environment and performance. However, we measured the benchmark of the performance used to compare the different input data and data flow. For example, the different amounts of workflow reflect the different data flow in the warehouse, but the benchmark gives us reliable data to measure different warehouse environments and workflows. In the next stage, the optimize step, we use a case study to illustrate our theory explained in the design and analyze phases.

20A.5 Perform

Perform is the last phase of our study, after previous experiments and design applications. The next stage is to prove the feasibility of our design by using design of experiments (DOEs) and optimize the system performance according to the current configurations.

20A.5.1 Optimize

Based on our theoretical model, we conduct experiments following the order of the optimum facility layout and cost-effective equipment. As a major part of statistical analysis in Six Sigma methodology, we use DOE as a possible improvement through process. In DOE, the effects of several independent factors (variables) can be considered simultaneously in one experiment without evaluating all possible combinations of factor levels. For experiment 1, two independent variables (factors) and one dependent variable were used. The two independent variables included the tag replacement and the number of antennae. The dependent variable was the readability of the tags. The observed results of the experiment had an effect on the independent variable:

$$\text{Readability} = f(\text{TP}, \text{AN}), \tag{20A.2}$$

where:
TP is the tag placement
AN is the number of antennae

For experiment 2, we needed to regulate some of the variables until the results achieved full-read efficiency. This was done to satisfy the customer's requirement of 100% readability. The model of the experiment is as follows:

$$\text{Readability} = f(\text{AP}, \text{PP}), \tag{20A.3}$$

where:
AP is the antenna's position
PP is the portal's position

20A.5.1.1 Factors and levels

When executing a full factorial experiment, a response is obtained for all possible combinations of the experiment. Because of the large number of possible combinations in full factorial experiments, two-level factorial experiments are frequently utilized. Experiment 1 was a two-factor, two-level experiment. As described in Table 20A.1, a total of four trials were needed (i.e., $2^2 = 4$) to address all assigned combinations of the factor levels.

The specific situations to which a DOE is applied affect how factors and levels are chosen. Factor levels can also take different forms. In this experiment, levels are quantitative. Experiment 1 should allow for a systematic observation of a particular behavior under controlled circumstances.

Therefore, two principles were conducted in the experiment:

1. Tag placement: top, side
2. Number of antennae: one antenna and two antennae (on each side of portal; used only in one trial within the experiment)

Table 20A.1 Factorial Design 2^k

Factors: 2	Replicates: 10
Base runs: 4	Total runs: 40
Base blocks: 1	Total blocks: 1
Number of levels: 2, 2	

Table 20A.2 Experiment Factors and Levels

Factors and designations		Levels	
		1	2
Experiment 1	A1: Tag placement (TA)	Top	Side
	B1: Antenna number (AN)	1	2
Experiment 2	A2: Antenna's position (AP)	Horizontal	Nonhorizontal
	B2: Portal's distance (PD)	5 ft	7 ft

For experiment 2, in order to test the readability of the tags and metrics performance, the same trials were performed as experiment 1, with the following variables:

- Position of antennae: We installed two antennae at the same height on each side and at two different heights (3 and 5 ft).
- The distance between each side of the portal: 5 and 7 ft.

The standardized timescale was 30 seconds in consideration of limited real-world data acquisition times. All of the specifications were conducted 10 times with three different replacements of tags and 10 items in each trial. The experiment factors and levels are summarized in Table 20A.2.

20A.5.2 *Verify*

We have several ways to optimize the current layout design to improve the reading accuracy: For experiment 1, the placement of the tags on the pallet or item can be classified in three ways: top, front, and side. The performance of each classification is different. To sum up, the best position of a tag on an item is on the side (face to the antennae) because of the polarization and magnetic field (polarization test report). The performance of the antennae and tags are totally different with these three classifications. The third classification, tag on the side, had the best performance. Compared with the other two classifications results, the readability of the tags, when they are on the side, can be up to 60% of full satisfaction. This classification is still not very satisfying. We also determined that a significant change occurred when the number of antennae was varied. We carried out the experiment using one antenna with 10 items first. The readability of the tags was only 80%, compared with almost 100% when using two antennae (average: 96%). The read rate graph is shown in Figure 20A.2.

For experiment 2, the influence of the variable *tag placement* in the model is the same as in experiment 1. The antenna position had different effects on the two experiments. The experiment hypothesized that the antenna, when placed in different horizontal planes, would have a positive influence on readability. The nonhorizontal antenna showed better results when distance between antennae increased. For example, the experiment conducted from 5–7 ft demonstrated better results at 7 ft than at 5 ft. *Portal dock's position* is an important factor that has an influence on reading accuracy, especially in experiment 2. The hypothesis was that the shorter the distance is, the better the read efficiency would be. The experiment simulates real-world circumstances and is designed for two hypothesized cases. The requirements are the distance between each side of the portal must be 5 ft with the antenna on the same horizontal line and the distance between the readers must be 7 ft with different heights. Therefore, the objective of the experiment was to verify the hypothesis about whether the performance of readability is better with "nonhorizontal line" orientation. If the hypothesis was proved to be true, the improvement on the antenna reading efficiency would increase by varying the height of the two antennae when other factors are fixed (Figure 20A.2).

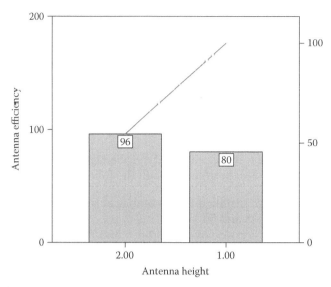

Figure 20A.2 Read rate graph.

Finally, the results of the experiment supported the aforementioned hypothesis. We identified the normal distance between each side as 7 ft, but the optimum and effective distance is ~5 ft. The factors of distance between each side and the nonhorizontal antenna both have an influence on the effectiveness of readability.

20A.6 Conclusion

The reaction time of the antenna on the tags was almost the same in these two cases. It can be determined that readability can achieve a full-read expectation when performed under the specification given below:

- The full-read range is 3–5 ft when the antennae are fixed at the same horizontal line on each side of portal.
- The full-read range is 6–8 ft when the antennae are fixed with different heights on each side of portal.
- The full-read requirement needs to have the tags on the sides of items or facing toward the antennae.

Frequency, distances and angles, type of tag, location and replacement, and influences of moisture, metals, and pallet patterns all played a part in the readability of the tags. The effective reading distance was analyzed in MATLAB (6.0, Release 13), for visualizing the results and provided documents for future research. The data points on the graph showed random variation, but the visualization graph gives a clue that the most effective scale for the antenna is between 2 and 3 ft around the middle line. The bar on the right side indicates the read rate that is based on our experiment specification of 10 tags per pallet. The reading rate can be reached at 90% or better under this specification (Figures 20A.3 and 20A.4).

To sum up, the total conditions for receiving >90% readability includes several considerations, which are as follows:

- The placement of tags
- The distance between antennas (if there is more than one antenna)

Reading Rate of RFID

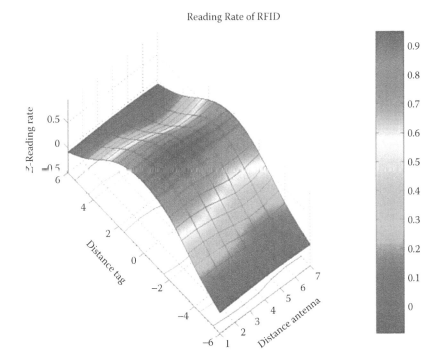

Figure 20A.3 Three-dimensional graph for effective reading rate.

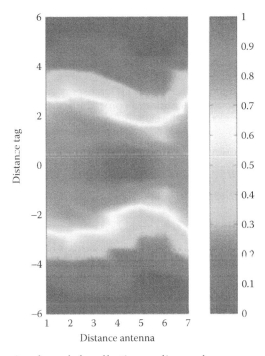

Figure 20A.4 Two-dimensional graph for effective reading scale.

- Appropriate stop by time when going through the portal (at least 3 seconds)
- The change of the position of the antenna when other limitations are fixed

Using Six Sigma methodology increases the efficiency of test procedures and validates the influence of real warehouse case layout scenarios.

Further Readings

Banerjee, P., Zhou, Y., and Montreuel, B. (1997). "Genetically assisted optimization of cell layout and material flow path skeleton." *IIE Transactions*, 29(4): 277–291.

Carbon, T. A. (2000). "Measuring efficiency of semiconductor manufacturing operations using data envelopment analysis (DEA)." *IEEE SEMI Advanced Semiconductor Manufacturing Conference*, September 2–14, Boston, MA.

Gaukler, G., Seifert, R., and Hausman, W. (2007). "Item-level RFID in the retail supply chain." *Production and Operations Management*, 17(6).

Gleixner, S. et al. (2002). "Teaching design of experiments and statistical analysis of data through laboratory experiments." *32nd ASEE/IEEE 2002 Frontiers in Education Conference*. Vol. 1. IEEE. November 6–9.

Gotsman, C. and Koren, Y. (2005). "Distributed graph layout for sensor networks." *Graph Drawing*. *Lecture Notes in Computer Science*, 3383. Heidelberg: Springer.

Jones, E. C., Volakis, J., and Verma, V. (2007). "How RFID reliability effects inventory control accuracy?" *Antennas and Propagation Society International Symposium*, IEEE, June 9–15, Honolulu, HI.

Kleijnen, J. P. C. (1995). "Verification and validation of simulation models." *European Journal of Operational Research*, 82(1): 145–162.

Lee, Y. M., Cheng, F., and Leung, Y. T. (2004). "Exploring the impact of RFID on supply chain dynamics." *Proceedings of the 36th Conference on Winter Simulation*. Winter Simulation Conference, Washington, DC.

Pan, J., Tonkay, G., and Quintero, A. (1999). "Screen printing process design of experiments for fine line printing of thick film ceramic substrates." *Journal of Electronics Manufacturing*, 9(3): 203–213.

Penttila, K., Sydeimo, L., and Kivikoski, M. (2004). "Performance development of a high-speed automatic object identification using passive WID technology." *Proceedings of International Conference on Robotics 8 Automation*, College Park, MA, June 20, IEEE, pp. 4864–4868.

Rao, K. V. S., Nikitin, P. V., and Lam, S. F. (2005). "Antenna design for UHF RFID tags: A review and a practical application." *IEEE Transactions on Antennas Propagation*, 53(12): 3870–3876.

Tompkins, J. A. et al. (1996). *Facility Planning*. 2nd edn. New York: John Wiley & Sons.

Wehking, K., Seeger, F., and Kummer, S. (2006). "RFID transponders: Link between information and material flows. How reliable are identification procedures?" *Logistics Journal*.

Zhang, Y., Liu, J., and Zhao, F. (2006). "Information-directed routing in sensor networks using real-time reinforcement learning." *Combinatorial Optimization in Communication Networks*, 18: 259–288.

Appendix 20B: A framework for effective Six Sigma implementation[*][†]

20B.1   Introduction

As competition gets more intense, customers demand higher quality products and/or services organizations look for the ways to improve their operational performance to address customer expectations (Hammer 2002). In the pursuit of improved operational

---

[*] By Erick C. Jones, Mahour Mellat Parastb, and Stephanie G. Adams.

performance and higher customer satisfaction, Six Sigma has been recognized as a systematic and structured methodology that attempts to improve process capability by focusing on customer needs (Dasgupta 2003; Harry 1998; Linderman et al. 2003). It has been described as an approach for organizational change, which incorporates elements of quality management and business process reengineering (Quinn 2003).

There has been a significant increase in the application of Six Sigma in industry over the past decade. According to Hoerl (1998), General Electric's (GE) operating margins increased from 13.8% to 14.5%, an increase valued at about $600 million, which stemmed from Six Sigma quality initiatives. In 2002, at least 25% of Fortune 200 companies claimed that they have the Six Sigma program (Hammer 2002). By focusing on customer needs and defining quantifiable measures for achieving specific goals, Six Sigma projects result in greater customer satisfaction and enhance organizational performance and profitability (Blakeslee 1999; Goh et al. 2003; Harry 1998; Kondo 2001).

The implementation of Six Sigma has produced mixed results. While companies such as GE and Motorola report huge savings from their Six Sigma initiatives (Pande et al. 2000), critics of Six Sigma argue that many quality-based initiatives (such as Six Sigma) will fail because of the intense business competitiveness (Stebbins and Shani 2002). As organizations are looking toward Six Sigma programs, there is a need to address the issue of effective implementation of Six Sigma projects. We believe that developing a framework for Six Sigma implementation will help scholars and practitioners to gain insight into effective implementation of Six Sigma projects. It also helps organizations to effectively utilize their resources so that they can benefit from their Six Sigma initiatives.

The purpose of this chapter is to present a framework for implementation of Six Sigma projects. We begin by defining Six Sigma methodology using DMAIC or DMADV, which are popular frameworks for Six Sigma projects (Breyfolge 2003; Pyzdek 2003). We utilize the PDCA cycle to relate Six Sigma initiatives to process improvement. We identify the key variables that affect successful implementation of Six Sigma. In reference to the organizational context, we identify factors that are crucial in effective implementation of Six Sigma. We believe that the proposed framework can be used as a guideline for further research in implementing Six Sigma while helping organizations to oversee their Six Sigma programs.

20B.2 Six Sigma methodology

According to Feld and Stone (2002), Six Sigma is a data-driven philosophy used to drive management decisions and actions across an organization. Caulcutt (2001) indicates that Six Sigma reduces waste, increases customer satisfaction, and improves processes with a considerable focus on financially measurable results. For the purpose of this chapter, Six Sigma is defined as a set of methodologies and techniques used to improve quality and reduce cost utilizing a structured and disciplined methodology for solving business problems. Our definition is consistent with that of Hammer (2002).

A popular framework for implementing a Six Sigma methodology is the DMAIC process. DMAIC, or define, measure, analyze, improve, and control, is a key process of a standard framework for a Six Sigma project and is shown in Figure 20B.1.

According to Jing and Li (2004), the psychology of this approach is that the key process input variables are narrowed down to a vital few, with the idea that having control of the vital few will allow for good control of the whole process. DMAIC is widely used when a product or process is already in existence but performing inadequately. DMAIC focuses on eliminating unproductive steps, developing and applying new metrics, and using technology to drive improvement (De Feo and Barnard 2004). Another popular

Figure 20B.1 DMAIC framework.

Figure 20B.2 DMADV framework.

approach associated with Six Sigma projects, DMADV, or define, measure, analyze, design, and verify, is used for developing new products/services (Figure 20B.2). While the focus of DMAIC is on eliminating waste and improving an existing process, DMADV is primarily utilized to develop new products/services.

Six Sigma has a strategic component aimed at not only developing commitment to it, but also an active involvement of higher management (Snee 2000). This strategic component is the responsibility of management to identify the key processes of their organization, measure their effectiveness and efficiency, and initiate improvements of the worst performing processes. It is suggested that firms should implement their Six Sigma initiatives by integrating them with their business strategy (Cheng 2007).

20B.3   Literature review

As evidenced by Linderman et al. (2003), academic research on Six Sigma is lagging behind its practice in the industry. While empirical research is needed to fill the gap between the theory and the practice of Six Sigma, few studies have been carried out to understand the effective implementation of Six Sigma projects.

Most studies on Six Sigma have been primarily focused on anecdotal evidence and case studies. Academic research on Six Sigma has been accelerated in recent years (Linderman et al. 2003; McAdam and Lafferty 2004; Schroeder et al. 2008). McAdam and Lafferty (2004) argue that successful implementation of Six Sigma requires attention to both process perspective (methodology) and people perspective (behavior). While early research on Six Sigma has been focused on the technical side of Six Sigma in terms of tools, techniques, and methodologies, recent studies have paid attention to the psychological, contextual, and human side of Six Sigma such as reward systems for Six Sigma (Buch and Tolentino 2006), goal setting (Linderman et al. 2006), organizational context (Choo et al. 2007a), and psychological safety (Choo et al. 2007b). Six Sigma has been traditionally focused on cost reduction and efficiency; however, recent studies show that it could be used as a methodology to increase profitability (Sodhi and Sodhi 2005) and it could drive creativity (Biedry 2001), enhance organizational learning (Wiklund and Wiklund 2002), and facilitate innovation (Byrne et al. 2007). In terms of performance variation, the human side of Six Sigma exhibits the highest level of variation between different groups in a company (Fleming et al. 2005). In addition, it requires top management commitment, a highly disciplined approach, and training (Hahn et al. 2000).

Different theoretical frameworks have been used to understand Six Sigma implementation. Building upon goal theory literature, Linderman et al. (2003) address the role of specifying challenging goals for Six Sigma projects, where Six Sigma projects with challenging goals result in a greater magnitude of performance. They also indicate that the use of a structured method (in Six Sigma projects) increases performance. In another study, Linderman et al. (2006) empirically show that goals can be effective when Six Sigma projects employ Six Sigma tools and methods. However, specifying unrealistic and very challenging goals are counterproductive, resulting in frustration and lack of motivation for team members.

From a knowledge management perspective, Choo et al. (2007a) develop a knowledge-based framework for Six Sigma projects. By focusing on two complementary sources of knowledge creation in Six Sigma projects—prescribed methodology and organizational context—they argue that Six Sigma projects that can maintain a balance between the effective implementation of prescribed methodology (e.g., tools and techniques such as quality control) and context (e.g., leadership, organizational culture, and black belt roles) can generate higher levels of knowledge. To the extent that organizations can manage such a balance, a sustainable quality advantage will be achieved.

With reference to Six Sigma implementation as a structured and systematic process improvement methodology, little has been said on the design and structure of Six Sigma projects. While companies implement Six Sigma in a variety of ways, it is important to determine how much of those efforts is done in a systematic and structured way.

According to Mader (2002), structured application of tools and techniques increases the rate of success in process improvement. Standardization of methodologies reduces variability in the processes so it is crucial for organizations to implement Six Sigma in a structured and systematic way so that they can benefit the most from their efforts (Choo et al. 2007a). It is important to note that the success of any quality improvement project (including Six Sigma) requires the coexistence of attention to both detail and innovation (Naveh and Erez 2004). While a structured method (such as PDCA) facilitates attention to detail, contextual variables such as the role of black belts and executive commitment are the key in providing a culture of innovation.

Previous research on Six Sigma has paid little attention to the role of black belts in Six Sigma projects and how they can facilitate successful implementation of Six Sigma. Linderman et al. (2003) argue that improvement specialists such as black belts serve as role models for Six Sigma improvement projects where their effective involvement increases the success of Six Sigma projects.

To address the above shortcomings in the literature, we have employed the PDCA framework to relate Six Sigma implementation to process improvement. The steps in Six Sigma projects (DMAIC or DMADV) are similar to those of PDCA (Choo et al. 2007a; Shewhart 1931, 1939). The PDCA cycle is a well-established framework for process improvement where it focuses on continuous learning and knowledge creation (Deming 1993), which is the key in the success of any quality improvement initiative (Karlsson and Wiklund 1997; Wiklund and Wiklund 2002). While companies may use different terminologies for their Six Sigma projects, by adhering to a unified and well-established framework such as PDCA we attempt to integrate Six Sigma projects under the PDCA cycle. We believe that framing Six Sigma projects within a PDCA cycle provides a more comprehensive view of Six Sigma implementation in organizations.

20B.1 Development of constructs and their relationships

Eight constructs have been defined for Six Sigma implementation. The constructs have been developed based on the review of the literature (Breyfogle 2003; Pyzdek 2003) along with other constructs developed by the researchers after observing several Six Sigma organizations. These constructs investigate variations in implementing Six Sigma. Choo et al. (2007a) argue that the quality of improvement projects is affected by both methods and psychological variables. In another study, the effect of contextual variables on Six

Sigma implementation has been addressed (Choo et al. 2007b). Our instrument addresses both the method aspect of Six Sigma and the financial and organizational aspect. The following is a list of the constructs along with their brief description.

1. *Black belt roles.* Black belts are used differently across organizations. The purpose of this construct is to measure the degree to which the roles of black belts are dedicated to Six Sigma or if they have to split their time between their Six Sigma responsibilities and other day-to-day management activities. The role of black belts in effective implementation of Six Sigma projects has been addressed in previous studies (Choo et al. 2007a; Hammer 2002; Linderman et al. 2003). Black belts bridge the gap between the top management and the Six Sigma project team (Schroeder et al. 2008).

2. *DMAIC versus DMADV.* These two Six Sigma processes are designed to be used for specific types of projects. The purpose of this construct is to measure whether they are used according to their intention as recommended by Breyfogle (2003).

3. *Plan.* Deming (1986) describes how quality projects are managed in the following process: plan–do–check–act. It is sometimes referred to as a PDCA cycle (Arveson 1998). For both DMAIC and DMADV, this is the "define" step. This construct contains the first steps to start a project such as project selection, project planning, and project scope and metrics. The purpose of this construct is to compare how organizations actually start projects with how DMAIC and DMADV recommend the starting projects (Breyfogle 2003; Pyzdek 2003).

4. *Do.* The purpose of this construct is to measure how companies perform the second step in the PDCA cycle. For both DMAIC and DMAVD, this is the "measure" stage in the Six Sigma process. Experiments and tests are performed to evaluate the current performance and improvements on a project. Examples of these tools are DOEs and failure modes and effects analysis (FMEA; Breyfogle 2003).

5. *Check.* The purpose of this construct is to measure the third step in the PDCA cycle to check if the new ideas will perform as expected. Both DMAIC and DMADV have "analyze" as the third step. The purpose of this construct is to determine whether the organizations check the data with statistical tools and which sources of variation are critical to the process. Another purpose is to determine whether judgments are made on the data before the statistical analyses are performed.

6. *Act.* The purpose of this construct is to measure the last step in the PDCA cycle which encompasses the last two steps in the Six Sigma processes. Although they are different in nature, the last two steps are used to set and implement plans to make sure that the changes or new ideas are successful.

7. *Financial responsibility.* The purpose of this construct is to measure how leaders of the project are held accountable for the reported benefits as well as the benefits to the leaders when the project goes well.

8. *Executive support.* The purpose of this construct is to measure how involved the executives are with the projects. The role of executives in supporting Six Sigma projects has been addressed in the literature (Byrne et al. 2007; Linderman et al. 2006; Schroeder et al. 2008). Executives not only need to support Six Sigma projects in terms of providing resources and their commitment but they also need to make a balance between Six Sigma projects and operational activities of the firm. It is believed that the balance between the innovation projects (such as Six Sigma) and the operational activities of company is the key to the success of the firm (Gottfredson and Aspinall 2005).

The above constructs capture both the methodological elements (e.g., PDCA) and the context element (e.g., executive commitment) of Six Sigma projects. From the knowledge-based view of quality improvement, creation of knowledge in Six Sigma projects requires the coexistence of both methodological and context elements (Choo et al. 2007a). Accordingly, we believe that the variables and the proposed framework are capable of addressing both aspects of knowledge creation in Six Sigma projects.

20B.5   The conceptual model for Six Sigma implementation

Figure 20B.3 shows the conceptual framework for Six Sigma implementation. The proposed framework for Six Sigma implementation has been developed with reference to both the contextual elements (e.g., leadership) and the methodological techniques (e.g., PDCA). In terms of the contextual variables, it has been argued that the existence of a clear vision about Six Sigma projects is the key to the success of Six Sigma initiatives, where by relating these improvement projects to the business strategy management can enhance the effectiveness of Six Sigma projects (Larson 2003).

It should be noted that the role of executive support for Six Sigma projects is facilitated by black belts, who are champions (or improvement specialists) that direct and manage Six Sigma initiatives (Schroeder et al. 2008). While top management is not directly involved in Six Sigma implementation, it is believed that through supporting and empowering black belts, top management can accelerate the progress of Six Sigma implementation. Accordingly, we hypothesize that

Pm: Executive support significantly affects the role of black belts in Six Sigma implementation.

The availability of the resources for Six Sigma projects is a key in the success of such initiatives (Choo et al. 2007a). It is a driving force for creativity and innovation for those who need to use resources for any improvement projects (Amabile and Gryskiewicz 1987). Executives not only need to provide adequate resources for Six Sigma projects but they also need to empower Six Sigma project team members to take responsibility for the initiative. From the goal theory perspective, specifying monetary goals (cost saving or revenue) motivates team members to achieve the desired outcome (Linderman et al. 2003). Accordingly, empowering Six Sigma project team members enables them to take necessary actions so that they can meet the goals. In this regard, the executive team needs to empower Six Sigma team members while making them financially

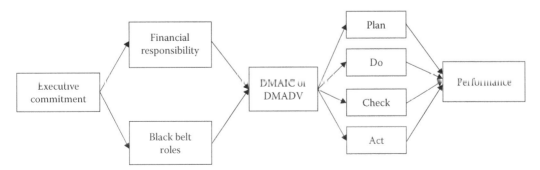

Figure 20B.3 Conceptual framework for Six Sigma implementation.

responsible for their actions and/or decisions regarding Six Sigma projects. Therefore, we hypothesize that

> Executive support significantly affects the financial responsibility of team members in Six Sigma implementation.

It has been argued that the success of Six Sigma projects requires utilizing specific tools and techniques (Choo et al. 2007a). Team members should be empowered to select the best methodology for dealing with Six Sigma projects. On the other hand, the autonomy of (Six Sigma) team members in making appropriate decisions in Six Sigma projects encourages them to select the best methodology for their process improvement initiative. If Six Sigma team members are held accountable financially, they will be motivated to utilize the appropriate methodology for their project. Accordingly, we hypothesize that

P4a: The selection of the appropriate methodology (DMAIC or DMADV) has a significant effect on the Plan phase of the PDCA cycle.
P4b: The selection of the appropriate methodology (DMAIC or DMADV) has a significant effect on the Do phase of the PDCA cycle.
P4c: The selection of the appropriate methodology (DMAIC or DMADV) has a significant effect on the Check phase of the PDCA cycle.
P4d: The selection of the appropriate methodology (DMAIC or DMADV) has a significant effect on the Act phase of the PDCA cycle.

It is believed that process improvement initiatives do not provide the desired results unless they are framed within a systematic process improvement structure. Elements such as vision, skills, resources, action plans, and incentives are necessary in effective Six Sigma implementation (Larson 2003). The lack or deficiency in any stages of the process improvement will have a negative effect on the desired outcome of the project. In the context of Six Sigma projects within the PDCA cycle, different phases of the PDCA cycle influence the outcome of the projects. Therefore, we propose that

P5: The Plan phase in the PDCA cycle has a significant effect on the performance of Six Sigma projects.
P6: The Do phase in the PDCA cycle has a significant effect on the performance of Six Sigma projects.
P7: The Check phase in the PDCA cycle has a significant effect on the performance of Six Sigma projects.
P8: The Act phase in the PDCA cycle has a significant effect on the performance of Six Sigma projects.

The proposed framework for Six Sigma implementation takes into account both the methodological aspect of Six Sigma and the organizational and contextual variables. In addition, it has the advantage of relating Six Sigma projects to the quality programs in an organization, where it relates Six Sigma initiatives to the PDCA quality improvement cycle.

As indicated by Sinha and van de Ven (2005), understanding Six Sigma concerns the integration of both the micro and the macro level of analysis within a firm. The success of Six Sigma projects requires multilevel understanding within an organization—a concept that is referred to it as Meso theory in organizational design (Daft 2001). In this regard, the proposed framework is capable of addressing both the macro (i.e., executive commitment) and the micro (e.g., PDCA cycle) level of analysis within a firm.

20B.6   Summary

The main purpose of this chapter was to develop a framework for effective Six Sigma implementation. Building upon previous studies in Six Sigma, we addressed the need for the utilization of a structured methodology for Six Sigma implementation. We also asserted the need for addressing contextual variables (e.g., executive commitment) in Six Sigma. With reference to the above findings, we specifically recognized the role of black belts in Six Sigma as well as the financial responsibility of the Six Sigma team. That, we believe, is the key in effective Six Sigma implementation. Six Sigma team members need to be empowered while held financially responsible. Our proposed framework can help the executives and managers to look at their Six Sigma initiatives from a broader perspective, linking the organizational variables with the methodological approach to Six Sigma.

Further research is needed to operationalize the constructs we have proposed in this study and test them using empirical data. Evidence from the industry shows that Six Sigma can be implemented to improve the performance of supply chain projects (Yang et al. 2007). Therefore, future research needs to address the implementation of Six Sigma in a supply chain environment.

References

Amabile, T. M. and Gryskiewicz, S. S. (1987). *Creativity in the R&D Laboratory*. Greensboro, NC: Center for Creative Leadership.

Arveson, P. (1998). "The Deming cycle." Retrieved from http://www.balancedscorecard.org/bkgd/pdca.html

Biedry, J. (2001). "Linking Six Sigma analysis with human creativity." *Journal for Quality and Participation*, 24(4): 36–38.

Blakeslee, J. (1999). "Implementing the Six Sigma solution: How to achieve quantum leaps in quality and competitiveness?" *Quality Progress*, 32(7): 77–85.

Breyfogle, F. W. III (2003). *Implementing Six Sigma: Smarter Solutions Using Statistical Methods*. 2nd edn. Hoboken, NJ: John Wiley & Sons.

Buch, K. and Tolentino, A. (2006). "Employee perceptions of the rewards associated with Six Sigma." *Journal of Organizational Change Management*, 9: 356–364.

Byrne, G., Lubowe, D., and Blitz, A. (2007). "Using a Lean Six Sigma approach to drive innovation." *Strategy & Leadership*, 35(2): 5–10.

Caulcutt, R. (2001). "Why is Six Sigma so successful?" *Journal of Applied Statistics*, 28: 301–306.

Cheng, J. (2007). "Six Sigma and TQM in Taiwan: An empirical study." *Quality Management Journal*, 14(2): 7–18.

Choo, A. S., Linderman, K. W., and Schroeder, R. G. (2007a). "Method and context perspectives on learning and knowledge creation in quality management." *Journal of Operations Management*, 25: 918–931.

Choo, A. S., Linderman, K. W., and Schroeder, R. G. (2007b). "Method and psychological effects on learning behaviors and knowledge creation in quality improvement projects." *Management Science*, 53: 437–450.

Daft, R. L. (2001). *Organizational Theory and Design*. 7th edn. Mason, OH: South-Western.

Dasgupta, T. (2003). "Using the Six Sigma metric to measure and improve the performance of a supply chain." *Total Quality Management*, 14: 355–366.

De Feo, J. and Barnard, W. (2004). *Juran Institute's Six Sigma: Breakthrough and Beyond*. New York: McGraw-Hill.

Deming, W. E. (1986). *Oct of Crisis*. Cambridge, MA: MIT Press.

Deming, W. E. (1993). *The New Economics, for Industry, Government, Education*. Cambridge, MA: MIT Press.

Feld, K. and Stone, W. (2002). "Using Six Sigma to change and measure improvement." *Performance Improvement*, 41(9): 20–26.

Fleming, J. H., Coffman, C., and Harter, J. (2005). "Manage your human sigma." *Harvard Business Review*, 83(718): 107–114.

Goh, T. N. et al. (2003). "Impact of Six Sigma implementation on stock price performance." *Total Quality Management & Business Excellence*, 14: 753–763.

Gottfredson, M. and Aspinall, K. (2005). "Innovation versus complexity: What is too much of a good thing?" *Harvard Business Review*, 83(11): 62–71.

Hahn, G. J., Doganaksoy, N., and Hoerl, R. W. (2000). "The evolution of Six Sigma." *Quality Engineering*, 12: 317–326.

Hammer, M. (2002). "Process management and the future of Six Sigma." *MIT Sloan Management Review*, 43(2): 26–32.

Harry, M. (1998). "Six Sigma: A breakthrough strategy for profitability." *Quality Progress*, 31(5): 60–64.

Hoerl, R. W. (1998). "Six Sigma and the future of the quality profession." *Quality Progress*, 31(6): 35–42.

Jing, G. and Li, N. (2004). "Claiming Six Sigma." *Industrial Engineer*, 36(2): 37–39.

Karlsson, S. and Wiklund, P. S. (1997). "Critical aspects of quality method implementation." *Total Quality Management*, 8(1): 55–66.

Kondo, Y. (2001). "Customer satisfaction: How can I measure it?" *Total Quality Management*, 12: 867–872.

Larson, A. (2003). *Demystifying Six Sigma: A Company Wide Approach to Process Improvement*. New York: American Management Association.

Linderman, K. W. et al. (2003). "Six Sigma: A goal theoretic perspective." *Journal of Operations Management*, 21: 193–203.

Linderman, K. W., Schroeder, R. G., and Choo, A. S. (2006). "Six Sigma: The role of goals in improvement teams." *Journal of Operations Management*, 24: 779–790.

Mader, D. P. (2002). "Design for Six Sigma." *Quality Progress*, 35(7): 82–86.

McAdam, R. and Lafferty, B. (2004). "A multilevel case study critique of Six Sigma: Statistical control or strategic change?" *International Journal of Operations & Production Management*, 24: 530–549.

Naveh, E. and Erez, M. (2004). "Innovation and attention to detail in the quality improvement paradigm." *Management Science*, 50: 1576–1586.

Pande, P. S., Neuman, R. P., and Cavanagh, R. R. (2000). *The Six Sigma Way: How GE, Motorola and Other Top Companies Are Honing Their Performance*. New York: McGraw-Hill.

Pyzdek, T. (2003). *The Six Sigma Handbook: A Complete Guide for Green Belts, Black Belts, and Managers at All Levels*. New York: McGraw-Hill.

Quinn, D. L. (2003). "What is Six Sigma?" In Bertels, T. (Ed.), *Rath & Strong's Six Sigma Leadership Handbook*. Hoboken, NJ: John Wiley & Sons, pp. 1–14.

Schroeder, R. G. et al. (2008). "Six Sigma: Definition and underlying theory." *Journal of Operations Management*, 26: 536–554.

Shewhart, W. A. (1931). *Economic Control of Quality of Manufactured Product*. New York: D. Van Nostrand.

Shewhart, W. A. (1939). *Statistical Method from the Viewpoint of Qualm Control*. Washington, DC: The Graduate School, Department of Agriculture.

Sinha, K. K. and van de Ven, A. H. (2005). "Designing work within and between organizations." *Organization Science*, 16: 389–408.

Snee, R. D. (2000). "Guest editorial: Impact of Six Sigma on quality engineering." *Quality Engineering*, 12(3): ix–xiv.

Sodhi, S. M. and Sodhi, N. S. (2005). "Six Sigma pricing." *Harvard Business Review*, 83(5): 135–142.

Stebbins, W. M. and Shani, A. (2002). "Eclectic design for change." In Docherty, P., Forslin, J., and Shani, A. B. (eds.), *Creating Sustainable Work Systems: Emerging Perspectives and Practice*. London: Routledge, pp. 213–225.

Wiklund, H. and Wiklund, P. S. (2002). "Widening the Six Sigma concept: An approach to improve organizational learning." *Total Quality Management*, 13: 233–239.

Yang, H. M. et al. (2007). "Supply chain management Six Sigma: A management innovation methodology at the Samsung Group." *Supply Chain Management: An international Journal*, 12: 88–95.

Applying and using Lean Six Sigma

chapter twenty-one

Using Six Sigma to evaluate radio frequency technologies at NASA

> You cannot inspect quality into the product; it is already there

> —W. Edwards Deming

21.1 Introduction

This case study is based on the research conducted at NASA between 2005 and 2006 and was funded by the Nebraska EPSCOR Grant. It describes how design for Six Sigma was utilized to conduct testing to determine the feasibility of using radio frequency identification (RFID) technologies for the International Space Station (ISS). We want to make sure that we acknowledge the student Tim Farhham for his involvement in the research.

21.2 Methodology: Integrated DFSS for research

The University of Nebraska's RFID and Supply Chain Logistics (RfSCL) has developed a new research strategy fusing traditional research methods with industry's new gold standard Six Sigma. This blend of methodologies helped the RfSCL laboratory use information received by the customers (NASA and Barrios) to better develop a solution that best meets the needs of these two organizations. Integrated design for Six Sigma (DFSS) uses the plan, predict, and perform methodology typically used in research but couples it with DFSS (Breyfogle 2003). DMADO (De-may-doh) is the acronym for design, measure, analyze, develop, and optimize (Breyfogle 2003). This new approach to research will help bridge the gap between academic organizations and industry.

21.2.1 Plan

The plan phase contains the first two steps in the DMADO process: define and measure. These steps lay down a roadmap for where the project needs to go and what resources are necessary.

21.2.1.1 Define

Define is where gathering information from the customer, suppliers, journals, and other sources are used to define the problem, scope of the project, and metrics to use. In our case, this involved meetings with NASA and Barrios employees, contacting the various suppliers used during the research, and studying the published literature on implementation of RFID in unique applications.

Quality function deployment (QFD) methods were used to determine the exact customer needs and areas to focus on to meet these customer requirements. The primary tool used was the house of quality (HOQ). This tool takes customer preferences and demands and turns them into technical requirements that can be quantified, measured, and analyzed.

The HOQ directed the focus of the research to be on the tag range and infrastructure requirements. This will be further discussed in Section 21.3 of this report.

21.2.1.2 Measure

After the HOQ focused the research, the tests and conceptualization models of the inventory management system started. The tests included read range, polarity, interference, orientation, and manufacturer reliability.

21.2.2 Predict

The predict phase is a repeating phase that contains analyze, design, and identify, and then leads back to measure from the plan phase. This allows researchers the flexibility to test many different types of systems and accurately compare each system.

21.2.2.1 Analyze

Measurements taken during the measure phase were then run through statistical software. The software then determined the factors that were significant in the model. This phase was crucial to the next phase where the analysis governed the design of possible systems that could use the hardware and concepts tested.

21.2.2.2 Design

In this phase, we used the analysis to plan out the design characteristics of the final model. These characteristics included processes, hardware, and software needs for a solution to the customer demands.

21.2.2.3 Identify

The identify phase begins with the design of the solution. It takes a critical look at the proposed solutions and identifies where there is need for improvement. These areas can be found by looking back at the HOQ and determining if the original focus was still maintained. Any gap in the system to deliver the needed output is then noted, and corrective actions can begin. This leads back to measuring the corrective actions and beginning the predict cycle again, until a design meets the original customer requirements.

21.2.3 Perform

Perform takes the place of the IC (improve and control) from DMAIC (define, measure, analyze, improve, and control). Due to the nature of research, implementation of the researched systems does not apply. Therefore, optimize and verify play an important role in giving the customer solutions.

21.2.3.1 Optimize

The optimize phase takes the final design and creates the detailed process and application map for the solution. This makes it possible for small changes to be made. These changes are detail oriented.

21.2.3.2 Verify

The verify phase is the last step in the process. Typically, it involves producing a prototype and delivering final process map for the customer to follow. Implementation plans and other such details are looked at to verify that customer demands were met.

21.3   Define

21.3.1   Project charter

21.3.1.1   Problem statement

The ISS currently uses the bar codes to audit and control inventory on board. However, recent events have uncovered the need to improve the current system. During a routine audit, it was discovered that supplies of food were less than expected.

The goals of the project were given by Barrios project director Amy Schellhase. The objectives were to test commercially available off the shelf (COTS) RFID hardware and develop possible inventory management solutions for the ISS. The project was not to exceed 6 months and deliver a final recommendation on hardware to implement and processes to develop.

21.3.1.2   Business case

Although NASA is not a business, there was a significant business case to develop an inventory management system using RFID to reduce the labor involved with auditing consumable materials on the ISS. Using simple calculations, we can determine the hourly cost of having an astronaut in space and calculate the labor cost of manual auditing. By reducing the auditing time, astronauts can spend more time running experiments that contribute to the objectives of their mission. This time savings can then be turned into a dollar value.

NASA estimates the cost of an astronaut in space as $1 million per day. This makes a manual inventory count very costly. Previous projects have looked into astronauts scanning their items as they use them. However, this would cause an incremental labor cost for every item scanned. On average, an astronaut uses 50 consumable items per day. This extra 10 seconds per manual scan with a bar code reader or other such devices would add $5787 per day per astronaut in labor alone. The station normally contains three or four crew members at any given time, making the total labor cost exceed $20,000 per day, almost $7.3 million per year.

Currently, astronauts spend 1 week every 6 months taking inventory. This current cost of labor is $22.4 million per year. By reducing this labor cost, much of NASA's money can be spent on research instead of inventorying the ISS.

21.3.1.3   Scope

The first step in the process was to identify shareholders in the project. These shareholders contain anybody affected by the outcome of the project. The shareholders for this project included the following:

- Barrios employees
- NASA employees
- Russian Space Agency
- European Space Agency
- Astronauts from all countries

21.3.1.4   Deliverable

There were two deliverables for the RISCL team. The first deliverable was the recommendation of hardware to install on the ISS. This deliverable consisted of testing multiple RFID systems and determined which systems could be used in NASA's specific application. The other deliverable was to develop processes to use the said hardware to manage the consumable inventory on board the ISS.

To understand the customer requirements and technical requirements of the project, an HOQ was constructed. Figure 21.1 shows how the HOQ was constructed.

Interactions (roof correlation matrix):

- Time to inventory: −1
- Cost of development: −1, −1
- Time of development: −1, −1, 2
- Cost of implementation: −1, −1, 1, 1
- Time of implementation: −1, −1, −1, −1, −1
- Life expectancy of tags: 1, 1, 1, 1
- Life expectancy of reader: 1, 1, 1, 1
- Power life of reader (recharging): −2, −2, −2, −1
- Range of tag: −2, −2, 1
- Size of tag: −2, −2, −1, −2

Customer requirements	Design requirements	Readability of tags in environment	Time to inventory	Cost of development	Time of development	Cost of implementation	Time of implementation	Life expectancy of tags	Life expectancy of reader	Power life of reader (recharging)	Range of tag	Size of tag		
		X	N	N	N	N	N	X	X	X	T	N		
Accurate inventory	5	9	3	9	9	3	3	9	9	3	9	1	335	27.92%
Low crew interaction	5	3	9	9	9	3	9	1	3	3	3		260	21.67%
Low infrastructure	4	3	3	3		9	9		3	3	3	1	148	12.33%
Radiation proof	5	9		3	3	3							90	7.50%
Low-power requirements	3		3	3	3	3	3	3	3	3	9	1	102	8.50%
Durability	4	9	3	3	1				3	1			80	6.67%
Nonintrusive (to packing)	3	3	3	3	3	3	1	3			9	9	111	9.25%
Short implementation time	2	1	3	3	9	9	9			3			74	6.17%

Figure 21.1 House of quality. N, nontechnical, T, technical, X, nonrelated.

The HOQ shows the main criteria that any implementation should focus on. The three top criteria are accurate inventory, low crew interaction, and low infrastructure. Because the implementation requirements are not known, the "basement" of the house was not included in the initial analysis but was used during testing.

21.3.1.5 Test requirements

The last step in the define phase was to define the test requirements by compiling a list of all forms of tests that needed to be performed and determine the metrics for which to test. For example, for interference testing, signal strength and read rate were used. For range testing, distance and orientation to antennae were parameters. For read rate, the amount of reads per second was used.

21.3.1.6 Obstacles

A key concern with the use of RFID tags is interference with the surrounding environment. The RfSCL laboratory investigated and documented many scenarios that would impede the collection of inventory data. Preliminary tests were used to define the key elements that needed to be further tested to best understand how RFID tags could be used.

Experiments in the laboratory showed that the tags directly mounted on metal containers did not read. This obstacle can be overcome by changing the Russian metal food canisters to NASA's plastic sealable bags.

21.3.1.7 Define acceptable hardware vendors

Due to the nature of RFID, vendor selection is very important. Many manufacturers make big claims but are frequently not able to follow through with their claims. The RfSCL contacted many vendors and narrowed the field to only a couple of vendors.

Selection criteria for vendors related directly to customer (NASA/Barrios) demands. Critical issues were longevity of vendor, such that it would be difficult to switch vendors should one vendor discontinue service for any reason. Another key issue was the vendor's ability to get tags on a regular basis and to deliver tags when needed. With these criteria in mind, the RfSCL laboratory chose a vendor that most closely met the requirements.

21.3.1.8 Application strategy

RFID needs an effective application strategy for it to be beneficial. The last part of the project is to determine an effective way to use the RFID equipment in order to meet the customer demand. Barrios will then implement the RFID application.

21.4 Measure

The test procedure used by the RfSCL laboratory closely followed what is considered to be a "noisy" test (Emigh 2005). Due to the ISS not being an anechoic environment, tests were not conducted in an anechoic chamber. The term "noisy" comes from the radio noise that is abundant in the RfSCL laboratory. This radio noise is caused by but not limited to telecommunication towers, motor vehicles, miscellaneous electrical devices, and ambient radio signals from television and radio transmission towers nearby.

Using wooden antenna mounts, the laboratory conducted a series of tests on each of the passive system. These tests included range, orientation sensitivity, obstacle, read rate, and accuracy.

21.4.1 Read range testing

The read range test determined the approximate range of RFID tags of each system. This metric is very important because of the nature of the RFIS systems. It is a limiting factor of what types of procedures can be implemented on a per tag basis.

21.4.1.1 Read range experiment procedure

Read range was tested in the RfSCL laboratory using set guidelines. The tags were placed 5 ft above the concrete/tile floor and mounted to a wooden stand. The floor was marked in 0.5-ft increments. Starting with a 1-ft separation, the tag was queried, and if the tag was read successfully, the separation was increased by 0.5 ft, and then requeried. This process was continued until the tag was no longer able to be read.

21.4.2 Orientation sensitivity testing

For some tags, as the orientation of the tag to the reader changed, the read range changed. We referred to this phenomenon as orientation sensitivity. After documenting this factor, a test was developed for quantifying the sensitivity. Table 21.1 shows the data collected.

21.4.2.1 Orientation sensitivity experiment procedure

Orientation sensitivity was conducted much in the same way as read range with one variation. This change was that the tags were rotated 45° in the antenna axis of rotation. The range was recorded for each orientation.

Table 21.1 Polarization Test Data

Interval	R0, A0	R45, A0	R90, A0	R0, A30	R45, A30	R90, A30	R0, A45	R45, A45	R90, A45	R0, A60	R45, A60	R90, A60
						Antennae						
1	14.8	15.17	16.3	29.29	25.09	25.19	19.28	21.7	22.39	0	14.02	15.63
2	14.8	15.27	16.28	29.42	25.23	25.31	19.06	21.65	22.35	0	13.92	15.27
3	14.6	15.04	16.25	29.22	25.5	24.96	19.45	21.6	22.37	0	14.45	15.36
4	14.75	14.87	16.27	28.68	25.37	25.59	18.91	21.52	22.47	0	14.83	15.12
5	14.74	15.19	16.32	28.78	25.27	25.43	18.96	21.52	22.53	0	14.31	15.16
						Antennae (polarized)						
1	29.77	28.76	26.76	14.04	12.45	31.27	16.26	12.83	12.84	0	0	10.11
2	29.32	28.89	26.64	13.94	12.46	31.21	16.05	12.87	13.27	0	0	10.41
3	29.39	28.01	27.33	14.2	12.37	31.09	16.01	12.75	13.71	0	0	10.38
4	28.96	27.78	27.2	14.15	12.45	31.18	16.33	12.58	13.48	0	0	10.62
5	29.45	28.03	27.3	14.03	12.64	31.42	16.28	12.92	13.02	0	0	10.67
						Antennae (nonpolarized)						
1	19.51	16.25	19.17	23.05	25.45	21.26	11.17	19.06	22.59	0.56	10.29	5.45
2	20	16.34	18.09	23.06	25.67	21.14	11.27	19.32	22.46	0.52	10.09	5.64
3	19.72	16.88	18.11	22.9	24.99	21.21	11.11	19.4	22.35	0.51	9.87	5.51
4	19.96	16.32	19.25	22.74	25.07	20.87	11.22	19.55	22.44	0.56	9.8	5.48
5	20.13	16.75	19.06	22.55	25.08	21.21	11.48	19.75	22.29	0.56	9.89	5.45

R, range; A, angle.

21.4.3 Read accuracy testing

Read accuracy is defined as scanned tags properly transferring their information. This test ensured that all tags, reader, and anticollision logarithms work properly.

21.4.3.1 Read accuracy experiment procedure

Read accuracy was tested by using the following procedure: First, 10 tags were placed on a nonconductive substrate (white printer paper). The tags were then placed 5 ft away from the antennae. The tags were then scanned for 10 seconds and the tag numbers received were recorded. Then the recorded tag numbers were compared with the actual tag numbers, and the accuracy was then recorded.

21.5 Analyze

21.5.1 Read range analysis

Using the data gathered in the measure phase, statistics were then used to determine the actual range of the tags. Using the customer specification of 99.9% accuracy, we formulated what range would give 99.9% accuracy. This was done by separating the factorial data by tag angle and RFID system. Descriptive statistics were then applied to each group of data and then used to predict the 99.9% read range, as demonstrated in Equation 21.1. Table 21.2 shows the data with the calculations of Equation 21.1.

$$99.9\% \text{ range}_{\text{system}} = \min(\mu_{0°} - Z_{0.005}\sigma_{0°}, \mu_{30°} - Z_{0.005}\sigma_{30°}, \mu_{60°} - Z_{0.005}\sigma_{60°}, \mu_{90°} - Z_{0.005}\sigma_{90°}). \quad (21.1)$$

Table 21.2 Read Range Analysis

| | Matrix | | | | | | | | | | | |
| | 4 × 4 | | | | 2 × 2 | | | | Alien–SamSys | | | |
Trial	0°	30°	60°	90°	0°	30°	60°	90°	0°	30°	60°	90°
1	12.50	16.00	15.00	16.00	9.00	5.50	6.00	7.00	5.50	5.00	4.50	4.50
2	12.50	16.00	15.00	16.50	9.00	5.50	6.00	7.00	5.50	5.50	4.50	4.50
3	12.50	16.00	16.00	16.00	9.00	5.50	6.00	6.00	5.50	5.50	4.50	4.00
4	12.50	16.00	16.00	16.00	9.00	5.50	6.00	7.00	5.50	5.50	5.00	4.00
5	12.50	16.00	16.00	16.00	8.00	5.50	6.00	6.00	5.50	5.00	5.00	4.50
6	12.50	16.00	16.00	16.00	8.00	6.00	6.00	7.00	5.50	5.00	4.50	4.50
7	12.50	16.00	16.00	16.00	8.00	5.50	6.00	7.00	5.50	5.00	5.00	4.50
8	12.50	16.00	16.50	16.00	8.00	5.50	6.00	7.00	5.00	5.00	4.50	4.50
9	12.50	16.00	16.00	16.00	8.00	5.50	6.00	6.00	5.50	5.00	5.50	4.50
10	12.50	16.00	16.00	16.50	8.00	5.00	6.00	6.00	5.00	5.00	5.00	4.00
Mean	12.5	16.0	15.9	16.1	8.4	5.5	6.0	6.6	5.4	5.2	4.8	4.4
SD	0.0	0.0	0.5	0.2	0.5	0.2	0.0	0.5	0.2	0.2	0.3	0.2
Z	3.1	3.1	3.1	3.1	3.1	3.1	3.1	3.1	3.1	3.1	3.1	3.1
99.9% range	12.5	16.0	14.4	15.4	6.8	4.8	6.0	5.0	4.7	4.4	3.7	3.6
Minimum range	**12.50**				**4.77**				**3.60**			

Table 21.3 ANOVA for Accuracy—Type III Sums of Squares

Source	Sum of square	df	Mean S	F	p
	Main effects				
A: Floor angle	2.66667	3	0.888889	32.00	.0000
B: Polarized	0.222222	1	0.222222	8.00	.0068
C: Tag angle	0.444444	2	0.222222	8.00	.0010
	Interactions				
AB	0.666667	3	0.222222	8.00	.0002
AC	1.33333	6	0.222222	8.00	.0000
BC	0.111111	2	0.0555556	2.00	.1465
ABC	0.333333	6	0.0555556	2.00	.0841
Residual	1.33333	48	0.0277778		
Total	7.11111	71			

Notes: As indicated by the *p*-values obtained from the data, all three cofactors of main effects were significant, as were the interactions between floor angle and polarization and floor angle and tag angle. The interaction between polarization and tag angle as well as the interaction between all three factors were less significant. F, F-test; df, degrees of freedom.

21.5.2 Orientation sensitivity analysis

Using an analysis of variance (ANOVA), on the average the significance of each factor was analyzed. The ANOVA is summarized in Table 21.3.

21.5.3 Read accuracy analysis

Read accuracy showed 100% readability. This test was conducted by verifying the tag number received by the computer was the tag that was scanned. This test was conducted by placing

20 tags in a cargo transfer bag (CTB) with crumpled paper and Styrofoam. The bag was then shaken and scanned for 10 seconds. All tags were read properly throughout the 20-scan study.

21.6   Develop

21.6.1   Smart shelf

The smart shelf system integrates shelving hardware with RFID antennas. This system would have dual capabilities. The main function of the smart shelf is to provide accurate inventory. However, the system also has the capability to locate items on board the ISS.

The unique infrastructure of this system would consist of blanketing the storage areas with mats that act as RFID antennae. These antennae read passively tagged consumable items and report their location within a couple feet (Figure 21.2).

21.6.2   Door tracking

Door tracking is a system that tracks the tagged items as they move through the ISS. This system allows for accurate item quantities to be recorded by recording when items were moved from storage and into trash areas. Location of items can also be found with this system by a query that reports which module an item is located in (Figure 21.3).

21.6.3   Waste tracking

Waste tracking is a unique form of inventory management because it measures items and quantities used instead of scanning the station for inventory. This concept exploits the centralization of waste on the ISS and therefore uses less infrastructure and less power than other systems. It also has the possibility of avoiding misreads due to interference or background reflection.

The process begins on the ground with packaging like items in bags. For the purpose of this report, we use 36 razor cartridges. Because a typical launch consists of two to four bags of these cartridges, all of one bag will be used before another bag is opened. Once one bag of razors are used and the Ziploc® bag is thrown away, the system

Figure 21.2 Smart shelf diagram.

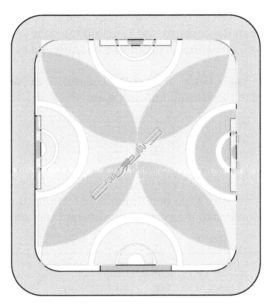

Figure 21.3 Door antennae placement.

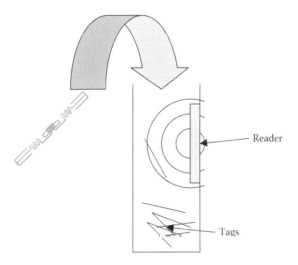

Figure 21.4 Waste tracking system.

registers that 36 razors have been used and need to be replenished during the next launch (Figure 21.4).

21.6.4   *Sensor active tag*

The sensor active tag (SAT) system has the potential to be the ideal inventory management system for ISS. The key function of the SAT system is that it can query consumables on board the ISS on a regular schedule as well as locate items on board the ISS. These audits can keep the ground crew constantly aware of what items are being used on a daily basis.

A SAT is an active tag that can read passive tags. These SAT devices are placed inside CTBs and periodically scan the contents and transmit the data to the central Information Management System (IMS) software. SATs can also be equipped to help the crew locate the bag using audible methods.

Development of this tag is being done at the University of Nebraska in a joint effort between the Department of Electrical Engineering and the Radio Frequency Supply Chain Logistics Laboratory.

21.7 Optimize

21.7.1 Implementation

A Delphi decision matrix was used to determine which application was used. This organized the different applications to the customer needs and ranked the equipment and the proposed systems.

Factors	Weight	System			
		Sensor active tag	Door tracking	Waste tracking	Smart shelf
Infrastructure	3	7	4	9	5
Intervention	3	9	9	9	9
Reliability	2	8	8	7	9
Location	1	9	7	5	9
Development	1	5	8	7	5
		78	70	80	74

Factors	Weight	Tag				
			Matrix			
		Alien	4×4	2×2	RF code	SAVI
Ability to get tags	2	10	8	7	8	6
Size/shape	2	7	7	9	8	3
Read distance	2	6	9	7	7	7
Durability	2	9	9	9	9	10
Ease of tagging	1	6	10	9	7	1
Cost	1	10	10	10	4	2
		80	86	83	75	55

The decision matrix shows that the 4×4 matrix tags used with the waste tracking inventory method would work best.

21.7.2 Cost analysis

There will be several costs associated with the implementation of the RFID waste tracking system. Section 21.7.2.1 outlines all of the costs associated with the system.

21.7.2.1 Fixed cost

The fixed costs include three parts. The first is the readers. These will be for the ISS, NASA ground crew, and Barrios. The second is the software cost of in-house labor for

Table 21.4 Fixed Costs for Waste Tracking System

Readers	$75,000.00
Software	$100,000.00
Installation	$4,500,000.00
Total	**$4,675,000.00**

Table 21.5 Annual Incremental Costs for Waste Tracking System

Tags	$15,000.00
Ground crew labor	$300,000.00
Maintenance	$150,000.00
	$465,000.00

Table 21.6 Annual Savings for Waste Tracking System

Year	Fixed cost ($)	Annual cost ($)	Annual savings ($)	Annual return ($)
1	4,675,000.00	465,000.00	22,400,000.00	17,260,000.00
2		465,000.00	22,400,000.00	39,195,000.00
3		465,000.00	22,400,000.00	61,130,000.00

NASA programmers to integrate the RFID system into their existing WMS software. The last is the installation cost that sends the equipment to the ISS and has the astronauts install the hardware (Table 21.4).

The incremental costs of the waste tracking system also include three parts. Table 21.5 outlines all the incremental costs.

The return on investment can be seen in Table 21.6. The table shows the annual savings of $20.4 million earnings before interest and taxes (EBIT) over 3 years.

21.8 Conclusion

RFID has the ability to help streamline the ISS inventory management systems. With less time spent on inventorying consumable items, they will be able to do more research than before. The initial cost of almost $5 million is returned in the first year with the labor savings.

References

Breyfogle, F. W. (2003). *Implementing Six Sigma*. 2nd edn. Hoboken, NJ: Wiley.

Carreau, M. (2001). "Space station crew endures food shortage." Houston Chronicle, December 1.

chapter twenty-two

Using Six Sigma to evaluate automatic identification technologies to optimize broken case warehousing operations

> You cannot inspect quality into the product; it is already there.
>
> —W. Edwards Deming

22.1 Introduction

This project was conducted by a team of professors and students from the University of Texas at Arlington as well as Tecnológico de Monterrey, Campus Querétaro. The principal investigator was Dr. Erick C. Jones, an industrial and manufacturing engineering professor at the University of Texas at Arlington as well as the director of radio frequency identification (RFID) and auto-identification (RAID) laboratories. This project was also supported by Dr. Beatriz Murrieta, an industrial engineering professor at Tecnológico de Monterrey, Campus Querétaro as well as the department chair. Support and project input were also given by Dr. Vettrivel Gnaneswaran, the assistant director of the RAID laboratories.

The project leads for this investigation include Jose Sanchez Gonzalez, Mackenzie Dacres, and Rayanne Macnee. Gonzalez is a graduate Master of Science in industrial engineering (MSIE) student at the University of Texas at Arlington. Dacres is an industrial engineering undergraduate student at the University of Texas. Macnee is a nursing undergraduate student. Support was also given by Walter Muflur and Cynthia Vinueza-Garcia, both undergraduate industrial engineer students at the University of Texas at Arlington.

22.1.1 RAID laboratory

The RAID laboratories were established in 2011 by the principal investigator Dr. Erick C. Jones. The vision of the laboratory is that everything will be tracked wirelessly in 10 years. The mission of the laboratory is to provide integrated solutions in logistics and other data-driven environments through automatic data capture, real-world prototypes, and analysis. The goal of the laboratory is to support the marriage of industries' supply chain needs for automated identification technology with academia's theoretical applications. The purpose of the facility is to support project initiatives such as RFID, Logistics, Manufacturing, and Information Technology. The RAID laboratory has had six major projects fully funded within the 1 year it has been established and has five more upcoming.

22.1.2 The National Science Foundation–International Research Experiences for Students

The National Science Foundation (NSF) is an independent federal agency that was established in 1950 by Congress. It has an annual budget of $6.9 billion, all of which goes toward the progression of science, national health, prosperity, and welfare. That $6.9 billion currently funds ~20% of all federally supported basic research. The International Research Experiences for Students (IRES) program is funded to support the development of worldwide research and interaction. It seeks to increase students' comfort level when working with people from different cultures by exposing them to the technological, cultural, economic, and sociopolitical aspects of Mexican society. The outcomes are as follows: an increase in research skills, cultural awareness, and a student's abilities in research methods and problem-solving skills.

22.2   Background

22.2.1   Company background

Research was conducted at a large food product distribution Center in Latin America.

22.2.2   Problem statement

The information associated with a stretch-wrapped full pallet license plate number (LPN) is lost when a "full" pallet has to be broken or taken apart to fulfill orders with individual cases on that pallet. The cases associated with the full pallet do not have the information such as lot, expiration date, or other relevant information that was included with the original LPN. This is causing customer returns of cases that are approximated at 15% of the outbound volume. It is expected that the volume of "broken" pallet case picking will be increasing in the near future which using the current methods may lead to increased labor hours, high turnover, and lower worker productivity.

22.3   Research objective

22.3.1   Research question

In developing a research objective, the team narrowed the focus of the project down to three specific research questions. These questions include the following: Can RFID impact the number of outbound cases without LPN information? Can engineering work process redesign (EWPR) impact the incident ratio (IR)? and can RFID and EWPR impact the proposed future labor hours? We will explore these questions through the course of this chapter.

22.3.2   Relationship to business problem

These research questions are closely tied to the business problem faced by employees of Company XYZ. By investigating RFID technologies, we will gain insights into the "broken pallet" case-picking process and will be able to evaluate the impacts of pallet LPN information transfer on cases as well as future workflow and workstation redesign for increased worker optimization and improved safety conditions if RFID is implemented and if broken pallet case volume increases.

22.3.3   Research hypothesis

In conducting our research, we developed three test metrics, which are listed below.

22.3.3.1   Test metrics

1. The number of outbound cases without LPN information (NCLPN)
2. The IR
3. The projected number of future labor hours (PNLH)

In addition to the three test metrics, four hypotheses were created.

22.3.3.2   Null hypothesis

H_{01}: RFID technologies do NOT reduce the NCLPN alternative hypothesis.
H_a: RFID technologies do reduce the NCLPN.
H_{02}: RFID technologies and EWPR do NOT reduce the IR.
H_{03}: RFID technologies and EWPR do NOT reduce the PNLH.

22.3.3.3   The rejection criteria

H_{01} is that if the NCLPN can be reduced by 10%, then we cannot reject the H_{01}.
H_{02} is that if the IR can be reduced by 10%, then we cannot reject the H_{02}.
H_{03} is that if the PNLH can be reduced by 10%, then we cannot reject the $H_{03.}$

22.4   Research methodology

We will employ design for Six Sigma research (DFSSR) approach, developed by Jones (2011). DFSSR is a research methodology focused on reducing variability, removing defects, and getting rid of wastes from processes, products, and transactions. The approach utilizes quality tools and optimization techniques with effective cost justification (Figure 22.1).

The DFSSR-based approach contains seven different levels, where the problem defined needs to be measurable followed by the definition of metrics.

22.4.1   Research methodology

This section describes the different actions that need to be done under each specific phase.

Figure 22.2 outlines the seven phases of DFSSR methodology and the corresponding steps with each phase. In the first stage "plan," we define and measure the problem. In the second stage "predict," we measure, analyze, design, and identify. In this project, the last phase "perform" that uses optimize and verify is not utilized due to time constraints.

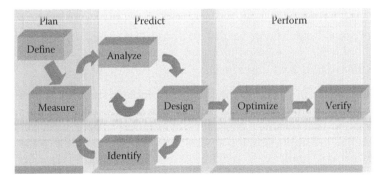

Figure 22.1 DFSS-R research methodology. (From Erick, J. C. and Christopher, C. A., *RFID and Auto-ID in Planning and Logistics. A Practical Guide for Military UID Applications*, CRC Press, Boca Raton, FL, 2011. With permission.)

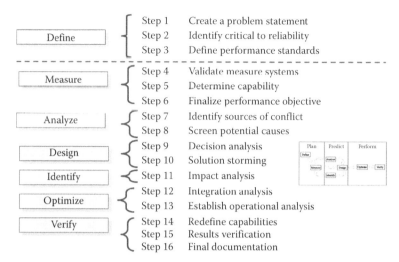

Define	Step 1	Create a problem statement
	Step 2	Identify critical to reliability
	Step 3	Define performance standards
Measure	Step 4	Validate measure systems
	Step 5	Determine capability
	Step 6	Finalize performance objective
Analyze	Step 7	Identify sources of conflict
	Step 8	Screen potential causes
Design	Step 9	Decision analysis
	Step 10	Solution storming
Identify	Step 11	Impact analysis
Optimize	Step 12	Integration analysis
	Step 13	Establish operational analysis
Verify	Step 14	Redefine capabilities
	Step 15	Results verification
	Step 16	Final documentation

Figure 22.2 Phases of DFSSR.

Table 22.1 elaborates specifically on the tools the team utilized for this project. Figure 22.3 provides an example of a sample flowchart and how it might be utilized to solve a problem. Figure 22.4 displays the organizational structure of a fishbone diagram. Figure 22.5 illustrates how by using Pareto analysis one can identify 20% of the causes to 80% of the problem. These tools were all utilized through the course of the project to come to our conclusions.

The team also conducted thorough data collection through plant visits, interviews, and observations.

22.4.2 Research approach

As mentioned earlier, the research approach used by the team was the DFSSR methodology. We utilized the define, measure, analyze, design, and identify phases. In the define phase, we identified the company business problem and narrowed the scope of the research project. Utilizing the results of the previous phase, we will now identify the metrics for the business problem statements in the measure phase. Using the results of the previous phase, we analyzed the impacts of the current operations on the metrics defined. We identified the problem and metrics and completed an analysis of each. We now identify how the RFID technology will impact the metrics. In the identify phase, we evaluated the cost benefit of the proposed scenario in the design phase.

22.4.3 DFSSR—Steps and description

The define state results in defining the business problem statements. The cases associated with the full pallet do not have the information such as lot, expiration date, or other relevant information that was included with the original LPN. This is causing customer returns of cases that are approximated at 15% of the outbound volume. It is expected that the volume of "broken" pallet case picking will be increasing in the near future which using the current methods may lead to increased labor hours, high turnover, and lower worker productivity.

Table 22.1 Tools Used for the DFSSR Methodology

Define	Measure	Analyze	Design	Identify	Optimize	Verify
Interview process/ observation	Hypothesis testing	Fishbone	Scenarios	Cost–benefit analysis	Factorial design of experiments	Control plans
Language processing	Analysis of variance	Pareto analysis	Fractional factorials	Visual systems	Theory of constraints	TMP
Prioritization matrix	Quality function deployment	Process mapping	Data mining	5-S	Response surface methodology	Multiple response optimization
System map	Flowchart	Multi-vari chart	Blocking	TPM	Blocking	Visual systems
Stakeholder analysis	Measurement system analysis	Chi-square	Response surface methodology	Mistake proofing	Data mining	SPC/APC
Thought process map	Graphical methods	Regression	Multiple response optimization	SPC/APC	Multiple response optimization	Mistake proofing
Value stream mapping	Process behavior charts	Buffered tolerance limits	Theory of constraints	ROI analysis	Fractional factorials	Continuous improvement

Note: 5-S, sort, set in order, shine, standardize, and sustain; APC, advanced process control; SPC, statistical process control; TPM, total productive maintenance.

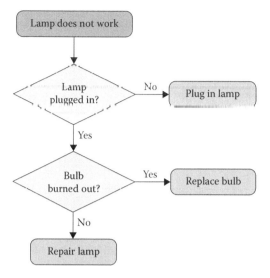

Figure 22.3 Sample flowchart.

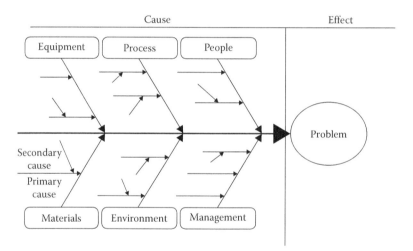

Figure 22.4 Sample fishbone diagram.

Figure 22.5 Pareto analysis.

After conducting the measure phase, the team was able to determine the following metrics:

Metric 1: NCLPN
Metric 2: IR
Metric 3: FNLH

After determining the metrics, the team then began the analyze phase of DFSSR. The results of the facility layout analysis were as follows: Two floor-level conveyors travel to the outbound shipping transportation area. The pallets are staged in front of the conveyor so that they can be "broken" pallet case picked. Then the workers unload the pallet onto the floor conveyor where the pallets have a designated location.

This would indicate that fixed location designs are possible. However, integrating bar codes or additional RFID labels would require additional manpower needs as well as equipment for the operation. Based on the observation of material handling techniques used, worker productivity may be impacted by the calculation of incidence rate.

The flowchart was also analyzed, which demonstrates that the current operation of integrating automation for transferring LPN information from pallets to cases will require new operation procedures. This future-state analysis will impact several areas of the current flowchart. These areas include managing forklift traffic, staging pallets to be picked, labeling operation, and staging operation. In addition, worker procedures need to be analyzed for excessive exertion and better ergonomic postures must be considered.

The results of the fishbone diagram are that the current LPN is linked only to the pallet label. The information from this LPN would have to be linked at the case level. This is due to the fact that the conveyor scanner does not determine non-LPN information for broken pallet cases. However, the conveyor scanner may be able to use RFID LPNs in the future with strategic location. The current bar codes on the cases can only be used for the final/end customer and are not being manipulated for distribution center use. Possible manipulation of this system may be a solution. Lastly, the current employee working capability may contribute to the overall work productivity. Educating employee on proper working postures may also impact worker productivity and overall distribution center efficiency.

The results of the Pareto analysis confirm that the biggest impact for the lost LPN information would come from a variety of solutions. Company XYZ should consider linking the LPN to the cases, utilizing the conveyor scanner to automate manual counts on the conveyor, evaluating the manipulation of the customer's bar code to connect supply chain information, and modifying working postures and lifting techniques employed by the workers, which may be modified to impact work productivity.

After thorough analysis, the team moved into the design phase where we came up with three different scenarios:

Scenario 1: Passive Gen 2 RFID System
Scenario 2: EWPR
Scenario 3: Passive Gen 2 RFID System and EWPR (REWPR)

22.4.3.1   Scenario 1

In this scenario, we can see where the employee is attaching the passive RFID levels to the cases. These labels contain LPN information that will help keep track of the cases. Fixed or handheld readers are used by the employees to write, read, and verify the information in the vendor management system (VMS) (Figure 22.6).

Figure 22.6 Passive tags.

Figure 22.7 Redesigning the process.

22.4.3.2   Scenario 2

In this scenario, we redesign some of the processes to improve worker productivity. It will imply the change of the conveyor's height. As a result of the adjustments, the employee working methods will be affected with the goal of optimizing the worker movements with training and minimal facility upgrades (Figure 22.7).

22.4.3.3   Scenario 3

In this scenario, we can RFID Gen 2 system and EWPR.

These changes will impact the way the employees load the cases on the conveyor belts. In addition, the organization will be able to keep track of the cases that are shipped from the distribution center (Figure 22.8).

22.4.4   Summary

Using RFID for case-level LPN tracking, we consider the facility changes for higher worker productivity: Scenarios 1–3.

Figure 22.8 Passive tags/EWPR.

22.4.5 Identify

In this phase, we evaluate the cost benefit of the proposed scenarios in the design phase. This will be done through the calculation of the return of investment as well as the payback period.

22.5 Results

We created three different scenarios in the identify phase of DFSSR phase:

- Scenario 1 involves the implementation of a passive Gen 2 RFID system. This would call for a large initial investment. With this scenario, every employee would be provided a barcode (RF) scanner. There is, however, an estimated rate of investment of 31.4% and an estimated payback period of 0.43 years.
- Scenario 2 involves the implementation of an EWPR. This scenario would call for an initial investment and would involve providing braces to increase the conveyor height. It is estimated that there will be an estimated savings return of 29.3% and an estimated payback period of 1.27 years.
- Scenario 3 involves both the implantation of a passive Gen 2 RFID system and EWPR (RfEWPR). This would call for an initial investment rate of investment of 31.4% and a payback period of 0.44 years.

22.6 Discussion

22.6.1 Rejection of hypothesis

In addition, we also rejected all three of our null hypotheses. We found that RFID technologies do reduce the number of outbound cases without LPN information; RFID technologies and EWPR do reduce the IR; and RFID technologies and EWPR do reduce the PNLH.

22.7 Limitation

The results and recommendations calculated during this project may not be accurate because of a variety of limitations we encountered. These limitations include the fact that the team was only able to observe operation for a limited time period given the duration of the research project. In addition, the team mostly used observations and limitations for the analysis. Also, a small sample set was used in all data collection. In order to validate results and give a serious evaluation, more data collection is needed. Lastly, it is recommended that the last phase of DFSSR including the optimize and verify steps is preformed to insure that the technologies recommended are tested and are the proper solution for the given environment.

22.8 Conclusion

In conclusion, we are grateful to the National Science Foundation for the opportunity to be a part of this NSF–IRES in Mexico University of Texas at Arlington program. Because of their funding, we had the opportunity to enhance our Spanish language skills and experience the Mexican culture.

We also appreciate the opportunity for working with a research team and learning a scientific research approach using RFID and Logistics technologies.

Further Readings

Company XYZ's official website. (2012). "Breakfast, snacks, recipes, cereal." Company XYZ's, n.d. Web. July 02. http://www.Company XYZ.com/en_US/home.html

Erick, J. C. and Christopher, C. A. (2011). *RFID and Auto-ID in Planning and Logistics. A Practical Guide for Military UID Applications*. Boca Raton, FL: CRC Press.

Ngai, E. W. T. et al. (2010). "RFID systems implementation: A comprehensive framework and a case study." *International Journal of Production Research*, 48(9): 2583–2612.

NSF. (2012). "Nsf.gov—National Science Foundation—US National Science Foundation." National Science Foundation, n.d. Web. July 02. http://www.nsf.gov/index.jsp

Smith, H. A. and Konsynsk, B. (2003). "Developments in practice X: Radio frequency identification (RFID)—An internet for physical objects." *Communications of the Association for Information Systems*, 12(19): 301–310.

University of Texas at Arlington. (2012). "Fast facts." UTA College of Engineering, n.d. Web. July 02. http://www.uta.edu/engineering/about/fast-facts.php

Ustundag, A. (2010). "Evaluating RFID investment on a supply chain using tagging cost sharing factor." *International Journal of Production Research*, 48(9): 2549–2562.

Whitaker, J. Mithas, S., and Krishnan, M. S. (2007). "A field study of RFID deployment and return expectations." *Production and Operations Management*, 16(5): 599–612.

chapter twenty-three

Using Six Sigma to implement RFID automation at an automotive plant

You cannot inspect quality into the product; it is already there

—**W. Edwards Deming**

23.1 Research problem and literature search

The research problem investigated for this project is stated as:

> How to identify and locate defects which happen in a 15-step manufacturing process.

Previous research and case studies have shown feasible solutions for using radio frequency identification (RFID) technologies in tracking material flow in manufacturing operations that involve high metallic components.

> Nowadays, even the high frequency (HF) RFID system has been successful in many applications, such as commercial payment, product management, personal security and access control, but the technology still faces severe obstacles in metal industries due to electromagnetic interference. The applicable, main and economic approach for solving this obstacle which is Placing Electromagnetic Interference (EMI) sheet for reducing of metal interference problem is still dominating even that Ferrox tag has been designed for solving the same issue, but due to the reason that needs to be thick in order to let sufficient flux for energizing the chip which result to less competitive than EMI. This paper is providing a measurement method for determining if EMI sheet useful in RFID metal tags based on wireless power transfer which can be qualified and quantified loss and return to the tag coil. Finally, a 0.4 mm-thick EMI sheet was created and it is able to cover almost 60% of the reading range. (Wang, 2008)
>
> Most of companies are using manual paper-based documents for tracking and managing the inspection in material tests. However, manual-based method is the easiest way to occur errors, and the unreliable and ineffective information documenting process which requires a significant amount of time and effort. RFID is an automated and user-friendly technology which is useful for quality management by providing visibility through all the systems for gathering, filtering, managing, monitoring and sharing quality data. Higher the efficiency

and flexibility of information flow is the important issues, and it can be performed by RFID-QIM (RFID-based Quality Inspection and Management) technology, mobile devices (PDAs) and web portals. Also, these technologies increase speed and accuracy of data entry, reduce cost savings simultaneously. (Home-Rotil, 2011)

RFID in inter-organizational will offer a range of direct and indirect benefits. Direct benefits such as read multiple tags and store information into the system in a short time without any labor effort, and lower labor and inventory handling costs. Indirect benefits, for example, make it easier to find misplaced items, reduce shrinkage from theft and reduce transaction errors. Combine these two types of benefits; RFID is able to improve internal efficiency of the organization, higher worker productivity, manage inventory control, reduce stock-outs, and decrease theft and fewer scanning errors.

As the costs perspective, direct costs are including implementation costs, hardware and software, installation costs, training costs, reader equipment costs and maintenance costs; indirect costs are such as the time employees spent on learning and manipulating the system, and the cost of RFID software integrated with the existing systems. (Industrial and Manufacturing Systems Engineering, 2011)

23.2   Research overview and objectives

In order to minimize defects and improve customer service, the team asked the Five W's and One H (What, Where, Why, When, Who, and How) according to the past data that attributed errors. From the cause-and-effect diagram (Figure 23.1), we determine that skipping operation and inspection steps in the manufacturing line is the main problem of causing quality issue. We will apply RFID technology on tracking if each of 15 steps has been done properly; in other words, RFID can help to avoid the chance of skipping steps. At the same time, it will be useful for time studies as well, such as when the work-in-process (WIP) inventory comes in the step, how long it takes to be done, and when it actually leaves to the next working step. We will also investigate the computer software/hardware for technology improvement.

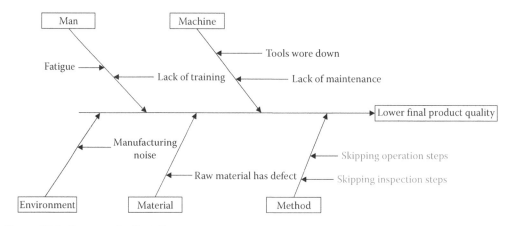

Figure 23.1 Cause-and-effect diagram.

Facility layout may be a possible solution for solving the skipping of manufacturing steps by the manufacturing cell's employees. A good facility layout will reduce detour routes, increase the space utilization, ease communication and support, decrease WIP moving time, and make production flow/process flow smoothly. We layout the project in the following aspects: (1) Be induced to the overall problem of the designated industry, (2) have a tour that will show us the facilities with the goal to assimilate and imagine how procedures cycle, (3) make a flowchart to predict possible issues in the industry, (4) obtain data from the industry for more accurate solutions, and finally (5) implement possible solutions.

The research objective is to demonstrate the capabilities of RFID technologies in ensuring that every manufacturing step is performed. Each manual process within the multi-step process and each computer-based process that affects manufacturing downtime will be evaluated for improvement. We will investigate these problems by analyzing the data that later will be part of conversion of procedures. We will also investigate the computer software/hardware for technology improvement. In return, an efficient flow of raw materials from station to station and an integrated automated system on which supervisors and employees will have an effective way of tracking incoming/outgoing materials at each station will be formulated. The overall goal is to use this project as a first step in reducing the excessive defects for the production line. The strategy to approach this problem is broken into the following aspects.

Specific objective 1. Identify the problems associated with the manual processes and evaluate operations to prevent workers from skipping steps within the manufacturing operation.

Each member of the group will cooperate with all the knowledge they have learned in school and real-life experience. Students will make sure that they combine skills such as verbal and visual knowledge of logistics. Students will discuss possible solutions as a team.

Specific objective 2. Identify possible engineering layouts that would support RFID technologies solutions and review implementation scenarios for RFID technologies.

As a team, we utilized Six Sigma very carefully to make sure what we are evaluating is a feasible solution. Utilizing flowchart diagrams, Pareto analysis, and work measurement the team analyzes the problems for possible opportunities for improvement.

23.3   Research method and approach

The plant has multiple champions who were utilizing the DMAIC method that stands for define, measure, analyze, improve, and control. Our team will employ the design for Six Sigma research (DFSS-R) approach to modify the manufacturing system (see Figures 23.2 through 23.4). DFSS-R is a research methodology aimed at reducing variability, removing defects, and getting rid of wastes from processes, products, and transactions.

According to the cause-and-effect diagram (Figure 23.1), we supposed that *skipping either operation or inspection steps* is the main factor causing lower final product quality that increases more and more customers' complaints. The fishbone diagram analysis identified two main problems:

1. Skipping operational steps
2. Skipping inspection steps

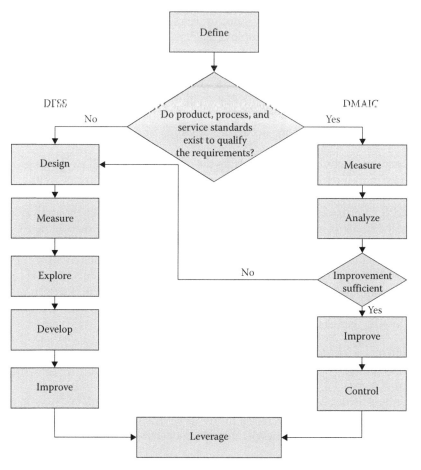

Figure 23.2 DFSS vs. DMAIC. (From Flowchart. Wikipedia, the free encyclopedia. http://en.wikipedia.org/wiki/Flowchart. With permission.)

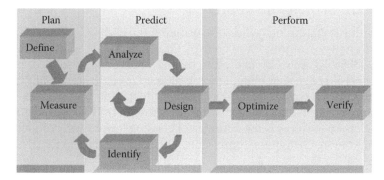

Figure 23.3 DFSSR. (From Flowchart. Wikipedia, the free encyclopedia. http://en.wikipedia.org/wiki/Flowchart. With permission.)

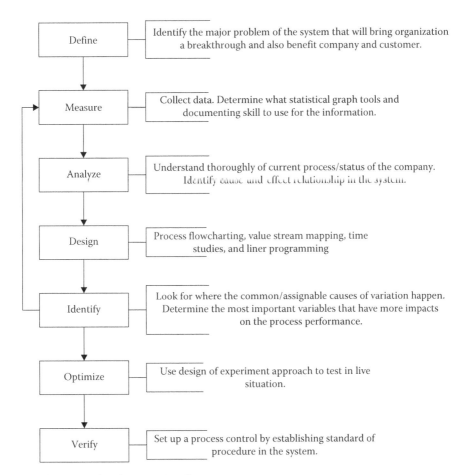

Figure 23.4 Six Sigma implementation flow.

23.3.1 Measure

In the measure step, we determine what metrics will be measured and what techniques will be used to obtain data for the identified characteristics described as metrics and techniques.

The number of steps skipped in operation (ST) is as follows:

- Manual observation
- Flowchart analysis
- Facility layout analysis

The correlation between the skipped steps and the defect rate (CSD) is as follows:

- Evaluate company data
- Pareto analysis

Reviewing metrics (ST) data to evaluate the skipped steps was our first step toward a solution. The data reviewed were given by COMPANY XYZ and developed by our team (Figure 23.5).

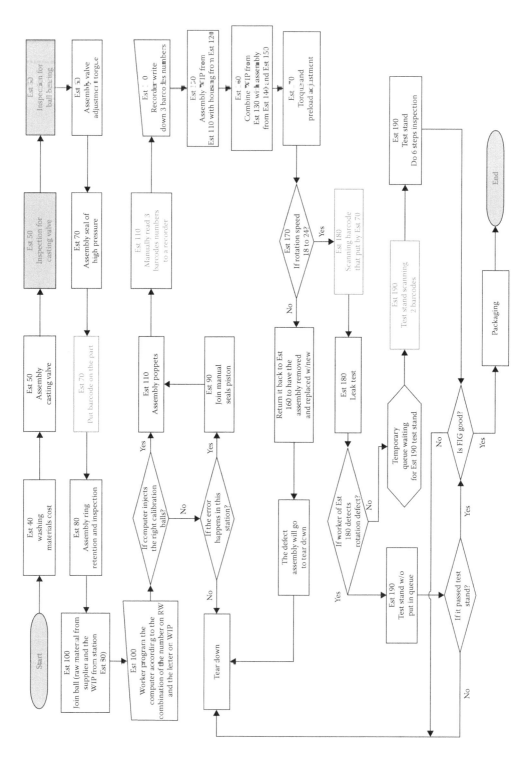

Figure 23.5 Example measure process flow.

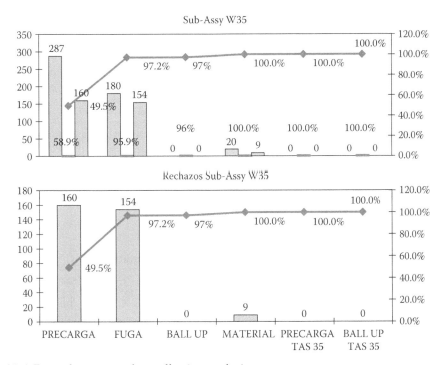

Figure 23.6 Example measure data collection analysis.

We then compared our walk and observed flowcharts with the companies' operational flowcharts. Metric 2 is the correlation between the skipped steps and the defect rate (CSD).

We analyzed the defect rate data in order to correlate the defects with stations with skipped processes by using Pareto analysis (Figure 23.6). In Pareto analysis, it identifies the top three stations that had defects and the rework because the worker puts the wrong combination number into the computer.

23.3.2 Design

In the design step, we utilize the information and insights that we acquired in the analysis stage to come up with possible options or solutions that are viable. RFID is a new generation of barcode. It has reader, antenna, tag, host computer, and middleware. Reader will scan the tag through the antenna and send the information that stored in the tag to host computer (Figure 23.7).

We suggest using RFID ultrahigh frequency (UHF) 433 MHz active tag or specialized passive tag (metal craft) because it is suitable to track metal and liquid, due to the fact that the manufacturing line is producing metal product (Table 23.1).

The purposes of applying RFID are as follows:

- Tracking all the steps that have to be done in each station. It will match our objective 1 and avoid workers' skipping steps.
- Tracking where and when the parts are in the plant. It will match our objective 2, providing the real-time location of WIP and finished goods (FG) and the status of WIP, FG, and rework, and also can easily do time study for each step in each station.

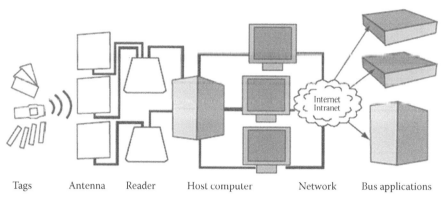

Tags   Antenna   Reader   Host computer   Network   Bus applications

Figure 23.7 Sample RFID system

Table 23.1 UHF Characteristics and Applications

Frequency	Key characteristics	Typical applications
UHF 433 and 860–960 MHz	• In use since late 1990s	• Supply chain and logistics at case and pallet level
	• Longer read ranges than HF tags (more than 4–5 feet)	• Inventory control
	• Very long transmit ranges for active 433 MHz (up to several hundred feet)	• Warehouse management
	• Gaining momentum due to worldwide retail supply chain mandates	• Asset tracking
	• Potential to offer lowest cost tags	
	• Incompatibility issues related to regional regulations	
	• Susceptible to interference from liquid and metal	

We develop three scenarios as possible solutions for solving problems mentioned in the define stage:

- Scenario 1: Handheld scanner scans location tags and active tag.
  - The tags will be put on the part.
  - Within a station, location tags will be put for each substep.
  - Workers have their own handheld scanners; they scan when every time they grab a new part and also scan the location tags within the station to make sure that they do complete all the substeps.
  - The use of RFID will track location and tracking of WIP, FG, and rework.
- Scenario 2: Active tag is scanned by each location reader's antenna (in each antenna's interrogation zone) and using cart system.
  - The tags will be put on the part.
  - Within a station, location reader's antenna will be put for tracking each substep.
  - Reader's antenna will be put on the carts.

- Every time part leaves the cart and returns to the cart, it will be scanned. Location reader's antenna has its own interrogation zone, so it will automatically scan the tag on the part to make sure worker completes all the substeps.
- This scenario could let workers concentrate on their works; they do not need to bother or be distracted by scanning tags.
- Scenario 3: Active tag is scanned by each location reader's antenna within the station and the reader on the conveyor belt.
  - Location reader's antenna is fixed in each station.
  - UHF 433 MHz active tag is used.
  - Workers have to do each step especially within a station in order to let the tag be scanned.
  - Raw material in/out at each station will be scanned automatically by the readers on the conveyor system.
  - Parts will flow with conveyor system to each station and any defects will stop conveyor from moving to the next step.

23.3.3 Identify

In the identify step, we utilize the information and insights that we acquired in the design stage to evaluate the solution.

RFID integrates from customer, shipping, production, track system, and quality control together into the host computer database. In that way, RFID will provide real-time information, real-time location system, and automatic data capture either in the internal organization or within the whole supply chain.

23.3.4 Results

In this section, we utilize the information and insights that we acquired in the identify step to evaluate the RFID costs, cost and benefit assumptions, and cost–benefit analysis (Table 23.2).

Table 23.2 RFID Components with Typical Cost Ranges and Purchasing Considerations

Component	Price range	Special consideration
Tags	Passive: $0.3–$0.5 Semiactive: up to couple of dollars Active: $10–$100 Special: varies	• Recurring cost • Mounting accessories and specialized packaging can double the cost of a tag. • Printer applicator and label encoders can cost anywhere from $3,500 to $20,000.
Readers and antennae	$1000–$5000	• Additional antennae can cost up to $500. • Handheld readers can cost anywhere from several hundred dollars to $1500.
Host computer system	$1,000–$100,000	• Not necessary unless new middleware or application cannot use existing system.
Middleware	$25,000–$1,000,000	• Often sold by site license • A reader appliance with built-in middleware can cost $8,000–$10,000 per unit. • Middleware can also be a feature of the application software.

23.3.4.1   *Benefit assumptions of using RFID technologies*
- Scenario 1: Handheld scanner scans location tags and passive tag.
  - Improve overall equipment effectiveness (OEE) rate from 70.0% to 73.5% or 3.5%
- Scenario 2: Passive tag is scanned by each location reader's antenna (in each antenna's interrogation zone).
  - Improve OEE rate from 70.0% to 75%, or 5%.
- Scenario 3: Active tag is scanned by each location reader's antenna within the station and the reader on the conveyor system.
  - Improve OEE rate from 70% to 77%, or 7%.

23.3.4.2   *Final recommendation*
- Based on the cost–benefit analysis, we suggest that scenario 1 is considered as the best alternative.
- The payback period is 2.666 years.
- Before implementing any option, we suggest that the optimize and verify future steps should be performed to validate and detail out the different scenarios.

23.4   *Conclusion*

From our project perspective:
- Finding a solution that will help them tracking if step in the operation in the manufacturing process is met.
  - We identified and evaluated three feasible solutions.
- Investigating RFID technology ability to reduce errors that are caused by skipping steps.
- We define the problem and determine the defects occurred somewhere in the operations and these problems encompass the inherent bottleneck and test stand. The theoretical methodology for solving these issues is DFSS-R.

From the National Science Foundation International Research Experiences for Students (NSF IRES) program perspective:

- We had an awesome cultural experience in Mexico.
- We learned to appreciate the language and the culture.
- We also experienced how business was performed first hand in Mexico.
- Flowchart analysis comparisons
- Next we reviewed data for metrics (CSD) to correlate the skipped steps and the defect rates within their line.
- We developed a Pareto analysis.
- We define the problem and determine the defects occurred somewhere in the operations and these problems encompass the inherent bottleneck and test stand. The theoretical methodology for solving these issues is DFSS-R.

23.5   *Gantt chart and project timeline*

The Gantt chart/project timeline is the overall timeline for implementation of Six Sigma or DFSS/DFSS-R (Table 23.3). It is the agreed upon schedule by all associated parties and keeps the project functioning on time and within budget.

Table 23.3 Typical Project Timeline for Six Sigma Implementation

Task	Week 1 (June 13–15)	Week 1 (June 1–18)	Week 2 (June 19–20)	Week 2 (June 21–25)	Week 3 (June 26–July 2)	Week 4 (July 3–6)
Introduction to companies and issues related	X					
Overview of Six Sigma and logistics topics		X				
Flowchart and data request		X				
Possible problems and solutions analysis			X			
Integration of logistics and Six Sigma into possible solutions				X		
Solution evaluation as team				X		
Final plant tours for final details					X	
Presentation and solution implementation						X

Acknowledgment

This case study utilized Six Sigma to use RFID to further automate manufacturing lines in an international automotive parts plant. The project was funded by the National Science Foundation International Research Program. The students involved in the project included Stan Ugoji and Jian Shin.

References

Bunduchi, R., Weisshaar, C., and Smart, A. U. (2011). "Mapping the benefits and costs associated with process innovation: The case of RFID adoption." *Technovation*, 31(9): 505–521. doi:10.1016/j.technovation.2011.02.004.

College of Engineering. The University of Texas at Arlington. Web. June 25, 2011. http://www.uta.edu/engineering/.

Flowchart. Wikipedia, the free encyclopedia. Web. June 19, 2011. http://en.wikipedia.org/wiki/Flowchart

Home-Rotil. Web. June 27, 2011. http://www.rotil.nl/communications/index.en.php

Industrial and Manufacturing Systems Engineering, Degrees in IE, Engineering Management, Systems Engineering, Logistics, IMSE Department, The University of Texas at Arlington. Web. June 25, 2011. http://www.uta.edu/ie/index.html

Jones, E. C. and Chung, C. A. (2007). *RFID in Logistics: A Practical Introduction*. Boca Raton, FL: CRC Press.

Kuo, S. K., Hsu, J. Y., and Hung, Y. H. (2011). "A performance evaluation method for EMI sheet of metal mountable HF RFID tag." *Measurement,* 44: 946–953.

Summers, D. C. S. (2007). *Six Sigma: Basic Tools and Techniques.* Upper Saddle River, NJ: Pearson Prentice Hall.

Wang, L. C. (2008). "Enhancing construction quality inspection and management using RFID technology." *Automation in Construction,* 17: 467–479.

WonderFox soft video converter factory series. Web. June 27, 2011. http://videoconverterfactory.com

Wright, P. (1982). "A user-oriented approach to the design of tables and flowcharts." In Jonassen, D. H. (ed.). *The Technology of Text: Principles for Structuring, Designing, and Displaying Text.* Englewood Cliffs, NJ: Educational Technology Publications, pp. 317–340.

You can be more. Web. June 27, 2011. http://youcanbemore.co.uk

chapter twenty-four

Using Six Sigma to evaluate using automated inventory tracking to reduce in processing food product shortages at an international food processing plant

> You cannot inspect quality into the product; it is already there.
>
> —W. Edwards Deming

24.1 Introduction

24.1.1 RAID laboratories

The Radio Frequency and Automated Identification (RAID) vision is that "Everything will be tracked wirelessly in 9 years." Their mission is to provide integrated solutions in logistics and other data-driven environments through automatic data capture, real-world prototypes, and analysis. Their objective is to attract recognized funding from notable federal agencies and nationally recognized organizations to provide a research facility that inspires future science, technology, engineering, and mathematics (STEM) researchers from K-12 as well as undergraduate students. A further goal of RAID is to attract national attention from academic rankings and research recognition. The RAID laboratory is one of the most developed laboratories at the University of Texas at Arlington (UTA). It has latest technology equipment that includes military grade fixed and mobile active radio frequency identification (RFID) systems (Savi Technologies, RF Code), industry grade high speed automated conveyor (Hytrol Conveyor), industry recognized RFID edgeware, enterprise resource planning (ERP) and warehouse management system (WMS) systems (Global Concepts), Wal-Mart/Department of Defense (DOD) mandated standard fixed and mobile passive RFID systems (Alien Technologies, Metrics), hospital tracking location systems (Ubisense Ultra-Wide Band), real-time location system, barcode readers, optical character recognition, satellite tags, 2-D barcode etching machines, and industrial antenna enclosures.

RAID's current projects include the National Science Foundation (NSF) International Research Experiences for Students (IRES) for undergraduate students at Monterrey Tech Queretaro in Mexico. This program is focused on research methodology, scientific approach, RFID technology, and application of these previously mentioned topics to companies based in Mexico.

24.1.2 The NSF and the IRES

The aspiration of the NSF and the IRES is to enhance students' research skills, increase students' comfort level, and motivate the awareness of cultural differences. The IRES program desires students to become acquainted with the Spanish language for the intention of evolving student's international capability. In addition, IRES requests students to develop their abilities in research methods and problem-solving skills in congruence with developing international skills. The advanced topics for industrial engineering classes and the program in Mexico were funded by the NSF program in Mexico in order to create an IRES program at Tecnológico de Monterrey, Campus Monterrey, Tecnológico de Monterrey Campus Querétaro, and the UTA.

Its purpose is to increase students' comfort level when working with people from different cultures by exposing them to the technological, cultural, economic, and sociopolitical aspects of Mexican society. Preceding projects of the NSF–IRES in Mexico of UTA include the continuous process improvement management using RFID technologies. This was addressed by investigating how automated tracking can confirm the impacts that specific manufacturing processes have on product defects. This project was with the TRW Automotive Plant in Querétaro. Another program was reducing backorder costs and increasing throughput using RFID inventory control. The problem resided in food manufacturing spoilage and loss of inventory which in turn has a major impact on throughput, backorders, and sales. Reducing backorders through automation was investigated for the Company X Corporation. The final previous project involved using a web-based web portal to reduce carrier errors. The need to use more than one transportation mode in order to have products delivered cost-effectively from suppliers to customers is more important than ever. With high gas prices, companies are seeking to reduce transportation costs in every possible way with a focus from Werner Global Logistics.

Some of the current projects include evaluating the impact of automation on subprocess variation in food process production for Company X Corporation and evaluating the impact of RFID technology on broken pallet case and customer return correlations for Kellogg's Distribution Center of Latin America.

24.2   Background

24.2.1   Business problem statement

The variations in subprocesses cause discrepancies in the planned amount of final food products. These discrepancies may increase the cost of operation because of expedited raw material supplier cost, as well as possibly reduce the projected final product amounts.

24.3   Research objective

The research objective of the project is to investigate the capabilities of RFID and the automation technologies so as to reduce subprocess variation.

24.3.1   Research question

Can RFID technologies identify, measure, and reduce variation in the food production subprocesses?

24.3.2   Relation to business problem

By investigating RFID, the team will gain insights into the subprocess variation components. From this, the team will be able to evaluate automation as well as other alternatives to improve the business problem.

24.3.3   Research hypothesis

Test metric: Subprocess variation will be measured by inventory inaccuracies.

Null hypothesis (H_0): RFID technologies do NOT reduce subprocess variation in the food production process.

Alternative hypothesis (H_a): RFID technologies do reduce subprocess variation in the food production process.

Rejection criteria: If the inventory accuracy can be reduced by 10%, then the team cannot reject the H_0.

24.4   Research methodology

24.4.1   Research laboratory methodology with steps

The approach taken will be that of design for Six Sigma research (DFSS-R) research methodology (Figure 24.1). DFSS-R was derived from Six Sigma methodology (DFSS). It was developed by Dr. Erick C. Jones in 2006 (Figure 24.2). This methodology consists of 7 stages which are DMADIOV and 16 steps are the steps within, shown in Figure 24.2.

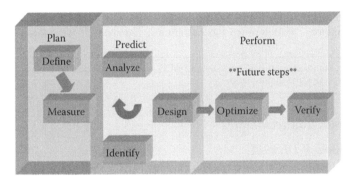

Figure 24.1 DFSS-R methodology.

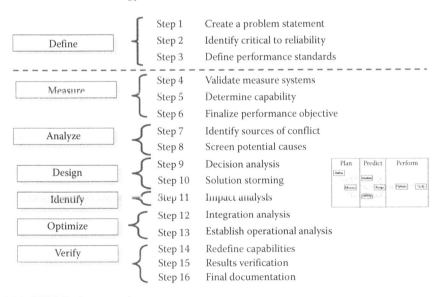

Figure 24.2 DFSS-R phases and steps.

Table 24.1 DFSS-R Data Collection Tools

Define	Measure	Analyze	Design	Identify	Optimize	Verify
Interview process/observation	Hypothesis testing	Fishbone	Scenarios	Cost–benefit analysis	Factorial design of experiments	Control plans
Language processing	Analysis of variance	Pareto analysis	Fractional analysis	Virtual systems	Theory of constraints	TPM
Prioritization matrix	Quality function deployments	Process mapping	Data mining	5-S	Response surface methodology	Multiple response optimization
System map	Flowchart	Multi-vari chart	Blocking	TPM	Blocking	Visual systems
Stakeholder analysis	Measurement system analysis	Chi-square	Response surface methodology	Mistake proofing	Data mining	SPC/APC
Thought process map	Graphical methods	Regression	Multiple response optimization	SPC/APC	Multiple response optimization	Mistake proofing
Value stream mapping	Process behavior charts	Buffered tolerance limits	Theory of constraints	ROI analysis	Fractional factorials	Continuous improvement

5-S, sort, set in order, shine, standardize, and sustain; APC, advanced process control; SPC, statistical process control; TPM, total productive maintenance.

24.4.2   *Data collection techniques*

24.4.2.1   *Tools of DFSS-R*

The team utilized numerous tools to collect data for DFSS-R data collection, which are shown within Table 24.1. Figure 24.3 shows the Gantt chart that highlights the steps necessary for the project to be completed on time and with the planned activities completed.

24.4.3   *Research approach*

24.4.3.1   *DFSS-R steps and description*

The DFSS-R methodology helps us to organize our information and use data interpretation tools correctly. In Figure 24.4, the team stated the tools that were used in each phase of the DFSS-R methodology. The tools used are explained in a detailed manner.

Flowcharting: It is a graphic representation of a process or series of steps and uses different shapes and arrows. An example is displayed in Figure 24.5.

Fishbone diagram: It is a method to summarize and illustrate the potential causes of a problem. It helps for reasoning and organizing the causes in an understandable manner. It has its name because of its fishbone shape where the main problem is stated in the head and the causes are stated in the bones of the body. An example is illustrated in Figure 24.6.

Pareto analysis: It is a phenomenon of which 20% of the phenomena (people) affect 80% of the outcome (wealth) used to explain multiple phenomena.

The RFID technology is based on wireless communication over the air that reads or writes information from a tag. The simplest systems comprise a tag, a reader, and an air

Calendar of event (Summer 2012)							
Task	Week 0	Week 1	Week 2	Week 3	Week 4	Week 5	Week 6
	5/23/2012	5/27/2012	6/3/2012	6/10/2012	6/17/2012	6/24/2012	7/1/2012
Familiarize to new country	▓						
Introduction to companies and issues related		▓					
Overview of Six Sigma and logistic topics		▓	▓				
Data request			▓	▓			
Possible problems and solution analysis			▓	▓	▓	▓	▓
Integration of logistics and Six Sigma into possible solutions					▓	▓	▓
Solution evaluation as a team			▓	▓	▓	▓	
Final plant tours						▓	
Presentation and solution implementation							▓

Figure 24.3 Gantt chart. The shaded cells are the durations of each task.

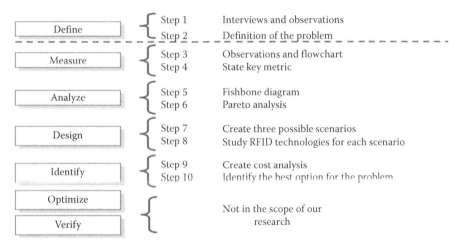

Figure 24.4 Tools of project phases.

intertace. How do RFID systems work? The antenna of the interrogator (reader) emits radio signals, where the electromagnetic (EM) field transmitted can be continuous. This signal is received with an antenna that comes in a variety of shapes and sizes. The antenna can be built-in or external, and circular polarization of reader antenna allows any tag versatile antenna orientation. A sample range for these antennas is anywhere 1 inch to >100 feet. Transponders (tags) respond with their unique code, contain a microchip and/or integrated circuit, antenna made of copper or aluminum coil, and normally an encapsulating

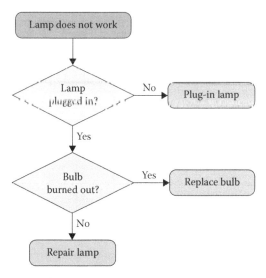

Figure 24.5 Flowcharting example.

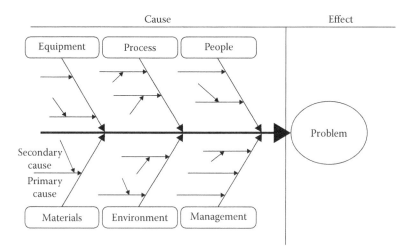

Figure 24.6 Fishbone example.

material of glass or polymer. The reader receives and decodes tag information and sends it to a computer via standard interfaces. Finally, software is available to filter data and monitor the network.

For further description of RFID tags, see Table 24.2. For a detailed cost of RFID, see Table 24.3.

The differences between RFID and barcode identification are detailed as follows. Barcode identification is where a scanner reads reflected light from barcodes and then discerns a sequence of numbers. The numbers are arranged according to a prescribed format, such as Universal Product Code (UPC) or European Article Number (EAN), which describes attributes about the item.

RFID relies upon power and command from a reader where the RFID tag emits data and then the reader discerns a sequence of numbers. The numbers are arranged according

Table 24.2 Types of RFID Tags

Passive	Active	Semiactive
• No on-board battery to power the tag; relies on the reader's energy to power the tag (interpret) and respond	• Older technology; uses on-board battery to power the tag and responds to a reader	• Newer technology combines best of both the above tag types
	• Battery could last several years based on the usage.	• On-board battery powers the tag, but reader energy is used to respond back
• Lowest range and accuracy of the three types	• Higher accuracy and range compared to passive tags	• Lasts longer than active tag and provides higher range and accuracy than passive tag
• Least costly	• Costliest	• Cheaper than active
• Specified by Wal-Mart, US Department of Defense, and Metro	• Used by the US Department of Defense for industrial applications	• Used in various toll collection systems

Table 24.3 RFID Cost Analysis

Component	Price range	Special considerations
Tags	Passive: $0.3–$0.5 Semiactive: up to couple of dollars Active: $10–$100 Special: varies	• Recurring cost • Mounting accessories and specialized packaging can double the cost of a tag. • Printer applicators and label encoders can cost anywhere from $3,500 to $20,000.
Readers and antennae	$1000–$5000	• Additional antennae can cost up to $500. • Handheld readers can cost anywhere from several hundred dollars to $1500.
Host computer system	$1,000–$100,000	• Not necessary unless new middleware or application cannot use the existing system
Middleware	$25,000–$100,000	• Often sold by site license • A reader appliance with built-in middleware can cost $8,000–$10,000 per unit. • Middleware can also be a feature of the application software.

to a prescribed format, such as Electronic Product Code (EPC)'s 96-bit, which also describes attributes about the item.

24.5   Results

24.5.1   DFSS-R approach

24.5.1.1   Define

In order to define the problem, the team used the following data collection techniques of interviewing and recording observations. An interview session was conducted. The interview involved asking people from human resources and those who work on the floor.

They were able to detail specifics that were not known to upper management and provide suggestions as to what the major problems were. This session also involved asking a production manager about what he or she believed was the problem. She discussed with us what she could ascertain from the inventory data that were available to her.

The result of the define phase was a clearly defined business problem statement. This statement is "Variations in sub-processes are causing discrepancies in the planned amount of final food products." With the attainment of this statement, the team can then proceed to focus the scope of the research.

24.5.1.2 Measure

Within the measure phase, the team employed the DFSS-R approach and tools to determine the metrics (Figure 24.7). A flowchart was used to identify the key metric of the variation of the final product and the involved subprocesses. This gave the team a way to discern the extent of the problem and whether a solution would successfully address the problem. The food process examination for Company X revealed phases that could be broken down in order to dissect each procedure better.

24.5.1.3 Analyze

In the analyze phase, the team used the following data collection approaches and DFSS-R tools to evaluate the current state of the subprocess variations. Data were obtained during the operational visits. A fishbone diagram and Pareto analysis were used to identify the contributing factors.

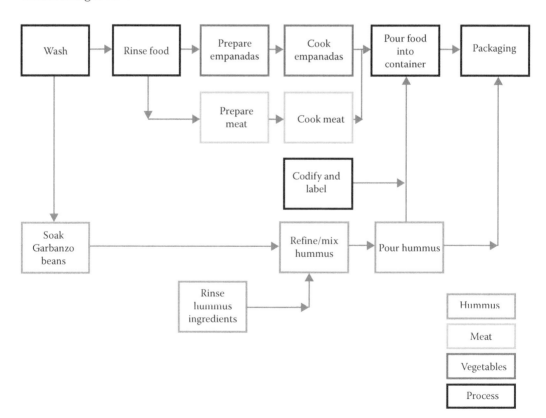

Figure 24.7 Food process mapping.

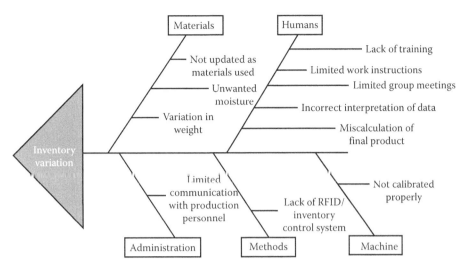

Figure 24.8 Cause-and-effect diagram for inventory variation.

The fishbone diagram identified 11 known causes for inaccuracies within the inventory (Figure 24.8).

The Pareto analysis determined the two most contributing factors to the inventory inaccuracies (Figure 24.9). It also determined 2 (20%) of the 11 known causes from the fishbone diagram creating 80% of the problems. The pivotal cause for error was the miscalculation/manual data entry error of raw material weights. The subordinate cause to this was the lack of RFID/inventory control system procedure enforcement.

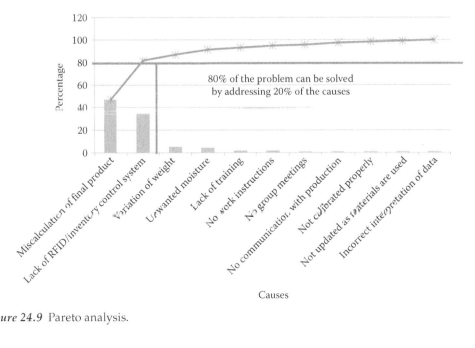

Figure 24.9 Pareto analysis.

24.5.1.4   Design

Given the results of our analyze phase, the team designed scenarios for continuous improvement. At least one scenario will test and evaluate the RFID technologies. The other scenarios are not related to this technology based on data collection, previous analysis, and facility layout.

Scenario 1 is simultaneous scanning procedure (SSP). This uses the existing bar-code system more efficiently. It would also eliminate wait by simultaneously scanning with materials movement. This option would require a procedural change but no new investment.

Scenario 2 is a passive RFID system (PRS). This option would update the existing bar-code system with passive RFID technologies. The movement of material would be automatically tracked at each location. Each tag would contain inventory accuracy data set about product. One of the greatest assets of this scenario is that it would minimize manual entry and documents the time of process.

Scenario 3 is an active humidity sensor RFID system (AHSRS). This system would include active tags with weight and humidity sensor. The bags of raw materials have sensor to monitor the quality of ingredients via a humidity sensor. This scenario would also include the final product container having a tag to monitor weight. The benefits would be that the scanning of tags will eliminate the need for manual entry and give the real-time weight and condition of product.

The purpose of the team is to consider eliminating manual entry, track all steps of materials though warehouse, and monitor the quality of the products. These goals would be met by utilizing handheld RF scanners using barcodes, passive tags that will be scanned at designated locations, and active tags that will relay qualities of material and products to host the system.

24.5.1.5   Identify

In the identify phase, the team evaluated the cost benefits of the proposed scenario in the design phase. This entails detailing the invested costs for each scenario and the projected returns (benefits and savings) for each scenario.

24.5.2   Return-on-investment findings

The team then calculated the return on investment (ROI) and identified the best scenario by the most useful ROI.

$$\text{ROI} = \frac{\text{Savings} - \text{investment}}{\text{Investment}}. \tag{24.1}$$

The team also calculated the payback period (PBP) for the different scenarios.

$$\text{PBP} = \frac{\text{Investment}}{\text{Savings}}. \tag{24.2}$$

The PBP would be 1.2 years as the period below is projected over a 5-year time period.

Scenario 2 (PRS) entails a passive tag being scanned by each location reader's antenna (in each antenna's interrogation zone). The estimated yearly savings (return) would decrease variation from 15.0% to 8.6%, or 6.4%. The ROI would be 85.8% and the PBP would be 0.54 years.

For Scenario 3 (AHSRS), the investment would be $2,800,000. This would entail an active tag being scanned by each location reader's antenna within the station. The estimated yearly savings (return) would decrease variation from 15.0% to 5%, or 10%. The ROI would be 36.3% and the PBP would be 1.4 years.

The results indicate that Scenario 2 (passive RFID system) has the best ROI (see Table 24.3).

24.6   Discussion

24.6.1   Rejection of hypothesis

Based on the assumptions, the team *rejected* the null hypothesis and concluded that RFID technologies do reduce subprocess variation in the food production process.

24.6.2   Selection of options

This is supported by the discernment within the identify phase to recommend the passive RFID system so that materials movement will be automatically tracked. The other scenarios that include a weight sensor may have the potential to identify the lost weight for the subprocesses to create the final food product. This possible solution could possibly eliminate the mishandling of weights recorded and input into the system. A further robust system may be created by combining a weight sensor with a humidity sensor. This system would be extremely robust in resolving the problem; however, more research would be desirable in order to justify the cost. Scenario 1 is a valid solution but is not as lucrative as Scenario 2. That being said, Scenario 1 would be a desirable improvement upon the existing system.

24.7   Limitations

The results and recommendations may not be accurate due to the limitations of this study. The team was critically hindered by limited observations. Our analysis was based on only three interactions where it would be more desirable to have on average 10 observations. This is in part due to the food process being quite extensive and not fully observed by two observation periods. This is a result of the team only being able to observe the operation for a limited time period given the duration of the research project. The team typically utilized observations for the analysis where it would have been more appropriate to have employed other methods as well. More date would be required for a superior recommendation and solution. Finally, the research methodology was not fully performed. The optimize and verify steps that include prototyping and verification of the proposed solution were not obtained. These processes would prove to be vital for the conclusion of a robust solution.

24.8   Conclusion

Although no final solution was reached, the team was able to confidently recommend scenarios for Company Xi to further pursue. The recommendation was based on research and would benefit from further investigation as to the full feasibility. Our test metric was that the subprocess variation will be measured by inventory inaccuracies. From this, the team was able to evaluate this metric employing the DFSS-R methodology with the tools: fishbone analysis, Pareto analysis, flowcharting, and cost–benefit analysis. The DFSS-R analysis proposes that RFID does reduce inventory variation.

Acknowledgments

This case study utilized Six Sigma to improve warehousing operations at international distribution center. The project was funded by the NSF International Research Program. The students involved in the project included Cinthya Vinueza-Garcia, Walter Mulflur, Erick C. Jones Jr., Juan Robles, and Mohammad Sodiqui.

We thank the NSF for the opportunity to be a part of this NSF IRES In Mexico UTA program. We are gracious for the opportunity to enhance our Spanish language skills and experience the Mexican culture. We are thankful to the Company XYZ Queretaro Plant for allowing us to have a hands-on engineering experience to work in an international work culture. Finally, we appreciate the opportunity for working with a research team and learning a scientific research approach using RFID and logistics technologies.

Further Readings

Chandler, C. M. D. (2007). "Formulation of Lean Six Sigma critical business processes for manufacturing facilities." Dissertations & Theses, The University of Texas at Arlington, Texas.

Gill, T. G. (2004). "Teaching flowcharting with FlowC." *Journal of Information Systems Education*, 15(1): 65–77.

Grahan, B. (2008). "Roots of the business process mapping." *BPTrends*. June 6, 2012.

Juran, J. (eds.) (1995). "The quality improvement process." In *A History of Management for Quality*. Milwaukee, WI: ASQC Quality Press.

Karuppusami, G. and Gandhinathan, R. (2006)."Pareto analysis of critical success factors of total quality management: A literature review and analysis." *The TQM Magazine*, 18(4): 372–385.

Lavengood, S. and Reeb, J. (2002). "Pareto analysis and check sheets." *Statistical Process Control*. June 7, 2012.

Majid, M. Z. A. and McCaffer, R. (1998). "Factors of non-excusable delays that influence contractors' performance." *Journal of Management in Engineering*, 14(3): 42.

Shah, R., Chandrasekaran, A., and Linderman, K. (2008). "In pursuit of implementation patterns: The context of Lean and Six Sigma." *International Journal of Production Research*, 46(23): 6679–6699.

Wong, K. C. (2011). "Using an Ishikawa diagram as a tool to assist memory and retrieval of relevant medical cases from the medical literature." *Journal of Medical Case Reports*, 5(1): 1–3.

chapter twenty-five

Using Six Sigma to improve trucking

You cannot inspect quality into the product; it is already there.

—W. Edwards Deming

25.1   Introduction

25.1.1   Research problem and literature review

A well-established trucking company in the United States that has a branch located in Mexico seeks to enhance their in-house transportation management system (TMS) software real-time vehicle tracking that utilizes global positioning system (GPS) and radio frequency identification (RFID) technologies. They believe that this will increase customer service, quality control, vehicle load quality, and route optimization, and reduce transportation costs.

Research problem. The data retrieval and entry process is very time consuming and has a large amount of errors associated with the process.

This process involves time for them and from their partner carriers (trucking companies). The process is the update and dispatch of the trailers as they cross the border. This process involves many individuals, which creates numerous opportunities for manual errors as data are inputted into the system.

The research question is

> What manual procedures are causing the greatest amount of time
> and errors in the data retrieval and entry process?

This will be investigated by reviewing the flowchart of the process to identify which process takes the most time and also by reviewing the data clerk's job tasks and the method of data retrieval and entry. Since the excess use of time is caused by manual process, web-implemented software will be used to reduce the time it takes to receive and enter the data. This new web solution will allow Company X to automatically update the data that Company X receives from the broker straight to the software.

The two relevant academic literature associated with research question involves optimizing the use of trailers after information is provided and optimizing the time and data, thus saving time and reducing errors. The relevant academic literature associated with research question involves optimizing the time and data. We utilize the following literature that investigates the manual process. The article "Combine automatic and manual process on web service selection and composition to support QoS" is describing how web-based software are taking over for manual process. The web-based software are going to overtake manual process because they are more efficient and save an enormous amount of time from manual process.

Next, we utilized the research from the Baskent University's Department of Industrial Engineering report on the multiple traveling salesman problem (MTSP; Bektas 2006) that describes problem solving between brokers and managers. This application with technologies such as RFID can be used to increase visibility of equipment. The application of MTSP solution procedure helps in evaluating the impact of using RFID devices during the research process.

25.2   Research overview and objectives

The research objective of the project is to identify the processes that are allocating a large amount of time. To meet this research objective, we will use the following *specific objectives*:

Specific objective 1. To identify the data entry processes collected by vendors that cause excessive time and errors that may be able to be automated with GPS and RFID. We suggest that conducting a time study investigation and flowcharting will provide us valuable data. This will help us identify the business efficiency.

Specific objective 2. To find a web-based software that will help them optimize the time and labor used during the data entry process at the company and to implement a web-based software in which the broker can input the data instead of inputting the data into Excel and sending the data via e-mail.

Specific objective 3. To investigate RFID technology ability to reduce errors and optimize the scheduling between the companies' routes. We will use the traveling salesman problem (TSP) to verify RFID.

The business challenge was a lengthened supply chain to and from Mexico, international multimode transportation. The solution they implemented was to place an account manager onsite to provide customers with a single point of contact for communication and support. This account manager is who we understand to be the broker that works onsite at Laredo, TX.

25.3   Research method and approach

We utilized the Six Sigma approach of design for Six Sigma research (DFSS-R) that has seven steps: define, measure, analyze, design, identify, optimize, and verify. For this project, we used all the steps excluding optimize and verify because we did not implement any solution in the company.

To meet our specific objective, we have divided into three tasks, which will be completed for this proposal. The first task goes with the first specific objective, the second task goes with the second objective, and the third task goes with the third specific objective. The team will be working on all three tasks at different times. The first task we are going to perform is a flowchart, the second task we are performing is a web-based software solution, and the third task is imbedding an RFID system.

25.3.1   Task 1—Flowchart

This section provides you with flowchart information. A flowchart is the first step taken to find the questions we have for a company. A flowchart is very important because it lays out the whole company process they are doing and we can get our data request forms from the flowcharts. For this proposal, we had two flowcharts: the first one was the internal flowchart and the second one is the trucking process. This flowchart will help us identify the process that is taking up much time. From the flowchart, we are able to identify that the data retrieval and entry process is causing the error.

25.3.2 Task 2—Web-based software solution

This section provides you with web-based software that can be used to input the information into the company's system. The most basic solution would be to design and make a web site where information can be entered and can be accessed live at the company's office and also the customers with their trailer information can see where their trailer is.

25.3.3 Task 3—Investigate RFID technologies for automating data capture

This section provides you with information on RFID research. Integration of RFID into vehicles to track their location, and this information can be used to optimize the routes that the vehicle is traveling. The RFID implementation can be verified using the TSP.

25.3.4 Define

In the define step, we seek to clarify the research objectives into a usable scope for our project.

In order to analyze the operations, we used a tradition definition tool. Flowchart analysis was used to determine the error and the excess use of time for data input.

Flowcharts were derived manually by the project team through documented data entry procedures, interviews, and walking the process. The data entry process was documented using interviews and the Intercambios (tractor owners) information was identified through company documents and spreadsheets.

25.3.4.1 Define—Results

The analysis of the team derived flowcharts described the following: It identified the clerical processes that take excessive time and have high potential for manual data entry error that allocates time/error and identified the Intercambios data collection processes that take excessive clerical input time and have high potential for manual data entry error.

25.3.5 Measure

In the measure step, we utilized our flowcharts created in the define step to determine what we want to measure for analysis and the means of measurement. After reviewing the flowcharts for accuracy, we decided to measure: the difference between the current-state flowchart tasks and the future-state flowchart tasks (CvF Task), data retrieval and entry methods (CvF DE), and the use of time caused by manual data entry (Tm DE).

25.3.6 Analyze

In the analyze step, we utilized our defined measurement criteria to analyze our data and measured the following characteristics with the previously mentioned techniques:

Metric 1: The difference between the current-state flowchart tasks and the future-state flowchart tasks (CvF Task). We estimated the amount of time utilized from clerical data entry demonstration.
The use of time caused by manual data entry (Tm DE) is the same information described earlier.
Metric 2: The use of time caused by manual data entry (Tm DE). Data entry errors can come from Intercambios entering data into the spreadsheet incorrectly. Studies show

that manual data entry ranges from 10% to 40% error rates and the potential of two types of manual data entry errors from both Company X and Intercambios provides the potential data error rates of 20%–80%.

Metric 3: The difference between the current-state flowchart manual tasks and the future state flowchart (CvF Task). Reducing steps and data inputted directly into the system by Intercambios will reduce some labor time and errors.

Metric 4: Data retrieval and entry methods (CvF DE). Eliminating steps will reduce labor time and errors and data inputted by RFID will reduce labor and all manual errors.

25.3.7 Design

In the design step, we utilize the information and insights that we acquired in the analysis stage to come up with possible options or solutions that are viable.

The solutions for the metrics are as follows:

Solution 1: A web-enabled front end database that can integrate into corporate AS400-based system that will reduce the time for CvF Task, CvF DE, and Tm DE.

Solution 2: An RFID-based data capture system for Company X trailers that integrates into Company X's corporate AS400-based system that will reduce the time for CvF Task, CvF DE, and Tm DE.

In the design step, we utilized our previous information and identified two possible solutions. Solution 1 addresses the excess time problem from clerical labor described in the analyze step and Solution 2 addresses the Intercambios data entry problem and the excess clerk time and potential errors from both data entry points.

25.3.7.1 Solution 1

An automated way to have the data entered into the system without excess labor is designed. We suggest a web site that is linked to database that can send data to the company's system. Note that a web site only cannot do this, but the database synchronization to the system is crucial and may require software consulting and access and permission from the corporate. The system that has required components initially collects the information that the Intercambios are e-mailing to the clerks located in an Excel file. The potential benefits include reducing duplication of effort from clerks, reducing potential clerical errors from excess data entry, no additional steps for Intercambios, and replacing e-mailed Excel sheet with web site entry.

25.3.7.2 Solution 2

An automation system to collect data and to automatically update the system without manual entry is implemented. The relevant technologies include satellite tags, GPS tags, and active RFID tags, and the active RFID is most feasible. The benefits include reducing all manual entry and also providing the ability to optimize the trailer returns so that the company can charge more money and keep more trailers rented. The challenge is that RFID is expensive but the payback for reduction of errors reduces the need for manual data entry from clerks and Intercambios. Database and Company X software integration is still a challenge because there are many other parts to look at.

25.3.7.3   Results

By using the DFSS-R methods, we make up with these results. In this project, we are finding solutions to reduce time and errors. Solution 1 is to design and make a web site that will help reduce time on data entry and errors. Solution 2 is to implement RFID so it will save a large amount of time and save on labor, and the error rate will decrease significantly.

25.3.8   Benefits of RFID

There are no errors and you save time and also there are automatic updates on system so there is no need to enter data twice into the system. The possible best benefit is labor saving and possibility of business growth because you can charge each Intercambios extra for the RFID equipment.

25.4   Web order entry

Web order entry allows designated personnel, including customers and vendors, a secure means to enter requests for shipments over the Internet for real-time execution. Web order entry is extensible, allowing customer-specific information about shipment requests to be captured, including multiple reference number types and custom shipment request properties. The application will validate location information ensuring consistency across shipments. Post office (PO) information can also be validated ensuring authentic shipments. Required fields can be set up ensuring that an important information is captured within each shipment request.

25.5   Tracking and tracing

Tracking and Tracing provides interested parties a method to obtain current information regarding status and location of existing freight movements at anytime through a secure Internet-based application. Information provided in tracking and tracing is customizable to the customer's need. For example, a distribution center (DC) could have visibility to all freight inbound to the facility displaying pertinent information to the DC. Multiple search criteria exist allowing custom searches by date, reference number, location, and other criteria. Other electronic delivery methods are also available for this tracking information.

25.6   Carrier selection

Carrier selection provides manual and systemic solutions for recommending and assigning carriers based on a diverse set of selection criteria. A completely automated solution is available whereby a carrier is automatically assigned to a load as soon as it is received. Once the carrier is assigned, the load is automatically tendered to the chosen carrier and electronic data interchange (EDI) shipment statuses will provide pickup and delivery information for completing the transaction.

25.7   Information exchange

EDI may be most easily understood as the replacement of paper-based load tenders with electronic equivalents. EDI saves unnecessary recapture of data, which leads to faster transfer of data, far fewer errors, less time wasted on exception handling, and hence a more stream-lined business process. EDI offers the prospect of easy and inexpensive communication of structured information with your business partners.

Electronic exchange of data can be in the form of EDI, flat files, e-mail, and a variety of other file types. Information exchange can be performed direct (peer-to-peer), without cost, or through a value-added network (VAN). Secured encryption for data transfer is available through AS1, AS2, pretty good privacy (PGP) encryption, and secure FTP (SFTP).

Information exchange with carriers includes load tenders, requests to move freight, acceptance or rejection of the tender, load status updates, location updates, appointment information, and pickup and delivery information. It can be used for functional acknowledgments, reconciliation to avoid orphaned transactions, and e-mail-based EDI services for carriers who do not have EDI capabilities.

Information exchange with customers/vendors includes load tenders directly from customer/vendors, purchase order information from the customer regarding vendor shipments, and shipment status updates to the customer. Freight bill will be electronically delivered to the customer.

25.8   Mapping

Our geographical information system (GIS) enables real-time mapping data of your integrated satellite network applications.

Mapping is used for time-sensitive shipments and a proactive measure to spot potentially late loads. Driver direction system, routing system, and engine monitor system to prevent breakdowns.

25.9   Automated paperless logs

Company X Enterprises introduced the first and only paperless driver log system authorized by the Federal Highway Administration, which promotes safety and saves hundreds of man-hours to maximize driver performance.

From this information, we can see that this would be the best solution for Company X in Mexico. The customer would be able to log on with their tractor number and see automatically where their product is and will also be notified automatically of any changes to the delivery, etc. This software will not only reduce the errors but also save the time that is used for data retrieval and error and customer service.

25.10   Gantt chart and project timeline

The Gantt chart/project timeline is the overall timeline for implementation of Six Sigma or DFSS/DFSS-R. It is the agreed upon schedule by all associated parties and keeps the project functioning on time and within budget.

Task	Week 1 (May 31–June 4)	Week 2 (June 5–11)	Week 3 (June 12–18)	Week 4 (June 19–25)	Week 5 (June 26–July 2)	Week 6 (July 3–6)
Company presentation and interview						
Overview of Six Sigma and logistics topics						

(Continued)

Task	Week 1 (May 31–June 4)	Week 2 (June 5–11)	Week 3 (June 12–18)	Week 4 (June 19–25)	Week 5 (June 26–July 2)	Week 6 (July 3–6)
Flowchart and data request forms						
Possible problems and solutions analysis (possible office visit)						
Integration of logistics and office time study						
Time study and problem solution						
Solution evaluation and decision						
Presentation and solution implementation						

25.11   Conclusion

In conclusion, we analyzed a business problem by investigating the research objectives of evaluating manual data entry processes that require a large amount of time. In order to adequately investigate this problem, we had to perform specific objectives and the summary of those investigations is to identify the data entry processes that are causing excessive time and errors. From the flowchart analysis, we identified these processes.

Finding a solution that will help them optimize the time and labor used during the data entry process. We identified and evaluated two feasible solutions, investigating automated technology ability to reduce errors and optimize the scheduling between the company's routes which we described a solution RFID that has a potential to combine trailers for Intercambios and reduce sending empty trailers and provide future revenue for the company. The web site and the RFID technology will eliminate the shaded areas in the Gantt chart.

Acknowledgments

This case study utilized Six Sigma to improve warehousing operations at international distribution center. The project was funded by the National Science Foundation International Research Program. The students involved in the project included Cinthya Vinueza Garcia, Walter Mulflur, Erick C. Jones Jr., Juan Robles, Mohammed Siddiqui, and Jithin Daniel.

Reference

Bektas, T. (2006). "The multiple traveling salesman problem: An overview of formulations and solutions procedures." *Omega*, 34(3): 209–219.

Further Readings

Barnes, R. M. (1958). Motion and time study. Vol. 84. New York: Wiley.

Company X case study, global implementation—Mexico study. http://www.CompanyX.com/assets/pdf/whitepapers/cs_mexico.pdf

Hosting option. June 25, 2011. http://www.numatek.com/web-hosting/prices.html

Jin, C. et al. (2008). "Combine automatic and manual process on web service selection and composition to support QoS." *12th International Conference on Computer Supported Cooperative Work in Design*. April 16–18, pp. 459–464. doi:10.1109/CSCWD.2008.4537022.

Oracle posts double-digit price hikes for database, middleware products. June 28, 2011. http://www.crn.com/news/applications-os/208800277/oracle-posts-double-digit-price-hikes-for-databasemiddleware-products.htm;jsessionid = By2ZM8KGz-BKZNqOJhavvw**.ecappj03

Website design price estimate. June 25, 2011. http://www.numatek.com/web-design/estimate.html

Wilson, J. M. (2003). "Gantt charts: A centenary appreciation." *European Journal of Operational Research*, 149(2): 430–437.

Wright, P. (1982). "A user-oriented approach to the design of tables and flowcharts." In Jonassen, D. H. (ed.). *The Technology of Text: Principles for Structuring, Designing, and Displaying Text*. Englewood Cliffs, NJ: Educational Technology Publications, pp. 317–340.

chapter twenty-six

Using Six Sigma logistics to optimize a city's supply chain inventory supplies

You cannot inspect quality into the product; it is already there.

—W. Edwards Deming

26.1   Introduction

Technical organizations often face the challenge of aligning their supply chain. The technical manager faces challenges in coordinating data collection and analysis efforts to evaluate the supply chain in a cost-effective manner. In some organizations, it may be prudent to utilize the current technical personnel to perform this analysis. Oftentimes, companies consider utilizing costly software and consultants prior to using their in-house resources. Allowing the technical manager to utilize an internal team to provide an analysis is more cost effective for several reasons.

1. Data collected will be utilized again if consultants are deemed necessary.
2. The in-house team will understand the implications of solutions that the model may provide and can make adjustments for reality.
3. Simplified assumptions can be agreed upon by internal stakeholders.
4. The project will prepare personnel for change.
5. The project provides a cost-effective solution.

Also, this study will reveal if your supply chain network may be too complex to model simply using the Excel Solver prior to investing in an extensive study. Though it is very important to perform supply chain analysis, many companies cannot justify the use of expensive software and consultant to perform these analyses continually. The technical manager can provide good solutions by creating this type of study.

Previously, a project team of students from the University of Nebraska–Lincoln and personnel from the city began a Six Sigma project to reduce obsolete inventory. The supply chain consisted of a network of warehouses, storerooms, suppliers, and the internal end user that represented the customer. During the Six Sigma process improvement study, the team determined that customer service needs were not being met, obsolete inventory was being driven by purchasing behavior, and the facility costs could be reduced with facility consolidations. The team analyzed the supply chain network of the city's Public Works Department using modeling techniques to recommend which warehouses could be consolidated. Based on the recommendations, 96,000 square feet could be reduced and gross of $3.5 million would be saved over 5 years not including taxes and depreciation. This represents a cost reduction of 25%.

26.2   Background

A technical manager's goal when locating facilities and allocating inventory is to maximize the overall profitability of the resulting supply chain network while providing customers with adequate service. Traditionally, revenues come from the sale of product and costs arise from facilities, labor, transportation, material, and inventory holding. Ideally, profits after tariffs and taxes should be maximized when designing a supply chain network. In this scenario, the city government does not pay taxes or collect revenues so their goals were to minimize the overall operating cost and still be responsive to the customer.

Trade-offs must be made by the technical managers during network design. For example, building many facilities to serve local markets reduces transportation cost and provides fast response time, but it increases the facility and inventory costs incurred by the firm. Technical managers can use network design models in two different situations. First, these models are used to decide on locations where facilities will be established and the capacity assigned to each facility. Second, these models are used to assign the current demand to the available facilities and identify lanes along which product will be transported. Managers must consider this decision at least on an annual demand basis, prices, and tariff charge. In both cases, the goal is to maximize the profit while satisfying the customer needs. The following information must be available before the design decisions can be made:

1. Location of supply sources and markets
2. Location of potential sites
3. Demand forecast by market
4. Facility, labor, and material costs by site
5. Transportation costs between each pair of sites
6. Inventory costs by site as well as a function of quantity
7. Sales price of product in different regions
8. Taxes and tariffs as product is moved between locations
9. Desired response time and other service factors (Chopra and Meindl 2004)

Given this information, a choice of model type can be made. Previous literature highlights some general models that have differing goals. Each model has differing objectives; the models that were considered for this study were the capacitated plant location model and the gravity location model. The capacitated plant location model seeks to minimize the total cost of the current supply chain network; the problem is formulated into an integer program. The gravity location model's goal is to locate an optimal location based on the cost inputs. Beyond optimization models, the technical manager could build a simulation of their supply chain (Chang 2004).

In this study, we chose to use the capacitated plant location model (Chopra and Meindl 2004) in order to determine the minimal number of facilities that could hold inventory and meet the customer demand. In our study, the city has chosen to consolidate warehouse facilities. Management is questioning whether all 12 facilities are necessary. They have assigned a supply chain team of University of Nebraska–Lincoln and city personnel to study the network for the public works operations and identify the warehouses that can be closed. The goal is to formulate the model to minimize the total costs taking into account costs, taxes, and duties by location. Given the taxes and duties do not vary between various locations, and that the city does not pay taxes, the team decided to use the existing facility locations and allocate demand to the open warehouses to minimize the total cost of facilities, transportation, and inventory.

26.3 Capacitated plant location model

The capacitated plant location network optimization model requires the following inputs:

N = number of potential locations
M = number of demand points
D_i = annual demand from market i
K_i = potential capacity of plant i
F_i = annualized fixed cost of keeping factory i open
C_{ij} = cost of producing and shipping one unit from factory i to marker j (cost includes production inventory, transportation, and duties)

and the following decision variables:

Y_i = 1 if plant is open, 0 otherwise
X_{ij} = quantity shipped from factory i to market j

The problem is formulated as the following integer program:

$$\min\left(\sum_{i=1}^{n} F_i Y_i + \sum_{i=1}^{n}\sum_{j=1}^{m} C_{ij} X_{ij} \right),$$

subject to

$$\sum_{i=1}^{m} X_{ij} D_{ij} \quad \text{for } j = 1, \cdots, m, \tag{26.1}$$

$$\sum_{j=1}^{n} X_{ij} \leq K_i Y_i \quad \text{for } i = 1, \cdots, n, \tag{26.2}$$

$$Y_i \in (0,1) \quad \text{for } i = 1, \cdots, n. \tag{26.3}$$

The objective function minimizes the total cost (fixed + variable) of setting up and operating the network. The constraint in Equation 26.1 requires that the demand at each facility market be satisfied. The constraint in Equation 26.2 states that no plant can supply more than its capacity. (Capacity is 0 if closed and K_i if it is open. The product of the terms $K_i Y_i$ captures this effect.) The constraint in Equation 26.3 enforces that each plant is either open ($Y_i = 1$) or closed ($Y_i = 0$). The solution will identify the plants that are to be kept open, their capacity, and the allocation of regional demand to these plants. The model is solved using the Solver tool in Excel (Chopra and Mcindl 2004).

26.4 Network modeling steps incorporated into a Six Sigma service project

The typical Six Sigma DMAIC approach was used with the addition of a network model within the analyze phase. DMAIC stands for define, measure, analyze, improve, and control. These are the steps in a standard improvement model for a Six Sigma-directed project.

26.4.1   Define

The main work in the define phase is for the team to complete an analysis of what the project should accomplish and to confirm its understanding with its sponsor(s). The team members should agree on the problem, understand the project's link to corporate the strategy and its expected contribution to return on invested capital (ROIC), agree on the project boundaries, and know what indicators or metrics will be used to evaluate success. The last two issues often prove important particularly in service environments (George 2003). The problem defined for this project was to reduce obsolete inventory.

26.4.2   Measure

One of the major advances of Six Sigma is its *demand* for data-driven management. Most other improvement methodologies tended to dive from identifying a project into improve without sufficient data to really understand the underlying causes of the problem. The measure phase is Six Sigma's stage for data collection and "measuring" the problem. This phase is generally broken into several steps, including establish baselines, observe the process, and collect data (George 2003). The measure of success was reducing the percentage of obsolete inventory in the supply chain.

26.4.3   Analyze

The purpose of the analyze phase is to make sense of all the information and data collected in the measure phase. A challenge to all teams is *sticking to the data*, and not just using their own experience and opinions to make conclusions about the root causes (George 2003). There are many tools available in the analyze phase, including network modeling. Network models provide a rich and robust framework for combining data, relationships, and forecast from descriptive models. They provide managers with broad and deep insights into effective plans, which are based on the company's decision options, goals, commitments, and resource constraints (Shapiro 2001). After using regression analysis and design of experiment analysis, the team chose to use supply chain optimization for a more robust solution.

The network model used within this project followed several steps including the following:

1. Collect input data and establish baseline
2. Set optimization constraints
3. Run alternatives with the capacitated plant location model (Chopra and Meindl 2004)
4. Show alternatives in revenue, savings, and customer service
5. Select an alternative

These steps lead to an alternative that minimized the cost of the supply chain. This alternative then directs the tasks within the improve stage.

26.4.4   Improve

The sole purpose of the improve phase is to make changes in a process that will eliminate the defects, waste, costs, and so on, which are linked to the customer needs identified in the define stage (George 2003). The improve stage differs for every Six Sigma project.

The common theme is that the improvements should be centered on the largest issues found in the analyze phase. The suggestions for consolidating facilities (the supply chain model recommended) and using a more robust criterion for eliminating outdated inventory were recommended for the improvement.

26.4.5   Control

The purpose of control is to make sure that any gains made will be preserved, until and unless new knowledge and data show that there is an even better way to operate the process. The team must address how to hand off what they learned to the process owner and ensure that everyone working on the process is trained in using any new, documented procedures. Six areas of control are critical: document the improved process, turn the results into dollars, verify the maintenance of gains continually, install an automatic monitoring system, pilot the implementation, and develop a control plan. Key performance indicators (KPIs) were identified to be tracked with statistical process control charts for the following year. This is further elaborated in the conclusions section.

The DMAIC process with the capacitated plant location model in the analyze phase was utilized to study the City of Houston's Public Works Warehousing Operations.

26.5   Case description

26.5.1   Organizational description

The organization used for this case study is a city in Southwestern United States, Public Works, Materials Management Branch (MMB). The MMB is responsible for the processing and coordination of all procurement and contract-related activities as well as warehousing and distribution of all general inventory items for the department.

The branch facilitates purchases ranging from pipes for restoration of sewer lines to computers and traffic signs. To promptly obtain goods and services, the department utilizes in excess of 800 commodity and service contracts. The branch is divided into three functional sections: procurement, contract management, and warehousing and distribution. This study was centered on the warehousing and distribution section.

The MMB has the responsibility for warehousing and distribution of general and automotive inventory items, from cradle to grave, for the department. Two central depots serve as staging locations for inventory that is distributed to a network of 10 general supply warehouses, 9 automotive warehouses, and many storerooms located throughout the city. The inventory consists of a variety of items, for example, pipe, valves, fittings, office, and janitorial supplies.

26.5.2   Project description

The MMB had been audited in previous years, and the audits identified the opportunities for improvement in the warehousing operations. The audits identified excess obsolete inventory, need to evaluate standard operating procedures, and labor productivity. Obsolete inventory is defined as inventory that has not had any requests for disbursement for over 1 year.

The current system contains 12 warehouses and 28 storerooms with an ongoing cost of $14.94 million. Upon inspection, it was estimated that the warehouses have a maximum of 30% space utilization. The current supply chain is shown in Figure 26.1. Public Works

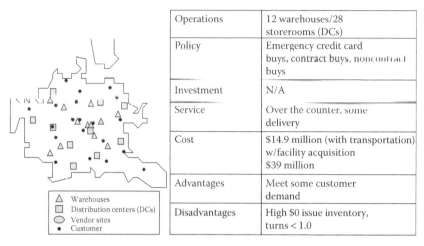

Operations	12 warehouses/28 storerooms (DCs)
Policy	Emergency credit card buys, contract buys, noncontract buys
Investment	N/A
Service	Over the counter, some delivery
Cost	$14.9 million (with transportation) w/facility acquisition $39 million
Advantages	Meet some customer demand
Disadvantages	High $0 issue inventory, turns < 1.0

Figure 26.1 Current supply chain description.

owned $10.1 million of inventory within the MMB warehouses. The inventory that was deemed as obsolete was valued at $3.6 million or 35% of the total inventory.

The modeling steps were followed to complete the analysis.

1. Collect input data and establish baseline
   The current supply chain information was collected to form the input data for the network model. The inputs included costs for electricity, gas, data lines, and labor. Also, holding and transportation costs were estimated for each facility. The warehouses do not pay taxes or water costs since they are in a city building, so information as to lost water sales and lost taxes were also captured and used in the cost equations.
2. Set optimization constraints
   The optimization constraints included the size limitations of each facility and the future demand at each facility. The facility size was collected from operations. The future demand on inventory was estimated to be the same as the last year's value.
3. Run alternatives with the capacitated plant location model (Chopra and Meindl 2004)
   The costs and data that were collected in steps 1 and 2 of the modeling were input into Chopra's model in Excel and the Solver Add-in was utilized to run the alternatives of the least cost model.
4. Show alternatives in revenue, savings, and customer service
   The different alternatives were then evaluated for revenue and savings with an return on investment (ROI) calculation assuming the project had a 5-year life. The customer service provided in each alternative was evaluated by a team from operations.
5. Select an alternative
   The optimal solution contains two warehouses, which are centrally located as shown in Figure 26.2. This gives a reduction of 96,000 square feet, which translates into $3.5 million over 5 years. This solution will increase the space utilization to 65% and reduce the obsolete inventory to 10% of the total value held within the warehouses.

As the city moves to the optimized model, the control phase of DMAIC will keep the improvements in place and running smoothly. The metrics that are given to continue the control are the KPIs given in Table 26.1. These should be measured and tracked

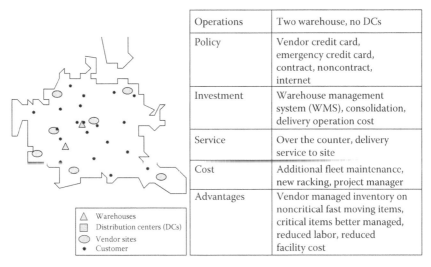

Operations	Two warehouse, no DCs
Policy	Vendor credit card, emergency credit card, contract, noncontract, internet
Investment	Warehouse management system (WMS), consolidation, delivery operation cost
Service	Over the counter, delivery service to site
Cost	Additional fleet maintenance, new racking, project manager
Advantages	Vendor managed inventory on noncritical fast moving items, critical items better managed, reduced labor, reduced facility cost

△ Warehouses
▢ Distribution centers (DCs)
⬭ Vendor sites
∗ Customer

Figure 26.2 Optimized supply chain.

Table 26.1 Key Performance Indicators

Category	Metric	Definition
Service	Turns	Annual dollars issued divided by average
Service	Percentage of obsolete inventory	Number of commodity codes (CCs) that have not been issued for over 1 year divided by the total number of CCs
Cost	Cost/pick	Total labor cost divided by the total number of picks
Cost	Cost/order	Total labor cost divided by the total number of orders
Asset management	Facility utilization	Number of pallet positions utilized vs. available
Future		
Transportation	Shipments/division	Track the number of deliveries to site for each division
Inventory	Velocity	Annual revenue divided by the daily overhead
Labor	Percentage of productivity by area/task	Actual labor hours divided by the efficiency standard for the task and track by the employee
Labor	Picks/hour	Number of pick issues divided by pick labor hours
Purchasing	PCard spend by category	PCard spend by contract, noncontract, and emergency usage vs. total PCard spend

utilizing statistical process control (SPC). This data could then be used to repeat an optimization in the future.

26.5.3 Lessons learned

The lessons learned included model complexity changes and challenges and limitations that could be better met. The model complexity was chosen to reflect a first look at the supply chain and a simple optimization. A more complex model may have been used if better original data had been available. The data that were available could not be validated because it came from the enterprise resource planning (ERP) system that was antiquated. The model was validated with site tours and sampling for volumetric data. We note that the model is only as good as the data provided. A common term is "garbage in" to a model and you get "garbage out." If a more complex model was utilized, software other than Excel as well as consultants would be needed.

It was difficult to reach an agreement between divisions on what part of the cost data can be incorporated as reduced costs. For example, the portion of overall utility costs can be equated to warehouse space versus the other uses of the buildings today, and the percentage of value can be used for holding cost because the city does not pay taxes and does not invest excess monies.

26.6   Implications for the technical manager

A technical manager should use this chapter to better understand his or her own first steps in supply chain optimization projects. This information should encourage the manager to begin to look at his or her information internally, before hiring a consulting team. The first look may give a viable answer that can be implemented for increased efficiencies and savings.

An technical manager can gather internal data and then use the methods in Excel Solver to create the appropriate supply chain model. The specific steps for using Excel can be found in different references; we recommend the steps outlined by Chopra. The technical manager can justify many good solutions for the supply chain by further using current technical personnel. This may provide another tool for the technical manager to justify his or her technical staff. If the internal team does not solve the problem, or the manager is ready for a second look at the issue, a consultant can be hired with very little lost to the company.

This study provided an overall method for performing continuous improvement projects using the Six Sigma methodology. Further, this study shows how in the analyze stage an technical manager can perform a supply chain analysis on current operations. The technical manager can use this study as a guide for both.

26.7   Conclusions

This chapter details a quick and relatively inexpensive way to perform an analysis for supply chain savings opportunities. The major benefit is that you utilize internal personnel who have been already budgeted for and may have a better understanding of operations than outside consultants. Also, the initial study may be modeled using existing spreadsheets before more costly software and consulting options are explored.

This simple analysis may not replace a complex supply chain analysis using some of the more advanced software that incorporates the CPLEX and ILOG modeling engines. These software companies allow for more constraints than the less powerful spreadsheet will allow. They have claimed to have 20%–50% more optimized results that may translate into more cost savings. This is contingent if consultants can better interpret data, future business strategies, and evaluate logistics networks including transportation traffic patterns.

From the case study application, we identified a workable set of challenges with lessons learned that can be valuable to organization when modeling the supply chain. The technical manager and his or her team can be a valuable asset when doing both continuous improvement projects and providing valuable supply chain modeling expertise.

References

Chopra, S. and Meindl, P. (2004). *Supply Chain Management*. Upper Saddle River, NJ: Pearson Prentice Hall.

George, M. L. (2003). *Lean Six Sigma for Service*. New York: McGraw-Hill.

Shapiro, J. F. (2001). *Modeling the Supply Chain*. Pacific Grove, CA: Duxbury.

Yang, Chang Ho, Yong Seok Choi, and Tae Young Ha. (2004). "Simulation-based performance evaluation of transport vehicles at automated container terminals." *OR Spectrum* 26(2): 149–170.

Further Readings

Bowersox, D. J., Closs, D. J., and Bixby, C. M. (2002). *Supply Chain Logistics Management*. Boston, MA: McGraw-Hill/Irwin.

George, M. L. (2002). *Lean Six Sigma*. Dallas, TX: George Group.

Johnson, J. C. et al. (1999). *Contemporary Logistics*. Upper Saddle River, NJ: Prentice Hall.

Kotnour, T. G. et al. (1998). "Understanding and leading large-scale change at the Kennedy space center." *Engineering Management Journal*, 10(2): 17–21.

Kotnour, T. G., Matkovich, J., and Ellison, R. (1999). "Establishing a change infrastructure through teams." *Engineering Management Journal*, 11(3): 25–30.

Leach, F. J. and Westbrook, J. D. (2000). "Motivation and job satisfaction in one government and development environment." *Engineering Management Journal*, 12(4): 3–9.

Liao, C.-J. et al. (2013). "Meta-heuristics for manufacturing scheduling and logistics problems." *International Journal of Production Economics*, 141(1): 1–3.

Michealson, H. B. (1982). *How to Write and Publish Engineering Papers and Reports?* Philadelphia, PA: ISI Press.

Turnquist, M. A. and Nozick, L. K. (2004). "A nonlinear optimization model of time and resource allocation decisions in projects with uncertainty." *Engineering Management Journal*, 16(1): 40–49.

Vof, S. and David L. (2003). *Woodruff, Introduction to Computational Optimization Models for Production Planning in a Supply Chain*. Berlin: Springer.

Yoon, C. and Makatsoris, H. (2004). "Supply chain modeling using simulation." *International Journal of Simulation*, 2(1): 24–30.

Nontraditional Lean Six Sigma and modern quality trends

chapter twenty-seven

Lean Six Sigma certification and belt levels

> Education is not the filling of a pail, but the lighting of a fire.
>
> **—William Butler Yeats**

27.1   Introduction

Lean Six Sigma (LSS) is one of the methodologies that have been popularized for the benefit of excelling in the cut-throat competition now prevalent due to rapid development in technology. The need for continuous improvement in various processes that drive every stage of the supply chain is evermore. We see more and more businesses striving to reduce aberrations and aim for high efficiencies. With the successful implementation by companies such as General Electric (GE) and Motorola, the top management of a variety of organizations is now more involved in equipping its task force with a wide range of skill sets in order to achieve its mission. Here, LSS has caught on as it is a powerful tool that combines and increases its umbrella of application along with its efficiency rate in cost cutting. This has been addressed in Chapters 9 through 14 and explains the inevitable pioneering in Six Sigma education.

Now, we are faced with two categories of receiving Six Sigma knowledge: (1) by training and (2) by certification.

Certification is almost always associated with exam and has a confirmation process at higher levels of skill. From the history of Six Sigma, we learn that the key roles have been assigned and structured according to their expert level determined in accordance with the effort put in to receive the required color of belt similar to karate training.

Value of such certifications is driven by a robust process to achieving the certification and success of certified students in many organizations. Certification can mean additional $20,000–$30,000 per year in salary. It paves way for greater possibilities in any student's career.

27.2   Executive level

This level leads teams by streamlining their efforts toward the overall focus pertaining to the culture of the organization.

27.2.1   Sponsorship information

To train your organization to completely align themselves to the objectives of LSS methodology, one must encourage at least company-level training for its employees and also be knowledgeable to understand the change implementations.

This would help them support the right kind of initiatives for the benefit of the futuristic vision for the organization as a whole. It must find good fit in the strategy for the company's goals. It will help them allocate resources that are cost effective, find optimal solutions in problem solving, manage projects, and decide among profitable projects for appropriate deployment.

27.2.2 Belt information

Six Sigma professionals are useful at all levels and are implemented in health care, manufacturing, service, and other type of industries (American Society for Quality 2012). There are trainings offered for black belts, master black belts, green belts, and yellow belts based on the karate color belt system as popularized by the "Godfather" of Six Sigma Mikel Harry. It determines their allotment for projects.

Let us discuss the responsibilities that come with appropriation for each belt.

1. Black belt
   a. Leads and trains the project teams consisting of green belt trainees to create problem-solving strategies
   b. Brings about dramatic transformational results
   c. Takes on taxing reengineering projects that create sustainability
2. Green belt
   a. Collects data for analysis in projects taken up by black belt leaders
   b. Assists in cross-functional application across projects and teams
   c. Implements base-level tools
3. Yellow belt
   a. A team participant that is just aware of Six Sigma processes conceptually
   b. Reviews team progress toward the goal of improvement

27.2.2.1 Master black belt

Consultation for organization and personnel equipped with green and black belt-level certification.

Certifications are offered by various quality bodies and a fee may be charged. However, three types of organizations are in the forefront of Six Sigma education:

- Company offerings: GE, 3M, and others offer certification with possibility of experience and no grading through exam. They offer blended black and green belt training.
- Universities: Educational institutions include University of Nebraska–Lincoln, University of California at Berkeley, Purdue University, and University of South Carolina. Exam can be given online.
- Organization (PRO) quality: Some of them are American Society for Quality (ASQ), Institute of Industrial Engineers (IIE), International Supply Chain Education Alliance (ISCEA), International Quality Federation (IQF), Juran Institute, and so on.
  - Yellow belt training is offered for 1 day.
  - Green and black belt training is offered for 1 week with a semester format system and green and black belt exams are considered to be prerequisites for further training.
  - On-site consulting similar to company certification process takes 3–4 months additionally with the project verification process.
  - Exam can be taken online.

27.2.2.2   Yellow belt
- Company offerings: Yellow belt training is offered for 1 day.
- Universities: They have a semester format system starting with a yellow belt.
- Organizations pro quality: They also offer training in yellow belt education for a period of 1 day.

27.2.2.3   Green belt
- Company offerings: Green belt training usually extends for a week and it is supposed to give $100,000 in savings to the company on its implementation.
- Universities: They have a two semester format system and yellow belt exam is considered a prerequisite for green belt training in the other semester.
- Organizations pro quality: They also offer training in yellow belt education for a period of 1 day.

27.2.2.4   Black belt
- Company offerings: Black belt level has offered training for 3 weeks and also the requirement to complete an application project within 4 months. LSS black belt gives $250,000 in savings.
- Universities: They have a two semester format system. Yellow and green belt exams are considered to be prerequisites for further training. Black belt additionally requires a project to be evaluated.
- Organizations pro quality: They have a confirmation process for attaining black belt level. A submission of affidavit review paper or report review of project along with a public presentation of the project work is a must.

27.3   ASQ certified Six Sigma black belt

The American Society of Quality Engineers was founded in 1946 to bring together employees and employers in pursuit of quality from organizations across the world. It represents the quality movement with about a million individual and organizational members in 150 countries (American Society for Quality 2012).

Their mission is to analyze data and perform statistical functions accurately to identify variable relationships and distribution of data for further analysis. The trainees must be able to validate any points of disparity for effective communication across teams.

They have the black belt training that transforms the employee into a professional that understands the underlying Six Sigma principles, which at the same time have a necessary awareness of tools required for Lean thinking.

However, the personnel with the LSS black belt must equip themselves with a thorough understanding of integrating Lean into the Six Sigma methodology and using various tools advantageously to ensure sustainability and improvement in quality by adding value economically as well as qualitatively. They offer an option of choosing the field of implementation to specialize on. There is also a supplemental overview of using Minitab for statistical calculation that is popular due to its ease of use in industry.

27.4   IIE BB certification

It was founded in 1948 and it is a professional body that plans to advance the skill set of industrial engineers in fields such as Lean, operations research, engineering management, quality control, and reliability (Institute of Industrial Engineers 2012). They provide training materials and conduct workshops and conferences for continuous enrichment of the body of knowledge in industrial engineering and its related divisions.

It has a four-week training program for an accredited certification in black belt for green belt professionals to hone their leadership skills.

27.4.1 Institute of industrial engineers

In this section, we include a report on Six Sigma and the role of industrial engineering by Larry Aft, Director of Continuing Education from the IIE.

In the 1960s, the definition of industrial engineering was adopted by the American IIE:

> Industrial engineering is concerned with the design, improvement, and installation of integrated systems of people, materials, equipment, and energy. It draws upon specialized knowledge and skills in the mathematical, physical, and social sciences together with the principles and methods of engineering analysis and design to specify, predict, and evaluate the results to be obtained from such systems.

Since its inception, industrial engineering has focused on improving operations. The role of the industrial engineering has evolved; as Pritsker points out (1990), the early industrial engineers worked on work simplification and methods improvement, work standards, and facilities layout. This evolved to a more comprehensive view of industrial engineering as a systems approach, with the objective naturally being to improve the operation and design of the system.

American business has consistently worked to improve quality and productivity, usually via a packaged program such as quality circles, total quality management (TQM), business process reengineering, and the latest adaptation, Six Sigma and Lean. All of these improvement strategies incorporate elements and tools of industrial and systems engineering. The philosophy of Six Sigma recognizes that there is a direct correlation between the number of product defects, the wasted operating costs, and the level of customer satisfaction. With Six Sigma, the common measurement index is defects per unit and can include anything from a component, a piece of material, or a line of code to an administrative form, time frame, or distance (Harry 1998). Six Sigma emphasizes identifying and avoiding variation. The use of Six Sigma changes the discussion of quality from one where quality levels are measured in percentages (parts per hundred) to a discussion of parts per million or even parts per billion (Pyzdek 1996).

The Six Sigma strategy involves a series of systematic steps that are specifically designed to lead the organization through the gauntlet of process improvement. These major steps include the following:

1. *Define*. This step involves the identification of critical process characteristics.
2. *Measure*. After the identification of critical quality characteristics, this step documents the existing process, performs necessary measurements, and estimates the process capability.
3. *Analyze*. The third step allows analysis of performance measures to determine the amount of improvement that might be possible to make the critical quality characteristic "best in class." This may involve process redesign.
4. *Improve*. This step guides the organization to specific product characteristics that must be improved to achieve the performance and financial goals. Once this is done,

the characteristics are diagnosed to reveal the major sources of variation. Key process variables are identified by way of statistically designed experiments. Optimum performance levels for each result from the experiments.

5. *Implement.* Improvements are implemented via a well-thought-out and documented project plan developed within the Six Sigma process.

6. *Control.* This step is for documentation, validation, and monitoring of the new process conditions via statistical process control methods. Anticipated improvements are verified.

Lean is a business improvement methodology that maximizes the shareholder value by achieving the fastest rate of improvement in customer satisfaction, cost, quality, process speed, and invested capital.

A Lean enterprise views itself as part of an extended value chain, focusing on the elimination of waste between you and your suppliers, and you and your customers. Lean has its origins in the teaching and writings of TQM and just in time (JIT), which espouse the idea of *delighting the customer through a continuous stream of value adding activities.* Specifically, it is an extension of the phrase *world class* as defined by Dr. Richard Schonberger as

> …adhering to the highest standards of business performance as measured by the customer.

In other words, value is always defined from the customer's perspective. Understanding your customer's needs is a prerequisite for driving Lean principles and methodologies.

A commonly held definition of Lean enterprise is

> a group of individuals, functions, and sometimes legally separate but operationally synchronized organizations.

The *value stream* defines the Lean enterprise. The objectives of the Lean enterprise are to

- Correctly identify and specify the *value to the ultimate customer/consumer* in all its products and services.
- Analyze and focus the value stream so that it does everything from product and service development and production to sales and service in such a way that activities that do not create value are removed and actions that do create value proceed in a continuous flow as pulled by the customer.

From the time a customer need is recognized until it is satisfied, the process and all its elements must add value for the "value stream" to be meaningful. The basic components of this Lean system are waste elimination, continuous flow, and customer pull.

As defined by John Krafcik, in his book, *The Machine That Changed the World,*

> Lean production is "Lean" because it uses less of everything compared with mass production: half the human effort in the factory, half the factory space, half the investment in tools, half the engineering hours to develop a new product in half the time. Also, it requires far less than half of the needed inventory on site. The expected results are fewer defects, while producing a greater and ever growing variety of products.

Lean applies to any organization type and can be applied to all areas within the business. Essentially, Lean is a three-pronged approach incorporating *a quality belief, waste elimination, and employee involvement* supported by a *structured management system*. Basically, we have taken simple processes and complicated them resulting in longer lead times, reduced flexibility, increased inventories, and the inability to meet customer demands.

Many organizations have used both Six Sigma and Lean successfully, whereas many others have not. The failures have occurred when individuals who are only trained in the tools attempt to use them without having a systems view of the organization's operations. This is why, under the leadership of qualified industrial and systems engineers, the methodologies are so powerful.

27.5   Summary

The appropriate application of correct Six Sigma and Lean methods to processes, regardless of the business or industry, can bring about significant savings and benefits. When used as part of an overall industrial and systems engineering process, benefits are significant. As organizations develop the capacity to use these methods within the overall improvement strategies, a number of lessons have been learned.

27.6   Lesson learned

- Through the use of Six Sigma techniques, teams have been able to learn and apply *formal and systematic methodology* for problem identification, evaluate potential solutions, and implement those solutions. In addition, the teams have been able to better prioritize the issues and solutions.
- Inclusion of the operators as part of the continuous improvement team improved morale of the people by empowering them to take ownership of the process improvements. This improvement in morale helped facilitate implementation of solutions.
- A better understanding of how important is the commitment of senior management to the Six Sigma process improvements.
- Lean and Six Sigma are more than tools. They are part of a systematic improvement process.

27.7   ISCEA LSS pathways to excellence

The Six Sigma series has several envisioned certification pathways. They are categorized as yellow belt programs, green belt programs, and black belt programs. Program prices are based on the average price from the approved course providers (http://www.iscea.net/SixSigma).

27.7.1   Executive LSS yellow belt—Invitation only

The Executive LSS Yellow Belt (ELSSYB) program is designed for certification of an executive at the "executive yellow belt" level. Executive yellow belts are able to understand the LSS initiatives to the point to provide executive top-line support to these programs. Understanding why an initiative is not working or requires strategic investment to succeed is crucial for these types of initiatives.

27.7.2 Certified LSS yellow belt

The Certified LSS Yellow Belt (CLSSYB) program is designed for certification of an individual at the "yellow belt" level. Six Sigma yellow belts are able to describe, support, and participate on Six Sigma projects. They are key to success of the program in that often they act as the key subject matter experts on projects and at many companies they may be the program sponsors who must invest resources into the projects.

27.7.3 Certified LSS green belt certification program

The Certified LSS Green Belt (CLSSGB) program is designed for certification of an individual at the "green belt" level. Six Sigma green belts are able to "lead" Six Sigma projects and apply their Six Sigma knowledge in order to produce bottom-line earning before interest and taxes (EBIT) savings for medium-sized companies. They participate in large-scale Six Sigma projects led by black belts in larger organizations.

27.7.4 Certified LSS black belt certification program

The Certified LSS Black Belt (CLSSBB) program is designed for certification of an individual at the "black belt" level. Six Sigma green belts are able to "lead, train, and manage" Six Sigma projects and apply their Six Sigma knowledge in order to produce bottom-line EBIT savings for all types of organizations. Demonstration of management abilities is required in the form of successfully leading Six Sigma project(s) based on experience levels.

Programs generally address the following topics:

1. Overview and history of LSS
   - Corporate: implementation and investment in LSS initiatives
2. Overview of Six Sigma certification and training programs
   - Define: identifying and planning for sponsorship investment to Six Sigma projects
3. Define: (BB) organizing, planning, and setting specific project goals
4. Measure I: identifying and validating measurements and measurement systems
   - Measure II: running initial data and changing measurement system strategies
5. Analyze: evaluating data baseline data and estimating new performance goals
   - Design: identifying the need to redesign the process or product for success
6. Improve/Lean: implementing solutions and fostering acceptance and buy-in
7. Control: implementing tools and dashboards for maintaining successful results
   - Exam preparation
   - Topics are only covered as they are the needs of the appropriate certification level

27.8 Company certifications

Companies such as GE, 3M, Motorola Institute, and IBM offer corporate trainings and parade Lean Sigma beliefs by following Lean and Six Sigma principles through all levels of business dealings.

Motorola has integrated the training with learning of solutions offered since 2008. This program continues to focus proper measurement and analysis of data for improving the quality of products and services sold to customers (Motorola 2012).

Companies such as IBM in 2012 and 3M (2012) offer similar training modules and opportunities to their employees (Byrne et al. 2012).

References

3M. (2012). "Building value for our customers." Retrieved April 2013. http://multimedia.3m.com/mws/mediawebserver?mwsId=66666UFbEVsSyXTt48TuoXTaEVlQEVs6EVs6EVs6E666666&fn=3M%20Six%20Sigma%20Nov%202004.pdf.

American Society for Quality. (2012). "Media room." December 16, 2012. http://www.asq.org/media-room/index.html.

Byrne, G., Lubowe, D., and Blitz, A. (2012). "Driving operational innovation using Lean Six Sigma." December 17, 2012. http://www-935.ibm.com/services/at/bcs/pdf/br-stragchan-driving-inno.pdf.

Harry, M. J. (1998). "Six Sigma: A breakthrough strategy for profitability." *Quality Progress,* 31(5): 60–64.

Institute of Industrial Engineers. (2012). "What is Institute of Industrial Engineers?" Accessed on April 10, 2013. http://www.iienet2.org/Details.aspx?id = 295.

Motorola Solutions Six Sigma® Program. (2012). December 17, 2012 http://www.motorola.com/web/Business/Solutions/Motorola-Training-Site/North-America-Training/Documents/StaticFiles/BlackBelt_Flyer_0412.pdf.

Pritsker, A. (1990). *Papers, Experiences, Perspectives.* West Lafayette, IN: Systems Publishing Corporation.

Pyzdek, T. (1996). *The Complete Guide to the CQE.* Tucson, AZ: Quality Publishing.

chapter twenty-eight

Lean Six Sigma practitioners, consultants, and vendors

> It is not enough to do your best; you must know what to do, and then
> do your best.
>
> —W. Edwards Deming

28.1 Introduction

We begin this chapter by describing the responsibilities of each level of practitioner. Businesses have many stakeholders including stockholders, customers, suppliers, company management, employees and their families, the community, and the society. Each stakeholder has a unique relationship with the business. Each stakeholder is both supplier and customer, former many closed-loop processes that must be managed, controlled, balanced, and optimized if the business is to thrive.

28.2 Champion

The most critical element of Six Sigma implementation is management support; without management support, the true potential of Six Sigma will never be realized. The management must participate in development of vision and strategy for Six Sigma deployment, manage resource development, and assist with project definition and alignment by coaching sponsors and black belts. Once the project run has been established, they must assure project tracking and reporting—including validated financial benefits and transfer of solutions—and assess performance to initiate actions to assure success.

28.3 Black belt

A black belt-level employee must give full-time commitment by acting as an initiator. By being an initiator, he prepares an initial project assessment to estimate benefits and identifies the needed project resources. The technologist perspective must be applied to implement Six Sigma tools effectively and appropriately to be used in the DMAIC (define, measure, analyze, improve, and control) process. The practitioner is also a leader that identifies the barriers to protect the progress of project while leading the team involved and inform the sponsor about project status. The employee must be able to deliver results on time.

28.4 Green belt

Green belts serve on project teams and may also lead projects. But they do not require full-time responsibility, which is the key difference compared to black belts.

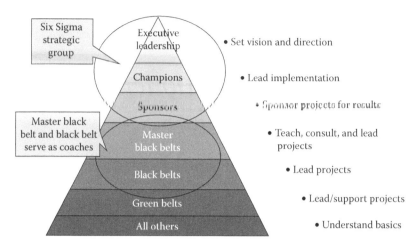

Figure 28.1 Key roles for success.

Figure 28.1 shows the level occupied by each of these practitioners along with their goals.

28.5   Current major vendors

Many companies such as IBM, Motorola, and GE offer solutions to businesses based on their type of industries. As they supply such solutions, they are termed as vendors. Most of the top companies listed by *Fortune 500* magazine integrate Lean and Six Sigma ideologies. Some of them are 3M (2012), Amazon.com, Bank of America (2012), Boeing Frontiers (2012), Caterpillar Inc. (2012), Dell, Eastman Kodak Company, General Electric, Honeywell, and so on (Fekete and Hancu 2010; Improvement and Innovation.com 2012; *USA Today* 2012).

28.6   Vendor selection model

Outsourcing for timely production and reduced cost has become a trend. In this scenario, a useful selection model is appropriate. We start by collecting as much data about each of the potential vendors of Lean Six Sigma Solutions. Information about its client base, success rate, resources, certifications, associations, and so on is to be collected and rated.

We also identify the key performance metrics that are aligned with our organization goals. Some of the metrics are as follows:

- Delivery time
- Resource utilization
- Customer service
- Quality
- Efficiency
- Flexibility in incorporating changes

Weights are allotted to each category based on the objective of organization. Scoring for every vendor is given based on the performance in each area of concern. An overall performance index ranking is calculated using a Pugh Matrix (Fekete and Hancu 2010) by multiplying the weight and the score. The one that has the highest rate is chosen.

A simpler process of evaluation is

- Analyzing the needs of the organization to adopt Lean Six Sigma (LSS).
- Requesting for quotations and evaluate them for feasibility and flexibility.
- Interviewing vendor and using scorecard.
- Negotiation.

(LSS Academy) Results of the scorecard give a direction in the selection process.

28.7 Consultants and integrators

28.7.1 Master black belt

This level is before the black belt level and the master black belts play a role in advising black belts during project execution, that is, consulting and expanding the knowledge base of employees by teaching classes. They also get to lead major projects and participate in program development. However, to achieve this status, additional training beyond the black belt is required.

Their responsibility is to ensure proper alignment of task force and integrate new methodologies in the existing methods within the company's culture.

Some of the consultant groups available now are Rath & Strong, Pivotal Resources, BMGI, Xerox (2009), Bain & Company, Six Sigma Qualtec, and iSixSigma (2009).

References

3M. (2012). "3M flexible circuit foundry." *Solutions at 3M*. December 17, 2012. http://solutions.3m.com/wps/portal/3M/en_US/FlexibleCircuits/Home/.

Bank of America. (2012). December 17, 2012. http://newsroom.bankofamerica.com/?s = speeches &item = 54.

Boeing Frontiers. (2012). Boeing. December 17, 2012. http://www.boeing.com/news/frontiers/archive/2003/july/i_ca1.html.

Caterpillar Inc. (2012). Caterpillar. December 17, 2012. http://www.cat.com/cda/components/fullArticleNoNav?ids = 89047&languageId = 7.

Fekete, Z. A. and Hancu, L. V. (2010). "A supplier selection model for software development outsourcing." Retrieved April 2013. http://anale.steconomiceuoradea.ro/volume/2010/n2/191.pdf.

Improvement and Innovation.Com. (2012). "Improvement and Innovation.com." December 17, 2012. http://www.improvementandinnovation.com/.

iSixSigma. (2009). "The 2005 Global 500." iSixSigma. December 17, 2012. http://www.isixsigma.com/implementation/financial-analysis/cost-quality-not-only-failure-costs/.

USA Today. (2012). "Feds may unleash Six Sigma on terrorism." December 17, 2012. http://usatoday30.usatoday.com/money/companies/management/2002-10-30-sixsigma_x.htm.

Xerox. (2009). "Lean Six Sigma and the quest for continuous improvement." http://www.xerox.com/downloads/usa/en/x/Xerox_Lean_Six_Sigma_Brochure.pdf.

Review Questions

Question 1: The relevant stakeholders in an important project would typically include all of the following EXCEPT:

(a) Owners and stockholders
(b) Potential suppliers

(c) Potential competitors
(d) Hourly employees

Question 2: In highly effective Six Sigma companies, most employees receive some training. What group is most likely to receive sponsorship training?

(a) Green belt candidates
(b) Senior management
(c) Black belt candidates
(d) Master black belt candidates

chapter twenty-nine

Six Sigma project management*
What to do after the Six Sigma sponsorship phase

The journey of a thousand miles begins with a single step.

—**Lao Tzu**

Overview

Steps	Substeps
Define	• Quality projects (sponsorship) • Define specific project (black belt)
Measure	• Identify critical characteristics • Clarify specifics and targets • Measure system validation
Analyze	• Determine baseline • Identify improvement objectives • Evaluate process inputs
Improve	• Formulate implementation plan • Perform work measurement • Lean execute the proposed plan
Control	• Reconfirm plan • Set up statistical process controls • Operational process transition

Source: PWD Group LLC.

Recommended Tools by Belt

		Define	Measure	Analyze	Improve	Control
Stage 1	Yellow belt	Interview process		X–Y map/QFD/ house of quality	EVOP	Control plans
		Language processing		Fishbone diagram		Visual systems
		System map (flowchart)				5-S
						SPC/APC
Stage 2	Green belt	Prioritization matrix	Hypothesis testing	FMEA	Design of experiments	TPM
		Thought process map	Flowdown	Multi-vari chart	Fractional factorials	Mistake proofing

(Continued)

* By B. Gray, E. C. Jones, and J. A. Perkins.

Recommended Tools by Belt (Continued)

		Define	Measure	Analyze	Improve	Control
		Stakeholder analysis	Measurement system analysis	Chi-square	Data mining	
		Thought process map	Graphical methods	Regression	Blocking	
			Analysis of variance	Pareto analysis		
Stage 3	Black belt	Value stream map	Process behavior charts	Buffered tolerance limits	Theory of constraints	
		Gantt chart development	QFD	Process capability analysis	Multiple response optimization	
		Project management			Response surface methodology	

Source: PWD Group LLC.

Notes: Black belts are expected to know the continued knowledge obtained throughout green belt and yellow belt certifications. The same goes for green belt and yellow belt. Empty cells represent that skill is not required for that belt stage; however, a general knowledge of the step is expected.

5-S, sort, set in order, shine, standardize, and sustain; APC, advanced process control; EVOP, evolutionary operations; FMEA, failure modes and effect analysis; QFD, quality function deployment; SPC, statistical process control; TPM, total productive maintenance.

29.1   Introduction

After a sponsor has identified for a Six Sigma project, an implementation plan has to be put together. This includes determining which project the sponsor wants to be worked first, what the project metrics are, who the customer is, what resources will be needed, and what time frame the project will need to be worked in. To ensure that all of these topics remain clear throughout the project, it is important to explicitly define as much of the project as possible. This chapter discusses the critical components for managing the project in order to enable a successful project (Figure 29.1).

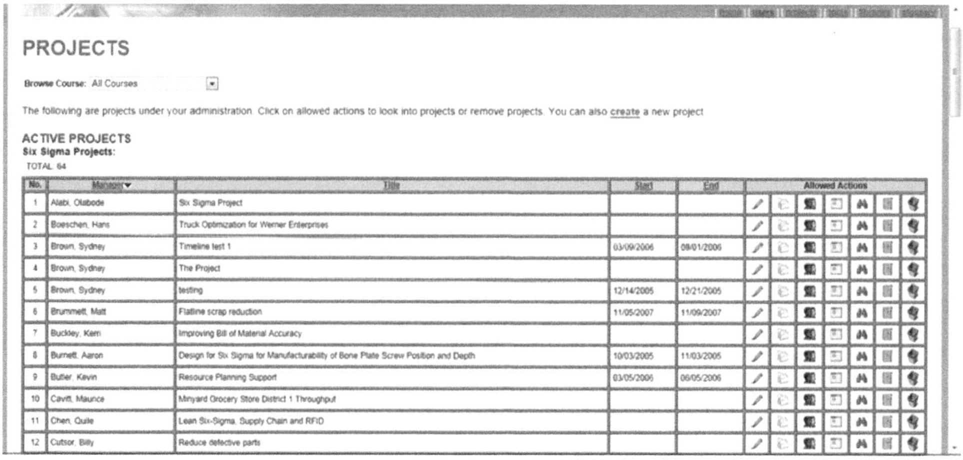

Figure 29.1 Project lists.

29.2   Project definition worksheet

The project definition worksheet (PDW) is a guide for further clarifying the project definition. A completed PDW also provides the key elements needed for the project charter. The purpose of the PDW is to ask the appropriate questions in order to clarify the scope and support for the project. Traditionally, not all the questions can be answered at this point of time. However, the goal is to direct whoever is filling out the form to narrow the scope of the issue and view the problem from many directions (Figure 29.2; Tables 29.1 through 29.5).

One way to clarify the scope and scale of the project is to explicitly state what the project includes and what it excludes. The "four W's and one H" are a useful guide in considering includes/excludes (Table 29.6):

- What—product line, production equipment
- Where—geographic region, storage condition, site
- When—stage of the benefits driven procurement (BDP), steps of the production process
- Who—channel, customer segment
- How—shipment method, order entry type

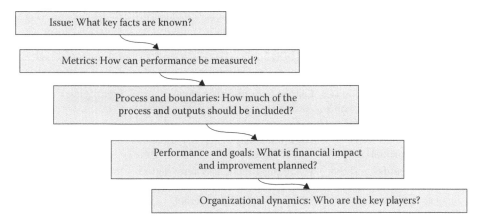

Figure 29.2 Process definition worksheet.

Table 29.1 PDW Key Facts

Action	Element	Definition	Actual
1	What is the specific problem you want to solve or defect you want to eliminate?	The specific condition that is affecting the success of the business	Crown area separation in BMT tires
2	What customer is most affected by this problem?	The entities most affected by this problem; can be internal or external customer	Dealers, fleet, end users
3	What is the output associated with this problem?	For example, a service, a transaction, a document, or a physical product	Bias truck tires
4	Where is the problem occurring?	For example, a geographical location or a part of the organization	India, Malaysia, Thailand, Taiwan
5	When was the problem observed?	Identify when the problem was observed in terms of production output	>10 years

Table 29.2 PDW Metrics

Action	Element	Definition	Actual
1	Identify the business metrics	The business metric, critical to satisfaction (CTS), or Big Y is the specific outcome of the process targeted by the project.	Number of crown area separation adjustments
2	Estimate current performance	The current performance is the quantitative level of business metric for a time prior to the start of the project. If possible, the reference period should be past 12 years of performance.	Fifty-three percent of 2400 tires adjustment through 3Q02 (9 months)
3	What are potential consequence metrics?	Any other business measurement that may be impacted negatively as a result of improving the primary metric	Product cost, plant, output, capacity
4	How do you know about the extent and duration of the business problem?	The primary data source you used to identify business problem. If possible, title, author and date of the report, and any other sources that can be used to support this claim should be included.	Asia adjustment data 3Q02 2002 GI350 ABFP 62056

Table 29.3 PDW Boundaries

Action	Element	Definition	Actual
1	What are the key activities that have an impact on the output of the interest?	List the activities that have influence on the output of the process	Identifying the system
2	Develop a high-level business process map.	High-level map with only three to seven major steps	Process map
3	List the includes and the excludes.	What: product line; Where: region, storage, condition; When: steps on the process; Who: channel, customer; How: shipment method	Includes/excludes
4	Determine the intended scope and scale of the project.	Project scope defines the part of the processes that are to be included in the project. The project scope statement should also consider any time, resources, equipment, and facilities that may be significant.	Project scope and scale
5	Which current and past projects are relevant to this project?	What other work are you aware of that might be closely related to the project?	Various changes over the past 5 years. Data/information is available
6	Write the problem statement.	This item is really a summary of information entered earlier in the worksheet. It should include answers to these questions. Problems? Where does it occur? How often does it occur? Magnitude of the problem?	Fifty-three percent of BMT adjustments are categorized as cost accounting standards (CAS). It is most prevalent in India and Malaysia. This impacts the good year in the region and therefore price.

Table 29.4 PDW Performance and Goals

Action	Element	Definition	Actual
1	Identify the cost categories that are impacted by this problem.	There may be financial or process impacts to a cost center that can be expressed in terms of items such as labor, material, or inventory.	Country p or L, region P or L, inventory working capital, net selling price
2	Identify the total financial impact of the problem	This refers to tangible financial impact. See guidelines for reporting the Six Sigma project report for more information	$1.68 million
3	What is the amount of improvement planned for this project?	This is a general statement of the project objective. This statement should be structured as follows: improve some metrics from some current metric levels to some goals by some time frames.	Goal statement
4	Estimate the potential financial benefits of the project.	It is a forecast of the potential values/benefit of this project. This is an estimate based on the current performance and the established goal.	Expected benefits
5	Estimate how long it will take to complete the project.	Be aggressive but realistic if a completion time of 6 months or less is not realistic, then the project should be rescoped.	Timeline: 6 months

Placing an element in the "excludes" category does not mean that it cannot be used for comparison or to benchmark an "included" category. Excluded elements may always be addressed in a follow-up "translation" or transfer project (Table 29.7).

29.3   Voice of the customer

Voice of the customers can be classified into the following topics:

- Customer identification
- Customer feedback
- Customer requirements

29.3.1   Customer identification

The concept of Six Sigma has evolved for the customer. In an organization, everything starts and ends with the customer. The customers' expectations and needs are set as quality. The customer expects performance, reliability, value for their money, and on-time delivery (Harry and Schroeder 2006). The customer is not necessarily the one who will ultimately buy the product. In Lean Six Sigma (LSS), the concepts evolved in such a way that the operator at the next station is the customer. The primary customer of the process will or should have the highest impact on the process (Pande et al. 2000).

Table 29.5 PDW Organization

Action	Element	Definition	Actual
1	Identify the process owners.	Person who has the primary responsibility of ongoing operation of the process	Plant business centers; plant and region Qtech designers and compounders; sales
2	Identify the stakeholders.	Identify the key individuals with the stake in this problem/process/project.	All the above as well as customers, sales force, country, and strategic business unit (SBU) management
3	Identify the project sponsor.	Who is the individual most suited in terms of authority? Influence the stake in the outcome to sponsor the process.	Technology directors
4	Which black belt should be selected to lead this project?	Consider the project scope. Impact and complexity and expertise likely should be required.	Black belt: To be determined
5	Identify the core project team members or function.	The core team has typically three to six people who have complementary skills and expertise likely to be required for the project success.	Team member/function;
6	Identify possible resources or constraints.	Identify resources that might be needed, for example, access to database and travel to customer plant sales sites. List potential constraints including limitation on use of response.	Possible resource constraint

Table 29.6 Includes/Excludes

	Includes	Excludes
What		
Where		
When		
Who		
How		

The define phase concentrates on defining the primary customers who contribute most of the revenues for the organization. Every business has many potential customers and the customers have their own business criteria. The organization considers cost, quality, features, and availability factors for weighing the potential customers. The cost, quality, functionality, and availability (CQFA) value grid helps the organization succeed in one way or the other. It is required to define the customers and to analyze the customer data and that obtaining wrong customer data can cause a flaw in the project (Pande et al. 2000). Hence customer data are vital. To have proper customer data, customer surveys are required which are free from bias, so the organization or the company can produce and design products for the right markets and the right customers. Therefore, in any market, the customers can constitute:

Table 29.7 Includes/Excludes Example

	Includes	Excludes
What	Crown area separation, bias medium truck tires	All other conditions and product types
Where	India (first), Malaysia (second), Indonesia, Thailand, Philippines, Taiwan	Other locations
When	Design, manufacturing, selling	Molds, logistics
Who	Fleet and replacement market	Other
How	Method of movement	Does not include other

- Current, happy customers
- Current, unhappy customers
- Lost customers
- Competitor's customers
- Prospective customers

Any organization must want to work with any of the above-mentioned customers to improve their market, which is nothing but customer retention and customer loyalty. In order to obtain the right customer data, the following methods should be used (Eckes and Harder 2001):

- Surveys
- Focus groups
- Interviews
- Complaint systems
- Market research
- Shopper programs

The traditional methods of obtaining customer data are as follows:

- Targeted and multilevel surveys
- Targeted and multilevel interviews
- Customer scorecards
- Data warehousing
- Customer audits
- Supplier audits
- Quality function deployment

As mentioned earlier, various levels (belts) are certified in order to define the methodology followed and used to implement Six Sigma methodologies. The challenge lies in getting as much valuable data as possible. Certified belt users help define and distinguish the internal and external customers who are important to a particular project in an organization. This allows the sponsors and the executive management to understand their customers better and work toward effective deliverables for the customers.

29.3.2   Internal customers

Internal customers are those in the company who are affected by the product or service as it is being generated. The internal customers are often forgotten in the process of concentrating on the processes used to satisfy the external customers. The concept of LSS includes involving every employee and making him/her responsible for the process. This affects employee satisfaction and involvement on a positive scale, which in turn affects the customer satisfaction. For employee satisfaction, sound communication is important and can be improved by the following (Lowenstein 1995):

- Company newsletters
- Story boards
- Team meetings
- Staff meetings

In order to involve employees in a better and more effective way, training and education by black belts is very important. The black belts and master black belts are vested with a crucial responsibility of selecting the right people for the Six Sigma project. This process is called stakeholder analysis. When Six Sigma projects are initiated, there might be resistance from the people involved for the proposed change. It is not always obvious why the resistance arises. No matter how brilliant the idea and obvious the benefits, any effort to make a change will trigger resistance. People have many different reasons to resist change; often these reasons are quite legitimate. Resistance may take many different forms, depending on the perspective, position, and personality of the person. Resistance to change is one of the most frequent reasons why projects ultimately fail. Sponsors and black belts are left with no option but to accept the resistance. Their success lies in using this resistance as an opportunity to improve the proposed methodology more effectively. The stakeholder analysis is performed by the following:

- Identify key stakeholders of the project
- Identify the current level of support/resistance
- Define the needed level of support
- In case of gaps, develop strategy to move each stakeholder to the needed level of support.

The first and foremost step is to identify the key stakeholders of the projects. The question "Who are the key stakeholders?" is very crucial and can possibly be answered from the following:

- Owners of the process
- Anyone contributing to the process
- Anyone affected by the process
- Those who benefit from process output
- All who see themselves as stakeholders

The next step is to assess the commitment of the stakeholders for the proposed project. If the stakeholders have a hidden agenda, it is important to resolve it before the commencement of the project (Figures 29.3 through 29.5).

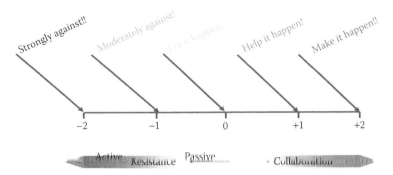

Figure 29.3 Stakeholder's commitment.

Key stakeholder	What is valued by the key stakeholder?	Commitment					Recommended actions
		−2	−1	0	+1	+2	
Executive						X◯	
TTL			X			◯	
BTM			X		◯		
Production director				X◯			
Purchasing manager				◯ X			
Finance manager			X◯				

X Current stakeholder position (present level).

◯ Where the stakeholder needs to be (required level).

Figure 29.4 Stakeholder commitment. BTM, business technology management; TTL, time-to-live.

The stakeholder analysis can be refined by performing two-dimensional stakeholder analyses. The refinement is done by adding the power/influence dimension. This gives the black belts a map showing where people stand regarding the project and helps in focusing on the efforts for project. Some people may be opposed to the change but have little influence on the success/failure of your project. The goal is to move everyone with success/failure influence toward a positive or at least neutral stance. The success/failure influence of those who remain stubbornly opposed must be minimized. The two-dimensional stakeholder analysis is illustrated in Figure 29.6.

29.3.3   External customers

External customers are not part of the organization but are impacted by it. External customers play a crucial role as they provide most of the money. The external customers can be end users, intermediate customers, and impacted parties. End users are the customers who buy the product for their own utility. Intermediate customers are those who buy the product and then resell. Retailers, distributors, and wholesalers are all some of the entities of intermediate customers. Impacted parties are those who do not actually buy the product but are impacted by it. The define phase concentrates mostly on identifying

TBM uptime—Stakeholder analysis

Key stakeholder	What is valued by the key stakeholder	Commitment −2	−1	0	1	2	Recommended actions
Plant manager	He gives strong support and he is also the sponsor of the project.					X ●	
Production manager	He will give support, because he will get benefit, but he may be influenced by the BT managers.			X	●		He will be involved during the next meetings.
Qtec manager	He will give support, because he will get benefit. In view of waste, but he, one of this technology managers, is involved in the project as black/green belt.				X ●		
Production control manager	His support is required if trials are required (later in the project). If we can improve the uptime, he will get benefit from the project.		X		●		"Some evening discussions"
BTM truck	If we improve the situation, he will get benefit. But he may lose the possibility to hide some of his own problems (dangerous).		X			●	Involve him before "starting" of the project; small meeting before project launch; always try to get informations about his feeling (during walk over to parking, etc.).

X   Present level

●   Required level

Figure 29.5 Stakeholder analysis—Example. BTM, toleranced by manufacturing.

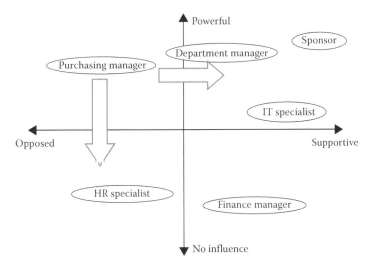

Figure 29.6 Two-dimensional analysis.

the external customers and it is more complex. In the following sections, the identification of the customers and the data analysis are discussed based on the belt certifications. The external customers in general can be identified by the following:

- Are the customers interested in the lowest possible price?
- Are the customers interested in the highest quality imaginable?
- Are the customers interested in sparing no expense?

29.4 *Project charter and teams*

The define step involves a team and its sponsors reaching an agreement on what the project and project goals are. Presuming that the project charter is already in place, the most important function of define is for the team to decide what has to be accomplished and confirm their understanding with the sponsorship. The sponsor and team should agree on the following:

- Decide and determine what the problem is and the agreement on the problem: what customers will be affected, what their "voices" are saying, how the current process or outcomes fail to meet their needs
- The link of the project between corporate strategy and its expected contribution to return of investment
- Decide on the boundaries of the project
- Know what indicators or metrics are used for the evaluation of the success of the project.

In service environments, the last two points that are mentioned above often prove particularly more important. When the process that has to be performed is studied and mapped out, deciding the start and end points of the project becomes a simple matter. Most services have not been mapped prior to improvement. There is often an argument between the team and the sponsors in the early stages of the project improvement as the team creates a supply, input, process, output, and customer (SIPOC) or value stream map and then the direction to identify exactly what they should include as part of their project and what they shouldn't.

29.4.1 Setting project boundaries

Projects that are too big will end up with floundering teams who have trouble finishing in a reasonable frame of time. Choosing a small or insignificant project, you will never convince anyone that Six Sigma is worth the investment. Projects that do not significantly contribute to financial payback will cause everyone from line managers to senior executives to quickly lose interest. Another concern to be considered when thinking about project boundaries is the level of the certification the employees possess. During their training period, black belts work on a project of limited scope only. The metrics for success in the define phase might be as follows:

- Customer satisfaction—to make sure that all customers segments are represented
- Speed/lead time
- The team to determine the defects and opportunities for improvement and to have a sigma-level improvement
- How will the above processes help financially?

There are two key issues in define phase that must be addressed:

1. It is important to make sure that the right people are on the project. The decision should be determined not only by the kind of people that are representative of the work area(s) affected by the project and that possess the knowledge, experience, and training that help the team reach the project goals but also by an evaluation of the dynamics of the team.
2. It is important to make sure that everyone in the team starts from the same point and works to the same project goals. This includes all the members of the team, belt certified professionals, champions, black belts, and the other staff who are working for the process but might not be part of the team.

Of course, granting authority to employees does not guarantee that people will work together or necessarily achieve all the lofty goals that are espoused in this approach. Many issues surround empowerment and teamwork that must be addressed. These issues range from operations and behavior to organizational design. For example, if the existing culture does not reward this type of activity, it is doubtful that participatory approaches will work until the cultural issues are resolved. However, using teams can lead to cultural changes that facilitate improvement. This chapter focuses on the issues related to managing projects and teams to help make the transition succeed.

From a behavioral perspective, empowerment is a tool used to enhance organizational learning. Organizational learning implies a change in organizational behavior in a manner that improves performance. This type of learning takes place through a network of interrelated components. These components include teamwork, strategies, structures, cultures, systems, and their interactions. Cooperative learning relies on an open culture where no one feels threatened to expose opinions or beliefs, which promotes a culture where individuals can engage in learning and questioning, and do not remain constrained by "taboos" or existing norms. This strategy includes continuous improvement projects as a governing principle for all team members.

29.4.2   Project charter

The project charter is represented in the following four areas:

1. Content of the charter
2. Negotiation of the charter
3. Project management
4. Project measures

29.4.2.1   Content of the charter

The development of a charter is a vital element for establishing an improvement team. The charter is a document that defines the teams' mission, the boundaries of the project, the consequences, and the time frame. It is usually the top management that creates the charter and presents it to the teams or the teams can also create a charter and present it to the top management and obtain their approval. Either way, the top management is responsible for giving the team the support and direction of work.

The charter begins with creating a purpose statement. This may be a single- or double-line statement explaining the purpose of forming the team. The purpose statement should be in correlation with the organization's vision and mission statements. The objective to be achieved should also be defined in this purpose statement.

The objective should be defined in such a way that it can be measured. The scope or the boundaries of the project should also be given in the project charter. This is to determine the organizational limit within which a team is permitted to operate. Time delaying and energy draining can be prevented if the boundaries of the project are defined.

Teams are more effective when they know what is expected of them. The team has the permission, authority, and blessing from the various levels of management to operate, conduct research, and consider and implement changes that may be of need for the process. A charter provides the following advantages:

- Eliminates confusion
- Determines the subject boundaries
- Areas that are not to be addressed
- Determines the deliverable
- Provides a basis for the team to set a goal
- Authorizes the team to collect relevant data

A team project charter should contain the following topics (Figures 29.7 and 29.8):

- Financial impact
- Problem statement
- Scope or boundary of the project
- Goal statement
- Role of team members
- Milestones or deliverables
- Resources required

29.4.2.1.1   Financial impact.   The financial impact is a short summary of the reasons for carrying out the project. It normally involves the quality, cost, or delivery of

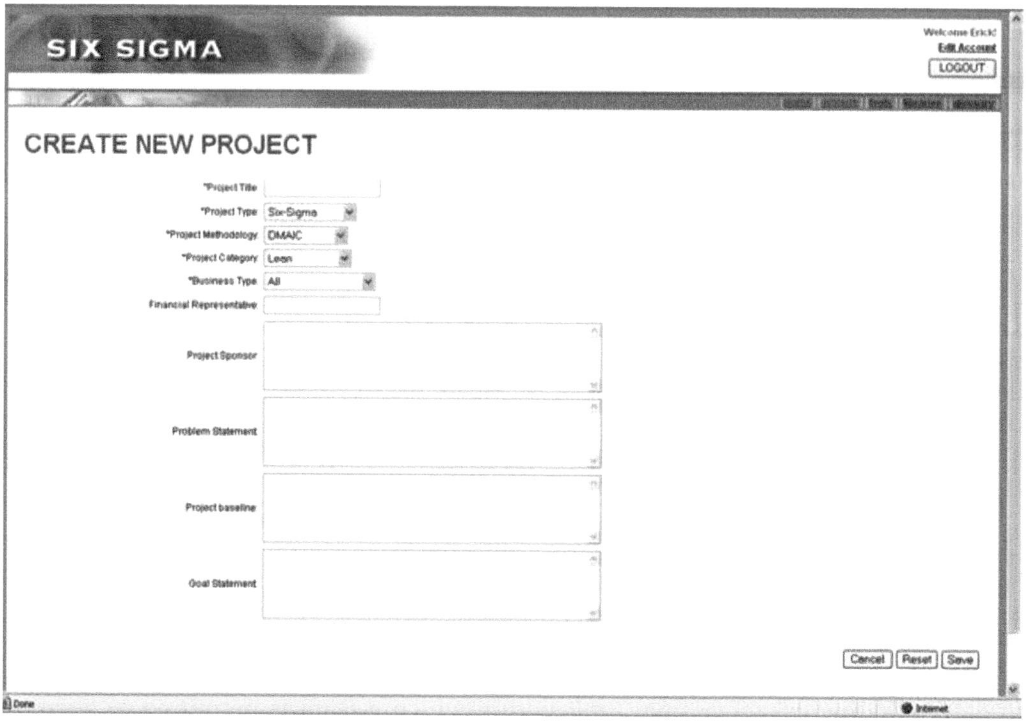

Figure 29.7 Project charter template.

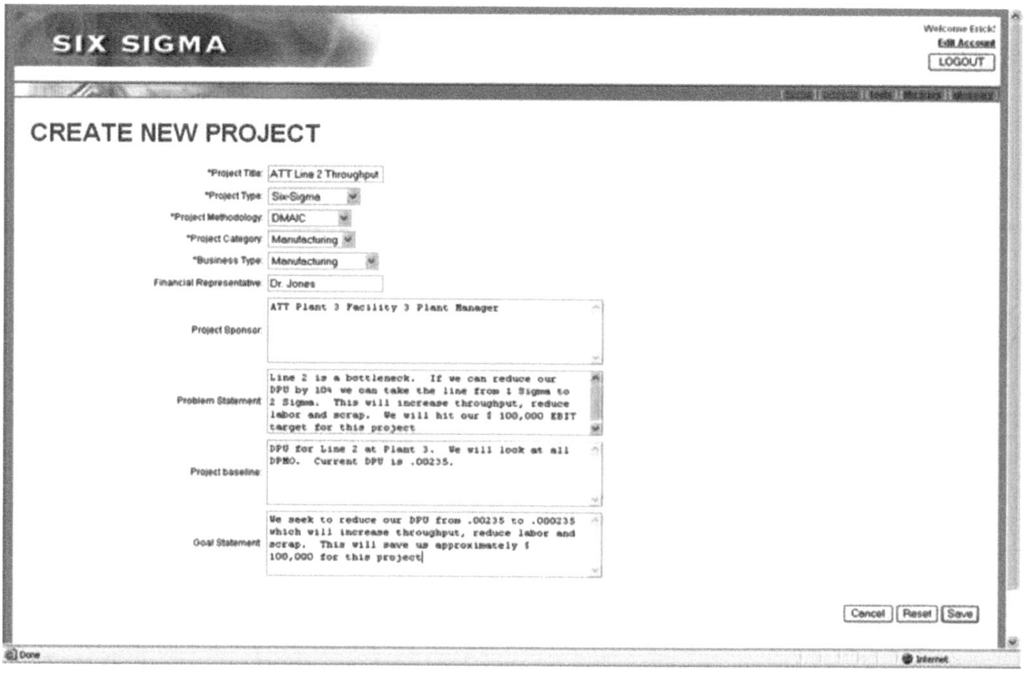

Figure 29.8 Sample project charter.

a product with a financial justification. There are four basic activities that will have a financial impact:

1. Design of a new product
2. Redesign of an existing product
3. Design of a new process
4. Redesign of an existing process

A common problem for most companies is a lack of measurement that shows the financial impact of a project. A project improvement team should carry out their project in accordance with financial justification guidelines (Eckes and Harder 2001). As an example, suppose that the existing quality defect rate is at 5000 defects per million opportunities (DPMO); a possible justification would show a reduction to 250 DPMO with a cost savings of $1,000,000. The advantages and shortcomings of the project should be looked into. There should be an involvement of the entire organization, if necessary, to determine the key costs and their resources for a successful project. Projects that do not provide financial augmentation should be eliminated right away.

29.4.2.1.2   Problem statement.   A problem statement gives a detailed statement on the issue that has to be improved. The problem statement should be crafted so that it fully describes the problem. Examples include how long a problem has existed, how it has affected the business, and what is the gap in performance and the measurable item that might be affected? The problem statement should be written in a manner that it does not cause one to jump to conclusions. A sample problem statement would be "The ABC Company, in 2007, has experienced a 25% drop in sales, with a 40% drop in net profit." The problem statement should contain a reference to a baseline measure for guidance. The collection of good data and process performance measurements will provide a picture of the areas in the company that need improvement the most. In addition, the foundation of the work will provide a measure for other teams working on other projects as well.

29.4.2.1.3   Goal statement.   The goal statement is created and agreed to by the team and the champion. The goal is anticipated to be attained in a 120- or 160-day period. According to the Six Sigma methodology, the project is required to realize a 50% reduction from the initial metrics. For example, reducing the collectibles from 120 to 60 days or reducing the scrap from 25% to 12.5%. One of the most efficient ways of formulating the goal statement is through the theory of constraints (TOC).

29.4.2.1.3.1   Step 1: Identify.   Concentrating on a nonconstraint resource would not increase the throughput (the rate at which money comes into the system through sales) because there would not be an increase in the number of orders fulfilled. There might be local gains, but if the material ends up waiting longer somewhere else, there will be no global benefit.
    In order to manage a constraint (bottleneck), it is first necessary to identify it. A constraint (bottleneck) is a resource whose capacity is less than the demand. This knowledge helps determine where an increase in "productivity" would lead to increased profits. To increase throughput, flow through the constraint must be increased.

29.4.2.1.3.2   Step 2: Exploit.   Once the constraint is identified, the next step is to focus on how to get more output within the existing capacity limitations. Because the constraint is what limits the system's throughput, we have to make it work to the maximum capacity.

29.4.2.1.3.3 *Step 3: Subordinate.*   Subordination usually involves significant changes to current (and generally long-established) ways of doing things at the nonconstraint resources. The non-bottlenecks to the system constraint is subordinated. All the other components of the system must work so as to guarantee full-speed functioning of the constraint.

29.4.2.1.3.4 *Step 4: Elevate.*   After the constraint is identified, the available capacity is exploited, and the nonconstraint resources have been subordinated, the next step is to determine if the output of the constraint is enough to supply market demand. If so, there is no need at this time to "elevate" because this process is no longer the constraint of the system. In that case, the market would be the constraint and the TOC thinking process should be used to develop a marketing solution. If, after fully exploiting this process, it still cannot produce enough output to meet demand, it is necessary to find more capacity by "elevating" the constraint.

29.4.2.1.3.5 *Step 5: Go back to Step 1.*   Once the output of the constraint is no longer the factor that limits the rate of fulfilling orders, it is no longer the constraint. Step 5 is to go back to Step 1 and identify the new constraint because there is always one. The five step process is then repeated.

29.4.2.1.4 *Milestones or deliverables.*   A well-organized project is bound to have set of short-term goals or deliverables that are used to keep the project on track and help bring it to completion. It has been pointed out that the initial team projects should be at the 120-day length. Only half of the project time is supposed to be allocated to define and measure the phase. Assigning teams and the right kind of people for the project is very important. The success rate of the project decreases as the length of time assigned to complete the task increases. A typical milestone chart might be the following:

- Day 0: start the team activities
- Day 1: start the define portion of the project
- Day 3: begin the measure portion of the project
- Day 80: start the analysis of the project
- Day 120: start the improvement phase of the project
- Day 160: bulk of project control elements in progress

Resources required for a project are very important and need to be noted and detailed. Typical resources might be as follows:

- Qualified people
- Machine time
- Machinery
- Laboratory or office space
- Phones and faxes
- Computer equipment
- Utilities

The define phase of Six Sigma should provide the top management the following information:

- Importance of carrying out the project
- Goal of the project
- Skills of the champion and other leaders
- Boundaries of the project
- The key process
- Metrics
- Customer requirements

29.4.2.2 Negotiation of the charter

The team presents the project charter to the top management. However, the project team might be closer to the actual facts through another approach toward the problem. So there are bound to be charter negotiations. They might be as follows:

- Objectives—change in the design or final product due to customer feedback
- Scope—increase or reduce the constraints for the work tasks to be performed by the team
- Resources—an accurate requirement of resources is difficult to provide. If there is a requirement for more resources, management may be required to prioritize certain resources beyond the team's control.
- Project transition—the transition of a project to normal company controls might require a time extension.
- Project closure—project closure date might be required to be moved up because of diverse events or changes in customer preference.

29.4.2.3 Project management and its benefits

Project managers are hired by most organizations to ensure that the complexities of a project move in an organized fashion and make the proper transitions in a timely and economical manner. The type of "matrix management" has proved very effective in providing deliverables on time. The project management roles and responsibilities are to include the following:

- Leading the cross-functional team
- Possessing excellent communication skills and ability to convey the message clearly
- Scheduling meeting to check progress at regular intervals
- Sustaining the team and its motivation
- Developing a detailed project plan
- Letting the team know the benefits of the project to all the share holders
- Tracking the progress of the tasks and deliverables
- Maintaining the flow of information between financial and information management

Projects need charters, plans, and boundaries. A project may be selected from a broad range of areas including the following:

- Customer feedback
- Improvement in process capabilities
- Cost reduction chances

- Defects reduction
- Employing Lean principles
- Growth in market share
- Reduction in cycle times
- Improvement in services

The project should be consistent with the strategies of the company for survival and/or growth and should be rather specific.

29.4.2.4   Project measures

It should be noted that the vital project measurements are hard to decide until the project charter and its associated processes are complete. The accurate selection of project measures ensures the overall success of a Six Sigma implementation. Since most projects deal with time and money issues, most project measures will also be related to time and money. The project measures provide information, which is required to analyze and improve the business process, as well as manage and evaluate the impact of the Six Sigma project. After a list of activities of the project is prepared, the budget of the project can be determined. During the project, the actual costs are collected as inputs and used to determine the estimated costs. The project manager or the team leader compares the revised estimated costs with the actual costs and determines the progress of the project. A project should be reasonable, attainable, and based on estimates of the tasks to be accomplished.

The various revenue factors included in the analysis are as follows:

- Income from additional sales generated due to the revised cost of the product and the changes in the quality, features, and availability to the customer
- Reduced amount of defects, scrap, returns, warranty claims, cost of poor quality, and poor market presence

The cost factors included in the budget are as follows:

- Labor cost
- Administrative expenses
- Equipment costs
- Subcontracted work
- Overheads
- Contingency funds

The timing of the revenues and costs also plays a very important role. Sometimes, the revenues projected might not be obtained because the funds were not available in the right time frame. The precision and detail of the project planning phase plays a very important role in the success of the project. The costs associated with each project are obtained based on historical data, quotes, standard rates, or similar activities performed previously. Estimates of project revenues are described based on four types of measurements:

- Budget—the plan of the total costs and cash inflows expressed in dollar amounts for the project. The plan also includes timing of the revenue and costs and a cost–benefit analysis
- Forecast—the predicted total revenues and costs, adjusting to include the actual information at the point of completion of project

- Actual—revenues and costs that have actually been occurred for the project
- Variance—the difference between the budgeted and actual revenues and costs. A positive variance shows that the project will be favorable, whereas a negative variance shows that a loss will be incurred.

29.4.3 Teams

There has been a move toward flattening hierarchies in organizations in order to empha-size team work and empowerment. Led by consultants such as Tom Peters, top managers have eliminated the layers of bureaucratic managers in order to improve communication and simplify work. Having many layers of management can have the effect of increasing the time required to perform work. For example, it has been reported that in the 1980s, one of the largest automobile manufacturers in the United States required 6 months to deter-mine its standard colors for office phones. This decision probably required many meetings and proper authorization. Although such decisions need to be made, the time required to make this decision was excessive.

Too many layers of management can also impede creativity, stifle initiative, and make empowerment impossible. With fewer layers of management, companies tend to rely more on teams. When Lee Iacocca took the reins at Chrysler Corp., one of his first acts was to eliminate several levels of management. Iacocca credited this move with making other needed changes easier within the organization.

29.4.3.1 Team leader roles and responsibilities

Quality professionals are in unanimous agreement that to be successful in achieving team-work and participation, strong leadership both at the company level and within the team is essential. However, what is not always clear is what it means to be an effective team leader. We know that leaders are responsible for setting the team direction and seeking future oppor-tunities for the team. Leaders reinforce values and provide a system for achieving desired goals. Leaders establish expectations for high levels of performance, customer focus, and continuous learning. Leaders are responsible for communicating effectively when evaluat-ing organizational performance and for providing feedback concerning such performance.

An important aspect of leadership is the organization's preparedness to follow the leadership. The best general is probably not going to be successful if the troops are not well trained or prepared. Hersey and Blanchard propose a theory called a situational leader-ship model that clarifies the interrelation between employee preparedness and effective-ness of leadership. According to Hersey and Blanchard, situational leadership is based on interplay among the following:

- The amount of guidance and direction a leader gives (task behavior)
- The amount of socioeconomic support a leader provides (relationship behavior)
- The readiness level that followers exhibit in performing a specific task, function, or objective

Therefore, if team members are trained and prepared so that they are "task ready," the leadership will be more effective. Readiness, in this context, is the "extent to which a fol-lower has the ability and willingness to accomplish a specific task." Readiness is a function of two variables. These are ability and technical skills and self-confidence in one's abilities. Therefore, effective leadership helps employees become competent and instills confidence in employees that they can do the job.

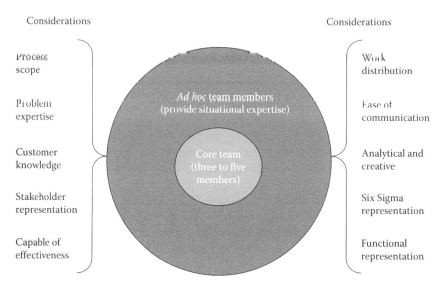

Figure 29.9 The core and *ad hoc* team.

As it relates to quality management, leadership is especially difficult. Leaders are told that they should empower employees. To many leaders, this implies laissez-faire or hands-off approach to management. In other words, many leaders feel that they are to provide resources but that they should not be involved in overly controlling employee behavior. Although the literature contains examples of companies that have been successful in delegating authority to this extent, quality management is not a vehicle by which leaders abdicate their responsibility.

In most organizations, employees want leaders who provide clear direction, necessary information, and feedback on performance, insight, and ideas. Skilled team leaders need to demonstrate this ability to lead. The single most important attribute of companies with failed quality management programs is a lack of leadership. A close second is poor communication, which is related to leadership. Effective leaders are people who are able to provide visions, ideas, and motivation to others in order to achieve the greater good.

Besides team leaders, there are a variety of roles that individuals occupy in teams. Team roles can be defined functionally. Teams often require different functional talents such as management, human resources, engineering, operations, accounting, marketing, and management information systems to name a few. In these cases, the managers overseeing the project help to identify the talents needed and then search for the team members to provide these talents (Figure 29.9).

29.4.3.2 *Team formation and evolution*
The way a team is formed depends, to an extent, on the objectives or goals of the team. Regardless of the type of team your firm employs, teams experience different stages of development. These stages include the following:

- Forming—where the team is composed and the objective for the team is set
- Storming—where the team members begin to get to know each other and agreements have not yet been made which facilitate smooth interaction between team members
- Norming—where the team becomes a cohesive develop

- Performing—where a mutually supportive, steady state is achieved
- Mourning (the final stage in successful projects)—where the team members regret the ending of the project and breaking up of the team

29.4.3.3   Team rules
During the norming stage, teams develop ground rules. Such ground rules can forestall conflict. It is often useful to establish ground rules for a team to be functional. If a team is functional, individual participation enhances the group's effectiveness. If the team is dysfunctional, such participation reduces the effectiveness of the group. Acts of commission include talking behind the backs of other team members or otherwise acting out one's feelings. There are also acts of omission in such passive aggressive behavior as forgetting to attend meetings or withholding information. Counteractive behavior improves the group's effectiveness by negating dysfunctional behavior. Counteractive behavior can be enacted by either the team, the facilitator, the team manager, or even the offending individual.

29.4.3.4   Types of teams
At the point, we will pause to define the various types of teams that are used in improving quality. Continuous process improvement often requires small teams that are segmented by work areas. Projects with multiple departments in a company require cross-functional teams. Large projects require teams with large budgets and multiple members. Smaller projects, such as "formulating a preventive maintenance plan for oiling the metal lathes," probably will require a much smaller team. A number of teams are listed and defined as follows:

1. Process improvement teams: These teams work to improve processes and customer service. These teams may work under the direction of management or may be self-directed. In either case, the process improvement teams are involved in some or all of the following activities: identifying opportunities for improvement, prioritizing opportunities, selecting projects, gathering data, analyzing data, making recommendations, implementing change, and conducting postimplementation reviews. Many process improvement teams are an outgrowth of quality-related training. These teams use the basic tools and plan–do–check–act (PDCA) cycle to effect change relating to processes.
2. Cross-functional teams: These teams enlist people from a variety of functional groups within the firm. In the real world, problems often cut across functional borders. As a result, problem-solving teams are needed which include people from a variety of functions. These cross-functional teams often work on higher level strategic issues that involve multiple functions. Such teams often work on macro-level, quality-related problems such as communication or redesigning companywide processes.
3. Tiger teams: These teams are high-powered teams assigned to work on a specific problem for a limited amount of time. These teams are often used in reengineering efforts or in projects where a specific problem needs to be solved in a very short period of time. The work is very intense and has only a limited duration.
4. Natural work groups: The teams are organized around a common products, customers, or services. Many times, these teams are cross-functional and include marketers, researchers, engineers, and producers. The objective of natural work groups include tasks such as increasing responsiveness to customers and market demand. In order to implement natural work groups, managers need to provide an enriched job design and work life for employees. The key elemental impact of natural work groups

is to improve service by focusing work units in an organization on the customer. A by-product is improved communication with customers. Often a natural work group will be established for a specific customer.

5. Self-directed work teams: These teams are chartered to work on projects identified by team members themselves. There is little managerial oversight except to establish the teams and fund their activities. Self directed teams are identified as either little s or big S teams. Little s self-directed work teams consist of employees empowered to identify opportunities for improvement, select improvement projects, and complete implementation. Big S self-directed teams are involved in managing the different functions of the company without a traditional management structure. These types of teams contain totally self-directed employees who make decisions concerning benefits, finances, pay, processes, customers, and all the other aspects of running the business. Often big S self-directed work teams hold partial ownership of the companies they work for so that they participate in the benefits of their teamwork.

6. Technology and teams: New tools for teamwork are constantly emerging. Also, team effectiveness is a precursor to project task performance. Integrated information systems include tools such as CAD/CAM and CIM. This aids in achieving improvement in efficiency and effectiveness. Process technology is used in helping to improve task performance. Process standardization methods such as the tools of quality and customer input methods complete the model. This model amplifies that more and more, team effectiveness is assisted by integrated tools and technologies, and the impact of technology should increase. As software becomes cheaper and easier to use, more tools will be used by everyone involved with the project.

7. Virtual teams: The term virtual teams are emerging as more companies become "virtual organizations," loosely knit consortia that produce products and services. Virtual teams are teams that rarely or never physically meet, except in electronic meetings using group decision software. Among virtual organizations, projects often cross-organizational boundaries. Today, Internet- and Intranet-based applications called teamware are emerging which allow us to access the World Wide Web and build a team, share ideas, hold virtual meeting, brainstorm, keep schedules, and archive past results with people in far-flung locations around the world. Hectic schedules and the difficulty in finding convenient times to meet to solve problems will make teams of this type more important in the future.

29.4.3.5   Implementing teams

The teams in our examples have something in common. The performance of the team is essential to their individual success, and in some cases, even lives hang in the balance. If the National Association for Stock Car Auto Racing (NASCAR) team performs ineffectively, the driver loses. If the Massachusetts General Hospital team is ineffective, people die. If the Navy SEALs do not function properly, lives are lost and the mission fails. How do we engender this sense of urgency in quality improvement teams? How do we create a momentum or team ethic that will help us beat the odds and be successful? Accomplishing this often requires facilitation and team building. Facilitation is helping or aiding teams by maintaining a process orientation and focusing the group. Team building is accomplished by following a process that identifies the roles for team members and then helps them to become competent in achieving those roles.

The role of the facilitator is very important in managing teams, particularly when team members have little experience with teamwork. The role of the facilitator is to make it easy for the group to know where it is going, know why it wants to get there, know how to get there, and know what it is going to do next. A facilitator focuses the group on the process it must follow. Successful facilitation does not mean that the group always achieves its desired results. The facilitator is responsible for ensuring that the team follows a meaningful and effective process to achieve its objectives.

The facilitator should plan how the group will work through a task, help the group stay on track and be productive, draw out quiet members, discourage monopolizes, help develop clear and shared understanding, watch body language and verbal cues, and help the group achieve closure. Again, facilitators must remain neutral on content. Facilitators cannot take sides or positions on important areas of disagreement. However, facilitators should help key members reach points of agreement.

29.4.4 Meeting management

Effective meeting management is an important skill for a facilitator of quality improvement teams. Often quality improvement involves a series of meetings of team members who meet to brainstorm, perform root-cause analysis, and carry out other activities. Tools for a successful meeting management include an agenda, predetermined objectives for the meeting, a process for running the meeting, and processes for voting, and development of an action plan using these tools requires outstanding communication skills as well as human relations skills. The steps required for planning a meeting are as follows (Figure 29.10):

- Defining an agenda
- Developing meeting objectives
- Designing the agenda activity outline
- Using process techniques

Structured processes, including a set of rules for managing meetings, work well in conducting meetings. It is paradoxical that structured processes are inhibiting, time consuming, and unnatural. These processes are used so that meetings stay focused, involve

```
Date:                                          Time:
In Attendance:
      Project Sponsor(s), Black Belt, Core Team Members, Master Black Belt
Agenda:
      – Review Business Case for the project
      – Review and discuss Project Charter
      – Develop team-behavior contract
      – Develop Project Plan, including team information and project schedule
Assignments:
What                                  Who            By When


Next Meeting:

```

Figure 29.10 Meeting notes.

deeper exploration, separate creative from evaluative activities, provide objective ground rules defensiveness, and separate the person from the idea.

Tools such as flipcharts, sticky dots, whiteboards, and sticky notes are used commonly in structured process activities. The focus of team meetings moves from generating ideas to clarifying ideas, to evaluating ideas, and to action planning. Some of the techniques, such as silent voting and idea writing, help team members reach consensus rapidly.

Another useful meeting management tool that was pioneered by Hewlett-Packard is the "parking lot." The parking lot is a flipchart or whiteboard where topics that are off the subject are parked with the agreement that these topics will be candidates for the next meeting's agenda. At the end of the meeting, the group agrees on the agenda for the following meeting, and the parking lot is erased.

29.4.5 Conflict resolution in teams

As people work closely together in teams, conflicts arise. Conflict resolution is a hugely important topic for team leaders and members. Conflicts are endemic to all kinds of team projects. Using team processes, assumptions are questioned, change is brainstormed, and cultures are challenged. This type of creative activity results in possible conflict. It is claimed that team leaders and project managers spend >20% of their time-resolving conflict. If this is true, then conflict resolution resounds as one of the very important under discussed topics in team building.

There are many sources of conflict. Some conflicts are internal, such as personality conflicts or rivalries, or external, such as disagreements over reward systems, scarce resources, lines of authority, or functional differentiation. Teams bring together individuals from a variety of cultures, backgrounds, and functional areas of expertise. Being on a team can create confusion for individuals and insecurity as members are taken out of their comfort zones. It is interesting to note that these are also some of the reasons why the teams are successful. Some organizational causes of differences are more insidious: faulty attribution, faulty communication, or grudges and prejudice. Four recognizable stages occur in the conflict resolution process:

1. Frustration—People are at odds and competition or aggression ensues.
2. Conceptualization and orientation—Opponents identify the issues that need to be resolved.
3. Interaction—Team members discuss and air the problems.
4. Outcome—The problem is resolved.

One of the things a leader must be able to do is to manage conflict in the organization. In order to foster a well-run workplace, leaders must be able to resolve conflict effectively in the organizations. Leaders resolve conflict in a variety of ways:

- Passive conflict resolution: Some managers and leaders ignore conflict. This is probably the most common approach to working out conflict. There may be positive reasons for this approach. The leader may prefer that subordinates work things out themselves. Or the conflict may be minor and will take care of itself over time. Leaders feel that some issues are small enough to not merit micromanagement.
- Win–win: Leaders might seek solutions to problems that satisfy both sides of a conflict by providing win–win scenarios. One form of this is called balancing demands

for the participants. This happens when the manager determines what each person in the conflict wants as an outcome and looks for solutions that can satisfy the needs of both parties.

- Structured problem solving: Conflicts can be resolved in a fact-based manner by gathering data regarding the problem and having the data analyzed by a disinterested observer to add weight to the claims of one of the conflicting parties.
- Confronting conflict: At times it is best to confront the conflict and use active listening techniques to help subordinates resolve conflicts. This provides a means for coming to a solution of the conflict.
- Choosing a winner: In some cases, where the differences between the parties in the conflict are great, the leader may choose a winner of the conflict and develop a plan of action for conflict resolution between parties.
- Selecting a better alternative: Sometimes there is an alternative; neither of the parties to the conflict has considered. The leader then asks the conflicting parties to pursue an alternative plan of action.
- Preventing conflict: Skilled leaders use different techniques to create an environment that is relatively free of conflict. These approaches are more strategic in nature and involve organizational design fundamentals. By carefully defining goals, rewards, communication systems, coordination, and the nature of competition in a firm, conflict can be reduced or eliminated. Conflicts are often the results of the reward systems in the firm. A systems approach will focus attention on organizational design rather than individual interactions.

29.5   Project tracking

Project management is an essential process in establishing a scheduled plan and allocating available resources judiciously. The strings of activities that constitute a project are to be monitored to prevent straying off the timeline and for proper implementation. This results in the objectives of a project being adhered to. By utilizing the allocated resources within time and cost constraints at the desired level of performance, specified goal can be achieved. The phases of project management include the following:

- Planning
- Scheduling
- Controlling

Tracking is in the controlling phase of project management to ensure that the requirements of the undergoing project are met. It involves an exercise of making rightful decisions (Figure 29.11):

- Identifying sources of bottlenecks
- Delegating duties
- Choosing the right trade-offs
- Measuring results against expected outcomes
- Applying timely corrective plans
- Establishing tolerance level in timeline
- Future developments to current work
- Efficient communication and continuous tracking

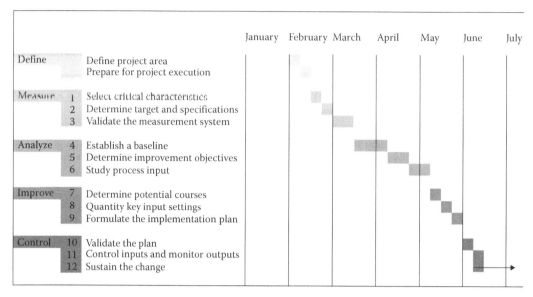

Figure 29.11 Project timeline.

29.5.1 Work breakdown structure

The work breakdown structure (WBS) is a descriptive document on the constitution of the project and the assigned responsibilities for each work group. Each responsibility is converted to a string of activities and further broken down to subtasks or elements that are set to motion by a team. From subtask to activity, each element has an associated time frame that should be flexible enough to accommodate variations that occur during implementation. Interrelationships between predecessor events and the following events are to be streamlined and scheduled. Offsetting is required to avoid losses by arranging for a safety stock of resources and to accommodate any increase in costs incurred. Deadlines are typically fixed by nature. For adherence to such fixed time requirements, resources have to be flexible and aplenty. There should be arrangements for coordinating between the different departments in the organization for sourcing their equipment, material, and manpower from the same pool. Seamless scheduling becomes essential for parallel tasks without slack time. Planning for such activities requires tools that can recognize the amounts of resources and time, and calculate the costs. These include the following:

- Project evaluation review technique (PERT)
- Critical path method (CPM)
- Gantt chart

The WBS is useful to define the relevant activities and the investment required. As is evident, these tools are helpful for the analyze phase in the Six Sigma process.

The following network planning rules are based on these tools:

- All activities must begin on completion of preceding events.
- Logicality of event precedence is implied using arrows.

- The direct event connection can be established through only one activity.
- Network must begin and end at single events.
- Event numbers must be unique.

The PERT and CPM tools are similar except that CPM is activity oriented and PERT is event oriented. Gantt charts are representative of activities or events against time. The purpose, frequency, method (written reports, summary, and forms), prioritization among random events, feedback loops, and contingency plans must be kept in mind when developing Gantt charts. The feedback loop includes the methods for contingency measures. It is an auto-generated control plan. Success or failure is to be measured against the following:

- Fulfillment of goals
- Adherence to time limits
- Boundary variance
- Utilization index of resources

A properly executed plan must have a high rate of success for all these measures. Contingency plans must be at no extra costs and waste of resources. Unanticipated events and complexities of the project that pose challenges in terms of better technologies are not acceptable excuses for occurrences of failure. Performance standards for results are set high and issues must be justifiably resolved by the team leader in order to maintain the schedule for the given budget levels.

Units of measurement for a timeline can be specified as days, weeks, months, and so on for convenience. Methods used for planning and monitoring include manual methods such as paper, graphs, and markers up to computer software.

Advantages of manual methods are as follows:

- Easy to use
- Apt for scheduling
- Reduced cost
- Customizable
- Flexible to needs
- Hands-on—feel of status

Advantages of computer software are as follows:

- Monitoring of random events
- Impact of alternatives
- Detailed project status reports
- Automatic calculation of time frames
- Easy reporting
- Quicker generation of summaries
- Plan is in real time
- Automated data collection activities

29.5.2   *Milestone reporting*

Significant events in the project timeline are termed as milestones. They help keep pace with the project schedule and in reporting the progress and status level of the project.

They act as a decision point to proceed with the project state. Discussions occur at these junctures to check and preemptively define a course of action to minimize both the occurrence and the effects of schedule constraints. The time is already set during the project planning phase as to which activity can be termed as an important milestone. For any well-managed project, a set of stages or milestones are used to keep the project on track and to help bring a project to completion. Only half of the project would be allocated to define and measure stages. Assigning teams a first project with lengths of >160 days will lower their success rate.

29.5.3　Project report

The project report is a progress summary that states the performance versus the anticipated benefits and costs incurred with the initial planned budget. Tasks completed against the milestones are also analyzed for status. The postmortem analysis is the next step that gives what went wrong and is used as a base for building and improving on further plans. The project report also defines the benefits that are obtained from more effective processes.

29.5.4　Document archiving

Document archiving is documenting the data, the source of materials, process parameters, and reports generated during the project. The organization of these files must be coherent and have special storage conditions (Figure 29.12).

These conditions include the following:

- Security with special access
- Retrievable
- Protection from damage
- Traceability of these documents using indicators
- Duplication of files for safety
- Using reliable mediums

Figure 29.12 Project records.

References

Eckes, G. and Harder, B. (2001). *Making Six Sigma Last: Managing the Balance between Cultural and Technical Change*. New York: Wiley.

Harry, M. J. and Schroeder, R. (2006). *Six Sigma: The Breakthrough Management Strategy Revolutionizing the World's Top Corporations*. New York: Random House Digital.

Lowenstein, M. W. (1995). *Customer Retention: An Integrated Process for Keeping Your Best Customers*. Milwaukee, WS: ASQC Quality Press.

Pande, P., Neuman, R. P., and Cavanagh, R. R. (2000). *The Six Sigma Way: How GE, Motorola and Other Top Companies Are Honing Their Performance*. New York : McGraw-Hill.

Further Readings

Chung, C. A. (ed.) (2003). *Simulation Modeling Handbook: A Practical Approach*. Boca Raton, FL: CRC Press.

Jones, E. C. (2003–2010). *Total Quality Management Using Six Sigma Techniques. Lecture notes.* Retrieved from https://elearn.uta.edu/.

Jones, E. C. and Chung, C. A. (2010). *RFID in Logistics: A Practical Introduction*. Boca Raton, FL: CRC Press.

Index

Note: Locators "*f*" and "*t*" denote figures and tables in the text

A

Active humidity sensor RFID system (AHSRS), 514
Actual time, 434
Advanced process control (APC), 407*f*
Allowances, 434
American Society for Quality (ASQ), 441, 538–539
American Society of Mechanical Engineers (ASME), 430
Analysis of variance (ANOVA), 301–303, 302*t*, 349, 416, 477
    box plot, 392*f*, 395*f*
    Fisher's LSD, 350
    one-way, 391–399, 391*f*
        dimple pattern, 393*f*
        dot plots, 395*f*
        general linear model, 398*t*
        golfers, 396*f*
        variables, selection, 391*f*
    p-value, 349–350
    residual plots, 392
        distance, 394*f*
        four-in-one, 392, 394*f*
        general linear model, 397*f*
        golfers, 396*f*
    Tukey's method, 350
    value plot, 392*f*
Analyze phase
    baseline, establishing
        discrete outputs, 325–332
        performance metrics, 324–325
    black belt
        multi-vari studies, 356–357
        process capability, 368–373
        regression, 357–368
    classifications, 322
    green belt
        ANOVA, 349–350
        chi-square test, 349
        entitlement, 342–344
        FMEA, 351–355
        hypothesis testing, 345–349
        X–Y process map, 350–351, 351*f*
    plan, 323*f*

    yellow belt
        cause-and-effect/fishbone diagrams, 337–338, 337*f*, 338*f*
        FTA, 338–340
        root cause, 340–342
    Z-scores, 322–324, 330*f*
        one-sided, 333*f*
        two-sided, 334*f*
Antenna number (AN), 456–457
Antenna's position (AP), 456–457
Apostles, 233
Attributes
    ethical, 203, 218
    performance, 203, 218
    process control charts, 291
        c/u, 219, 317
        np, 219, 316–317
        p, 218–219, 316
    sensory, 203, 218
    structural, 203, 218
    temporal, 203, 218
    types, 315

B

Balanced scorecard (BSC), 287
Baldrige criteria, 17*f*
Baseline process metrics
    characteristic/variable, measure, 324
    continuous outputs, 325, 333*f*
        normal/potential, capability, 333–337
        statistics, 332*f*
    defects/defectives
        counting, 324, 327–328
        inspection, tires, 327*f*
        lots, 330*t*
    discrete outputs, 325
        binomial/Poisson, capability, 331–332
        metrics, 329*f*
        statistics, 326*f*
        test request form, 328*t*
    estimation, 325, 325*f*
    primary, 324

Benchmarking, 379
   benefits, 381
   continuous, 258
   effectiveness, 258
   process, 381–382, 381f
   purpose, 380
   sequence, 258–259
   standards, comparison, 381
   types, 380
      compliance, 380
      continuous improvement, 380
      effective, 380
      perception, 380
Benefits driven procurement (BDP), 551
Binary regression model, 366
Black belt, analyze phase
   multi-vari studies, 356–357
      advantages, 357
      characteristics, 357, 358f, 359f
      example, 358f
      measurement, 358f
   process capability
      contingency tables, chi-square, 372–373
      indices, 370
      metrics, 371–372
      normal distribution, 369
      performance, 370–371, 371f
      RTY, 372
   regression
      fitted line plot, 360–362, 361f
      least squares, method, 365–366, 365f
      logistic, 366–368, 367f, 368t
      model, simple linear, 364, 364f
      multiple, 362–363
      output, 362f
      residuals, 362, 363f
      scatter plot, 360, 361f
      statistical tools, 359f
      study time/performance, 360t
Black belt, measure phase
   charts
      attributes process, 315–317
      control, 311–312, 312f
      c/u, 317
      implementation, 312–313
      median, 315
      np, 316–317
      p, 316
      variable/attribute control, 312t
      $\bar{x}$/R, 313–314, 313f, 314f
      $\bar{x}$/s, 315
   instrument selection, 306–307
Block, 388
Bottleneck, 385
Brainstorming, 180t, 185
Business process re-engineering (BPR), 58, 101, 103
Business technology management (BTM),
      557f, 558f

C

Cargo transfer bag (CTB), 478, 480
Cause-and-effect
   inventory variation, diagram, 513f
   relation, 428
Caveats, 13
Central limit theorem (CLT), 205
Certified LSS Black Belt (CLSSBB), 543
Certified LSS Green Belt (CLSSGB), 543
Certified LSS Yellow Belt (CLSSYB), 543
Chronocyclegraph, 431
Commodity codes (CCs), 531
Concept, design, optimize, and verify (CDOV), 447–448
Continually apply companywide quality control
      (CWQC), 27–29
Control chart
   analysis, 417
   classification, 417f
   components, 418, 418t
   jump-in-process, 418–419
   outside limits, 419
   recurring cycles, 419
   selection, 417
   trends, 418
   variability, lack, 419
Control phase
   black belt
      kanban pull, 422–423
      mistake proofing, 420–422
      SPC, 415–419
   green belt
      Kaizen, 411–412
      5S program, 408–411
      TPM, 412–415
   hierarchy, 407f
   natural erosion
      continuous improvement, 407, 407f
      standardization, 407, 408f
   steps
      inputs/outputs, 406
      plan, validation, 405–406, 405f
   Z-score, 405
Cost, quality, features, and availability (CQFA),
      228–229, 554
Cost accounting standards (CAS), 552
Cost–benefit analysis, 187, 191
Cost of good quality (COGQ), 441
Cost of poor quality (COPQ), 441
Cost of quality (COQ)
   caveats, 13
   justification, 13
   traditional, 13
Critical path method (CPM), 181, 574–575
Critical to process (CTP), 282
Critical to quality (CTQ), 182, 182f, 237, 281–282
   creation, 237
   VOC, translation, 256–257, 256t

Critical to satisfaction (CTS), 282, 448, 552
Crosby, Philip. B., 4, 10, 15, 24
Cross-tabulations, 349
Customer
    attributes, 254, 256
    data, 228–229
        collection, 234, 234*t*
        methods, 228–229
    data analysis, 235–236
        control charts, 235, 235*f*
        line graph, 235, 235*f*
        matrix diagram, 236, 236*f*
    external, 232
    identification, 228–229
    internal, 229–231
    metrics
        process performance, 233*f*
        selection, 233
    needs, 254, 282
    perception, 254
    retention/loyalty, 232–233
    service, 232
    surveys, 234
Customer relationship management systems
    (CRMs), 37

D

Data collection
    accuracy, LSS
        check sheet, 183
        GR&R, 183
    green belt
        checklist, 298, 298*t*
        check sheets, 297, 297*t*
        guidelines, 296
        measurement scales, 295, 295*t*
        meeting process chart, 298*t*
        sampling, types, 296
    yellow belt
        check sheets, 291, 291*t*
        control charts, 291
Defect output, 326–329
Defects per million (DPM), 445
Defects per million opportunities (DPMOs), 87,
    326–328, 329*f*, 563
Define, characterize, optimize, and verify (DCOV),
    447–448
Define, measure, analyze, design, and verify
    (DMADV), 19, 446–449, 461–466
Define, measure, analyze, identify, design, optimize,
    and verify (DMADIOV), 450
Define, measure, analyze, improve, and control
    (DMAIC), *see* DMAIC
Define phase, 486
    black belt
        affinity diagram, 266–268, 267*f*, 268*f*

        groupings, identification, 268*f*
        interrelationship diagram, 269–270, 270*f*
        Kano model, 262–263, 263*f*
        QFD, 263–266
        quality, house, 264*f*, 265*f*, 266*f*
    customer, voice, *see* Voice of customer (VOC)
    green belt
        benchmarking, 258–259, 259*f*
        brainstorming, 262
        Pareto diagrams, 257
        thought process map, 259, 261–262, 261*f*
    project
        charter/teams, *see* Project charter/teams
        definition worksheet, 224–227, 224*f*
        includes/excludes, 227*f*
        management, 250
        report, 253
        selection, 222–223, 223*f*
        timeline, 251*f*
        tracking, 250
    yellow belt
        SIPOC, 254
        SMART, 253–254, 253*t*
Deming, William. E., 8–10, 13–14, 23–24, 27
    prize, 16
        categories, 28
        CWQC, 29
        guidelines, 30
        3 phases, 29
        TQC, 27–29
        TQM, 28–29
    value chain, 8, 9, 14, 24
Design for Six Sigma (DFSS), 19, 66, 128, 176, 177*f*,
    445–449, 452, 471
Design for Six Sigma research (DFSSR), 447, 449–453,
    485, 485*f*, 487*t*
    DMADIOV, 450
    3P's
        methodology framework/substeps, 449*f*, 450*f*
        phase description, 450*f*
        statistical tools, methodology, 451*f*
        theoretical model, 453
Design of experiment (DOE), 70, 151, 185, 187, 189,
    387–388, 451, 455–456, 464
    advantages, 387
    perform, examples
        one factor, multiple levels, 391–399
        planning experiments, 399–401, 401*t*
    terminologies, 387–390
Design phase, 486
Distribution center (DC), 521, 530–531
DMADO (De-may-doh), 452
DMAIC, 19, 66, 80, 87, 89, 91, 94, 128, 151, 153, 156–157,
    176–177, 177*f*, 179, 404, 472, 545
    enterprise-wide deployment
        challenges, 426–427
        flowchart, 428*f*
        pyramid approach, 426

DMAIC
    Lean/Six Sigma
        combinatory approach, 426
        comparison, 426*t*
        integration, 427*f*
        organizational uses, 425–426
    phase, tools, 180*t*
    quality circle, 178*f*
    system/process, 427–428
DMAIC *versus* DMADV, 464
Document archiving, 253
Dow Chemical, 119–120, 123
Drift, 305

E

Earnings per share (EPS), 88
Earnings before interest and taxes (EBIT), 92, 96, 98,
        481, 543
    comparison, 95*t*
    sensitivity, 96*t*
Electromagnetic Interference (EMI), 493
Electronic data interchange (EDI), 521–522
Electronic Product Code (EPC), 511
Empirical methods, population variability
    frequency distribution, 194
    graphical methods, 198–202
    histogram, 194–198
Engineering management (EM), 151
Engineering work process redesign (EWPR), 484,
        490*f*, 491*f*
Enterprise resource planning (ERP), 50, 60, 62, 532
Enterprise-wide deployment, 426–427, 428*f*
Entitlement, 342
Entity, 382
Error, 328
    types, 342–344
European Article Number (EAN), 510
European quality award, 17, 17*f*, 38, 38*f*
Evolutionary operation (EVOP), 388
Executive LSS Yellow Belt (ELSSYB), 542
Experimental error, 388

F

Failure modes, effects and criticality analysis
        (FMECA), 352
Failure modes and effect analysis (FMEA), 175–176,
        180*t*, 184, 338, 353*f*
    categories, 352
    potential, 352*f*
    process steps, 353–354, 353*f*
    risk assessment/priority number, 354
    types, 354
Family error rate, 350
Fault tree analysis (FTA), 338–340
    FMEA, comparison, 339
    symbols, 339

    event, 339, 339*f*, 340*f*
    gate, 339, 340*f*
    uses, 338
Feigenbaum, Armand V., 10, 15, 24–25
Finished goods (FG), 499
Fishbone diagram, 175*t*, 180*t*, 185, 191, 508, 510*f*
Flowcharting, 508, 510*f*
    engineering economics metric,
        identification, 95*f*
    symbols, 213*f*
Focus/prioritization tools, LSS
    cause-and-effect matrix, 184
    failure modes/analysis, 184
    Pareto chart, 184
Food product shortages
    background, 506–507
    DFSSR approach
        analyze, 512
        define, 511
        design, 514
        identify, 514
        measure, 512, 512*f*
    limitations, 515
    research methodology
        approach, 508
        DFSSR, 507*f*
        phases/steps, 507*f*
        RFID cost, 511*t*
        tools, 508, 508*t*, 509*f*
    ROI, 514

G

Gage repeatability/reproducibility (GR&R), 183
Gantt chart, 509*f*
Garvin's approach, 25
    global, 3
    manufacturing-based
        Crosby, Philip. B., 4
        engineering specifications, 4
    product-based
        attributes, 3–4
        quality trail, 4*f*
    user-based
        Juran, Joseph. M., 4
        user's expectation, 4
    value-based
        aesthetic, 5
        conformance, 5
        durability, 5
        feature, 4–5
        perceived quality, 5
        performance, 4–6
        reliability, 5
General Electric (GE), 60, 537–538, 543
Geographical information system (GIS), 522
Global approach, 3
Global positioning system (GPS), 517–518, 520

Graphical methods, population variability
  box/whisker plot, 201–202
  scatter diagram, 198
  stem/leaf diagram, 199
Green belt, improve phase
  benchmarking, 379–383
  SWOT analysis, 383–385
  TOC
    drum buffer rope, 386–387, 387f
    implementing, methods, 386
Green belt, measure phase
  data
    collection, 295–298
    types, 298–299
  graphical methods
    boxplots, 308, 309f
    histograms, 310, 311f
    relationship table, 310t
    scatter diagrams, 308–310
    stem/leaf plots, 310
    trend charts, 310
  measurement methods
    ANOVA, 301–303, 302t
    calibration, 307–308
    correlation, *see* Measurement correlation
    enterprise, 306–308
    guidelines, 301f
    precision/accuracy, 299, 299f
    range, 300–301
    types, 306
  Pareto analysis, 294–295

H

High performance companies (HPCs), 113, 122t
High performance organization (HPO)
  HPT, 116
  list, 122t
  outcomes, 116–117
  traditional organization
    comparison, 127t
    differences, 125–127
    structure *versus*, 126f
High performance team (HPT), 116, 127
Histogram
  minitab edit/entry, 196t, 197t
  shapes, 197–198
  table, 195t
House of quality (HOQ), 471–474
Hurdle rate, 117
Hypothesis tests, 342
  alternative, 346
  general, 346f
  null, 346
  translation, 346–349
  uses, 345–346
  US legal system, 345

I

Identify, define, evaluate, and activate (IDEA), 447, 449
Identify, design, optimize, and validate (IDOV), 19,
    447–448
Identify phase, 491
Improve and control (IC) phase, 472
Improve phase
  architecture, 376–379
    new process, 377
    nonvalue-adding, identification, 377
    relationship, 377–378, 379f
  black belt
    DOE terminology, 387–390
    guidelines, design, 390–391, 390t
  categories, 376
  green belt
    benchmarking, 379–383
    SWOT analysis, 383–385
    TOC, 385–387
  models/transfer equations, 379f
  plan, 378f
Industrial engineering (IE)
  academic interference, 152t
  projects, 151f
  Six Sigma role, 63f, 151f
  specialized knowledge, 153f
Information Management System (IMS), 480
Inner array, 389
Institute of Industrial Engineers (IIE), 538–540
Internal rate of return (IRR), 440
International Automotive Task Force (IAFT), 37
International Organization for Standardization
    (ISO), 37
International Quality Federation (IQF), 538
International Research Experiences for Students
    (IRES), 484, 505
International Space Station (ISS), 471, 473, 475,
    478–481
International Supply Chain Education Alliance
    (ISCEA), 538, 542
Invent, innovate, develop, optimize, and verify
    (IIDOV), 447–448
Inventory, 385
Iron Byron, 391
Ishikawa, K., 15, 25
ISO 9000
  ISO 9000:2000 standard, 36–37
  performance, 36
  purpose, 35–36
ISO/TS 16949 standard, 37

J

Japanese Union of Scientists and Engineers (JUSE),
    16, 27
Juran, Joseph. M., 4, 8, 10–11, 13–14, 24
Just in Time (JIT), 16, 36, 71, 541

K

Kaizen
    blitz/event/workshop, 411–412
    defined, 411
    strategy, 411
Kanban pull
    card, 423f
    material flow, 423
    methods, 423
Kano model, 262–263, 263f
    Beijing Olympics, 262
    delighters, 262–263
    dissatisfiers, 262
Key performance indicators (KPIs), 529–530
Key process input variable (KPIV), 378, 378f, 416
Key process output variable (KPOV), 416

L

Lean
    enterprise, 71–72
    Japanese quality movement, 71
    manufacture
        cellular, 73
        elements, 72–73
    organizational uses, 425
    pro-company, 75
    structured management system
        employee involvement, 542
        quality belief, 542
        waste elimination, 542
    success, 74
    tools, 73
    VSM, 73–74
    waste, sources, 72
Lean Six Sigma (LSS), 23–24, 26, 58, 79, 119, 228, 547,
        553, 556
    consultants
        integrators, 547
        master black belt, 547
    control, 188–190, 189t
    COQ, 13
    define phase
        collate information, methods, 182
        project chartering, 179–181, 181f
        team, assignments, 181–182
    DFSSR, phase
        analyze, 191
        define, 190
        design, 191
        identify, 191
        measure, 190–191
        optimize, 191
        verify, 192
    economic tools, 439–440
    engineering education
        green/black belts controversy, 150
        Six Sigma engineers, 150

    evaluation models
        company performance, calculation, 118–119
        inflation discount, 118
        MARR estimation, 118
    evolution, 430
    history, 79
    hypothetical statements, 178–179
    improvement, continuous
        action, monitoring/recommendation, 432–433
        alternative, evaluation, 432
        possible solutions, 432
        problem definition/analysis, 432
    improve phase, 186–187, 187f
    integration
        ISO 9001, 156
        logistics, 156–157
        Malcolm Baldrige award, 153–154
    Japanese contributions, 16
    JIT, 16
    measure phase
        data collection/accuracy, 183
        focus/prioritization tools, 184
        generating ideas, 185
        process description, 183–184
        quantifying/describing variation, 184
    motion study
        history, 430–431
        SMED, 429
        techniques, 438
    overview, 176–178
    practitioners
        black belt, 545
        champion, 545
        green belt, 545–546
    principles, 88–89
    QM, integrating, 19
    quality costs, evaluation
        good, 441
        poor, 441–442
        total, 442
    real difference, 79–81
    roles
        industrial, 151, 151f
        SS–IE IF, 151–153
    supply chain, 20
    techniques, 3, 8, 10–12, 18–19
    testing models, 113–117
    time study
        allowances/standard time, determination,
            437–438
        challenges, 439
        cycles, number, 437
        definitions, 434–435
        elements, 435
        equipment, 433–434
        human factors, 439
        observation sheet, 436
        operator, rating/recording, 435, 437
        types, 438

vendors
  major, 546
  selection model, 546–547
Lean Six Sigma: Combining Six Sigma with Lean Speed, 79
Least significant difference (LSD), 350
License plate number (LPN), 484
Long-term variation, 334, 335*f*, 336
Lower control limit (LCL), 314
Low performance companies (LPCs), 113, 122*t*
L-type matrix, 234, 234*t*

M

Malcolm Baldrige National Quality Award
    (MBNQA), 16–17
  human resource/workforce management,
      33–34, 156
  leadership, 155
  LSS integration, 153–154
  model
    full integration, 33, 153–154
    Honeywell, 34
    Six Sigma/quality, 30–31, 153–154
  process, 155*f*
  self-assessment, 31
  Six Sigma, 31
    implementation, 32–33, 33*f*
    integration, 155*f*
    PDCA, 31
    plus, 34
  SSPIM, 30
  strategic planning, 34, 156
  workforce, 153, 156
Manufacturing-based approach
  Crosby, Philip. B., 4
  engineering specifications, 4
Materials Management Branch (MMB), 529–530
Matrix management, 565
Mean time between failure (MTBF), 5
Mean time to first failure (MTFF), 5
Measure, analyze, improve, and control
    (MAIC), 125
Measurement correlation
  bias, 303–304, 304*f*
  linearity, 304, 304*f*
  precision/tolerance, 305–306
  stability, 305
Measure phase, 489
  black belt
    attributes control charts, 316–317
    chart, implementation, 312–313
    control charts, 311–312, 312*f*
    median charts, 315
    variables/attribute charts, 312*t*
    $\bar{x}$/R charts, 313–314, 313*f*, 314*f*
    $\bar{x}$/s charts, 315
  classifications, 280, 282*f*
  critical characteristics, 281–283

flow down process, 283*f*
  construction, 283–284
  "little y," characteristics, 284–285
  project, 281*f*
green belt
  data, types, 298–299
  data collection, 295–298
  graphical methods, 308–311
  measurement methods, 299–308
  Pareto analysis, 294–295
quality, stage, 281*f*
sponsorship level, 287
targets/specifications
  determination, 285–286
  needs, 286–287
two dimensions, 285, 285*f*
yellow belt, 287–294
  data collection, 290–292
  flowchart, 288–289, 289*f*, 290*f*
  process maps, 288–290, 288*f*, 291*f*
  waste/variation analysis, 292–294, 292*f*
Methods-time measurement (MTM), 438
Minimum acceptable rate of return (MARR),
    88, 117–118, 440
Minitab, 356
Mistake proofing, 175*t*, 180*t*, 191–192
  characteristics, 420
  examples, 422
  method, 421, 422*t*
  system, 420*f*
  technique, 421, 422*t*
  warning mechanism, 421
Muda, 72, 74
Multicolinearity, 363
Multimedia Video Task Analysis (MVTA), 437

N

NASA, radio frequency technologies
  analyze
    orientation sensitivity, 477
    read range/accuracy, 476–477
  develop
    door/waste tracking, 478–479
    SAT, 479–480
    smart shelf, 478
  DFSS, methodology
    analyze, 472
    perform, 472
    plan, 471
    predict, 472
  measure
    orientation sensitivity, 475–476
    read accuracy, 476
    read range, 472, 475
  optimize
    cost analysis, 480–481
    implementation, 480
  project charter, define phase, 473–475

National Aeronautics and Space
   Administration (NASA), 471
National Association for Stock Car Auto Racing
   (NASCAR), 570
National Institute of Standards and Technology
   (NIST), 16, 17
National Science Foundation (NSF), 489, 505
Net present value (NPV), 440
Nippon Telephone and Telegraph, 15
Noisy, 475
Nonbottleneck operation, 386
Nonconforming parts per million (NCPPM), 63
Nondestructive evaluation (NDE), 307
Nondestructive testing (NDT), 307
Normal pace, 434
Normal time, 434
Nuisance factor, 395
   golfer, 393, 395
   randomized block design
      OFAT, 397, 398f, 399f
      SWAW, 397, 398f
   types, 395

O

Off fixture displays (OFDs), 60, 64
One-factor-at-a-time (OFAT), 397, 398f
One-sided test, 347
Operating characteristic (OC), 37
Operation expenses, 386
Operation Six Sigma (OSS), 176, 446–447, 449
Organizational performance
   customer
      relationship management, 8
      service surveys, 8
   360-degree evaluation, 10
   Deming, William. E., 8
   job analysis, 10
   LSS initiatives, 8
   processes, 105t
      behavior, 104
      change, 104
      work, 104
   QM thinking, 6
   quality perspective
      engineering, 6, 7f
      executive management, 7, 8f
      financial, 8–9, 9f
      functional, 6
      human resources, 10
      marketing, 8, 8f
      operations, 6
      systems, 6, 7f
   strategy, 7
Outer array, 389
Output y, 283
Overall equipment effectiveness (OEE), 502

P

Pareto analysis, 486, 488f, 508, 513f
Parking lot, 249, 572
Passive RFID system (PRS), 514
Payback period, 440–441
Perform phase
   graph, read rate, 458
   optimize
      factors/levels, 456, 457t
      TP, 456
   verify, 457
Plan-Do-Check-Act (PDCA), 28, 73,
      152–153, 153f, 446, 449, 451, 461,
      463–466, 569
Plan phase
   define, 453
   measure, 454
Poka-yoke, *see* Mistake proofing
Portal's distance (PD), 457
Post office (PO), 521
Pretty good privacy (PGP), 522
Price-earnings (PE) ratio, 88
Process optimization, 428
Process scorecard, 334
Product-based approach
   attributes, 3
   quality trail, 4f
Proficiency testing, 303
Project charter/teams, 239, 244
   advantages, 239
   boundaries, setting, 238
      cooperate learning, 238
      organizational learning, 238
   content, 239
   core and *ad hoc*, 246f
   evolution, 246
   financial impact, 240
      activities, 240
      problem statement, 240
   formation, 246
   goal statement, 240–241
      elevation, 241
      exploitation, 241
      identification, 240–241
      subordination, 241
   implementation, 247–250
      conflict resolution, 249–250
      facilitation, 248
      meeting notes, 248, 248f
   leader roles/responsibilities, 244–245
   management
      benefits, 242–243
      roles and responsibilities, 242–243
   measures, 243–244
      costs and revenue factors, 243–244
      types, 244

milestones/deliverables, 241–242, 252–253
negotiation, 242
rules, 246
types
    cross-functional, 246–247
    natural work groups, 247
    process improvement, 246
    self-directed work, 247
    technology, 247
    tiger, 247
    virtual, 247
Project definition worksheet (PDW), 224–227, 224*f*, 551–554
boundaries, 225*t*
key facts, 224*t*
metrics, 225*t*
organization, 226*t*
performance and goals, 226*t*
Project evaluation review technique (PERT), 181, 574–575
Project management, Six Sigma
charter and teams, 559–573
    conflict resolution, 572–573
    meeting, 571–572
customer, voice
PDW, 551
    boundaries, 552
    key facts, 551
    performance and goals, 553
tracking
    document archiving, 576
    milestone reporting, 575–576
    report, 576
    work breakdown structure, 574–575
VOC
    external, 557–559
    identification, 553–555
    internal, 556–557
Pugh Matrix, 546

Q

Quality
alliance, 34
assurance, 12
awards/standards
    Deming prize, 16, 27–30
    European, 17, 17*f*, 37–38
    ISO 9000, 34–36
    JUSE, 16
    MBNQA, 16–17, 17*f*, 30–34
conformance, 24
control, 11, 65
    units, 38
definition, 3, 29
dimensions, 233

Garvin's approach, 3–6, 25
global, 3
manufacturing-based, 4
product-based, 3–4
user-based, 4
value-based, 4–6
gurus
Crosby, Philip. B., 15, 24
Deming, William. E., 14, 23–24
Feigenbaum, Armand. V., 15, 24–25
Garvin, D. A., 3–4, 15, 25
Ishikawa, K., 15, 25
Japanese contributions, 15–16, 25–26
Juran, Joseph. M., 14–15, 24
Taguchi, G., 15, 25
house, 264*f*, 265*f*
implementation
LSS initiatives, 11
Six Sigma, 12
improvement, 65
inspection, 26
JIT, 57
justification, 12
Keiretsu partnerships, 23
LSS techniques, 3, 10, 13, 15, 19–20
management, 3, 8, 10–12, 18–19
measurement, 12–13
organizational training
implementation, 11
integrated approach, 11
QA, 12
QC, 11–12
perspective
engineering, 6, 7*f*
executive management, 7, 8*f*
financial, 8–9, 9*f*
functional, 6
human resources, 10
marketing, 8, 8*f*
operations, 6
systems, 6, 7*f*
planning, 65
points, 14
product, 59, 64
QM, 3, 8, 18–19
standards, 38
ISO 9000, 18
ISO 14000, 18, 38
ISO 14001, 18
QS 9000, 38
strategic planning
supplier parterning, 36–37
value chain, 36
three spheres, 11, 11*f*, 14
total management, 10, 18–19
trade, 3, 18
Quality assurance (QA), 11, 24

Quality control (QC), 11–12, 24–25
Quality Control Handbook, 13
Quality control unit (QCU), 38
Quality function deployment (QFD), 59, 81, 263–266, 471
Quality management (QM), 24
    LSS techniques, 19
    principles, 18
Quality philosophers
    Crosby, Philip. B., 15
    Deming, William. E., 14
    Feigenbaum, Armand. V., 15
    Ishikawa, K., 15
    Juran, Joseph. M., 14–15
    Taguchi, G., 15
Quality Systems (QS), 38
Quest Diagnostics, 124–125

R

Radio frequency (RF), 454–455
Radio Frequency and Automated Identification
    (RAID), 505
Radio Frequency and Supply Chain Logistics
    (RfSCL), 452, 471, 473–475
Radio frequency identification (RFID), 449, 452–455,
    459, 471, 473–476, 478, 480–481, 483,
    517–521, 523
    components, 501*t*
    real-world testing, 452
        perform, 455–456
        plan, 453–454
        predict phase, 454–455
    tags, 510, 511*t*
Research experiences, NSF, 483–484
Residual error, 388
Return on assets (ROA), 440
Return on equity (ROE), 88
Return on invested capital (ROIC), 528
Return on investment (ROI), 12, 73, 191, 440, 514, 528, 530
RFID automation, implementation
    Gantt chart/project timeline, 502–503
    method/approach
        costs, 501
        defect rate, 497
        design, 499
        DFSSR, 495, 496*f*
        DMAIC, 495, 496*f*
        identify, 501
        manufacturing system, modification, 495
        measure, 497, 498*f*, 499*f*
        Pareto analysis, 499
    overview/objectives
        cause-and-effect diagram, 494*f*
        software/hardware, investigation, 494
        technology, 494
    Six Sigma, 503*t*
RFID-based Quality Inspection and Management
    (RFID-QIM), 494

Risk priority number (RPN), 354, 355*f*
Rolled throughput yield (RTY), 372
Round robin testing, 303

S

Scatter plot, 360, 361*f*
Science, technology, engineering, and mathematics
    (STEM), 505
Secure FTP (SFTP), 522
Sensor, 454–455, 460
Sensor active tag (SAT), 479–480
Shop Management, 430
Short-term variation, 334–337, 336*f*
    rational subgrouping, 335, 335*f*
    standard deviation, estimation, 336
Simultaneous scanning procedure (SSP), 514
Single-minute exchange of die (SMED), 429
Situational leadership model, 245
Six Sigma, 445–454, 456, 458, 460–468
    advantages, 90
    analyze phase, 322–373
    approach
        application, 89, 90*f*, 90*t*
        cost, 88
    background, 58–59
    Baldrige model, 31–33, 33*f*
    baseline, 94
    business process improvement, 101
    case study, 96
    challenges, 61–62, 61*f*
        champion belt infrastructure, 61–62, 61*f*
        corporate-wide metrics, 61, 61*f*
    companies
        examine/compare failure factors, 122–125
        finalized list, 119*t*–120*t*
        high/low performing, 122*t*
        identification, 114–117
        performance, 92–93, 118–119
        returns, rate, 121*t*
        trends, 123*f*
    comparison, 94–96
    constructs, development, 463–465
    control phase, 404–423
    corporate support, 60
    critical factors, 93–94
    data
        availability/finalization, 114
        collection, 114
    defects/levels, 64*f*
    define phase, 221–270
    deliverables, 351*f*
    DMAIC, 19
    dual structures, 102
    EBIT growth rate, 98*t*
    engineering education, 150
        economics metric, 94, 95*f*
        green/black belts controversy, 62–63, 150

industrial roles, 63*f*, 151*f*, 153*f*
    OFD, 63–64
    projects, 91–94
evaluation, 96
firm
    performance, 103, 105–106, 107*f*
    technological innovation, 107*t*
framework
    conceptual model, 465–467
    DMADV, 462
    DMAIC, 462
goal settings, 102
house belt system, 61–62, 149
HPO, 114–117, 115*t*, 116*t*
IE
    projects, 151*f*
    specialized knowledge, 153*t*
impact, 92
implementation
    critical factors, 119, 124*t*
    flow, 497*f*
    success factors, 124*t*
improve phase, 376–401
industries, 60
innovation, 106–107
integration
    human resources management, 156
    leadership, 155–156
    Malcolm Baldrige, 153–154, 155*f*
interrelationship diagram, 122*f*
Juran approach
    six phases, 65
    trilogy, 65
Lean, 15, 19
    integration strategy, 128
    quality decision matrix, 91
    separation, 58
    testing/evaluation models, 113–119
MARR estimation, 118
measure phase, 279–317
methodology, 31, 94
    different, 65–66
    Juran approach, 65
network modeling
    analyze, 528
    control, 529
    define, 528
    improve, 528–529
    measure, 528
non-LSS projects, 119–128
operation/design
    DMAIC: OSS, 447
    methodology, comparison, 447
    research, 449
organizational uses, 425–426
PDCA, 152–153, 153*f*
philosophy, 79, 81
plus, 34

principles, 89
process, 20
    innovation, 106–107
    metrics, 324*f*
    organization, 104–105, 105*t*
    variability, 80
productive work, criteria, 117
projects
    justification, 91
    scenarios, 97*t*
quality
    control/management, 101–102
    gurus, 58
    movement, 102
roots, 59–60
three rules, 63
sample calculation, 96–97
sensitivity, 96
special, 102–103
speed improvement, 79
statistics, 63
    tools, 66
success factors, 124*t*
supply chain, logistics
    capacitated plant location model, 527
    case description, 529–532
    technical manager, implications, 532
thinking, 57–58, 58*f*
three-sigma rule, 63
TPS approach, 58
TQM, retention/change, 59
trucking improvement
    carrier selection, 521
    Gantt chart/project timeline, 522–523
    information exchange, 521–522
    mapping, 522
    paperless logs, automated, 522
    research method/approach, 518–521
    research overview/objectives, 518
    tracking/tracing, 521
    web order entry, 521
warehousing operation, 483–492
Six Sigma-IE interface framework (SS-IE IF), 151–153
Six Sigma process improvement method (SSPIM),
        30, 153–154
Specific, measureable, achievable,
        relevant, and timed (SMART)
        objective, 179
Sponsorship phase, 549
5S program
    4S/5S+1S, 409
    steps
        Seiketsu (standardize), 410
        Seiri (sort), 409–410
        Seiso (scrub), 410
        Seiton (straighten), 410
        Shitsuke (sustain), 410–411
    workplace organization, 409*f*

Stakeholders, 550, 554, 556–558, 568
    analysis
        example, 231*f*
        two-dimensional, 231*f*
    commitment, 230, 230*f*
    key, 230
Statistical models, population variability
    ANOVA
        Fisher's LSD, 212
        p-value, 212
        Tukey's method, 212
    binomial distribution, 202–203
        skewed, 203*f*
        symmetric, 202*f*
    CLT, application, 205
    hypothesis testing
        alternate, 206
        null, 206
        translation, 207–208
        US legal system, 206
    Poisson/normal distribution, 203–205
    regression
        least squares, method, 209–210
        multiple, 211
        residuals, 210–211
        simple linear model, 209
Statistical process control (SPC), 152, 188, 531, 549–550
Statistics/quality tools
    cause-and-effect diagram, 213, 213*f*
    check sheets, 213–214, 214*t*
    control charts
        attribute process, 218
        implementing process, understanding,
            215–216, 216*f*
        median, 216–217
        variable/attribute process, 215
        $\bar{x}$/R, 216, 217*f*
        x/s, 217–218
    definitions, 193–194
        data, 194
        parameter, 193
        population, 193
        sample, 193
        statistic, 194
    flowcharting, 212–213
    Pareto diagrams, 214–215
Stem/leaf diagram
    Minitab, 199–200
    percentiles/percentile rank, 200–201
    plot, 199*f*
    quartiles/interquartile range (IQR), 200
Stick-with-a-Winner (SWAW), 397, 398*f*
Strategic business unit (SBU), 554
Strengths, weaknesses, opportunities, and
            threats (SWOT) analysis, *see* SWOT analysis
Success, key roles, 546
Supplier development program, 37

Supplier relationship management system
            (SRMS), 37
Supply, input, process, output, and customer (SIPOC),
            254, 559–560
Supply chain management, 124
Survey research, 255
SWOT analysis, 383–385
    external environment
        competitive, 384
        economic, 384
        social, 384
        sociopolitical, 384
    opportunities/threats, 384–385
    strengths, 383–384
    weaknesses, 384
System, 427

T

Tag placement (TP), 456
Taguchi, G., 15, 25
Takt time, 72
Target, 285
Team(s)
    conflict resolution
        process, 572
        ways, 573
    formation/evolution, 568–569
    leader roles/responsibilities, 567–568
    rules, 569
    types
        cross-functional, 569
        natural work groups, 569–570
        process improvement, 569
        self-directed work, 570
        technology, 570
        tiger, 569
        virtual, 570
Terrorist, 233
Theory of constraints (TOC), 563–564
Throughput, 385–386
Time-to-live (TTL), 557
Total productive maintenance (TPM), 73, 549–550
    autonomous activities, 415
    benefits, 414
    features, 412
    implementing steps, 415
    loss, 412–413, 413*t*
    maintainability, design, 413–414
    metrics, 414
    total, 412
Total quality control (TQC), 24
Total quality management (TQM), 26, 57–59, 87–88,
            92, 103, 445, 540–541
    customers, 29
    definition, 28
    Deming award, 10

integrative approach, 10
  United States gurus, 10
Total return price (TRP), 88
Toyota Production System (TPS), 58, 71, 79
Traditional organization, 125–127
Transactional process, 328
Transportation management system (TMS), 517
Two-sided test, 347

U

Ultrahigh frequency (UHF), 199, 500f, 500t
Universal Product Code (UPC), 510
University of Tokyo, 15
Upper control limit (UCL), 314–315
User-based approach
  Juran, Joseph. M., 4
  user's expectation, 4

V

Value-added network (VAN), 522
Value-based approach
  aesthetic, 5
  conformance, 5
  durability, 5
  feature, 4–5
  perceived quality, 5–6
  performance, 4
  reliability, 5
  serviceability, 5
Value stream, 541
Value stream mapping (VSM), 73, 259–260, 260f
Vendor management inventory (VMI), 37
Vendor management system (VMS), 489, 490f
Voice of customer (VOC), 128, 228–237
  data collection, 234, 234t
  external, 232
  hearing, methods, 236
    active techniques, 236
    passive techniques, 236
  identification, 228–229
  internal, 229–231
  life cycle, 232
  metrics selection, 233–234
  retention/loyalty, 232–233
  service, 232
  surveys, 234
  understand, process, 236–237
    deploy, 237
    gather data, 237
    plan, 236

yellow belt, 254–257
  CTQ, translation, 256–257, 256t
  qualitative/quantitative research, 255
  sources, 255

W

The Wall Street Journal, 92
Warehouse management system (WMS), 505, 531
Warehousing operation
  background, 484
  research methodology
    approach, 486
    DFSSR steps, 486–490
    phases, DFSSR, 485–486, 486f, 488f
    scenarios, 489–490
  research objective
    hypothesis, 484
    test metrics, 485
Waste/variation analysis
  activities, types, 293f
  set in order, 293
  shine, 293
  sort, 292–293
  standardization, 293
  sustain, 293
  value, 293–294
Whisker plot, 308
Work breakdown structure (WBS), 181, 251–252, 574
  advantages
    computer software, 252
    manual methods, 252
  CPM, 252
  Gantt charts, 252
  PERT, 251–252
Work-in-process (WIP), 72, 386, 410, 423, 494

Y

Yellow belt, analyze phase
  cause-and-effect/fishbone diagrams, 337–338, 337f, 338f
  FTA, 338–340
  root cause
    corrective actions, 340t
    subjective/analytical tools, 341
    5 whys, 341
    5 Ws and H, 341

Z

Z_{shift}, 336